T0186823

High Performance Visualization

Enabling Extreme-Scale Scientific Insight

Chapman & Hall/CRC Computational Science Series

SERIES EDITOR

Horst Simon
Deputy Director
Lawrence Berkeley National Laboratory
Berkeley, California, U.S.A.

AIMS AND SCOPE

This series aims to capture new developments and applications in the field of computational science through the publication of a broad range of textbooks, reference works, and handbooks. Books in this series will provide introductory as well as advanced material on mathematical, statistical, and computational methods and techniques, and will present researchers with the latest theories and experimentation. The scope of the series includes, but is not limited to, titles in the areas of scientific computing, parallel and distributed computing, high performance computing, grid computing, cluster computing, heterogeneous computing, quantum computing, and their applications in scientific disciplines such as astrophysics, aeronautics, biology, chemistry, climate modeling, combustion, cosmology, earthquake prediction, imaging, materials, neuroscience, oil exploration, and weather forecasting.

PUBLISHED TITLES

PETASCALE COMPUTING: ALGORITHMS AND APPLICATIONS
Edited by David A. Bader

PROCESS ALGEBRA FOR PARALLEL AND DISTRIBUTED PROCESSING
Edited by Michael Alexander and William Gardner

GRID COMPUTING: TECHNIQUES AND APPLICATIONS
Barry Wilkinson

INTRODUCTION TO CONCURRENCY IN PROGRAMMING LANGUAGES
Matthew J. Sottile, Timothy G. Mattson, and Craig E Rasmussen

INTRODUCTION TO SCHEDULING
Yves Robert and Frédéric Vivien

SCIENTIFIC DATA MANAGEMENT: CHALLENGES, TECHNOLOGY, AND DEPLOYMENT
Edited by Arie Shoshani and Doron Rotem

INTRODUCTION TO THE SIMULATION OF DYNAMICS USING SIMULINK®
Michael A. Gray

INTRODUCTION TO HIGH PERFORMANCE COMPUTING FOR SCIENTISTS AND ENGINEERS
Georg Hager and Gerhard Wellein

PERFORMANCE TUNING OF SCIENTIFIC APPLICATIONS
Edited by David Bailey, Robert Lucas, and Samuel Williams

HIGH PERFORMANCE COMPUTING: PROGRAMMING AND APPLICATIONS
John Levesque with Gene Wagenbreth

PEER-TO-PEER COMPUTING: APPLICATIONS, ARCHITECTURE, PROTOCOLS, AND CHALLENGES
Yu-Kwong Ricky Kwok

FUNDAMENTALS OF MULTICORE SOFTWARE DEVELOPMENT
Edited by Victor Pankratius, Ali-Reza Adl-Tabatabai, and Walter Tichy

INTRODUCTION TO ELEMENTARY COMPUTATIONAL MODELING: ESSENTIAL CONCEPTS, PRINCIPLES, AND PROBLEM SOLVING
José M. Garrido

COMBINATORIAL SCIENTIFIC COMPUTING
Edited by Uwe Naumann and Olaf Schenk

HIGH PERFORMANCE VISUALIZATION: ENABLING EXTREME-SCALE SCIENTIFIC INSIGHT
Edited by E. Wes Bethel, Hank Childs, and Charles Hansen

High Performance Visualization

Enabling Extreme-Scale Scientific Insight

Edited by

E. Wes Bethel
Hank Childs
Charles Hansen

CRC Press
Taylor & Francis Group
Boca Raton London New York

CRC Press is an imprint of the
Taylor & Francis Group, an **informa** business
A CHAPMAN & HALL BOOK

CRC Press
Taylor & Francis Group
6000 Broken Sound Parkway NW, Suite 300
Boca Raton, FL 33487-2742

Printed in the United States of America on acid-free paper
Version Date: 2012924

International Standard Book Number: 978-1-4398-7572-8 (Hardback)

Visit the Taylor & Francis Web site at
http://www.taylorandfrancis.com

and the CRC Press Web site at
http://www.crcpress.com

Foreword

Wes Bethel, Hank Childs, and Chuck Hansen have developed an eminently readable and comprehensive book. It provides the very first in-depth introduction to the interaction of two highly important and relevant topics in computational science: high performance computing and scientific visualization. The book provides a broad background on both topics, but more importantly, for the first time in book form, they describe some of the most recent developments in scientific visualization as we move from the Petascale era to Exaflops computing.

It has been exactly a quarter century since the 1987 publication of the ground breaking report by McCormick, DeFanti, and Brown that defined the field of scientific visualization. That report set in motion the development of a field that has, by now, become an integral part of computational science. As a community, we have come to accept the notion that scientific visualization is the tool to "see the unseen" in the vast amount of data being produced by numerical simulations. From understanding the behavior of subatomic particles in QCD to watching the explosion of supernovae or the evolution of galaxies, we can now "see" these phenomena, as if they are truly happening in front of our eyes. The scientific visualization community has worked diligently on refining algorithms, exploring new display technologies, thinking about the challenge of distributed visualization, and developing large software frameworks, and by now has become a fairly mature scientific activity.

In the same time frame, high performance computing has seen even more dramatic developments. In 1987 we were still thinking of Cray vector computers and frame buffers when it came to HPC and visualization. In the 1990s computing technology made a dramatic transition to MPPs using commodity hardware and the MPI programming model, while increasing performance by a factor of one million from the Gigaflops to the Petaflops level in 2012. Today, were are close to yet another transformation of the HPC field as GPUs and accelerators become integrated, while the amount of parallelism seems to be ever increasing.

In the context of this potential rapid transformation of the high performance computing field, the book by Bethel, Childs and Hansen arrives exactly at the right time. It succeeds perfectly and solidly combines the two almost parallel threads of development in scientific visualization and high performance computing all into one single volume for the first time. It will provide a solid foundation for anyone who considers using the most recent tools for visualization in order to understand complex simulation data or to understand the ever increasing amount of experimental data. I highly recommend this timely book for scientists and engineers. It closes an important gap in the available literature on computational science and it is a great addition to the CRC Press series on Computational Science. It provides a solid reference as the community embarks on the Exascale adventure.

Horst Simon
Lawrence Berkeley National Laboratory and University of California, Berkeley

May 2012

Preface

The field of scientific visualization is changing as rapidly as the computational landscape. Tools, techniques, and algorithms that were born in the 1970s and 1980s—the era of the serial processor—have undergone a rapid evolution over the past two decades to make effective use of emerging multi- and many-core computational platforms and to accommodate an explosive growth in data size and complexity.

This book—*High Performance Visualization*—focuses on the subset of the broader field of scientific visualization concerned with algorithm design, implementation, and optimization for use on today's largest computational platforms. The editors—Bethel, Childs, and Hansen—along with chapter contributors, are leaders in the field of high performance visualization, having produced some of the field's seminal works, including algorithms and implementations that run at the highest levels of concurrency ever published and that are in the hands of today's scientific researchers worldwide for day-to-day use.

This book is motivated by, and is the outgrowth of and expansion upon, Bethel's PhD dissertation also entitled *High Performance Visualization*. Many of the ideas presented in this book are, in fact, from the contributing authors' PhD dissertations. Interestingly, all of the ideas presented here have also been reduced to practice in the form of production-quality visualization software applications that run on today's largest computational platforms and operate on today's largest scientific data sets.

Collectively, the chapters in this book aim to organize and articulate a large and diverse body of computer science research, development, and practical application. The chapter themes reflect major conceptual thrust areas within the field of high performance visualization.

E. Wes Bethel, Hank Childs, and Charles Hansen

May 2012

Preface

Contributor List

Sean Ahern
Oak Ridge National Laboratory
Oak Ridge, TN, USA

Jim Ahrens
Los Alamos National Laboratory
Los Alamos, NM, USA

Marco Ament
Universität Stuttgart
Stuttgart, Germany

Utkarsh Ayachit
Kitware, Inc.
Clifton Park, NY, USA

E. Wes Bethel
Lawrence Berkeley National Laboratory
Berkeley, CA, USA

Kathleen Biagas
Lawrence Livermore National Laboratory
Livermore, CA, USA

Peer-Timo Bremer
University of Utah
Salt Lake City, UT, USA

Carson Brownlee
University of Utah
Salt Lake City, UT, USA

Eric Brugger
Lawrence Livermore National Laboratory
Livermore, CA, USA

David Camp
Lawrence Berkeley National Laboratory
Berkeley, CA, USA

Hank Childs
Lawrence Berkeley National Laboratory
Berkeley, CA, USA

Cameron Christensen
University of Utah
Salt Lake City, UT, USA

John Clyne
National Center for Atmospheric Research
Boulder, CO, USA

David E. DeMarle
Kitware, Inc.
Clifton Park, NY, USA

Marc Durant
Tech-X Corporation
Boulder, CO, USA

Thomas Ertl
Universität Stuttgart
Stuttgart, Germany

Jean Favre
Swiss Center for Scientific Computing
Lugano, Switzerland

Tom Fogal
University of Utah
Salt Lake City, UT, USA

Randall Frank
Computational Engineering International, Inc.
Apex, NC, USA

Steffen Frey
Universität Stuttgart
Stuttgart, Germany

Christoph Garth
University of Kaiserslautern
Kaiserslautern, Germany

Berk Geveci
Kitware, Inc.
Clifton Park, NY, USA

Sebastian Grottel
Universität Stuttgart
Stuttgart, Germany

Attila Gyulassy
University of Utah
Salt Lake City, UT, USA

Charles Hansen
University of Utah
Salt Lake City, UT, USA

Cyrus Harrison
Lawrence Livermore National Laboratory
Livermore, CA, USA

Thiago Ize
University of Utah
Salt Lake City, UT, USA

Mark Howison
Lawrence Berkeley National Laboratory
Berkeley, CA, USA
Brown University
Providence, RI, USA

Kenneth I. Joy
University of California, Davis
Davis, CA, USA

Scott Klasky
Oak Ridge National Laboratory
Oak Ridge, TN, USA

Hari Krishnan
Lawrence Berkeley National Laboratory
Berkeley, CA, USA

Michael F. Krogh
Computational Engineering International, Inc.
Apex, NC, USA

Sidharth Kumar
University of Utah
Salt Lake City, UT, USA

Kwan-Liu Ma
University of California, Davis
Davis, CA, USA

Jeremy Meredith
Oak Ridge National Laboratory
Oak Ridge, TN, USA

Mark Miller
Lawrence Livermore National Laboratory
Livermore, CA, USA

Kenneth Moreland
Sandia National Laboratories
Albuquerque, NM, USA

Christoph Müller
Universität Stuttgart
Stuttgart, Germany

Paul Návratil
Texas Advanced Computing Center
Austin, TX, USA

Alan Norton
National Center for Atmospheric Research
Boulder, CO, USA

Manish Parashar
Rutgers University
Piscataway, NJ, USA

Valerio Pascucci
University of Utah
Salt Lake City, UT, USA

John Patchett
Los Alamos National Laboratory
Los Alamos, NM, USA

Tom Peterka
Argonne National Laboratory
Argonne, IL, USA

Sujin Philip
University of Utah
Salt Lake City, UT, USA

Norbert Podhorszki
Oak Ridge National Laboratory
Oak Ridge, TN, USA

Prabhat
Lawrence Berkeley National Laboratory
Berkeley, CA, USA

David Pugmire
Oak Ridge National Laboratory
Oak Ridge, TN, USA

Oliver Rübel
Lawrence Berkeley National Laboratory
Berkeley, CA, USA

Allen Sanderson
University of Utah
Salt Lake City, UT, USA

Karsten Schwan
Georgia Institute of Technology
Atlanta, GA, USA

Giorgio Scorzelli
University of Utah
Salt Lake City, UT, USA

Brian Summa
University of Utah
Salt Lake City, UT, USA

Gunther Weber
Lawrence Berkeley National Laboratory
Berkeley, CA, USA

Daniel Weiskopf
Universität Stuttgart
Stuttgart, Germany

Brad Whitlock
Lawrence Livermore National Laboratory
Livermore, CA, USA

Matthew Wolf
Georgia Institute of Technology
Atlanta, GA, USA

Jon Woodring
Los Alamos National Laboratory
Los Alamos, NM, USA

Kesheng Wu
Lawrence Berkeley National Laboratory
Berkeley, CA, USA

Hongfeng Yu
Sandia National Laboratories
Livermore, CA, USA

Fan Zhang
Rutgers University
Piscataway, NJ, USA

List of Figures

1.1 The visualization pipeline. Abstract data is input to the pipeline, where the pipeline transforms it into readily comprehensible images. 2
1.2 An image of two vortex cores merging. 2

2.1 A simple parallelization strategy for a parallel visualization framework. 14
2.2 A parallel visualization framework embedded into a client–server application. 17
2.3 Determining constraints and optimizations in a data flow network using contracts. 19
2.4 Reading the optimal subset of data for processing 20

3.1 The three most common visualization pipeline partitionings in remote and distributed visualization. 27
3.2 Three potential partitionings of a remote and distributed visualization pipeline performance experiment. 34
3.3 The three potential partitionings have markedly different performance characteristics. 35
3.4 Visapult's remote and distributed visualization architecture. 37
3.5 Chromium Renderserver architecture diagram for a two-tile DMX display wall configuration. 39
3.6 Chromium Renderserver running a 6-way parallel visualization of a molecular docking application on a 3×2 tiled display system. 41

4.1 Rendering taxonomy based on Molnar's description. 51
4.2 Ray casting for volume rendering. 54
4.3 Volume rendering overview. 54
4.4 Ray cast of a 316M triangle isosurface from timestep 273 of an RM instability calculation 60
4.5 Ray cast of the 259M triangle Boeing data set using HD resolution. 61
4.6 Frame rate when varying cache size and number of nodes in the ray casted RM and Boeing data sets. 63
4.7 Total memory used per node for the RM data set. 64

4.8 Display process scaling. 65

5.1 Sort-last image compositing. 73
5.2 Example of the direct-send algorithm for four processes. . . 75
5.3 Tree-based compositing. 75
5.4 Example of the binary-swap algorithm for four processes and two rounds. 76
5.5 Example of the 2-3 swap algorithm for seven processes in two rounds. 78
5.6 Example of the radix-k algorithm for twelve processes, factored into two rounds of $\vec{k} = [4, 3]$. 79
5.7 Performance comparing binary-swap with radix-k for an image size of eight megapixels. 82
5.8 Target k-values for two different machines. 83
5.9 Performance comparing optimized binary-swap with radix-k shows overall improvement in volume rendering tests of core-collapse supernovae simulation output. 83
5.10 Core-collapse supernova volume rendered in parallel and composited using the radix-k algorithm. 84

6.1 A stream surface visualizing recirculating flow in a vortex breakdown bubble. 92
6.2 Streamlines showing outflows in a magnetic flow field computed by an astrophysics simulation code. 93
6.3 Integral curve test data sets. 96
6.4 Parallelization over seeds versus parallelization over data algorithms. 97
6.5 Astrophysics test case scaling data. 104
6.6 Fusion test case scaling data. 105
6.7 Data structure as a 3D/4D hybrid of time–space decomposition. 107
6.8 The cumulative effect on end-to-end execution time. 109

7.1 A sample bitmap index. 121
7.2 Serial performance for computing conditional histograms and evaluating bin-based histogram queries using FastBit. 123
7.3 Comparison of line-based and histogram-based parallel coordinates and extensions of parallel coordinates to visualize the temporal evolution of a selected feature. 126
7.4 Visualizations illustrating the application of segmentation of query results. 128
7.5 Query-driven analysis of network traffic data to detect distributed network scan attacks. 131
7.6 Query-driven analysis of halo particles in a electron linear particle accelerator simulation. 134

7.7 Query-driven exploration of a large 3D plasma-based particle accelerator simulation using parallel coordinates to define and validate data queries. 137
7.8 Visualization of the relative traces of the particles of a particle beam illustrating the injection and initial acceleration of the particles by the plasma wave. 138

8.1 The Z-order, or Morton, space-filling curve maps multi-dimensional data into one dimension while preserving the locality of neighboring points. 148
8.2 PDA based on multiresolution. 152
8.3 Analysis, reconstruction, and synthesis of a signal, c_j. 157
8.4 A test signal with 1024 samples and a multiresolution approximation at $1/8^{th}$ the resolution. 160
8.5 The CDF 9/7 biorthogonal wavelet and scaling functions. . 163
8.6 A single pass of the 2D DWT and resulting decomposition after two passes of the DWT. 164
8.7 Direct volume rendering of an enstrophy field. 168

9.1 Comparing timings for a turbulent combustion simulation and its corresponding *in situ* visualization over many concurrency levels, up to 15,360 cores. 176
9.2 Image produced from *in situ* visualization of the CH_2O field from a turbulent combustion simulation 176
9.3 Diagram of the adaptor that connects fully featured visualization systems with arbitrary simulation codes. 177
9.4 Example of co-processing *in situ* visualization using adaptors, showing the VisIt system with the GADGET-2 simulation code. 182
9.5 Diagram of a concurrent *in situ* system performing data processing, monitoring and interactive visualization 188
9.6 Illustration of hybrid data staging architecture for *in situ* co-processing and concurrent processing. 189
9.7 An example of locality-aware, data-centric mapping of two interacting applications onto two multi-core nodes. 190
9.8 Image produced from an *in situ* visualization of a turbulent combustion simulation, showing the interaction between small turbulent eddies. 192

10.1 A comparison of data parallel visualization and streaming. . 201
10.2 Synchronous and asynchronous streaming in a data flow visualization network. 203
10.3 In push and pull pipelines, the source or sink direct the pipeline to process the next portion. 206
10.4 Culling unimportant data when slicing. 207

10.5 Metadata must remain accurate, without processing the data in the portion. 208
10.6 Assuming the cache can fit the results, the cache module will prevent upstream re-executions when the camera moves but, otherwise, the data remains unchanged. 209
10.7 Culling, prioritizing, and multiresolution streaming in a pull pipeline. 213

11.1 Simplified graphics pipeline based on OpenGL. 225
11.2 GPGPU architecture with multiple compute programs called in sequence from the CPU, but running each in parallel on the GPU. 228
11.3 Interactive sort-first volume rendering of the Visible Human. 235
11.4 Visualization of a laser ablation simulation with 48 million atoms rendered interactively. 238
11.5 Schematic overview of the abstraction layers of the CUDASA programming environment. 241
11.6 A rear-projection tiled display without photometric calibration. 245

12.1 4608^2 image of a combustion simulation result, rendered by hybrid parallel MPI+pthreads implementation running on 216,000 cores of a Cray XT6 system. 263
12.2 P^Hvolume rendering system architecture. Image courtesy of Mark Howison, E. Wes Bethel, and Hank Childs (LBNL). 265
12.3 For ghost zone data, the P^Hvolume renderer requires less memory and performs less interprocessor communication than the P^Timplementation. 270
12.4 Charts comparing P^H and P^T ray casting and total render performance. 271
12.5 Comparison of the number of messages and total data sent during the fragment exchange in the compositing phase for $P^H$and P^Truns. 273
12.6 Total render time in seconds split into ray casting and compositing components and normalized to compare P^T and P^H performance. 274
12.7 A streamline from computational thermal hydraulics. 275
12.8 Comparison of $P^T$and P^Himplementations of the *parallelize over seeds* algorithm. 276
12.9 Comparison of $P^T$and P^Himplementations of the *parallelize over blocks* algorithm. 278
12.10 Comparison of $P^H$and P^Tperformance for the *parallelize over seeds* algorithm. 281

12.11 Gantt chart showing a comparison of integration and I/O performance/activity of the *parallelize over seeds* P^T and P^H versions for one of the benchmark runs. 283
12.12 Performance comparison of the P^H and P^T variants of the *parallelize over blocks* algorithm. 284
12.13 Gantt chart showing a comparison of integration I/O, `MPI_Send`, and `MPI_Recv` performance/activity of the *parallelize over blocks* P^T and P^H versions for one of the benchmark runs. 285

13.1 Contouring of two trillion cells, visualized with VisIt on Franklin using 32,000 cores. 294
13.2 Plots of execution time for the I/O, contouring, and rendering phases of the trillion cell visualizations over six supercomputing environments. 296
13.3 Contouring of replicated data (one trillion cells total), visualized with VisIt on Franklin using 16,016 cores. 299
13.4 Rendering of an isosurface from a 321 million cell Denovo simulation, produced by VisIt using 12,270 cores of JaguarPF. 301
13.5 Volume rendering of data from a 321 million cell Denovo simulation, produced by VisIt using 12,270 cores on JaguarPF. 302
13.6 Volume rendering of one trillion cells, visualized by VisIt on JaguarPF using 16,000 cores. 303

14.1 Comparison of Gaussian and bilateral smooth applied to a synthetic, noisy data set. 312
14.2 Three different 3D memory access patterns have markedly different performance characteristics on a many-core GPU platform. 313
14.3 Using GPU-specific features can produce a 2× performance gain. 315
14.4 Filter performance has a 7.5× variation depending upon the settings of tunable algorithmic parameters. 315
14.5 Chart showing how filter runtime performance on the GPU varies as a function of CUDA thread block size. 316
14.6 Parallel ray casting volume rendering performance measures on the GPU include absolute runtime, and L2 cache miss rates. 321
14.7 Examples showing how different transfer functions produce differing visible and performance results in parallel volume rendering. 323
14.8 Performance gains on the GPU using Z-ordered memory increase with increased concurrency. 324

15.1 Workflow blending *in situ* and post-processing elements for processing combustion and fusion simulations. 340
15.2 A spectrum of scientific data model elements. 342

16.1 Diagram of VisIt programs and their communication. 361
16.2 Images from VisIt scalability experiments 365
16.3 Recent covers of the SciDAC Review Journal created using VisIt. 367

17.1 An example rendering of a cubic volume by eightprocesses. . 376
17.2 Image interlacing in IceT. 377
17.3 An IceT-assisted rendering of an isosurface from a Richtmyer–Meshkov simulation. 380
17.4 Results from a simulation of objects in a crosswind fire. . . 380
17.5 An example of using IceT to simultaneously render both surfaces and transparent volumes on a multitile display. 380

18.1 ParaView being used in the analysis of flow patterns associated with magnetic flux ropes. 385
18.2 A Python script demonstrating the ParaView scripting interface used to set up a visualization pipeline. 391
18.3 An application from the Computational Model Builder (CMB) suite based on ParaView's client–server framework being used to develop a suitable mesh of Chesapeake Bay for a surface water simulation. 392
18.4 The ParaView Co-processing Library generalizes to many simulations by using adaptors that map simulation data structures to data structures it can process natively. 392
18.5 A prototype web application based on ParaViewWeb for interactive analysis of 3D particle-in-cell (PIC) simulation results for space scientists. 394
18.6 The fragmentation pattern of a high-speed aluminum ball hitting an aluminum brick. 395
18.7 ParaView's fragment extraction can be used to validate that the fragments in a simulated explosion (left) match collected debris (center) and observed damage (right) from experiments. 395
18.8 A visualization of magnetic reconnection from the VPIC project at Los Alamos National Laboratory. 396
18.9 Finding and tracking of flux ropes in global hybrid simulations of the Earth's magnetosphere using ParaView 398

19.1 The architecture of the ViSUS software framework. 403
19.2 The first five levels of resolution of 2D and 3D Lebesgue's space-filling curve. 404
19.3 Parallel I/O strategies. 405

19.4 The LightStream data flow used for analysis and visualization of a 3D combustion simulation (Uintah code). 406
19.5 A ViSUS application running on both an iPhone and a Powerwall. 407
19.6 The ViSUS software framework visualizing and processing medical imagery. 408
19.7 Remote climate visualization with ViSUS. 409
19.8 ViSUS being used for large panorama image processing. . . 410
19.9 Remote visualization and monitoring of simulations. 411

20.1 Exploration of the current field of a 1536^3 MHD simulation. 421
20.2 Volume rendering of an isolated ROI showing the magnitude of the current field from a 1536^3 simulation. 422
20.3 Direct volume rendering of reduced MHD enstrophy data. . 423
20.4 Pathlines from the time-varying velocity field of a simulation. 424
20.5 Pathline integration of five randomly seeded pathlines using reduced storm simulation data. 425

21.1 The basic Ensight client–server configurations. 431
21.2 EnSight parallel rendering for CAVE and planar tiled displays. 431
21.3 EnSight rendering with parallel compositing. 432
21.4 The EnSight 10 GUI includes dynamically generated silhouette edges for selections and targets, hardware accelerated object picking and context sensitive menus. 436
21.5 Supernova simulation data visualized with Ensight. 438
21.6 Ensight visualization of air flow simulation results. 439
21.7 Ensight visualization of inertial confinement fusion simulation results. 441

List of Tables

5.1 Theoretical lower bounds for compositing algorithms. 81

6.1 Parallel IC performance benchmarks. 110

8.1 Biorthogonal wavelet filter coefficients for the CDF 5/3 (top) and 9/7 (bottom) wavelets 163

9.1 Summary of *in situ* processing techniques 173
9.2 Performance results for co-processing *in situ* processing with an adaptor based approach. 182
9.3 Timing results for *in situ* visualization by proxy 191

10.1 Taxonomy of streaming. 202

12.1 Comparison of memory usage at MPI Initialization for P^H and P^T volume rendering implementations. 269

13.1 Characteristics of supercomputers used in a trillion cell performance study. 295
13.2 Performance for visualizing trillion cell data set while varying over supercomputing environment. 296
13.3 Performance for visualizing trillion cell data sets while varying I/O pattern. 298
13.4 Performance for visualizing trillion cell data sets while varying data generation pattern. 298
13.5 Performance results for an isosurfacing weak scalability study. 300
13.6 Performance results for a volume rendering weak scalability study. 302
13.7 Table comparing volume rendering performance at extreme scale with and without an algorithmic option designed to minimize data movement. 303
13.8 Table demonstrating the impact of all-to-one communication at extreme scale. 304

14.1 Percent variation in runtime across block sizes 320

15.1 Expected exascale architecture parameters compared to current hardware. 333

16.1 VisIt's five primary user interface concepts. 363

Contents

Foreword v

Preface vii

Contributor List ix

List of Figures xiii

List of Tables xxi

Acknowledgments xxxiii

1 Introduction **1**
E. Wes Bethel
1.1 Historical Perspective 1
1.2 Moore's Law and the Data Tsunami 2
1.3 Focus of this Book 3
1.4 Book Organization and Themes 3
1.5 Conclusion 6

I Distributed Memory Parallel Concepts and Systems **7**

2 Parallel Visualization Frameworks **9**
Hank Childs
2.1 Introduction 9
2.2 Background 11
2.2.1 Parallel Computing 11
2.2.2 Data Flow Networks 12
2.3 Parallelization Strategy 13
2.4 Usage 16
2.5 Advanced Processing Techniques 17
2.5.1 Contracts 18
2.5.2 Data Subsetting 19
2.5.3 Parallelization Artifacts 19
2.5.4 Scheduling 21
2.6 Conclusion 22

3 Remote and Distributed Visualization Architectures **25**
E. Wes Bethel and Mark Miller
3.1 Introduction 26
3.2 Visualization Performance Fundamentals and Networks . . . 26
3.3 Send-Images Partitioning 28
3.4 Send-Data Partitioning 30
3.5 Send-Geometry Partitioning 31
3.6 Hybrid and Adaptive Approaches 32
3.7 Which Pipeline Partitioning Works the Best? 33
3.8 Case Study: Visapult 36
3.8.1 Visapult Architecture: The Send-Geometry Partition . 37
3.8.2 Visapult Architecture: The Send-Data Partition . . . 38
3.9 Case Study: Chromium Renderserver 39
3.10 Case Study: VisIt and Dynamic Pipeline Reconfiguration . . 42
3.10.1 How VisIt Manages Pipeline Partitioning 43
3.10.2 Send-Geometry Partitioning 43
3.10.3 Send-Images Partitioning 44
3.10.4 Automatic Pipeline Partitioning Selection 44
3.11 Conclusion 45

4 Rendering **49**
Charles Hansen, E. Wes Bethel, Thiago Ize, and Carson Brownlee
4.1 Introduction 49
4.2 Rendering Taxonomy 50
4.3 Rendering Geometry 52
4.4 Volume Rendering 53
4.5 Real-Time Ray Tracer for Visualization on a Cluster 56
4.5.1 Load Balancing 57
4.5.2 Display Process 58
4.5.3 Distributed Cache Ray Tracing 58
4.5.3.1 DC BVH 59
4.5.3.2 DC Primitives 60
4.5.4 Results 61
4.5.5 Maximum Frame Rate 62
4.6 Conclusion 66

5 Parallel Image Compositing Methods **71**
Tom Peterka and Kwan-Liu Ma
5.1 Introduction 72
5.2 Basic Concepts and Early Work in Compositing 72
5.2.1 Definition of Image Composition 73
5.2.2 Fundamental Image Composition Algorithms 74
5.2.3 Image Compositing Hardware 77
5.3 Recent Advances 77
5.3.1 2-3 Swap 77

5.3.2 Radix-k 78
5.3.3 Optimizations 80
5.4 Results 81
5.5 Discussion and Conclusion 82
5.5.1 Conclusion 83
5.5.2 Directions for Future Research 85

6 Parallel Integral Curves **91**

David Pugmire, Tom Peterka, and Christoph Garth

6.1 Introduction 92
6.2 Challenges to Parallelization 94
6.2.1 Problem Classification 94
6.3 Approaches to Parallelization 95
6.3.1 Test Data 96
6.3.2 Parallelization Over Seed Points 98
6.3.3 Parallelization Over Data 99
6.3.4 A Hybrid Approach to Parallelization 99
6.3.5 Algorithm Analysis 103
6.3.6 Hybrid Data Structure and Communication Algorithm 106
6.4 Conclusion 111

II Advanced Processing Techniques **115**

7 Query-Driven Visualization and Analysis **117**

Oliver Rübel, E. Wes Bethel, Prabhat, and Kesheng Wu

7.1 Introduction 118
7.2 Data Subsetting and Performance 119
7.2.1 Bitmap Indexing 120
7.2.2 Data Interfaces 122
7.3 Formulating Multivariate Queries 124
7.3.1 Parallel Coordinates Multivariate Query Interface . . 125
7.3.2 Segmenting Query Results 127
7.4 Applications of Query-Driven Visualization 129
7.4.1 Applications in Forensic Cybersecurity 129
7.4.2 Applications in High Energy Physics 132
7.4.2.1 Linear Particle Accelerator 133
7.4.2.2 Laser Plasma Particle Accelerator 135
7.5 Conclusion 139

8 Progressive Data Access for Regular Grids **145**

John Clyne

8.1 Introduction 146
8.2 Preliminaries 146
8.3 Z-Order Curves 147
8.3.1 Constructing the Curve 149

8.3.2 Progressive Access 149
8.4 Wavelets 151
8.4.1 Linear Decomposition 153
8.4.2 Scaling and Wavelet Functions 154
8.4.3 Wavelets and Filter Banks 156
8.4.4 Compression 158
8.4.5 Boundary Handling 159
8.4.6 Multiple Dimensions 164
8.4.7 Implementation Considerations 164
8.4.7.1 Blocking 165
8.4.7.2 Wavelet Choice 165
8.4.7.3 Coefficient Addressing 166
8.4.8 A Hybrid Approach 166
8.4.9 Volume Rendering Example 167
8.5 Further Reading 167

9 *In Situ* Processing **171**

Hank Childs, Kwan-Liu Ma, Hongfeng Yu, Brad Whitlock, Jeremy Meredith, Jean Favre, Scott Klasky, Norbert Podhorszki, Karsten Schwan, Matthew Wolf, Manish Parashar, and Fan Zhang

9.1 Introduction 172
9.2 Tailored Co-Processing at High Concurrency 174
9.3 Co-Processing With General Visualization Tools Via Adaptors 175
9.3.1 Adaptor Design 178
9.3.2 High Level Implementation Issues 178
9.3.3 In Practice 179
9.3.4 Co-Processing Performance 181
9.4 Concurrent Processing 183
9.4.1 Service Oriented Architecture for Data Management in HPC 183
9.4.2 The ADaptable I/O System, ADIOS 184
9.4.3 Data Staging for *In Situ* Processing 185
9.4.4 Exploratory Visualization with VisIt and Paraview Using ADIOS 186
9.5 *In Situ* Analytics Using Hybrid Staging 187
9.6 Data Exploration and *In Situ* Processing 190
9.6.1 *In Situ* Visualization by Proxy 190
9.6.2 *In Situ* Data Triage 191
9.7 Conclusion 193

10 Streaming and Out-of-Core Methods **199**

David E. DeMarle, Berk Geveci, Jon Woodring, and Jim Ahrens

10.1 External Memory Algorithms 200
10.2 Taxonomy of Streamed Visualization 202
10.3 Streamed Visualization Concepts 204

10.3.1 Data Structures 204
10.3.2 Repetition 205
10.3.3 Algorithms 205
10.3.4 Sparse Traversal 207
10.4 Survey of Current State of the Art 209
10.4.1 Rendering 209
10.4.2 Streamed Processing of Unstructured Data 210
10.4.3 General Purpose Systems 211
10.4.4 Asynchronous Systems 212
10.4.5 Lazy Evaluation 214
10.5 Conclusion 215

III Advanced Architectural Challenges and Solutions 221

11 GPU-Accelerated Visualization 223
Marco Ament, Steffen Frey, Christoph Müller, Sebastian Grottel, Thomas Ertl, and Daniel Weiskopf
11.1 Introduction 224
11.2 Programmable Graphics Hardware 225
11.2.1 High-Level Shader Languages 226
11.2.2 General Purpose Computing on GPUs 227
11.2.3 GPGPU Programming Languages 228
11.3 GPU-Accelerated Volume Rendering 229
11.3.1 Basic GPU Techniques 229
11.3.1.1 2D Texture-Based Rendering 229
11.3.1.2 3D Texture-Based Rendering 230
11.3.1.3 Ray Casting 231
11.3.2 Advanced GPU Algorithms 231
11.3.3 Scalable Volume Rendering on GPU-Clusters 233
11.3.3.1 Sort-Last Volume Rendering 233
11.3.3.2 Sort-First Volume Rendering 234
11.4 Particle-Based Rendering 235
11.4.1 GPU-Based Glyph Rendering 236
11.4.2 Large Molecular Dynamics Visualization 238
11.4.3 Iterative Surface Ray Casting 239
11.5 GPGPU High Performance Environments 240
11.5.1 New Challenges in GPGPU Environments 240
11.5.2 Distributed GPU Computing 241
11.5.3 Distributed Heterogeneous Computing 242
11.6 Large Display Visualization 243
11.6.1 Flat Panel-Based Systems 243
11.6.2 Projection-Based Systems 244
11.6.3 Rendering for Large Displays 246

12 Hybrid Parallelism **261**

E. Wes Bethel, David Camp, Hank Childs, Christoph Garth, Mark Howison, Kenneth I. Joy, and David Pugmire

12.1 Introduction . . . 262
12.2 Hybrid Parallelism and Volume Rendering . . . 264
12.2.1 Background and Previous Work . . . 264
12.2.2 Implementation . . . 265
12.2.2.1 Shared-Memory Parallel Ray Casting . . . 266
12.2.2.2 Parallel Compositing . . . 266
12.2.3 Experiment Methodology . . . 267
12.2.4 Results . . . 268
12.2.4.1 Initialization . . . 268
12.2.4.2 Ghost Data/Halo Exchange . . . 269
12.2.4.3 Ray Casting . . . 269
12.2.4.4 Compositing . . . 272
12.2.4.5 Overall Performance . . . 272
12.3 Hybrid Parallelism and Integral Curve Calculation . . . 275
12.3.1 Background and Context . . . 275
12.3.2 Design and Implementation . . . 276
12.3.2.1 Parallelize Over Seeds . . . 276
12.3.2.2 Parallelize Over Blocks . . . 277
12.3.3 Experiment Methodology . . . 278
12.3.3.1 Factors Influencing Parallelization Strategy . 278
12.3.3.2 Test Cases . . . 279
12.3.3.3 Runtime Environment . . . 279
12.3.3.4 Measurements . . . 280
12.3.4 Results . . . 280
12.3.4.1 Parallelization Over Seeds . . . 280
12.3.4.2 Parallelization Over Blocks . . . 282
12.4 Conclusion and Future Work . . . 283

13 Visualization at Extreme Scale Concurrency **291**

Hank Childs, David Pugmire, Sean Ahern, Brad Whitlock, Mark Howison, Prabhat, Gunther Weber, and E. Wes Bethel

13.1 Overview—Pure Parallelism . . . 292
13.2 Massive Data Experiments . . . 293
13.2.1 Varying over Supercomputing Environment . . . 296
13.2.2 Varying over I/O Pattern . . . 297
13.2.3 Varying over Data Generation . . . 298
13.3 Scaling Experiments . . . 299
13.3.1 Study Overview . . . 299
13.3.2 Results . . . 300
13.4 Pitfalls at Scale . . . 301
13.4.1 Volume Rendering . . . 301
13.4.2 All-to-One Communication . . . 303

13.4.3 Shared Libraries and Start-up Time 304
13.5 Conclusion 305

14 Performance Optimization and Auto-Tuning 307

E. Wes Bethel and Mark Howison

14.1 Introduction 308
14.2 Optimizing Performance of a 3D Stencil Operator on the GPU 310
14.2.1 Introduction and Related Work 310
14.2.2 Design and Methodology 312
14.2.3 Results 313
14.2.3.1 Algorithmic Design Option: Width-, Height-, and Depth-Row Kernels 313
14.2.3.2 Device-Specific Feature: Constant Versus Global Memory for Filter Weights 314
14.2.3.3 Tunable Algorithmic Parameter: Thread Block Size 314
14.2.4 Lessons Learned 317
14.3 Optimizing Ray Casting Volume Rendering on Multi-Core GPUs and Many-Core GPUs 317
14.3.1 Introduction and Related Work 317
14.3.2 Design and Methodology 319
14.3.3 Results 320
14.3.3.1 Tunable Parameter: Image Tile Size/CUDA Block Size 320
14.3.3.2 Algorithmic Optimization: Early Ray Termination 323
14.3.3.3 Algorithmic Optimization: Z-Ordered Memory 324
14.3.4 Lessons Learned 325
14.4 Conclusion 326

15 The Path to Exascale 331

Sean Ahern

15.1 Introduction 332
15.2 Future System Architectures 332
15.3 Science Understanding Needs at the Exascale 335
15.4 Research Directions 338
15.4.1 Data Processing Modes 338
15.4.1.1 *In Situ* Processing 338
15.4.1.2 Post-Processing Data Analysis 339
15.4.2 Visualization and Analysis Methods 341
15.4.2.1 Support for Data Processing Modes 341
15.4.2.2 Topological Methods 342
15.4.2.3 Statistical Methods 343
15.4.2.4 Adapting to Increased Data Complexity . . . 343

15.4.3 I/O and Storage Systems 344
15.4.3.1 Storage Technologies for the Exascale 345
15.4.3.2 I/O Middleware Platforms 346
15.5 Conclusion and the Path Forward 347

IV High Performance Visualization Implementations 355

16 VisIt: An End-User Tool for Visualizing and Analyzing Very Large Data 357

Hank Childs, Eric Brugger, Brad Whitlock, Jeremy Meredith, Sean Ahern, David Pugmire, Kathleen Biagas, Mark Miller, Cyrus Harrison, Gunther H. Weber, Hari Krishnan, Thomas Fogal, Allen Sanderson, Christoph Garth, E. Wes Bethel, David Camp, Oliver Rübel, and Marc Durant, Jean M. Favre, Paul Navrátil

16.1 Introduction 358
16.2 Focal Points 359
16.2.1 Enable Data Understanding 359
16.2.2 Support for Large Data 360
16.2.3 Provide a Robust and Usable Product for End Users . 360
16.3 Design 361
16.3.1 Architecture 361
16.3.2 Parallelism 362
16.3.3 User Interface Concepts and Extensibility 363
16.3.4 The Size and Breadth of VisIt 364
16.4 Successes 364
16.4.1 Scalability Successes 365
16.4.2 A Repository for Large Data Algorithms 365
16.4.3 Supercomputing Research Performed with VisIt 366
16.4.4 User Successes 366
16.5 Future Challenges 368
16.6 Conclusion 368

17 IceT 373

Kenneth Moreland

17.1 Introduction 373
17.2 Motivation 374
17.3 Implementation 374
17.3.1 Theoretical Limitations ... and How to Break Them . 375
17.3.2 Pixel Reduction Techniques 376
17.3.3 Tricks to Boost the Frame Rate 377
17.4 Application Programming Interface 378
17.4.1 Image Generation 378
17.4.2 Opaque versus Transparent Rendering 379
17.5 Conclusion 379

18 The ParaView Visualization Application 383
Utkarsh Ayachit, Berk Geveci, Kenneth Moreland, and John Patchett, Jim Ahrens
18.1 Introduction . . . 384
18.2 Understanding the Need . . . 384
18.3 The ParaView Framework . . . 386
18.3.1 Configurations . . . 387
18.4 Parallel Data Processing . . . 387
18.5 The ParaView Application . . . 390
18.5.1 Graphical User Interface . . . 390
18.5.2 Scripting with Python . . . 391
18.6 Customizing with Plug-ins and Custom Applications . . . 391
18.7 Co-Processing: *In Situ* Visualization and Data Analysis . . . 392
18.8 ParaViewWeb: Interactive Visualization for the Web . . . 393
18.9 ParaView In Use . . . 394
18.9.1 Identifying and Validating Fragmentation in Shock Physics Simulation . . . 394
18.9.2 ParaView at the Los Alamos National Laboratory . . . 396
18.9.3 Analyzing Simulations of the Earth's Magnetosphere . 397
18.10Conclusion . . . 398

19 The ViSUS Visualization Framework 401
Valerio Pascucci, Giorgio Scorzelli, Brian Summa, Peer-Timo Bremer, Attila Gyulassy, Cameron Christensen, Sujin Philip, and Sidharth Kumar
19.1 Introduction . . . 402
19.2 ViSUS Software Architecture . . . 402
19.3 Applications . . . 408

20 The VAPOR Visualization Application 415
Alan Norton and John Clyne
20.1 Introduction . . . 415
20.1.1 Features . . . 416
20.1.2 Limitations . . . 417
20.2 Progressive Data Access . . . 417
20.2.1 VAPOR Data Collection . . . 418
20.2.2 Multiresolution . . . 419
20.3 Visualization-Guided Analysis . . . 420
20.4 Progressive Access Examination . . . 422
20.4.1 Discussion . . . 423
20.5 Conclusion . . . 424

21 The EnSight Visualization Application 429
Randall Frank and Michael F. Krogh
21.1 Introduction . . . 429
21.2 EnSight Architectural Overview . . . 430

21.3 Cluster Abstraction: CEIShell 432
21.3.1 Virtual Clustering Via CEIShell Roles 433
21.3.2 Application Invocation 434
21.3.3 CEIShell Extensibility 434
21.4 Advanced Rendering 435
21.4.1 Customized Fragment Rendering 435
21.4.2 Image Composition System 438
21.5 Conclusion 440

Index **443**

Acknowledgments

Any work of this magnitude would not have been possible without the support, encourgament, and assistance of a large number of people. The editors wish to acknowledge:

- Mary C. Hester for her invaluable assistance and hard work with technical editing of this manuscript.
- The support staff and editors at CRC Press, including Ms. Randi Cohen, for assistance in preparation and production of this manuscript.
- Horst Simon for encouraging us to create this work.
- The program managers at the agencies that have funded the research that produced this work. The specific funding sources and offices are listed in the following subsections.
- The Visualization Group at Lawrence Berkeley National Laboratory for help with reviewing and editing manuscript copy.
- Our friends and families for their unwavering support during the preparation of this manuscript.

Chapter 2–Parallel Visualization Frameworks

This work was supported by the Director, Office of Science, Office and Advanced Scientific Computing Research, of the U.S. Department of Energy under Contract No. DE-AC02-05CH11231.

Chapter 3–Remote and Distributed Visualization Architectures

This work was supported by the Director, Office of Science, Office and Advanced Scientific Computing Research, of the U.S. Department of Energy under Contract No. DE-AC02-05CH11231.

Chapter 5–Parallel Image Compositing Methods

We gratefully acknowledge the use of the resources of the Argonne and Oak Ridge Leadership Computing Facilities at Argonne and Oak Ridge National Laboratories. This work was supported by the Office of Advanced Scientific Computing Research, Office of Science, U.S. Department of Energy, under Contract DE-AC02-06CH11357. Work is also supported by DOE with agreement No. DE-FC02-06ER25777.

This research was supported in part by the U. S. Department of Energy through the SciDAC program with Award No. DE-FC02-06ER25777, DE-SC0005334, and DE-SC0005373, and the National Science Foundation with Award No. OCI-0850566 and through the PetaApps Program with Award No. OCI-0749227 and OCI-0905008.

Chapter 6–Parallel Integral Curves

This work was supported by the Director, Office of Science, Office and Advanced Scientific Computing Research, of the U.S. Department of Energy under Contract No. DE-AC02-05CH11231 through the Scientific Discovery through Advanced Computing (SciDAC) program's Visualization and Analytics Center for Enabling Technologies (VACET).

This research used resources of the National Energy Research Scientific Computing Center (NERSC), which is supported by the Office of Science of the U.S. Department of Energy under Contract No. DE-AC02- 05CH11231; used resources of the Oak Ridge Leadership Computing Facility (OLCF), which is supported by the Office of Advanced Scientific Computing Research, Office of Science, of the U.S. Department of Energy under Contract No. DE-AC05-00OR22725; and used resources of the Argonne Leadership Computing Facility at Argonne National Laboratories, which is supported by the Office of Advanced Scientific Computing Research, Office of Science, U.S. Department of Energy, under Contract DE-AC02-06CH11357, and also supported by DOE with agreement No. DE-FC02-06ER25777.

Chapter 7–Query-driven Visualization and Analysis

This work was supported by the Director, Office of Science, Office and Advanced Scientific Computing Research, of the U.S. Department of Energy under Contract No. DE-AC02-05CH11231 through the Scientific Discovery through Advanced Computing (SciDAC) program's Visualization and Analytics Center for Enabling Technologies (VACET).

This research used resources of the National Center for Computational Sciences at Oak Ridge National Laboratory, which is supported by the Office of Science of the U.S. Department of Energy under Contract No. DE-AC05-00OR22725.

Chapter 8–Progressive Data Access for Regular Grids

The author would like to thank Pablo Mininni for the provision of the Taylor Green turbulence data set and many insightful discussions on computational fluid dynamics. The author would also like to thank Andrea Clyne for her help with editing. This work was supported by the Computational Information Systems Laboratory (CISL), at the National Center for Atmospheric Research (NCAR).

Chapter 9–*In Situ* Processing

This work was supported by the Director, Office of Science, Office and Advanced Scientific Computing Research, of the U.S. Department of Energy under Contract No. DE-AC02-05CH11231.

Further, this research was supported in part by the U.S. Department of Energy through the SciDAC program with Award No. DE-FC02-06ER25777 and DE-SC0005373, and the National Science Foundation with Award No. OCI-0850566 and through the PetaApps Program with Award No. OCI-0749227 and OCI-0905008.

At Georgia Tech, the technical contributors to their work include both full-time researchers, Greg Eisenhauer, Karsten Schwan, and Matt Wolf, and many current and former PhD students. For the latter, they acknowledge the research conducted by Hasan Abbasi, Jianting Cao, Jai Dayal, Jay Lofstead, Fang Zheng, and several MS and undergraduate students from Georgia Tech.

At Rutgers, the technical contributors would like to acknowledge the contributions of Ciprian Docan and Tong Jin to the research presented in this chapter. The research presented in this chapter was supported in part by the Department of Energy ExaCT Combustion Co-Design Center via subcontract number 4000110839 from UT Battelle, by the Scalable Data Management, Analysis, and Visualization (SDAV) Institute via the grant number DE-SC0007455, and via DOE grant number DE-FG02-06ER54857. The research and was conducted as part of the NSF Cloud and Autonomic Computing (CAC) Center at Rutgers University.

Chapter 10–Streaming and Out-of-Core Methods

Kitware is an open-source solutions provider specializing in advanced research and development in the areas of visualization, medical imaging, computer vision, quality software process, data management, and informatics.

Further, work was supported by the Advanced Simulation and Computing program of the U.S. Department of Energy's NNSA. Los Alamos National Laboratory, an affirmative action/equal opportunity employer, is operated by Los Alamos National Security, LLC, for the National Nuclear Security Administration of the U.S. Department of Energy under contract DE-AC52-06NA25396. LANL LA-UR-12-21714

Chapter 12–Hybrid Parallelism

This work was supported by the Director, Office of Science, Office and Advanced Scientific Computing Research, of the U.S. Department of Energy under Contract No. DE-AC02-05CH11231 through the Scientific Discovery through Advanced Computing (SciDAC) program's Visualization and Analytics Center for Enabling Technologies (VACET).

This research used resources of the National Center for Computational Sciences at Oak Ridge National Laboratory, which is supported by the Office of Science of the U.S. Department of Energy under Contract No. DE-AC05-00OR22725.

The authors acknowledge the Texas Advanced Computing Center (TACC) at the University of Texas at Austin for providing the GPU cluster Longhorn that contributed to the research results reported within this paper.

Chapter 13–Visualization at Extreme Scale Concurrency

This work was supported by the Director, Officce of Advanced Scientific Computing Research, Offfice of Science, of the U.S. Department of Energy under Contract No. DE-AC02-05CH11231 through the Scientific Discovery through Advanced Computing (SciDAC) program's Visualization and Analytics Center for Enabling Technologies (VACET).

This work used resources at: (1) the National Energy Research Scientific Computing Center, which is supported by the Office of Science of the U.S. Department of Energy under Contract No. DE-AC02-05CH11231; (2) the Livermore Computing Center at Lawrence Livermore National Laboratory, which is supported by the National Nuclear Security Administration of the U.S. Department of Energy under Contract DE-AC52-07NA27344; (3) the Center for Computational Sciences at Oak Ridge National Laboratory, which is supported by the Office of Science of the U.S. Department of Energy under Contract No. De-AC05-00OR22725.

Chapter 14–Performance Optimization and Auto-tuning

This work was supported by the Director, Office of Science, Office and Advanced Scientific Computing Research, of the U.S. Department of Energy under Contract No. DE-AC02-05CH11231 through the Scientific Discovery through Advanced Computing (SciDAC) program's Visualization and Analytics Center for Enabling Technologies (VACET). This work used resources of the National Center for Computational Sciences at Oak Ridge National Laboratory, which is supported by the Office of Science of the U.S. Department of Energy under Contract No. DE-AC05-00OR22725, and of the National Energy Research Scientific Computing Center (NERSC), which is supported by the Office of Science of the U.S. Department of Energy under Contract No. DE-AC02-05CH11231. This work was also supported by Brown University through the use of the facilities of its Center for Computation and Visualization.

Chapter 16–VisIt

This work was supported by the Director, Office of Science, Office and Advanced Scientific Computing Research, of the U.S. Department of Energy under Contract No. DE-AC02-05CH11231.

Chapter 17–IceT

Sandia National Laboratories is a multi-program laboratory managed and operated by Sandia Corporation, a wholly owned subsidiary of Lockheed Martin Corporation, for the U.S. Department of Energy's National Nuclear Security Administration under contract DE-AC04-94AL85000.

Chapter 18–ParaView

Kitware is an open-source solutions provider specializing in advanced research and development in the areas of visualization, medical imaging, computer vision, quality software process, data management, and informatics.

Sandia National Laboratories is a multi-program laboratory operated by Sandia Corporation, a wholly owned subsidiary of Lockheed Martin Corporation, for the U.S. Department of Energy's National Nuclear Security Administration.

Work supported by the Advanced Simulation and Computing program of the U.S. Department of Energy's NNSA. Los Alamos National Laboratory, an affirmative action/equal opportunity employer, is operated by Los Alamos National Security, LLC, for the National Nuclear Security Administration of the U.S. Department of Energy under contract DE-AC52-06NA25396. LANL LA-UR-12-21714.

Chapter 20–The VAPOR Visualization Application

We would like to thank Pablo Mininni and Mel Shapiro for the provision of the MHD and Erica data sets, respectively, and for their guidance on VAPOR's development. VAPOR is a product of the Computational and Information Systems Laboratory, at the National Center for Atmospheric Research. VAPOR was developed with support of the National Science Foundation (NSF 03-25934 and 09-06379).

Disclaimer

Chapter 1

Introduction

E. Wes Bethel

Lawrence Berkeley National Laboratory

1.1 Historical Perspective 1
1.2 Moore's Law and the Data Tsunami 1
1.3 Focus of this Book 3
1.4 Book Organization and Themes 3
1.5 Conclusion 6
References 6

1.1 Historical Perspective

As we are taught from an early age, the scientific process includes the following steps: ask a question, perform background research to see whether others have found answers, formulate a hypothesis designed to answer the question, design and conduct an experiment to answer the question, analyze data collected from the experiment, draw conclusions, revise the hypothesis and repeat if necessary. Visual data exploration and analysis, or just visualization, plays a central role in this process: looking at or analyzing data from experiments is a crucial part of the scientific discovery process.

Galileo Galilei's improvements to early telescope designs first opened up the heavens, the satellites of Jupiter, sunspots, and even the rotation of the Sun. He proved the Copernican heliocentric model of the solar system: it is the Sun, not the Earth, that is the center of the solar system. Thus, the telescope became the first device to make the "unseeable" "seeable."

In 1987, a landmark report from the computer graphics and visualization community [3] coins the term "visualization in scientific computing" and amplifies on the subject of its central role in the scientific discovery process. Through a series of operations, or processing steps, known as the *visualization pipeline* (Fig. 1.1), visualization transforms abstract data into readily comprehensible images (e.g., Fig. 1.2). Today, scientific visualization plays a central role and indispensable role in contemporary science.

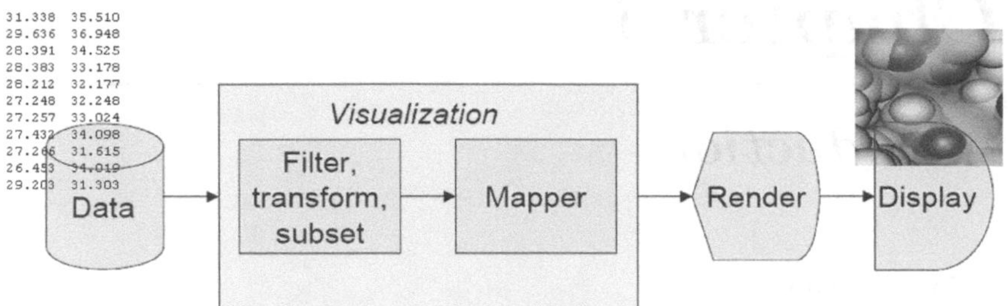

FIGURE 1.1: The visualization pipeline. Abstract data is input to the pipeline, where the pipeline transforms it into readily comprehensible images. Image courtesy of E. Wes Bethel (LBNL).

FIGURE 1.2: This image shows the merging of two vortex cores, something not possible to see by examining tables of numbers. Image courtesy of David Pugmire (ORNL), sample AMR data courtesy of Phil Colella (LBNL) and the SciDAC Applied Partial Differential Equations Center.

1.2 Moore's Law and the Data Tsunami

For most of the history of scientific computation, Moore's law [4] predicts the doubling of transistors per unit of area and cost every 18 months. The impact of this effect has been faster and smaller central processing units (CPUs), increased memory capacity, cameras having higher resolution, etc.

A direct result of this effect is that our ability to generate, collect, and store data vastly outpaces our ability to gain insight from it. We are overwhelmed by a deluge of data, a data tsunami. One of the major challenges faced by modern science is the explosion of information. It is widely accepted one of the most significant bottlenecks in modern science is the combination of managing an ever-increasing amount of information, and deriving knowledge or insight from massive data [1].

The field of computing has evolved rapidly over the past few decades from single-processor platforms having a few kilobytes of memory to those having almost a million processors and hundreds of terabytes of memory with no end in sight. Computational power is measured in floating point operations per second (FLOPs). Petascale computing (10^{15} FLOPS) is now prevalent and exascale computing (10^{18} FLOPS) is predicted to arrive in the next decade. The software tools created for these machines, whether for solving equations or for analyzing data, have, by necessity, had to evolve with technology and must continue to evolve.

1.3 Focus of this Book

Over the years, the field of visualization has also rapidly evolved to now include many different and diverse research focus areas: scalar, vector, and tensor field visualization and analysis; non-spatial information visualization; geometric modeling and analysis; virtual environments and display technology; software environments; perceptual issues; etc. Survey works like Hansen and Johnson, 2004 [2] provide an overview of the diversity of research in our community.

The main focus of this book is on topics that lie at the nexus between a rapidly evolving computational landscape, and visualization and analysis algorithmic research that targets this fast moving target. In this nexus are many other topics, which are covered by other volumes in this CRC Computational Science Series. While it is impossible to provide coverage of all relevant high performance visualization topics in the space of a single volume, we have endeavored to provide coverage of the major thematic areas.

1.4 Book Organization and Themes

Part I: Distributed-Memory Parallel Concepts and Systems. These chapters introduce concepts that are fundamental to parallel visualization.

- Chapter 2, *Parallel Visualization Frameworks*, introduces the concepts of parallel visualization algorithms and frameworks.
- Chapter 3, *Remote and Distributed Visualization Architectures*, examines the use of distributed computing resources for the purposes of performing visualization.

- Chapter 4, *Rendering*, discusses a crucial part of the visualization pipeline, namely the transformation of raw data and visualization processing results into an image.

- Chapter 5, *Parallel Image Compositing Methods*, presents techniques for assembling a final image from multiple partial images, such as those produced by parallel rendering components on large-scale computational platforms.

- Chapter 6, *Parallel Integral Curves*, shows the complexities of parallelizing a staple visualization algorithm, integral curve calculation, which forms the basis of a family of techniques for visualizing and analyzing complex flow fields.

Part II: Advanced Processing Techniques. Over the years, a substantial amount of research has focused on ways of accelerating visualization and analysis operations on HPC platforms. These chapters describe approaches that are complementary to parallelization strategies, but that are part of an overall strategy for achieving high performance visualization.

- Chapter 7, *Query-Driven Visualization and Analysis*, accelerates visualization and analysis by focusing processing on subsets of extreme-scale data deemed to be "scientifically interesting" using a combination of HPC computational platforms, advanced index/query technology, and effective user interfaces.

- Chapter 8, *Progressive Data Access for Regular Grids*, presents methods for the progressive access to, and processing of, structured mesh-based data, where a wavelet transformation enables both progressive and multiresolution access to and representation of scientific data.

- Chapter 9, *In Situ Processing*, is about performing visualization and analysis processing at the same time the data is computed, to avoid costly I/O.

- Chapter 10, *Streaming and Out-of-Core Methods*, presents techniques for visualization processing when the problem, or data set, is too large to fit entirely into memory.

Part III: Advanced Architectural Challenges and Solutions. The chapters in this part look forward into the future at alternative platforms and architectures. Topics here include using GPUs for various types of visualization (and analysis) processing, hybrid-parallelism and its impact on high-concurrency visualization, building and running a visualization application at extreme scale concurrency, performance optimization of fundamental visualization algorithms on modern multi- and many-core platforms, and a look forward into the exascale.

- Chapter 11, *GPU-Accelerated Visualization*, reviews the fundamental principles of modern graphics hardware and summarizes the latest research in GPU-based visualization techniques for stand-alone and cluster-based systems.
- Chapter 12, *Hybrid Parallelism*, discusses a blend of shared- and distributed-memory parallelism, and its application to two staple visualization algorithms on large HPC systems.
- Chapter 13, *Visualization at Extreme Scale Concurrency*, presents the results of experiments designed to better understand scalability limitations of a visualization application on large HPC platforms when applied to some of the largest-ever scientific data sets.
- Chapter 14, *Performance Optimization and Auto-tuning*, presents two studies aimed at increasing performance of two types of visualization algorithms on multi and many-core platforms through a combination of auto-tuning, use of algorithmic optimizations, and use of device-specific features.
- Chapter 15, *The Path to Exascale*, presents a summary of issues facing the visualization and analysis community as we transition from petascale to exascale class computational platforms.

Part IV: High Performance Visualization Implementations. This section is a survey of contemporary, high performance visualization implementations.

- Chapter 16, *VisIt*, and Chapter 18, *ParaView* are open source, production-quality, petascale-capable visualization and analysis frameworks.
- Chapter 17, *IceT*, the Image Composition Engine for Tiles (IceT) is a high-performance sort-last parallel rendering library, designed for use in large-scale, distributed-memory rendering applications.
- Chapter 19, *ViSUS*, is an environment that allows interactive exploration of very large scientific models on a variety of hardware, including geographically distributed platforms.
- Chapter 20, *VAPOR*, is an open source visual data analysis package that uses a wavelet-based progressive data model to provide a highly interactive, platform independent, desktop exploration environment, capable of handling some of the largest numerical simulation outputs, yet requiring only commodity computing resources
- Chapter 21, *EnSight*, is a full-featured, interactive, high performance visualization tool capable of scaling to the largest data sets; it has the ability to effectively leverage advanced computer systems both at the desktop/display and in computational clusters.

1.5 Conclusion

This book describes the state of the art at the intersection of scientific visualization, large data, and trends in high performance computing. It will prepare the reader to apply its concepts and perform additional research in this space.

References

[1] Richard Mount (ed.). The Office of Science Data-Management Challenge. Report from the DOE Office of Science Data-Management Workshops. Technical Report SLAC-R-782, Stanford Linear Accelerator Center, March–May 2004.

[2] Charles Hansen and Christopher Johnson. *Visualization Handbook*. Academic Press, Orlando, FL, USA, 2004.

[3] B. H. McCormick, T. A. DeFanti, and M. D. Brown (eds.). Visualization in Scientific Computing. *Computer Graphics*, 21(6), November 1987.

[4] Gordon E. Moore. Cramming More Components Onto Integrated Circuits. *Electronics*, 38(8), April 1965.

Part I

Distributed Memory Parallel Concepts and Systems

Chapter 2

Parallel Visualization Frameworks

Hank Childs

Lawrence Berkeley National Laboratory

2.1	Introduction	9
2.2	Background	11
	2.2.1 Parallel Computing	11
	2.2.2 Data Flow Networks	12
2.3	Parallelization Strategy	13
2.4	Usage	16
2.5	Advanced Processing Techniques	17
	2.5.1 Contracts	18
	2.5.2 Data Subsetting	19
	2.5.3 Parallelization Artifacts	19
	2.5.4 Scheduling	21
2.6	Conclusion	21
	References	23

Parallelization is the most common way to deal with the large data sets regularly generated by simulations or captured through experiments. This chapter seeks to answer key questions about frameworks for parallelizing visualization algorithms: What is the nature of these frameworks? How are they used? How do they parallelize processing? What problems result from parallelization? And how can optimizations be incorporated?

2.1 Introduction

Parallel visualization frameworks exist to deal with "large data." Of course, the notion of what is "large" is relative. Here, "large data" is defined as data that is too large be processed, in its entirety, all at one time, because it exceeds the available memory. This definition has three important criteria: (1) in its entirety, (2) all at one time, and (3) exceeds the available memory. It is not surprising that the approaches dealing with large data address one or more of these three criteria. Parallel visualization frameworks approach this problem through the third criterion; they use parallel resources with enough memory to store the data, as well as any derived data generated from it. This approach is popular because it has proven capable of dealing with virtually all visualization use cases, from applying simple algorithms to complex combinations of algorithms and from exploration to presentation.

Alternatives to parallelization address the large data problem through the other two criteria. The first criterion—processing the data in its entirety—can be approached in multiple ways. One technique is data subsetting: to process only the salient portions of the data set and ignore the portions that do not affect the final picture. An example of such a technique is query-driven visualization, which is discussed in Chapter 7. Another technique is multiresolution processing, which views coarse versions of the data, by default, and only processes data at finer resolutions when necessary (see Chap. 8). The streaming technique attacks the problem through the second criterion, processing all of the data at one time. This technique, instead, treats the data set as being composed of multiple pieces and processes data one piece at a time (see Chap. 10). Note that these techniques and parallel visualization frameworks are not mutually exclusive. Parallel visualization frameworks are flexible; they do not require that data be read in its entirety or that all data is processed at one time. As discussed in 2.5, parallel visualization frameworks can be used to provide a parallel foundation for any of the data subsetting, streaming, or multiresolution techniques.

A parallel visualization framework is like any software framework. It provides abstractions for key concepts, such as visualization algorithms or data representations, that are easily extended. It provides infrastructure code that dictates how modules in the framework interact and manages the flow of control within the framework. These approaches have been well borne out from a lineage of data flow networks (discussed further in 2.2.2).

Data flow networks have played a major role in visualization and analysis software since the early 1990s [11, 9, 1, 7], as they are so effective in rapid application development for solving a variety of visualization problems. They provide an execution model, a data model (i.e., a way to represent data), and algorithms to transform data. It is somewhat surprising that these frameworks can solve such a wide range of problems, since the data access patterns and nature of visualization algorithms vary widely. Parallel visualization frameworks extend data flow networks to operate in a parallel setting. It is even more surprising that their extension to the parallel world also has been so successful, since data access patterns are even more varied in a parallel setting. At the heart of this success is the commonality between visualization algorithms: data loading, data transformation, and data presentation. By focusing on these abstractions, parallel visualization frameworks are able to successfully support myriad algorithms.

This chapter gives an overview of parallel visualization frameworks. It introduces concepts for parallel computing (2.2.1) and data flow networks (2.2.2). It also describes the basic approach for data parallel processing (2.3), the usage of parallel visualization frameworks (2.4), and, finally, how advanced processing techniques can be incorporated into the design (2.5).

2.2 Background

2.2.1 Parallel Computing

The typical *supercomputer* consists of multiple *nodes*, a network for communicating between the nodes, and a file system that allows for data to be written to a disk for long-term storage and later retrieval. Each individual node often contains multiple *cores*, with that node's memory being shared by the cores. Finally, each instance of a parallel program is called a *task* (or sometimes a *Processing Element* or PE).

Distributed-memory parallelism techniques are required to coordinate between the nodes of a supercomputer, as the memory for each node is private to the others. The most common paradigm for this coordination is *message passing* and the the *de facto* standard for the this approach is the Message Passing Interface (MPI) library [10]. A pure distributed-memory program has one task for each core on each node of the computer.

Shared-memory parallelism techniques are used to achieve increased performance within a node. With this approach, there are fewer tasks than cores on a node and *threads*, which are lightweight programs controlled by a task, run on the remaining cores. Threads can share memory amongst themselves and between the main thread associated with the task, allowing for optimizations not possible with distributed-memory parallelism. POSIX threads (or pthreads) [8] and Open Multi-Processing (OpenMP) [2] are common APIs that enable shared-memory programming. A pure shared-memory program has one task on one node and uses multiple threads to make use of the node's cores.

Some programs use *hybrid parallel* techniques: distributed-memory parallelism across nodes and shared-memory parallelism within nodes. The most common hybrid parallel configuration has one task per node, with this task controlling threads on each of the node's cores. (Hybrid parallelism is discussed further in Chap. 12.) This chapter focuses on the simpler case where there is one task for each core (i.e., one instance of the program on each core) and the coordination between all cores—even those on the same node—comes through message passing.

Consider a distributed-memory parallel computer with eight quad-core nodes. A pure distributed-memory parallel program would have 32 tasks running and none of these tasks would make use of shared-memory techniques (even though some cores would reside on the same node). A pure shared-memory parallel program would only be able to use one of the nodes and would use threads to access its four cores. A hybrid configuration, however, could have: eight tasks, each running with four threads; sixteen tasks, each running with two threads; or even, configurations where the number of tasks and threads per node varies.

Problem decomposition, which defines the unit of parallelization, is an important aspect of programming parallel computers. The two main classes of problem decomposition are *task parallelism* and *data parallelism.* Task parallelism focuses on the tasks, with each task having a unique role, like an assembly line. In the context of visualization, a task parallel approach typically has data passing from task to task in a fixed sequence, with each task performing a new transformation of the data. Data parallelism, on the other hand, focuses on the data, with each task responsible for a piece of the whole. In the context of visualization, a data parallel approach typically has each task loading a portion of a large data set and processing that piece. For large data visualization tasks, data parallelism is more widely used. Data parallelism can be easily adapted to work with an arbitrary number of tasks where task parallelism requires that each task have a unique role, tying the parallelism to the work being performed. This chapter, as well as many of the following ones, describes data parallel solutions.

2.2.2 Data Flow Networks

Data flow networks are frameworks, often in the form of libraries, that provide an execution model, a data model (i.e., a way to represent data), and algorithms to transform data. The design has proven popular. There are multiple instances in use by the community: the Application Visualization System (AVS) [11], Open Data Explorer (OpenDX) [1], SCIRun [5], and the Visualization ToolKit (VTK) [9]. Their systems produce a desired visualization or analysis by connecting modules to form a *pipeline* through which data flows. The modules' types are *sources*, *sinks*, and *filters.* Sources produce an output, sinks take an input, and filters both take an input and produce an output.

A simple example of a data flow network has a file reader source (which would read data from a file and produce a mesh with a scalar field, such as pressure), followed by an "isosurface" filter (which would calculate the triangle mesh corresponding to isosurfaces of pressure), and finish with a rendering module sink (which would transform the triangle mesh into an image). In this example, the file reader creates a data set, which flows to the isosurface filter and the isosurface filter, in turn, makes a new data set that flows to the rendering module.

Data flow networks have many strengths:

- The design hides complexities from developers ranging from moving data through the network to synchronization to scheduling and execution. This allows them to focus on developing individual modules, such as a data loader for a specific file format or such as an algorithm to create "hedgehog" glyphs that show vector orientation.
- The modules interoperate, allowing for users to combine functionality in

unforeseen ways. For example, users can combine an isosurfacing filter and a slicing filter to calculate "isolines," even though the filter developers did not anticipate the need to combine these modules. In fact, users frequently construct custom pipelines for their specific analyses that involve dozens of modules.

- The design is highly extensible. Since the frameworks define abstract notions for sources, sinks, filters, and data flow, new, concrete instances of these modules can be incorporated into the framework at any time and with minimal intrusion.

Of course, data flow networks also have key weaknesses:

- Typically, the execution of the modules occurs in stages, with filters operating one at a time. This is not a cache-efficient approach.
- The output from each filter or source is a new data set, increasing the total memory footprint.

Data flow network libraries manage *pipeline execution* using either *push* or *pull* designs. OpenDX uses the push approach. It employs an executive that analyzes a pipeline and determines which modules need to execute and in what order. Further, it manages a cache of previous computations, which saves some modules from redundantly rederiving the same results on subsequent executions. VTK uses a pull approach, also referred to as demand-driven processing. VTK has two distinct phases: *update* and *execute*. The update phase begins at a sink in a pipeline and asks this module to get up-to-date. The sink then asks its input(s) to update. This update propagates up the pipeline, until it reaches a source. The source produces its output, for example by reading from a file, if it is a data loader, and this begins the execute phase. Then, the first filter in the pipeline executes, and the execute phase propagates down the pipeline. VTK's caching is automatic. Each source and filter maintain explicit knowledge of their output and they only recalculate that output if one of their attributes changed or, in the case of filters, one of their inputs changed.

2.3 Parallelization Strategy

The most common data processing strategy for parallel visualization frameworks is *pure parallelism*. This strategy is the brute force approach: data parallelism where the data is processed in its entirety and all at one time. Pieces of the whole data set are partitioned over the tasks. Each task reads its pieces and processes them, typically using data flow networks. Typically, each task instantiates an identical data flow network and the tasks differ only in the pieces of the whole data set they operate on. After each task finishes executing its data flow network, the data is rendered. The details of the

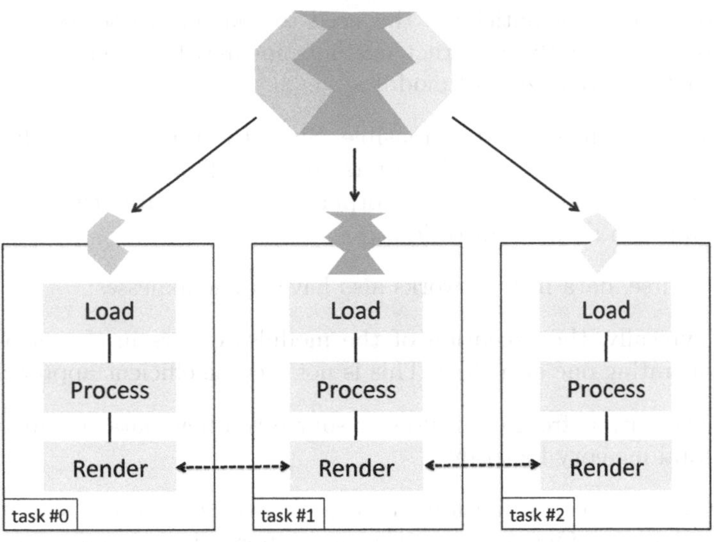

FIGURE 2.1: A parallel visualization framework. The data set for this simple example has three pieces, which are being distributed over three tasks. Parallel coordination is required for the rendering and may be required in the processing and loading as well.

rendering, which require parallel coordination, are described in Chapter 4. The most common case is for the data to be rendered in parallel on the tasks, although other use cases are described in 2.4. Figure 2.1 shows a diagram of this process.

Parallel resources bring benefits beyond increased memory to store data. Through increased parallelism, the visualization software has access to more I/O bandwidth (to load data faster) and more compute power (to execute its algorithms and render data more quickly). The efficacy of the pure parallelism approach has been repeatedly demonstrated. Chapter 13 describes an extreme example: it details an experiment where a parallel visualization framework scaled up to tens of thousands of tasks and processed data sets with trillions of cells.

The conventional data retrieval mechanism for parallel visualization frameworks is to have each task read a piece of the whole data set. This strategy is intuitive and easy to implement, since it does not require any parallel coordination. But the strategy is somewhat at odds with the complexities of I/O infrastructures for supercomputers. Generally, tasks do *not* have direct, dedicated access to the file system. Instead, their I/O requests are routed through a shared access point; for example, getting transparently forwarded by the operating system to a dedicated I/O node that performs I/O on behalf of other nodes. This means their requests are subject to contention and also, needlessly dividing large buffer reads in many smaller reads. (File systems perform best with few large reads compared to many small reads.) Some recent approaches build in explicit support for the details of the I/O infrastructure. This added complexity creates opportunities to aggregate read requests and gain significantly improved performance [6].

At their simplest, parallel visualization frameworks follow the "scatter-gather" design. Pieces of the data set are scattered over the tasks during the read phase. After scattering, the processing can often occur in isolation. Many visualization algorithms are embarrassingly parallel, meaning each task can execute on its piece of the data without coordinating with the other tasks. Examples of algorithms that fit this embarrassingly parallel model are isosurface calculation, slicing, and glyphing. Then, the parallel rendering phase serves to gather the results into a single location.

Even with embarrassingly parallel algorithms, artifacts can occur at the boundaries of pieces, unless special care is taken. The prototypical example of these artifacts are the broken isosurfaces that can occur at boundaries from inconsistent interpolations. These artifacts are typically remediated with ghost data, which are redundant layers of cells around each piece. More discussion of ghost data can be found in 2.5.3.

Non-embarrassingly parallel algorithms—algorithms that require coordination between the tasks—also can be accommodated within parallel visualization frameworks. These algorithms simply pass messages between tasks as they execute. Of course, care must be taken to coordinate with the overall execution of the framework. If, for example, each task executes once for each piece and if the number of pieces varies per task, then a collective communication pattern would deadlock. But, parallel visualization frameworks can be adapted to deal with this issue, and, in general, can fit any of the communication patterns that non-embarrassingly parallel algorithms employ. More discussion about incorporating non-embarrassingly parallel algorithms into parallel visualization frameworks can be found in 2.5.4.

Non-embarrassingly parallel visualization algorithms are common. Integral curve calculation, which is used to generate streamlines and other flow-based techniques, can divide its work in multiple ways, each employing different communication patterns. More description of these approaches can be found in Chapter 6. Identification of connected components, i.e., which regions of a data set are connected to other regions, requires parallel communication when the data is partitioned over tasks. Harrison et al. describe a technique for solving this problem, which uses all-to-all communication [4]. Of course, rendering gathers the results from all tasks, which requires parallel communication. Chapters 4 and 5 describe approaches for the parallel rendering of surfaces and volumes and how to composite the resulting images.

2.4 Usage

The most common usage of parallel visualization frameworks is via end-user applications. Part IV of this book describes the implementations of high performance visualization technologies, including four end-user applications. Three of these four—VisIt (discussed in Chap. 16), ParaView (Chap. 18), and EnSight (Chap. 21)—incorporate parallel visualization frameworks. In all three cases, the applications employ a client–server design where a parallel visualization framework is embedded into the server portion. Their servers execute in parallel on remote resources, enabling direct access to the data and sufficient compute power to calculate desired surfaces or images. The results are then transferred via a network, to their client, and displayed. Figure 2.2 shows an example. This client–server design is well suited for remote visualization for several reasons:

- Running a client on a local machine enables interactivity, both for the display of data and for manipulating the user interface. Further, in some situations, it enables the usage of GPUs on the local machine to accelerate rendering performance.
- When users have to move data between machines, they often cite it as the largest bottleneck in the overall visualization process. By running the server where the data resides, the data does not need to be transferred.
- Data often resides on the machine that generated it; the originating machines almost always have sufficient computational resources (e.g., processors, I/O, memory) to process this data.

The client–server design is not the only way to use parallel visualization frameworks, however. Some hardware configurations have their parallel resources directly connected to one or more displays, allowing visualizations to be directly displayed. A single connection to a display is common with dedicated visualization machines that connect one of their tasks to a user's monitor. End-user applications for this environment have only the connected task perform the user interface portions and rendering. The other tasks process data and communicate their results to the connected task, which may process data as well. Multiple connections to displays are common with "powerwalls" (high-resolution displays made up of tiles of traditional displays). In this case, each tile is connected directly to a task and each task is responsible for displaying the results for its tile.

When there is no connection to a display, a common usage of parallel visualization frameworks is with "windowless" applications. These applications process data in parallel, produce results, and write the output—typically images—to the file system for later study. Finally, some applications use parallel visualization frameworks on remote machines and have one of their tasks display the results, as if the user was directly connected to the task. In this

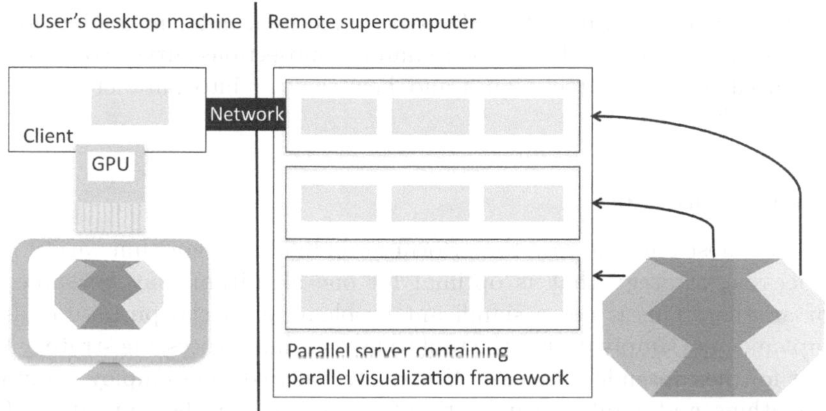

FIGURE 2.2: The client runs on the user's desktop machine, enabling interactivity for the program's user interface. In this example, the client directs the server via a direct network connection to one of the server's tasks. The server uses parallel visualization frameworks to parallelize the processing of the data. The resulting surface can then be transferred via the network connection back to the client, with accelerated rendering via the GPU. Alternatively, the server can render in parallel and transfer images to the client over the network. Chapter 3 describes this architecture and related ones in more depth.

case, the results are captured by the operating system where the task is running, sent to where the user is running, and displayed there. This effect is typically accomplished through "ssh-tunneling" on Unix-based systems. A key distinction between this configuration and a client–server design is that a client–server design manages the transfer of results explicitly, creating opportunities for image compression and other optimizations.

Other portions of this book describe more sophisticated software architectures that also use parallel visualization frameworks. Chapter 3 describes distributed architectures that are optimized for rendering performance. Chapter 9 describes *in situ* processing techniques, which eliminate I/O costs by processing data while it is being generated. These *in situ* techniques incorporate a parallel visualization framework in different ways: sometimes directly into the simulation code, sometimes via dedicated applications that run concurrently, and sometimes both.

2.5 Advanced Processing Techniques

This chapter has, up until now, described a basic form of parallel visualization frameworks. This section focuses on more complex forms, however, specifically how these frameworks can be adapted to support advanced processing techniques. First, 2.5.1 describes a mechanism for managing advanced processing techniques. Then, the remaining subsections survey some common, advanced processing techniques and how they fit into parallel visualization frameworks.

2.5.1 Contracts

The most efficient way to perform an individual algorithm varies. A data processing strategy that is optimal for one algorithm may be sub-optimal for another. This is not a significant problem for special purpose tools. The implementers simply design the tool to use the data processing strategy that is best for their intended use case. But richly featured tools employ many varied algorithms, and parallel visualization frameworks form the foundation of these tools. So, parallel visualization frameworks must be flexible; they must offer multiple ways of partitioning data over tasks and multiple ways to drive the flow of execution.

Contracts [3], are a mechanism for bringing flexibility to parallel visualization frameworks. Contracts describe the possible constraints and optimizations for program execution. This description, in turn, allows for better control over data partitioning, flow of execution, and algorithm selection. The contract itself is simply a data structure, with data members that represent an enumerated set of known constraints and optimizations.

The contract is finalized through an iterative process that gives each filter in the pipeline a chance to modify its data members, to reflect that filter's particular constraints, and to reflect possibilities for optimization. The initial version of the contract has default values: no constraints for execution and no optimizations. The iterative process starts with the final filter in the pipeline, then the second-to-last filter in the pipeline, all the way up to the first filter. This ordering allows every filter to understand the constraints of the filters that will consume its output(s) and also to communicate its constraints to the filters that produce its input(s). Once the contract is finalized, execution can begin, informed by this final contract's constraints and optimizations. Figure 2.3 illustrates this process.

In terms of implementation, contracts fit seamlessly into demand-driven data flow. As the *Update* cycle goes from filter to filter, a contract can be passed along. Each time a filter receives an *Update* call, the filter modifies the contract and produces a new version.

Contracts can be used effectively with a very large number of filters, which is important since they are used with richly featured tools that have hundreds to thousands of different filters. By focusing on the constraints and optimizations, specific knowledge of a filter is not encoded in the contract. This allows extensions to be added without touching core code, an important property for a framework. Further, the design allows for multiple optimizations to be combined. If one filter eliminates some portion of the data from consideration

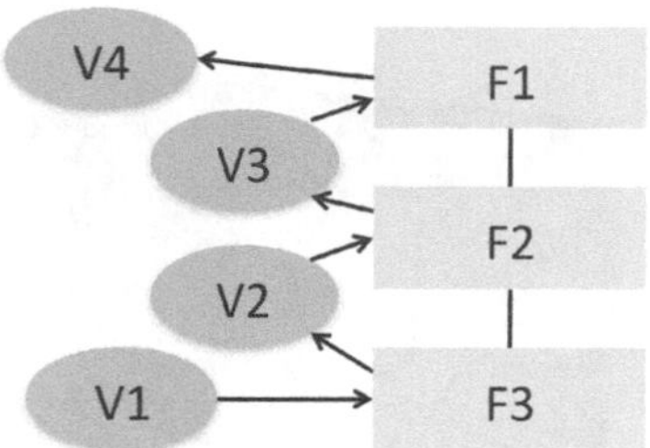

FIGURE 2.3: A contract summarizes the possible constraints and optimizations for the pipeline execution. Each filter has a chance to modify this contract. The initial version of the contract, $V1$, has default values. The final filter, $F3$, is the first to modify the contract, making a new version, $V2$. The next filter, $F2$, then modifies $V2$, to make a new version, $V3$. The first filter to execute, $F1$, is the last to modify $V3$, making the final version, $V4$. This version reflects the constraints and optimizations from all of the filters.

and another filter eliminates a different portion, then the final contract will prescribe processing only the data at the intersection of their remainders.

2.5.2 Data Subsetting

Data subsetting refers to only processing the portion of a data set that will affect the final result. Consider the example of slicing a 3D data set by a plane (see Fig. 2.4). Many of the pieces will not intersect the plane and reading them in would be a wasted effort. In fact, if P is the total number of pieces, then the number of pieces intersected by the slice is typically $O(P^{2/3})$.

It is possible to eliminate pieces from processing before ever reading them, if metadata is available. By accessing the spatial extents for each piece, the slice filter can calculate which pieces have bounding boxes that intersect the slice and only process those pieces. Note that false positives can potentially be generated by considering only the bounding box. In terms of the contracts described in 2.5.1, each piece would have its own Boolean field in the contract, indicating whether or not that piece should be processed. The Boolean for each piece would be *true* by default, but the slice filter would set it to *false* during its contract modification if the spatial metadata indicated it did not intersect the slice.

2.5.3 Parallelization Artifacts

As mentioned earlier, the approach of independently operating on pieces of a larger data set can introduce artifacts at the exteriors of the pieces. Even

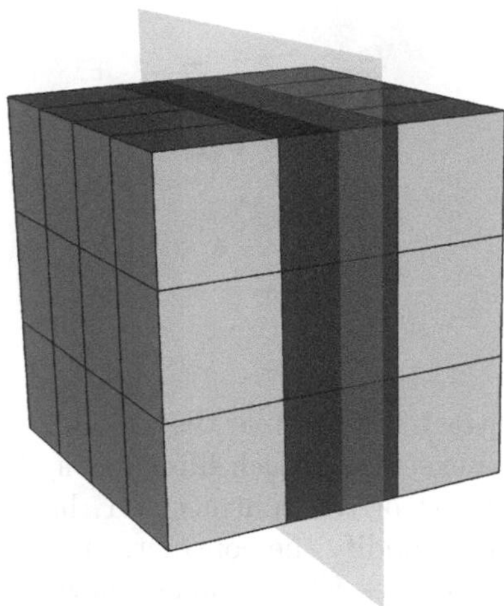

FIGURE 2.4: A data set decomposed into 36 pieces. Each piece is outlined with black lines and is colored light or dark gray. To create the data set sliced by the transparent plane, only the dark gray pieces need to be processed.

though the cause of these artifacts can differ, parallel visualization frameworks can be adapted to detect their causes and prevent them.

Some visualizations render the external faces of a volume—possibly after clipping away a region and giving the user a look inside a volume. However, calculating these external faces is difficult when operating on one piece at a time; faces that are external to a piece can be internal to the whole data set. And, if these faces are not eliminated, they can have multiple negative impacts. One negative impact is that the number of triangles being drawn can increase by an order of magnitude. Another negative impact is that these extra faces will be visible when the surface is transparent and result in an incorrect image. The solution to this problem is straightforward: identify faces that are external to a piece but internal to the data set. These faces are called *ghost faces.* In some cases, especially with unstructured grids, this identification requires significant additional memory, possibly in the form of a Boolean array for every face in the piece.

Interpolation is another cause of artifacts. If the data must be interpolated from cell centers to node centers—in anticipation of applying a Marching Cubes isosurface algorithm, which depends on nodal data, for example—then inconsistent values will get assigned to the nodes that are duplicated at the boundary of two pieces. These inconsistent values can ultimately in broken isosurfaces. The common solution to this problem is to introduce *ghost cells.* Ghost cells are an extra layer of cells around each piece that are used for interpolation, but discarded before rendering. The cost for this approach is even more significant. The piece's size is increased (more cells and more values) and additional memory must be allocated for understanding which cells are ghost and which are not, so the ghost cells can be removed before rendering.

Once again, contracts can be used to solve this problem. Each filter can modify the contract to identify their constraints, i.e., whether ultimately they will be calculating external faces or whether they will be performing interpolation. If they do, filters can be inserted to mark ghost faces or add ghost cells. If not, then no additional action is taken. In this case, this coordination allows the parallel visualization framework to prevent the conservative policy of always generating unnecessary ghost data; the correct picture can always be calculated at minimum cost.

2.5.4 Scheduling

Parallel visualization frameworks accommodate multiple strategies for scheduling the processing of pieces. Further, for a given execution, they can dynamically choose a strategy based on pipeline properties.

Dynamic load balancing is a scheduling strategy that assigns one task to play the role of *master* and the rest to be *slaves*. Any time a slave is idle, the master assigns it a piece for processing, until there is no more work to do. This strategy is particularly effective when the pieces take a different amount of time to process, as it naturally adapts to balance load. Further, the approach typically uses streaming (see Chap. 10), meaning that the memory requirements are lessened because only one piece at a time is processed by a task. But it does not work for all algorithms: some algorithms require all of the data to be immediately available and this strategy does not make some pieces available until later in the execution.

Static load balancing is a scheduling strategy that partitions the pieces over the tasks before execution. This strategy is prone to load imbalance when the time to process pieces is variable, impacting performance because the overall execution time is only as fast as the slowest task. However, this strategy has all of the data in memory at one time (spread out over the tasks), enabling certain classes of algorithms.

Again, contracts enable choosing the optimal strategy. Consider the two scheduling strategies just described. If an algorithm needs to have all of the data available to proceed, it can indicate this by setting a Boolean in the contract. Then, prior to execution, the parallel visualization framework can decide which strategy to employ. If any algorithm needs the entire data set available (and hence, set the Boolean in the contract), then the contract would select static load balancing. If not, it could choose the faster, dynamic load balancing.

2.6 Conclusion

Parallel visualization frameworks address the large data problem by enabling the usage of additional parallel resources, which increases available memory, compute power, and I/O bandwidth. Their parallelization strategies divide the data into pieces over their tasks, with individual algorithms executing in either embarrassingly or non-embarrassingly parallel manners. These frameworks are then used to build applications with sophisticated architectures, such as a client–server model with parallel visualization frameworks built into the server. Finally, these frameworks serve as the foundation for richly featured visualization tools, since they are flexible in usage, enabling optimizations for scheduling strategies, data subsetting, and many more.

References

[1] Greg Abram and Lloyd A. Treinish. An Extended Data-Flow Architecture for Data Analysis and Visualization. Research report RC 20001 (88338), IBM T. J. Watson Research Center, Yorktown Heights, NY, USA, February 1995.

[2] Barbara Chapman, Gabriele Jost, and Ruud van der Pas. *Using OpenMP: Portable Shared Memory Parallel Programming (Scientific and Engineering Computation).* The MIT Press, 2007.

[3] Hank Childs, Eric S. Brugger, Kathleen S. Bonnell, Jeremy S Meredith, Mark Miller, Brad J Whitlock, and Nelson Max. A Contract-Based System for Large Data Visualization. In *Proceedings of IEEE Visualization 2005*, pages 190–198, 2005.

[4] Cyrus Harrison, Hank Childs, and Kelly P. Gaither. Data-Parallel Mesh Connected Components Labeling and Analysis. In *Proceedings of EuroGraphics Symposium on Parallel Graphics and Visualization*, pages 131–140, April 2011.

[5] C.R. Johnson, S. Parker, and D. Weinstein. Large-Scale Computational Science Applications Using the SCIRun Problem Solving Environment. In *Proceedings of the 2000 ACM/IEEE Conference on Supercomputing*, 2000.

[6] Wesley Kendall, Jian Huang, Tom Peterka, Rob Latham, and Robert Ross. Visualization Viewpoint: Towards a General I/O Layer for Parallel Visualization Applications. *IEEE Computer Graphics and Applications*, 31(6), 2011.

[7] Konstantinos Konstantinides and John R. Rasure. The Khoros Software Development Environment for Image and Signal Processing. *IEEE Transactions on Image Processing*, 3:243–252, 1994.

[8] Bradford Nichols, Dick Buttlar, and Jacqueline Proulx Farrell. *Pthreads Programming.* O'Reilly & Associates, Inc., Sebastopol, CA, USA, 1996.

[9] William J. Schroeder, Kenneth M. Martin, and William E. Lorensen. The Design and Implementation of an Object-Oriented Toolkit for 3D Graphics and Visualization. In *VIS '96: Proceedings of the Conference on Visualization '96*, pages 93–ff. IEEE Computer Society Press, 1996.

[10] Marc Snir, Steve Otto, Steven Huss-Lederman, David Walker, and Jack Dongarra. *MPI-The Complete Reference, Volume 1: The MPI Core.* MIT Press, Cambridge, MA, USA, 2nd. (revised) edition, 1998.

[11] Craig Upson, Thomas Faulhaber Jr., David Kamins, David H. Laidlaw, David Schlegel, Jeffrey Vroom, Robert Gurwitz, and Andries van Dam. The Application Visualization System: A Computational Environment for Scientific Visualization. *Computer Graphics and Applications*, 9(4):30–42, July 1989.

Chapter 3

Remote and Distributed Visualization Architectures

E. Wes Bethel

Lawrence Berkeley National Laboratory

Mark Miller

Lawrence Livermore National Laboratory

3.1	Introduction	26
3.2	Visualization Performance Fundamentals and Networks	26
3.3	Send-Images Partitioning	28
3.4	Send-Data Partitioning	30
3.5	Send-Geometry Partitioning	31
3.6	Hybrid and Adaptive Approaches	32
3.7	Which Pipeline Partitioning Works the Best?	33
3.8	Case Study: Visapult	36
	3.8.1 Visapult Architecture: The Send-Geometry Partition	36
	3.8.2 Visapult Architecture: The Send-Data Partition	38
3.9	Case Study: Chromium Renderserver	38
3.10	Case Study: VisIt and Dynamic Pipeline Reconfiguration	42
	3.10.1 How VisIt Manages Pipeline Partitioning	43
	3.10.2 Send-Geometry Partitioning	43
	3.10.3 Send-Images Partitioning	44
	3.10.4 Automatic Pipeline Partitioning Selection	44
3.11	Conclusion	45
	References	46

The term *remote and distributed visualization* (RDV) refers to a mapping of visualization pipeline components onto distributed resources. Historically, the development of RDV was motivated by the user's need to perform analysis on data that was too large to move to their local workstation or cluster, or that exceeded the processing capacity of their local resources. RDV concepts are central in high performance visualization, where a visualization application is often a parallel application consisting of a collection of processing components working together to solve a large problem. This chapter presents RDV architectures and case studies that illustrate how these architectures are used in practice.

3.1 Introduction

Whereas many of the chapters in this book focus on architectures that run on a single, large, parallel platform, RDV concepts extend this idea, without loss of generality, to configurations consisting of more than one computational platform. Historically, this type of architecture was motivated by the reality of a user needing to perform visualization on a data set that was either too large to move across the network or that exceeded the capacity of their local computational platform. The RDV solution to this problem was the idea that part of the visualization pipeline would be executed "close to" or colocated with the large data, and then a smaller subset of data or visualization results would be sent to the user's local machine for additional processing and display.

From a high level, there are three fundamental types of bulk payload data that move between components of the visualization pipeline: "scientific data," visualization results (geometry and renderable objects), and image data [5, 18]. In some instances and applications, the portion of the pipeline that moves data between components is further resolved to distinguish between raw, or unprocessed, and filtered data [24, 11, 10]. For simplicity, these three partitioning strategies are referred to as *send images, send data,* and *send geometry,* as shown in Figure 3.1. The first few sections of this chapter discuss each of these strategies, as well as *hybrid* variants that are not strictly one of these three, but a combination of aspects from more than one of these three partitionings. The question of which pipeline partitioning works best is the subject of 3.7.

In reality, each or any of these individual stages of the visualization pipeline may be comprised of distributed or parallel resources. Three different case studies will be presented that explore alternative configurations of parallel components, comprised of high performance, remote and distributed visualization architectures. The Visapult architecture, which uses a hybrid partitioning combined with custom network protocols and that the Supercomputing Conference's Network Bandwidth Challenge three years in a row, is the subject of 3.8. Chromium Renderserver, which consists of a distributed-memory parallel scalable back-end rendering infrastructure that can deliver imagery to a remote client through an industry-standard protocol, is the subject of 3.9. Dynamic pipeline configuration across distributed resources in the VisIt application is the subject of 3.10.

3.2 Visualization Performance Fundamentals and Networks

Since RDV is about distributing components of the visualization pipeline, the overall performance of a distributed system is influenced in part by the following factors:

- *Component performance*: for a given amount of input data, each of the

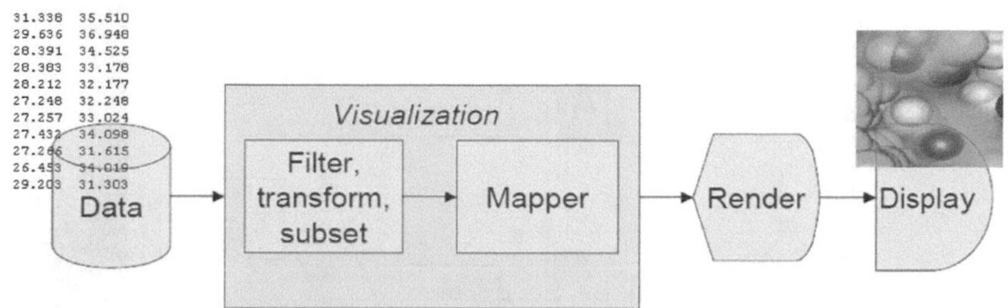

Figure 1.1 The visualization pipeline. Abstract data is input to the pipeline, where the pipeline transforms it into readily comprehensible images. Image courtesy of E. Wes Bethel (LBNL).

FIGURE 1.2 This image shows the merging of two vortex cores, something not possible to see by examining tables of numbers. Image courtesy of Dave Pugmire (ORNL), sample AMR data courtesy of Phil Colella (LBNL) and the SciDAC Applied Partial Differential Equations Center.

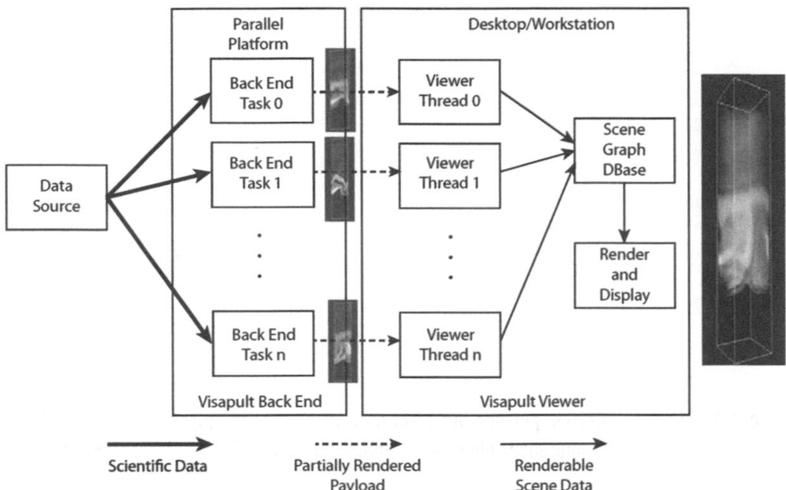

FIGURE 3.4 Visapult's remote and distributed visualization architecture. Image courtesy of E. Wes Bethel (LBNL).

FIGURE 3.6 An unmodified molecular docking application, run in parallel, on a distributed-memory system using CRRS. Here, the cluster is configured for a 3×2 tiled display setup. The monitor for the "remote user machine" in this image is the one on the right. Image source: Paul et al., 2008.

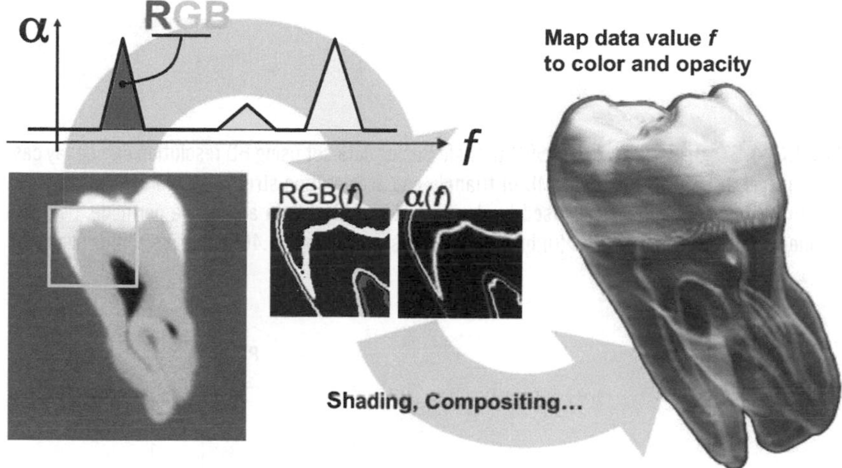

FIGURE 4.3 Volume rendering overview. The volume is sampled, samples are assigned optical properties (RGB, Alpha). The different properties are shaded and composited to form the final image.

FIGURE 4.4 On 60 nodes, the system can ray cast (left image) a 316M triangle isosurface from timestep 273 of a Richtmyer-Meshkov (RM) instability calculation (LLNL), in HD resolution, at 101fps, with replicated data, 21,326MB of triangles and acceleration structures, and at 16fps if the DC is used to store only 2,666MB per node. Using one shadow ray or 36 ambient occlusion rays per pixel (right image) the system can achieve 4.76fps and 1.90fps, respectively. Image source: Ize et al., 2011.

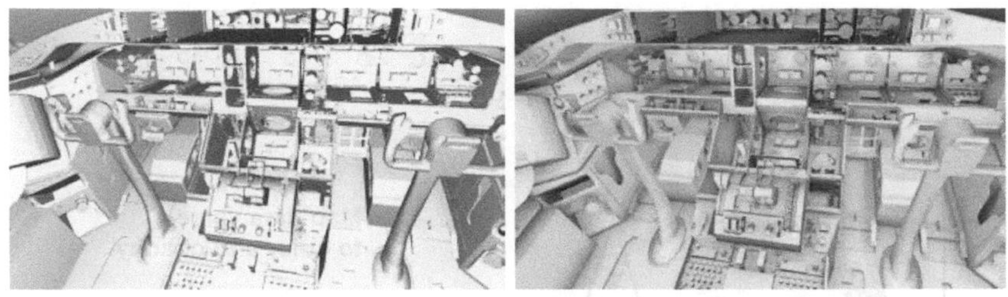

FIGURE 4.5 On 60 nodes, the 259M triangle Boeing data set using HD resolution can be ray cast (left image) at 96fps if all 15,637MB of triangle and acceleration structure data are replicated on each node and at 77fps if DC is used to store only 1,955MB of data and cache per node. Using 36 ambient occlusion rays per pixel (right image) the system achieves 1.46fps. Image source: Ize et al., 2011.

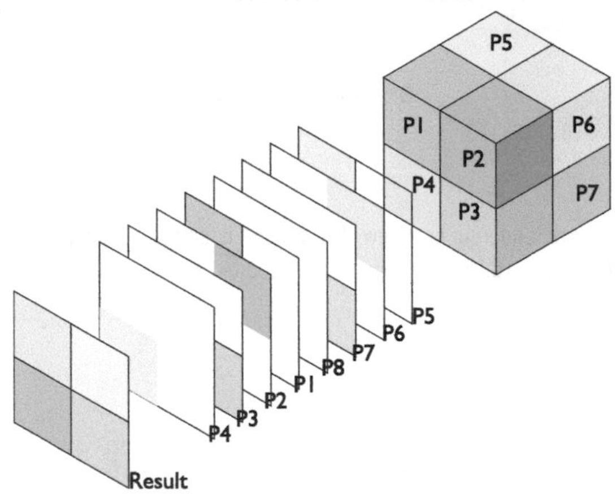

FIGURE 5.1 Sort-last image compositing.

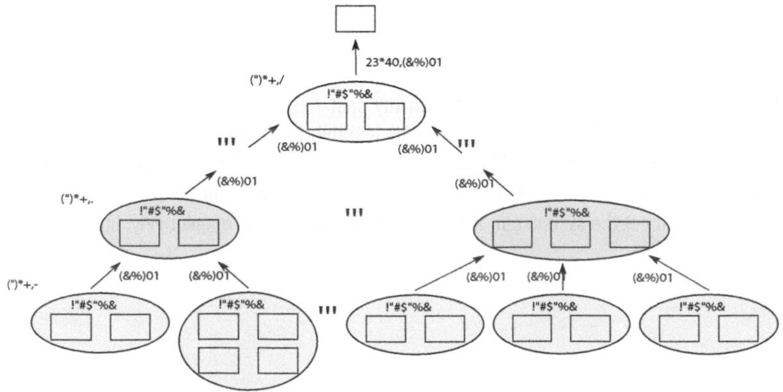

FIGURE 5.3 Tree-based compositing.

Table of target k-values for Intrepid BG/P and Jaguar XT5

Processes \ Megapixels	Intrepid !""	Intrepid #"	Intrepid $%"	Intrepid &'"	Jaguar !"	Jaguar #"	Jaguar $%"	Jaguar &'"
#"	#"	#"	#"	#"	!"	!"	!"	!"
$%"	$%"	$%"	$%"	$%"	!"	!"	!"	!"
&'"	&'"	&'"	&'"	&'"	$%"	#"	$%"	$%"
%!"	%!"	%!"	%!"	%!"	$%"	$%"	$%"	$%"
$'#"	%!"	$'#"	$'#"	$'#"	#"	#"	#"	#"
'(%"	%!"	$'#"	$'#"	$'#"	$%"	#"	#"	#"
($'"	%!"	$'#"	$'#"	$'#"	$%"	&'"	#"	#"
$")"	%!"	$'#"	$'#"	$'#"	%!"	&'"	&'"	#"
'")"	&'"	$'#"	$'#"	$'#"	#"	&'"	&'"	&'"
!")"	&'"	&'"	&'"	&'"	#"	$%"	&'"	%!"
#")"	&'"	&'"	&'"	&'"	!"	#"	&'"	%!"
$%")"	&'"	&'"	&'"	&'"	!"	&'"	&'"	#"
&'")"	&'"	&'"	&'"	&'"	$%"	$%"	%!"	#"

FIGURE 5.8 Target k-values for two different machines are shown. Optimizations such as active pixel encoding enable the use of higher k-values than before. In the original algorithm, $k = 8$ was used, but the tables above show that with active-pixel encoding, k-values as high as 128 are optimal, depending on the machine.

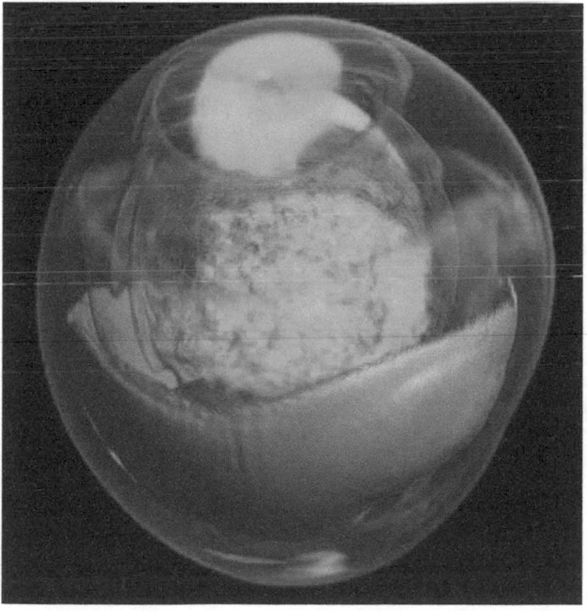

FIGURE 5.10 Core-collapse supernova volume rendered in parallel and composited using the radix-k algorithm.

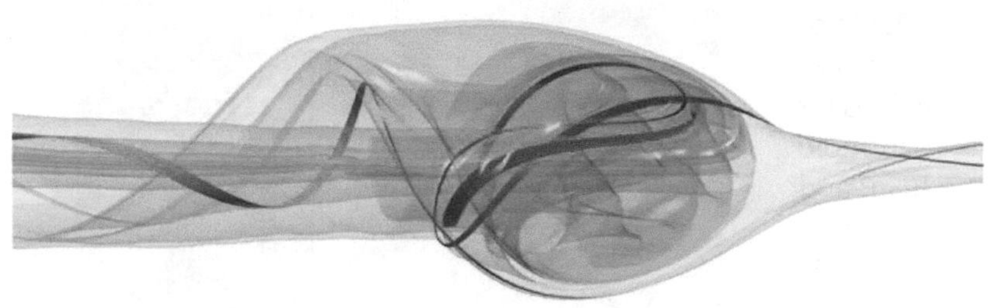

FIGURE 6.1 A stream surface visualizes recirculating flow in a vortex breakdown bubble. The surface, computed from high-resolution adaptive, unstructured flow simulation output, consists of several million triangles and was rendered using an adaptive approach for its transparency, designed to reveal the flow structure inside the bubble. Two stripes highlight the different trajectories taken by particles encountering the recirculation. Image courtesy of C. Garth, University of Kaiserslautern, data courtesy of M. Rütten, German Aerospace Center, Göttingen.

FIGURE 6.2 Streamlines showing outflows in a magnetic flow field computed by an Active Galactic Nuclei astrophysics simulation (FLASH). Image courtesy of D. Pugmire (ORNL), data courtesy of Paul Sutter (UIUC).

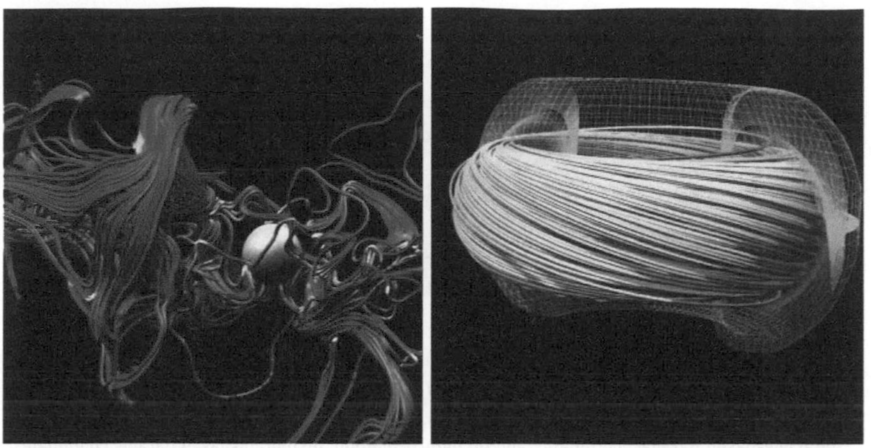

FIGURE 6.3 Data sets used for *IC* tests. Integral curves through the magnetic field of a GenASiS supernova simulation (left), data courtesy of Anthony Mezzacappa (ORNL). Integral curves through the magnetic field of a NIMROD fusion simulation (right), data courtesy of Scott Kruger (Tech-X Corporation).

TABLE 6.1 Performance Benchmarks

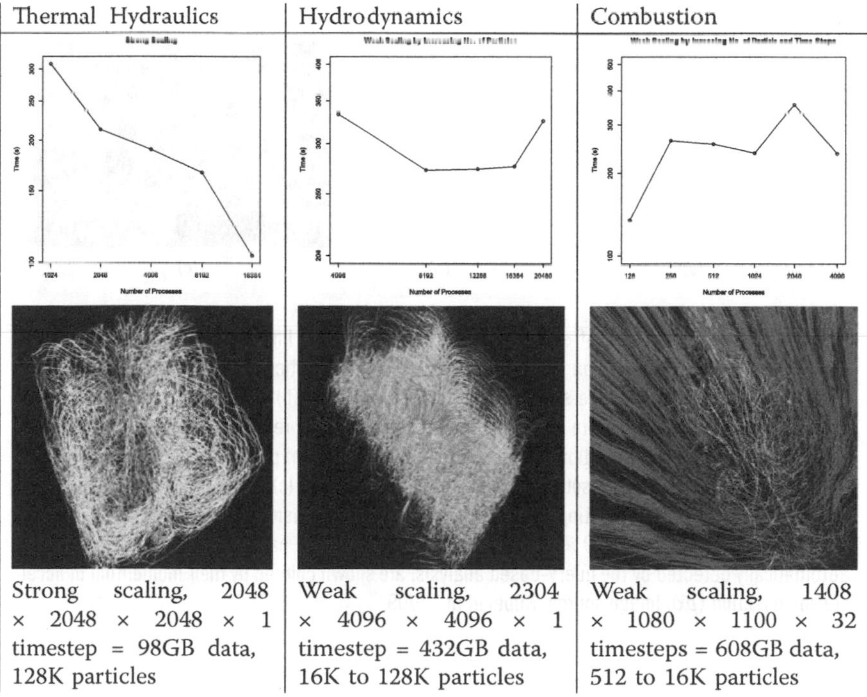

Thermal Hydraulics	Hydrodynamics	Combustion
Strong scaling, 2048 × 2048 × 2048 × 1 timestep = 98GB data, 128K particles	Weak scaling, 2304 × 4096 × 4096 × 1 timestep = 432GB data, 16K to 128K particles	Weak scaling, 1408 × 1080 × 1100 × 32 timesteps = 608GB data, 512 to 16K particles

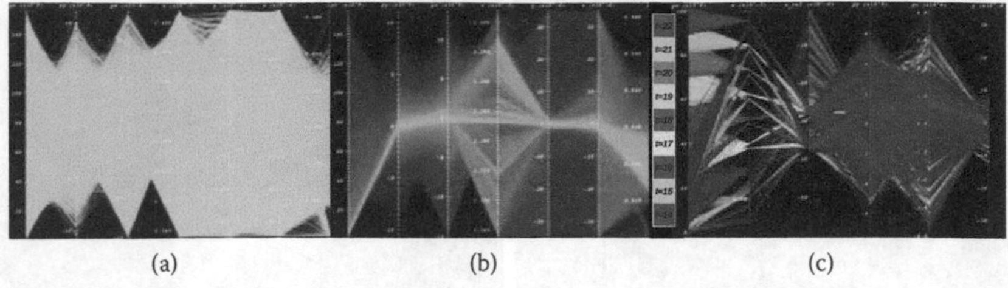

(a) (b) (c)

FIGURE 7.3 The left images, (a) and (b), show a comparison of two different parallel coordinate renderings of a particle data set consisting of 256,463 data records and 7 variables using: (a) traditional line-based parallel coordinates and (b) high-resolution, histogram-based parallel coordinates with 700 bins per data dimension. The histogram-based rendering reveals many more details when displaying large numbers of data records. Image (c) shows the temporal histogram-based parallel coordinates of two particle beams in a laser-plasma accelerator data set, at timesteps $t = [14; 22]$. Color is used to indicate the discrete timesteps. The two different beams can be readily identified in x (second axis). Differences in the acceleration can be clearly seen in the momentum in the x direction, px (first axis). Image source: Rübel et al., 2008.

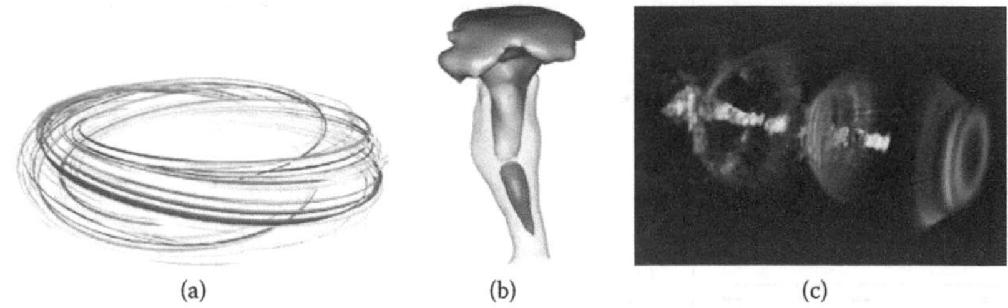

(a) (b) (c)

FIGURE 7.4 Applications of segmentation of query results. Image (a) displays the magnetic confinement fusion visualization showing regions of high magnetic potential colored by their connected component label. Image source: Wu et al., 2011. Image (b) shows the query selecting the eye of a hurricane. Multivariate statistics-based segmentation reveals three distinct regions in which the query's joint distribution is dominated by the influence of pressure (blue), velocity (green), and temperature (red). Image source: Gosink et al., 2011. Image (c) is the volume rendering of the plasma density (gray), illustrating the wake of the laser in a plasma-based accelerator. The data set contains approximately 229×10^6 particles per timestep. The particles of the two main beams, automatically detected by the query-based analysis, are shown colored by their momentum in acceleration direction (px). Image source: Rübel et al., 2009.

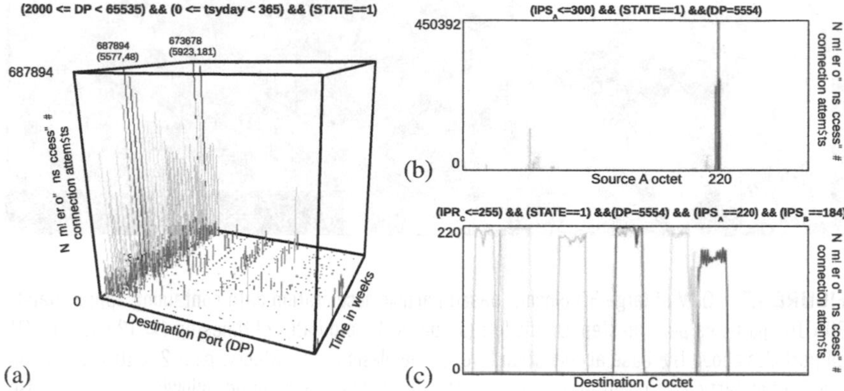

FIGURE 7.5 Histograms showing the number of unsuccessful connection attempts (with radiation excluded): (a) on ports 2000 to 65535 over a 42-week period, indicating high levels of activity on port 5554 during the 7th week; (b) per source A octet during the 7th week on port 5554, indicating suspicious activity from IPs with a 220 A octet; (c) per destination C octet scanned by seven suspicious source hosts (color), indicating a clear scanning pattern. Image source: Bethel et al., 2006.

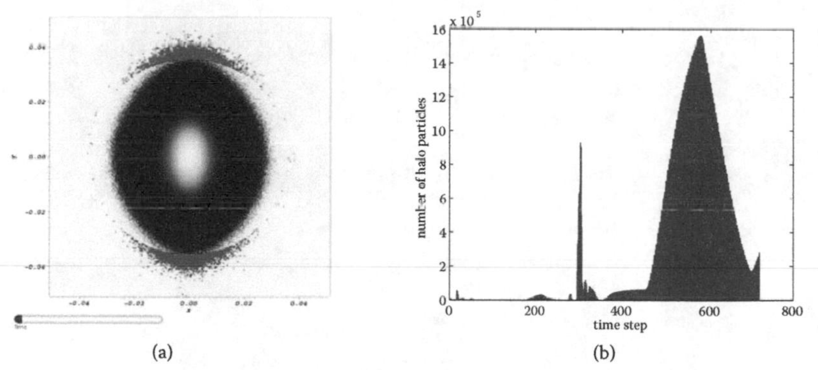

FIGURE 7.6 (a) Particle density plot (gray) and selected halo-query particles (red) for timestep 20 of the simulation. (b) Plot showing the number of halo particles per timestep in a 50TB electron linear particle accelerator data set. Image source: Chou et al., 2011..

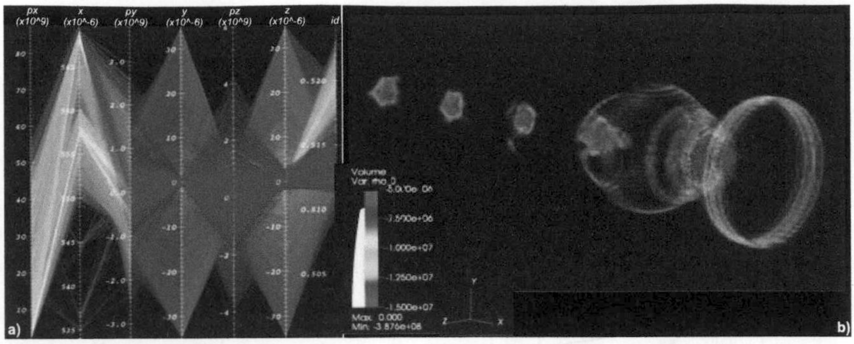

FIGURE 7.7 QDV of large 3D plasma-based particle acceleration data containing approximately 90 x 10^6 particles per timestep. On the left (a), parallel coordinates of timestep $t = 12$ showing: (1) all particles above the base acceleration level of the plasma wave (Query: $px > 2 \times 10^9$) (gray) and (2) a set of particles that form a compact beam in the first wake period following the laser pulse (Query: ($px > 4{:}856 \times 10^{10}$) AND ($x > 5{:}649 \times 10^{-4}$))(red). On the right (b), volume rendering of the plasma density illustrating the 3D structure of the plasma wave. The selected beam particles are shown in addition in red. Image source: Rübel et al., 2008.

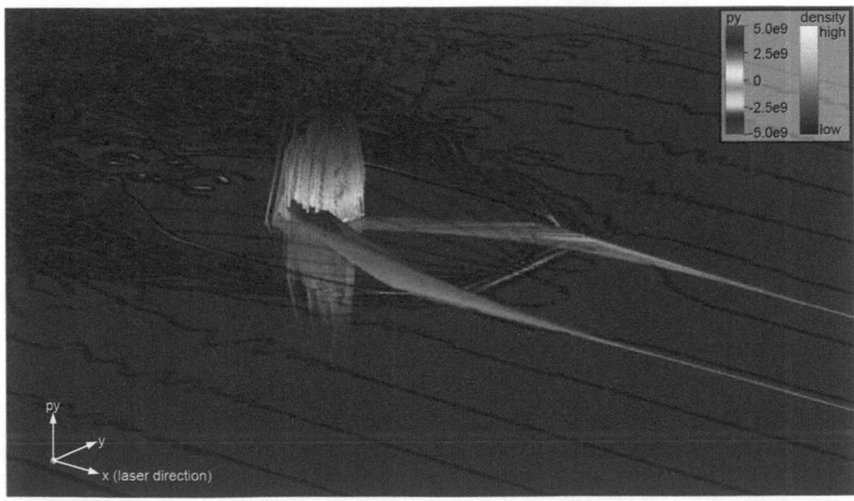

FIGURE 7.8 Visualization of the relative traces of a particle beam in a laser plasma accelerator. The traces show the motion of the beam particles relative to the laser pulse. The *xy*-plane shows iso-contours of the particle density at the timepoint when the beam reaches its peak energy, illustrating the location of the beam within the plasma wave. The up-axis and color of the particle traces show the particle momentum in the transverse direction, *py*. The image shows particles being injected from the sides and oscillating while being accelerated. Image source: Rübel et al., 2009.

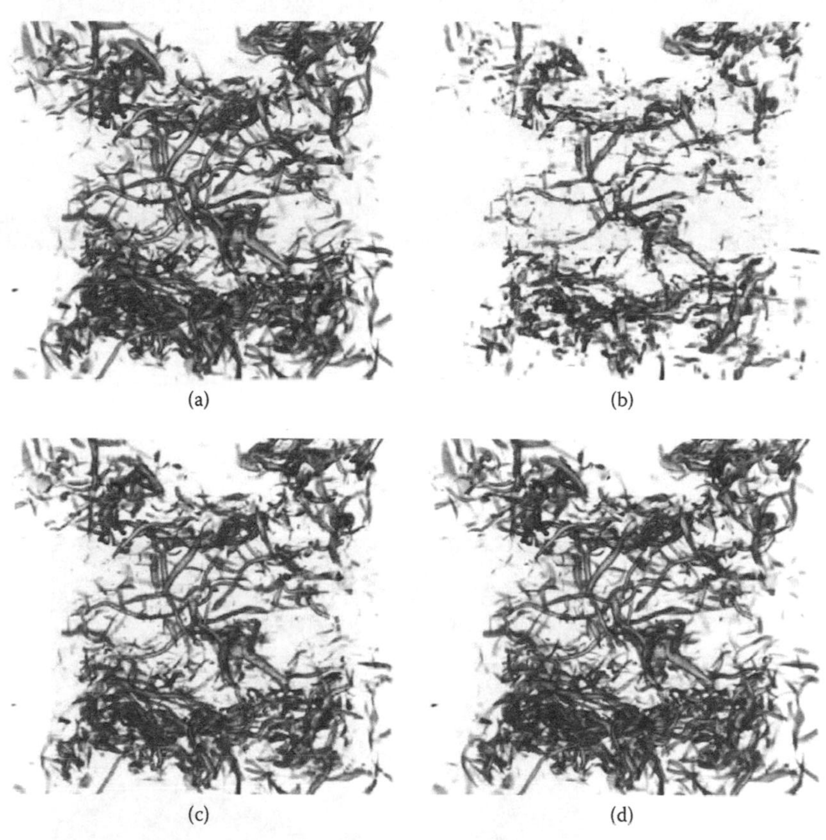

FIGURE 8.7 Direct volume rendering of an enstrophy field: original data (a), and shown in (b){(d), respectively, are results after reduction of the number of expansion coefficients used in reconstruction by factors of $\frac{1}{500}$, $\frac{1}{100}$, and $\frac{1}{10}$.

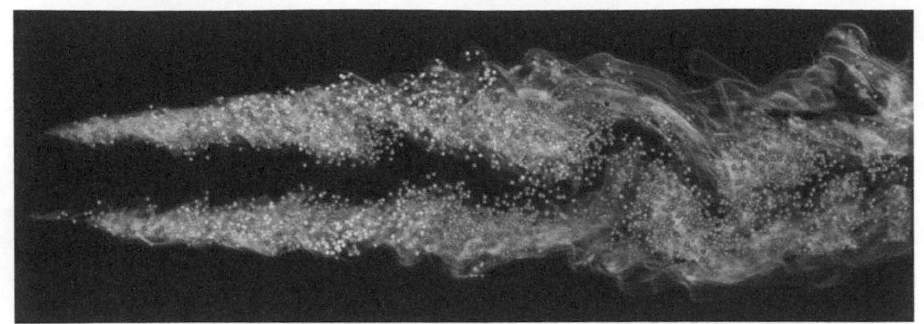

FIGURE 9.2 Visualization of the particles and CH_2O field.

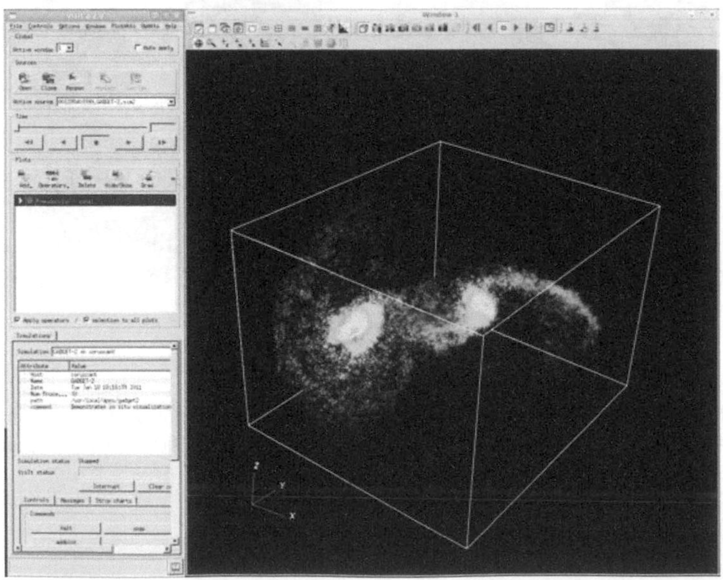

FIGURE 9.4 VisIt client connected to GADGET-2 instrumented with *Libsim*. Image source: Whitlock et al., 2011.

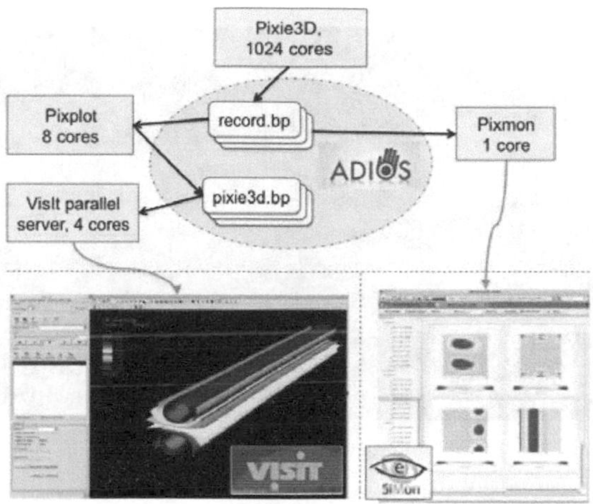

FIGURE 9.5 This example has four separate program — the Pixie3D — simulation, Pixplot, Pixmon, and VisIt — strung together as a workflow on the same supercomputer. A fifth program, the staging server (ADIOS), serves the read and write requests of these programs through its API.

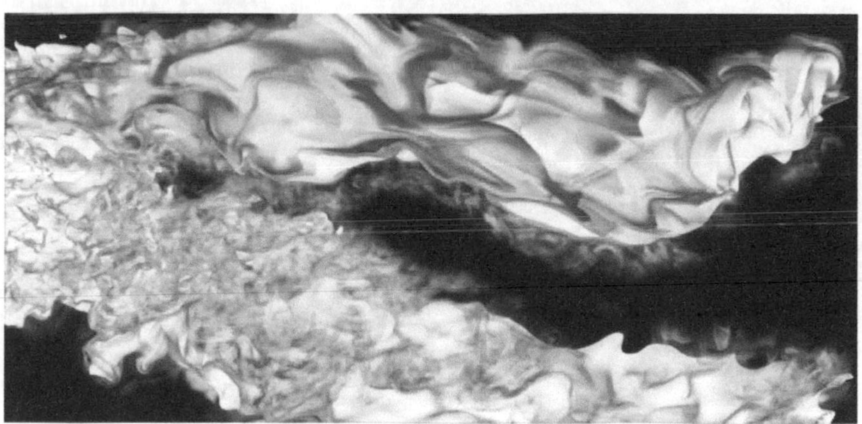

FIGURE 9.8 Visualization uncovering the interaction between the small turbulent eddies and the preheated layer of flame.

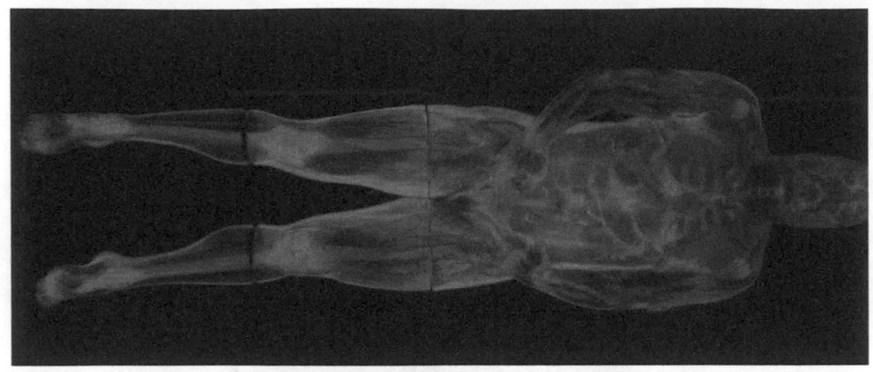

FIGURE 11.3 Interactive sort-first volume rendering of the Visible Human (2048 × 1024 × 1878). Image source: Moloney et al., 2011, data courtesy of the National Library of Medicine.

FIGURE 11.4 Visualization of a laser ablation simulation with 48 million atoms rendered interactively on a standard workstation computer. Image source: Gröttel et al., 2010.

FIGURE 11.6 A rear-projection tiled display without photometric calibration.

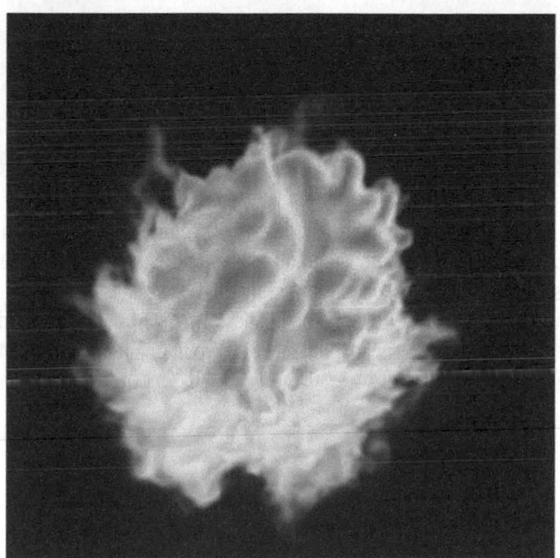

FIGURE 12.1 4608^2 image of a combustion simulation result, rendered by a hybrid parallel MPI+pthreads implementation running on 216,000 cores of the JaguarPF supercomputer. Image source: Howison et al., 2011. Combustion simulation data courtesy of J. Bell and M. Day (LBNL).

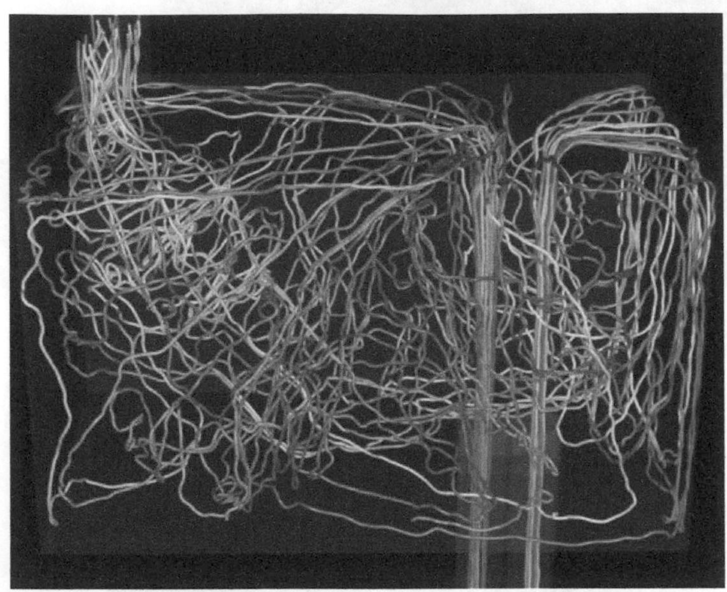

FIGURE 12.8 Integral curve computation lies at the heart of visualization techniques like streamlines, which are very useful for seeing and understanding complex flow-based phenomena. This image shows an example from computational thermal hydraulics. Algorithmic performance is a function of many factors, including the characteristics of the flow field in the underlying data set. This data set has different characteristics than those shown in Chapter 6, Figure 14. Image courtesy of Hank Childs and David Camp (LBNL).

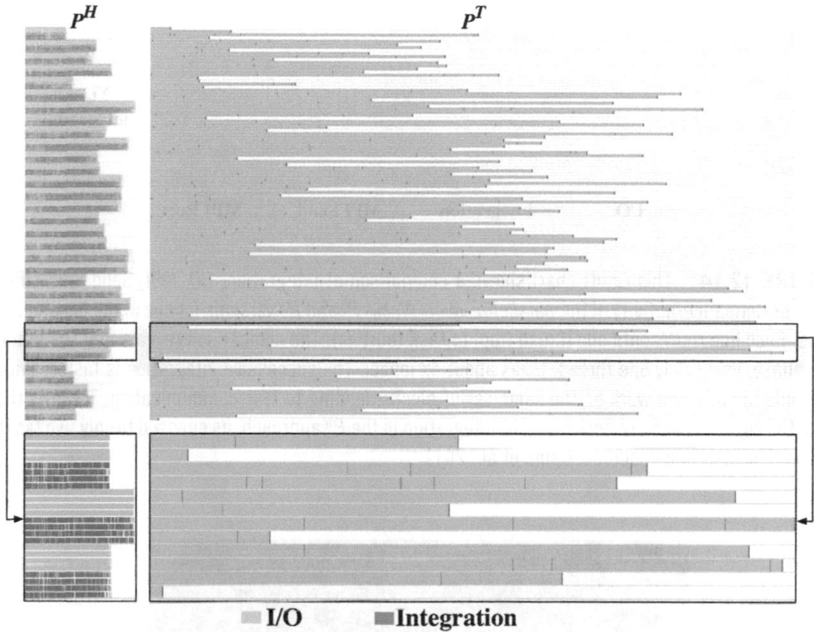

FIGURE 12.12 This Gantt chart shows a comparison of integration and I/O performance/activity of the *parallelize-over-seeds* P^T and P^H versions for one of the benchmark runs. Each line represents one thread (left column) or task (right column). The P^H approach outperforms the P^T one by about 10×, since the four I/O threads in the P^H can supply new data blocks to the four integration threads at an optimal rate. However, work distribution between nodes is not optimally balanced. In the P^T implementation, the I/O wait time dominates the computation by a large margin, due to redundant data block reads, and work being distributed less evenly. This can be easily seen in the enlarged section of the Gantt chart. Image source: Camp et al., 2011.

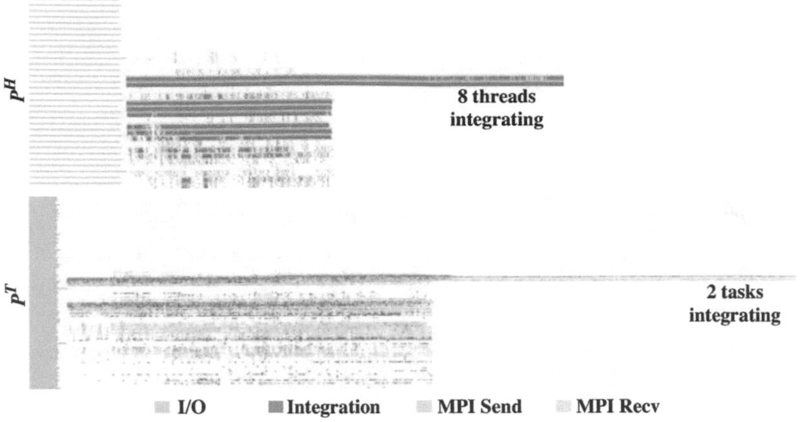

FIGURE 12.14 This Gantt chart shows a comparison of integration, I/O, MPI_Send, and MPI_Recv performance/activity of the *parallelize-over- blocks* P^T and P^H versions for one of the benchmark runs. Each line represents one thread (top) or task (bottom). The comparison reveals that the initial I/O phase, using only one thread, takes about 4x longer. The successive integration is faster, since multiple threads can work on the same set of blocks, leading to less communication. Towards the end, the eight threads are performing IC integration in the P^H approach, as opposed to only two tasks in the P^T model. Image source: Camp et al., 2011.

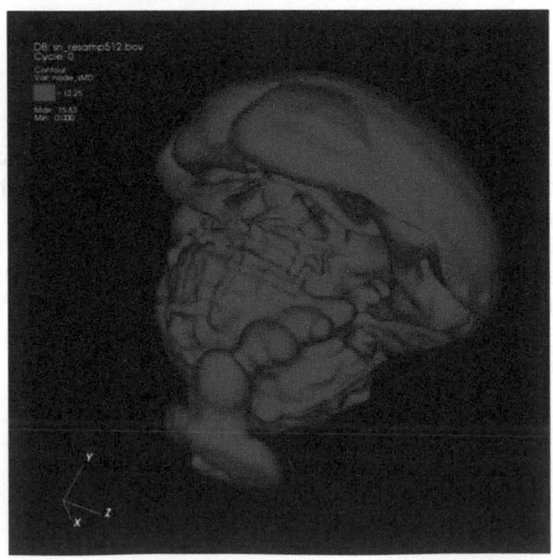

FIGURE 13.1 Contouring of two trillion cells, visualized with VisIt on Franklin using 32000 cores. Image source: Childs et al., 2010.

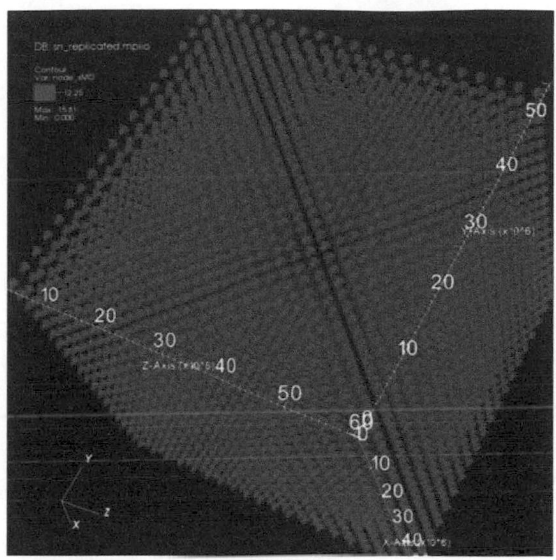

FIGURE 13.3 Contouring of replicated data (one trillion cells total), visualized with VisIt on Franklin using 16,016 cores. Image source: Childs et al., 2010.

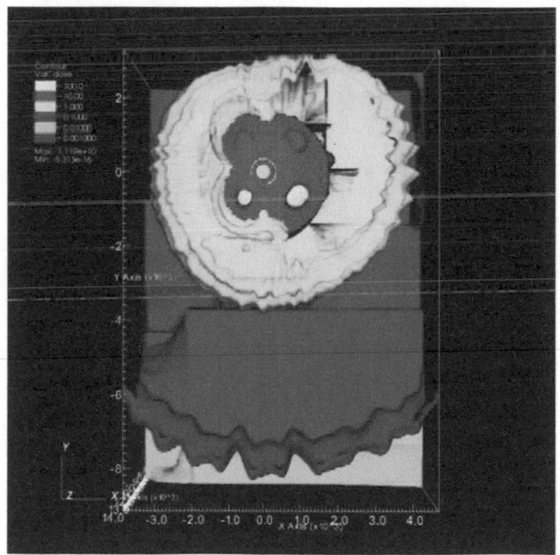

FIGURE 13.4 Rendering of an isosurface from a 321 million cell Denovo simulation, produced by VisIt using 12,270 cores of JaguarPF. Image source: Childs et al., 2010.

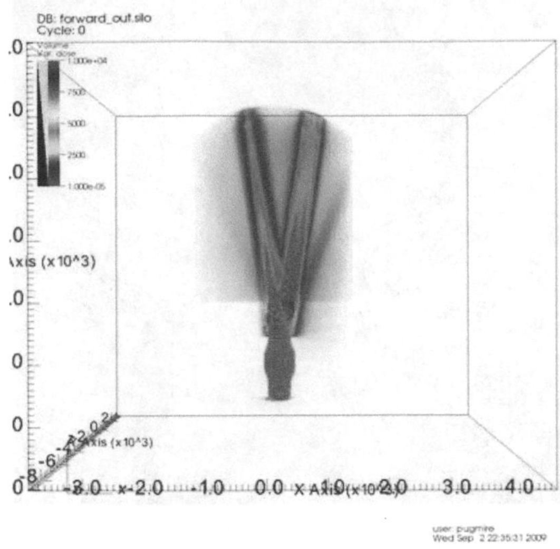

FIGURE 13.5 Volume rendering of data from a 321 million cell Denovo simulation, produced by VisIt using 12,270 cores on JaguarPF. Image source: Childs et al., 2010.

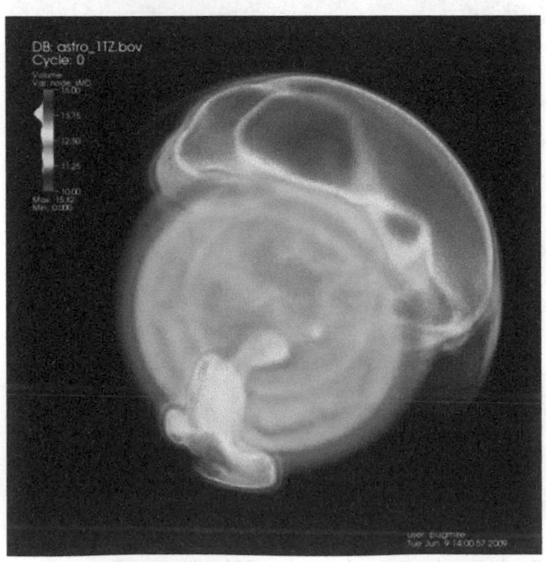

FIGURE 13.6 Volume rendering of one trillion cells, visualized by VisIt on JaguarPF using 16,000 cores. Image source: Childs et al., 2010.

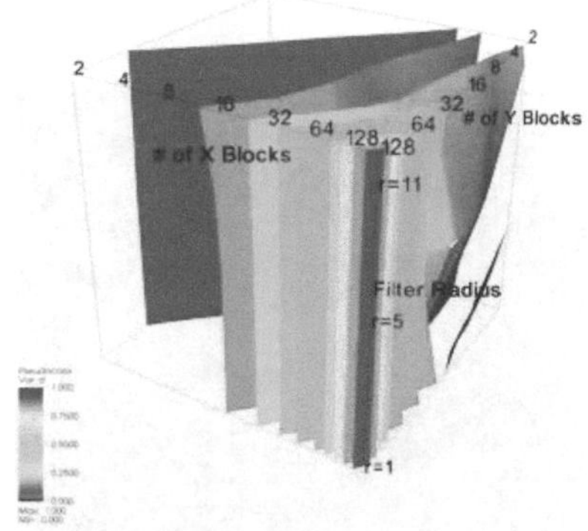

(a) Runtimes normalized by maximum highlight the poorest performing configurations.

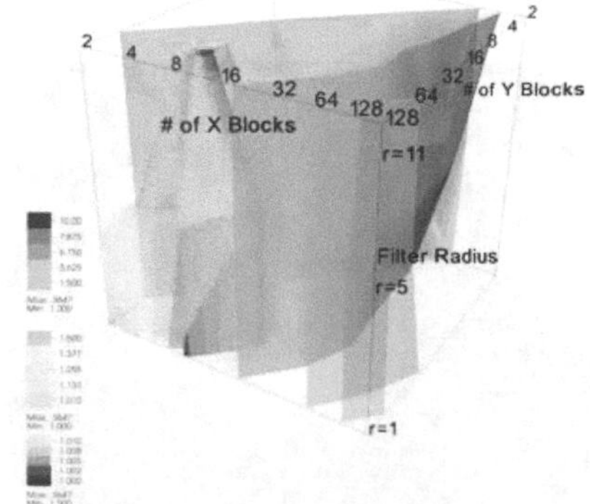

(b) Runtime normalized by minimum highlight the best performing configurations.

FIGURE 14.5 Visualization of performance data collected by varying the number and size of the GPU thread blocks for the 3D bilateral filter are shown at three different filter sizes, $r = \{1, 5, 11\}$. In (a), the performance data (normalized to the maximum value) highlights the poorest performing configurations; the red and yellow isocontours are close to the viewer. In (b), the performance data (normalized to the minimum value) highlights the best performing configurations. These appear as the cone-shaped red/yellow isocontours. Image source: Bethel, 2009.

		Array Order					Z Order				
		1	2	4	8	16	1	2	4	8	16
NoERT	1			1.00	0.77	0.74			0.90	0.58	0.44
	2		0.98	0.71	0.63	0.68		0.86	0.51	0.35	0.30
	4	1.04	0.70	0.59	0.61	0.74	0.89	0.51	0.32	0.26	0.33
	8	0.81	0.63	0.61	0.68	0.73	0.56	0.34	0.25	0.27	0.32
	16	0.83	0.71	0.72	0.68	0.71	0.42	0.30	0.27	0.27	0.32
ERT	1			0.92	0.70	0.66			0.83	0.53	0.40
	2		0.90	0.64	0.57	0.60		0.79	0.47	0.32	0.27
	4	0.95	0.64	0.53	0.54	0.65	0.82	0.46	0.29	0.23	0.29
	8	0.73	0.56	0.54	0.60	0.63	0.51	0.31	0.23	0.24	0.28
	16	0.74	0.62	0.62	0.59	0.61	0.38	0.27	0.24	0.24	0.28

(a) Runtime (s)

		Array Order					Z Order				
		1	2	4	8	16	1	2	4	8	16
NoERT	1			50	115	481			25	28	42
	2		36	57	238	632		17	17	22	67
	4	52	59	179	517	907	14	14	16	43	240
	8	139	243	527	839	873	14	17	43	182	202
	16	636	724	909	827	834	29	63	176	166	166
ERT	1			43	95	401			22	24	35
	2		29	48	193	530		15	15	19	56
	4	38	47	144	429	773	12	12	14	37	198
	8	106	192	436	711	741	12	14	37	153	163
	16	506	596	765	696	701	24	53	148	137	130

(b) L2 cache misses (millions)

FIGURE 14.6 Parallel ray casting volume rendering performance measures on the NVIDIA/Fermi GPU include absolute runtime (a), and L2 cache miss rates (b), averaged over ten views for different thread block sizes. Gray boxes indicated thread blocks with too few threads to fill a warp of execution. Surprisingly, the best performing configurations do not correspond to the best use of the memory hierarchy on that platform. Image source: Bethel and Howison, 2012.

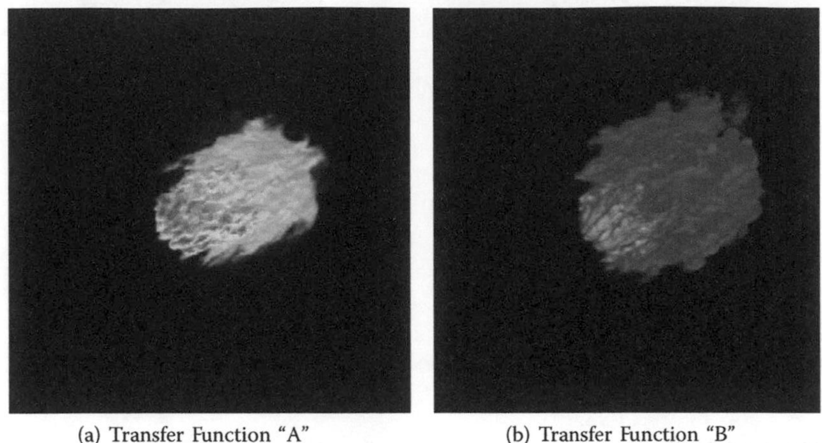

(a) Transfer Function "A" (b) Transfer Function "B"

FIGURE 14.7 Two different transfer functions have different benefits from early ray termination. They also yield images that accentuate different features of the underlying data set. The performance tests in this study use transfer function "A." Image source: Bethel and Howison, 2012.

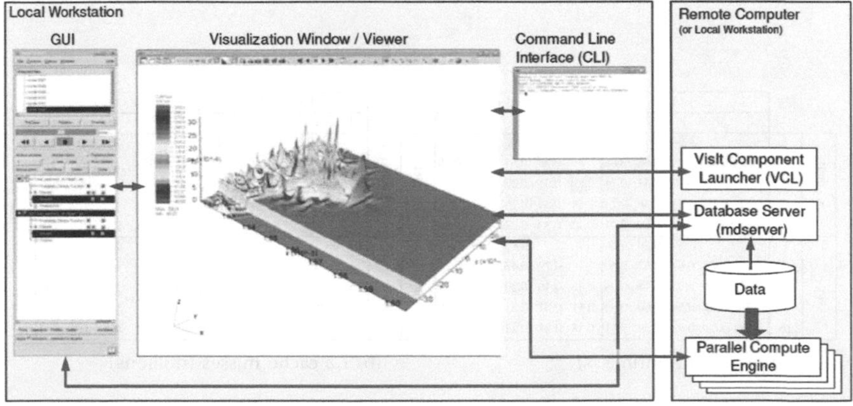

FIGURE 16.1 Diagram of VisIt programs and their communication. Image source: Childs et al., 2011.

FIGURE 16.3 Recent covers of the SciDAC Review Journal created using VisIt.

FIGURE 17.3 An IceT assisted rendering of an isosurface from a Richtmyer–Meshkov simulation. The detailed surface is represented by 473 million triangles.

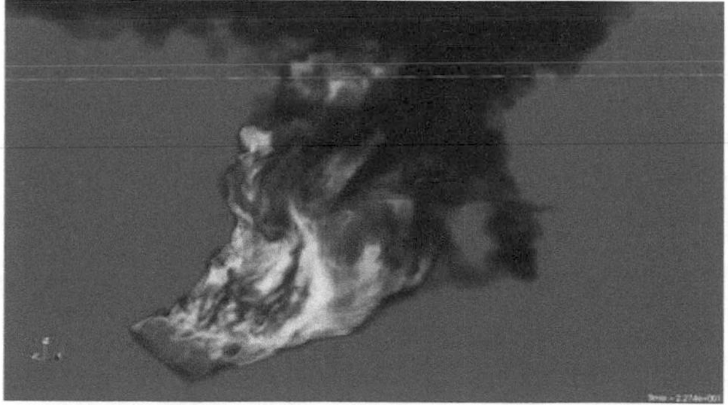

FIGURE 17.4 Results from a coupled SIERRA/Fuego/Syrinx/Calore simulation of objects in a crosswind fire. These 10,000,000 unstructured hexahedra are rendered using IceT's ordered compositing mode.

FIGURE 17.5 An example of using IceT to simultaneously render both surfaces and transparent volumes on a multitile display.

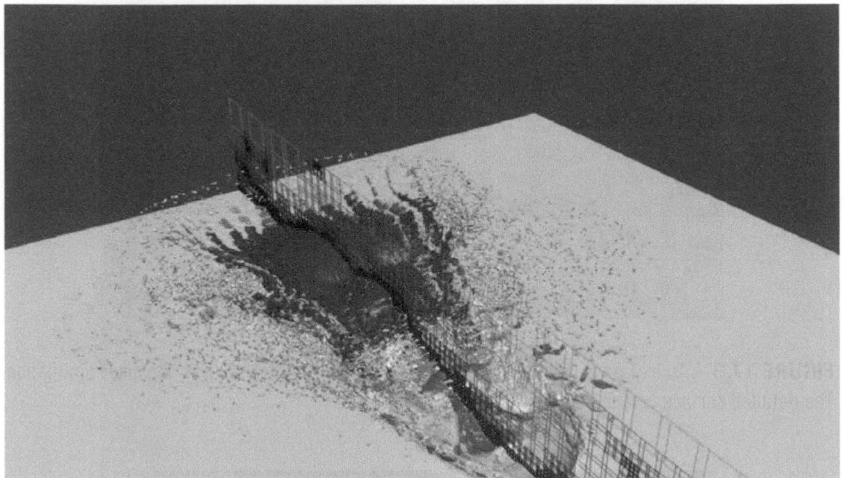

FIGURE 18.6 The fragmentation pattern of a high-speed aluminum ball hitting an aluminum brick.

FIGURE 18.7 ParaView's fragment extraction can be used to validate that the fragments in a simulated explosion (left) match collected debris (center) and observed damage (right) from experiments.

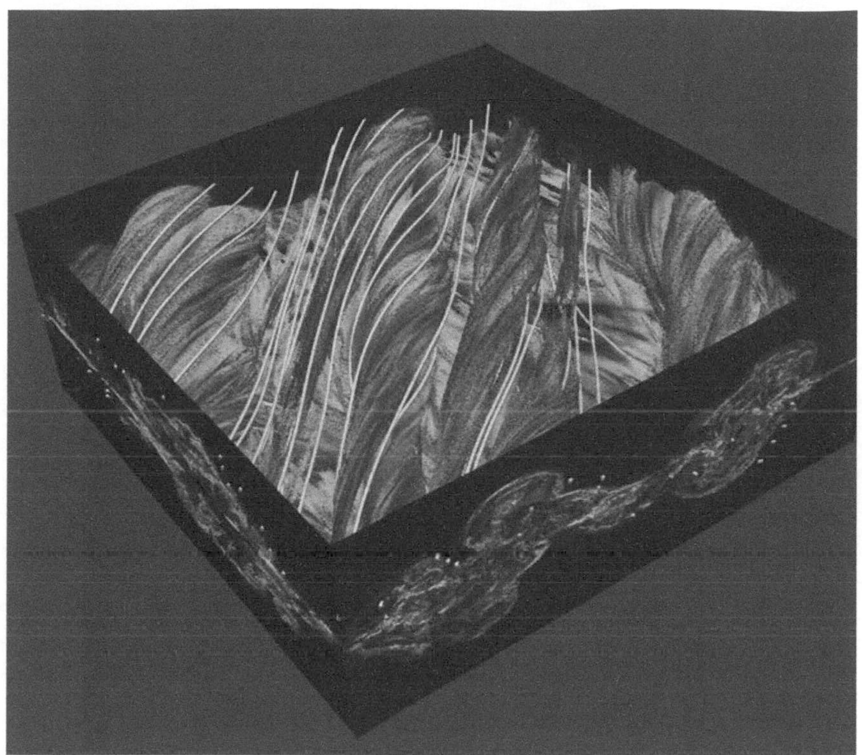

FIGURE 18.8 A visualization of magnetic reconnection from the VPIC project at Los Alamos National Laboratory. The visualization is generated from a structured data containing 3.3 billion cells with two vector fields and one scalar field, produced using 256 cores running ParaView in the interactive queue on the Kraken supercomputer at the National Institute for Computational Sciences (http://www.nics.tennessee.edu/).

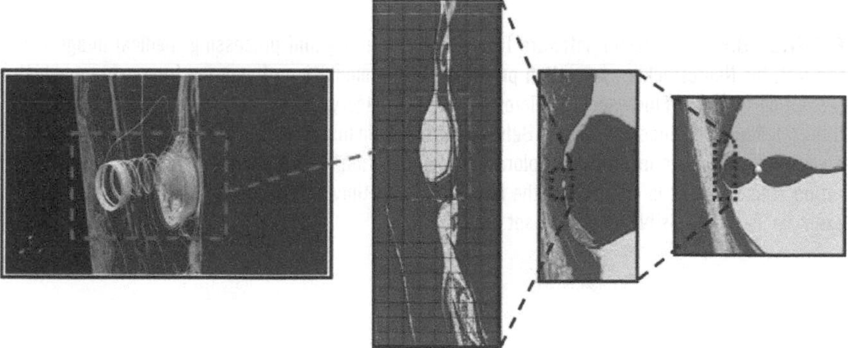

FIGURE 18.9 Finding and tracking of flux ropes in global hybrid simulations of the Earth's magnetosphere using ParaView.

FIGURE 19.5 The same application and visualization of a Mars panorama running on an iPhone 3G mobile device (left) and a Powerwall display (right). Data courtesy of NASA.

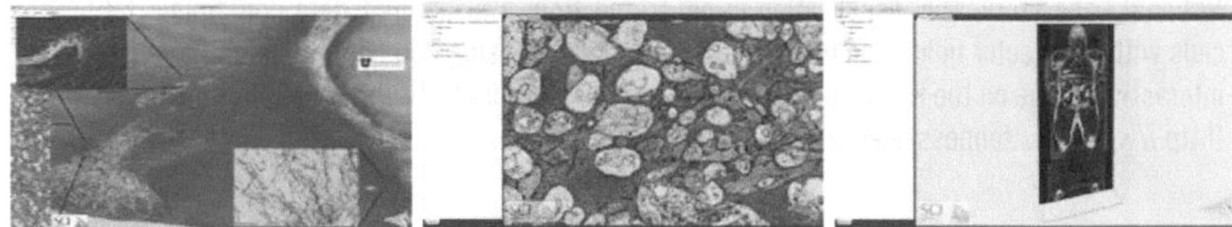

FIGURE 19.6 The ViSUS software framework visualizing and processing medical imagery. On the left, the Neurotracker application providing the segmentation of neurons from extremely high-resolution Confocal Fluorescence Microscopy brain imagery. This data courtesy of the Center for Integrated Neuroscience and Human Behavior at the Brain Institute, University of Utah. In the center, an application for the interactive exploration of an electron microscopy image of a slice of a rabbit retina. This data set is courtesy of the MarcLab at the University of Utah. On the right, 3D slicing example using the Visible Male data set.

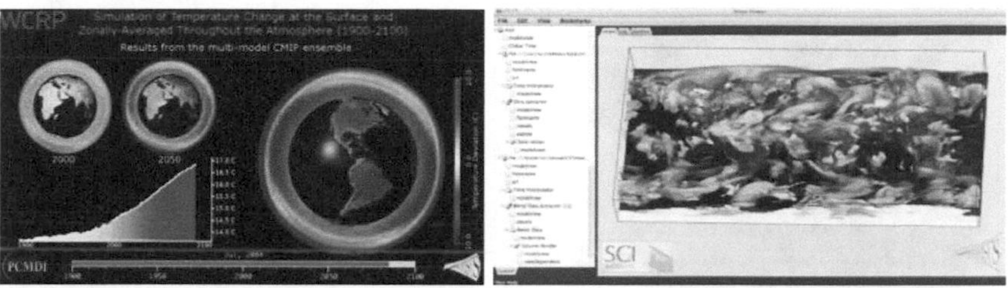

FIGURE 19.7 Remote climate visualization with ViSUS. The ViSUS framework providing a visualization for a temperature change ensemble simulation for the Earth's surface for the December 2009 climate summit meeting in Copenhagen (left). A visualization of global cloud density for a more recent climate simulation (right).

FIGURE 19.8 500 megapixel visualization of a 2D panorama data set of Mount Rushmore (left). The color shift between images in a panorama mosaic (middle). An application using a LightStream data flow to provide approximate gradient domain solution as a user interacts with the data (right). Data set courtesy of City Escapes Photography.

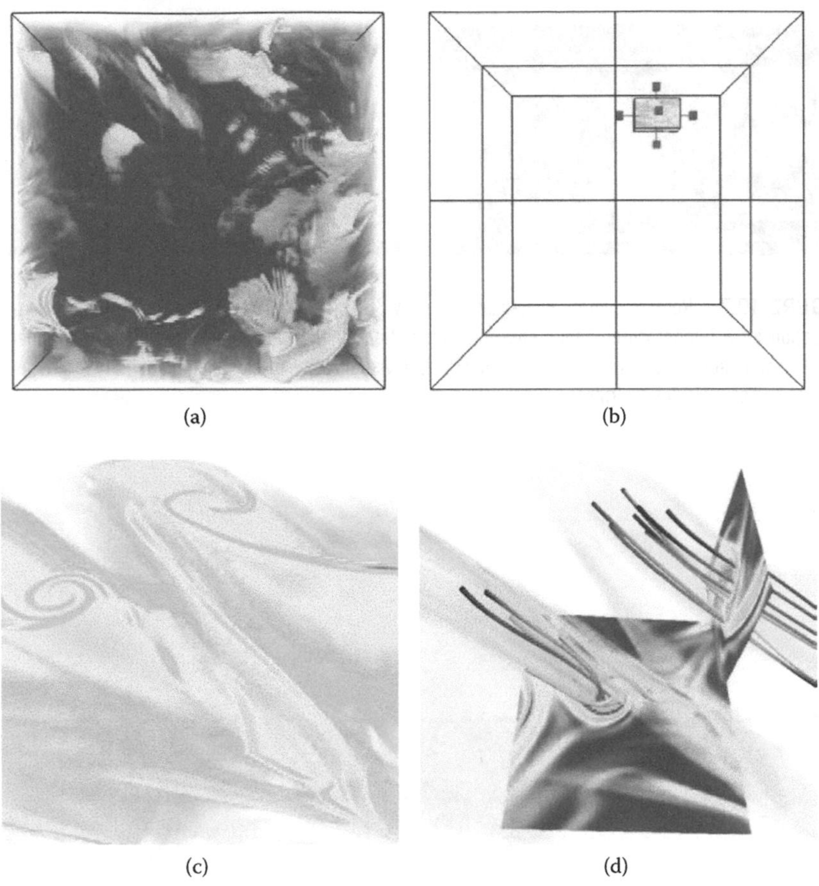

FIGURE 20.1 Exploration of the current field of a 1536^3 MHD simulation. A volume rendering of current magnitude generated using reduced data (a), an isolated ROI exhibiting a current "roll-up" (b), a close up of the phenomenon (c), and magnetic field lines passing through the center of the roll-up (d). The images in (b)–(d) were generated with the highest refinement level and LOD data.

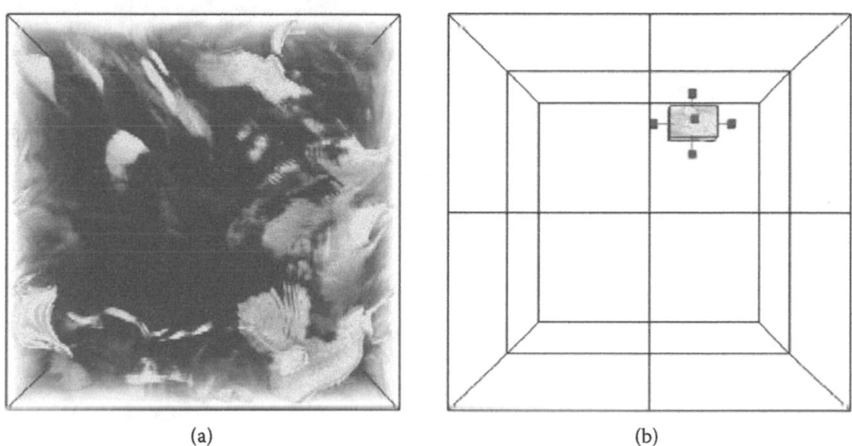

(a) (b)

FIGURE 20.2 Volume rendering of an isolated ROI showing the magnitude of the current field from a 1536^3 simulation. The original, unreduced data are shown (left) along with data reduced by a combination of both LOD and resolution (right). The corresponding reduction factors for LOD and resolution coarsening are 10:1 and 8:1, respectively.

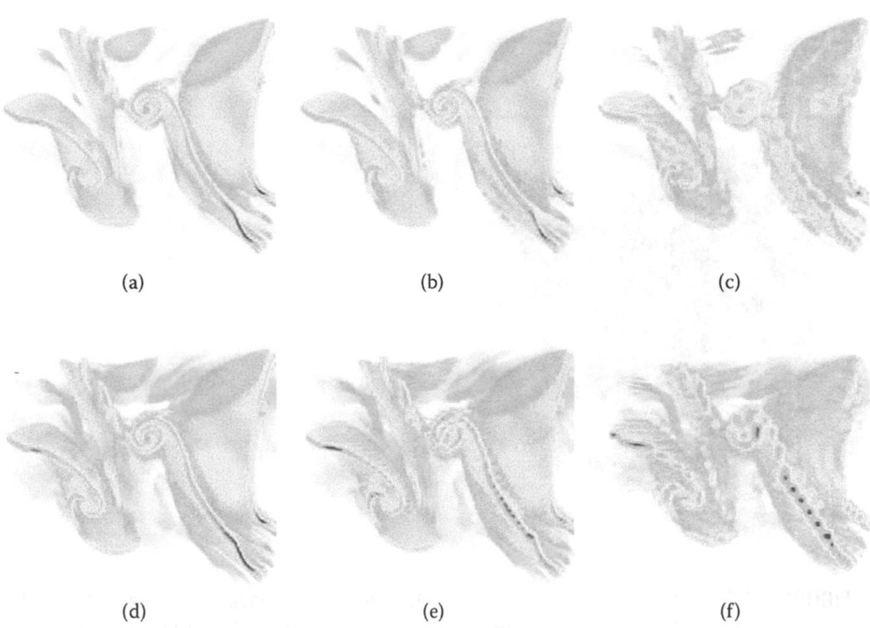

FIGURE 20.3 Direct volume rendering of reduced MHD enstrophy data (volume rendering of original data shown in Figure 20.2a). The images in the top row were produced with the native grid resolution, but varying the LODs with reduction factors of 10:1 (a), 100:1 (b), and 500:1 (c). The bottom row used the highest LOD for all images, but varies the grid resolutions with reduction factors of 8:1 (d), 64:1 (e), and 512:1 (f). Reduced LOD primarily benefits I/O performance, while reduced grid resolution primarily benefits memory, computation, and graphics.

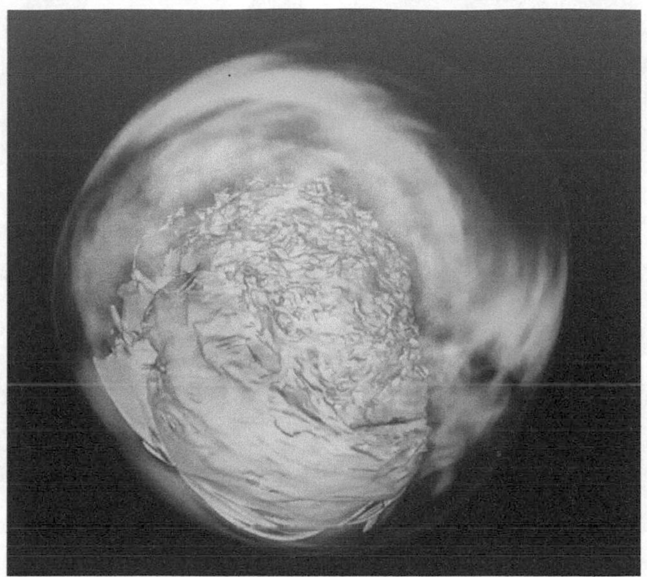

FIGURE 21.5 Supernova remnant density field rendering, demonstrating distributed structured volume rendering, with embedded polygonal surfaces. Data courtesy of Dr. John Blondin, North Carolina State University.

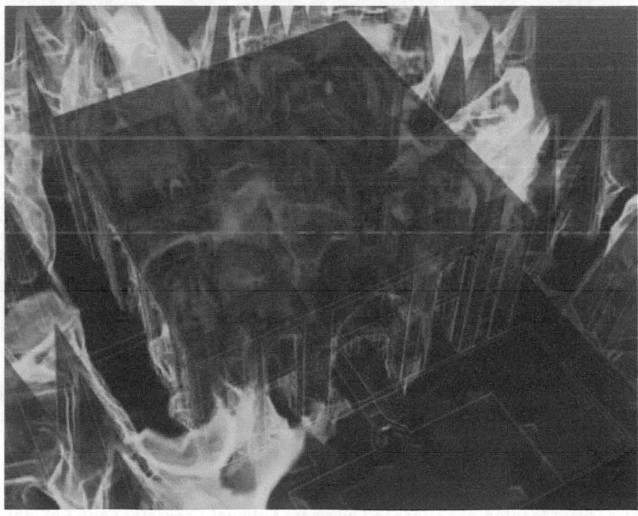

FIGURE 21.6 Simulated air flow through the Kashan–Iran Fin Garden building and surrounding vegetation. Unstructured volume rendered velocity through semi-transparent architectural geometry. Data courtesy of Studio Integrate.

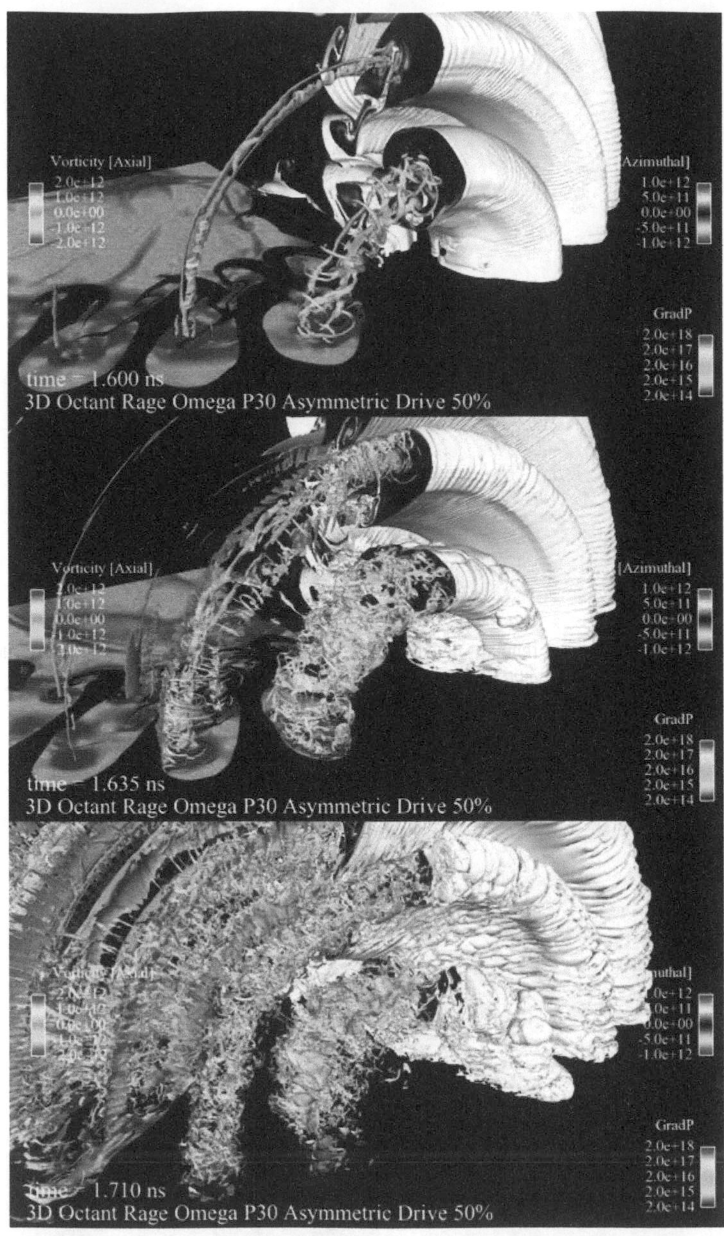

FIGURE 21.7 Inertial Confinement Fusion implosion visualization. The images were computed using EnSight servers located at Lawrence Livermore National Laboratory and EnSight clients running on a rendering cluster located at Los Alamos National Laboratory. Images courtesy of Robert Kares, Los Alamos National Laboratory.

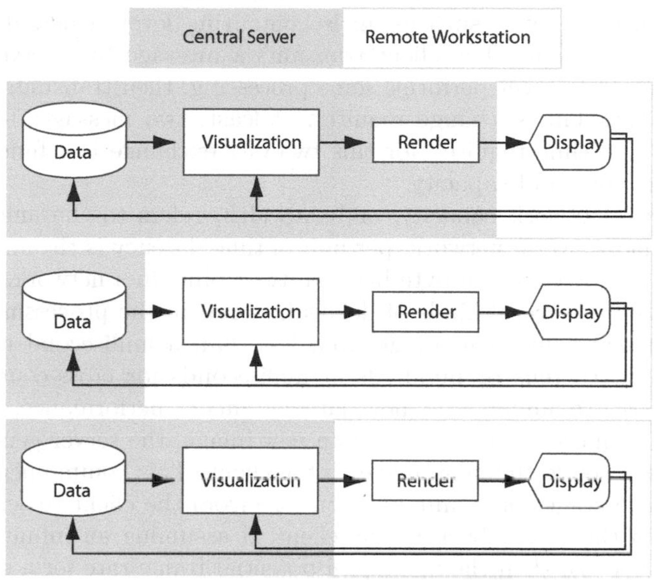

FIGURE 3.1: The three most common visualization pipeline partitionings in remote and distributed visualization. From top to bottom: send images, send data, and send geometry. Components in grey are located at the central computing facilities, and those in white are on the remote client. Image courtesy of E. Wes Bethel (LBNL).

different components in the pipeline will likely have different performance characteristics. Some components are more computationally expensive than others.

- *Inter-component communication*: moving data between components incurs a cost as well. The cost, or time required, is typically greater across wide-area than local-area networks.
- *Data characteristics and visualization algorithms*: some visualization algorithms, such as isocontouring, are sensitive to the characteristics of input data. For example, running an isocontouring algorithm on a "noisy" data set will result in a significantly greater amount of output than when run on a "smooth" data set of the same size. In extreme cases, the isocontouring algorithm can produce more output than input.
- *Data size*: the overall pipeline performance is influenced by the amount of data to be processed.

Of these factors, the chapter's focus is on the cost of inter-component communication, since it can have a surprising impact on overall pipeline performance. Consider the following use scenario, where the visualization pipeline is split across two machines in some unknown partitioning. First, a client (user) at a remote machine requests a new image, due, perhaps to changing a visualization parameter, such as an isocontouring level, a new data set, or a change in viewpoint. The client transmits a message to a server requesting an update. The server performs some processing, then transmits results back to the client. This exchange requires, at least, two messages to transit the network. The time required for this two-way exchange is a function of both network latency and capacity.

Whereas network capacity, or bandwidth, refers to the amount of data that can move over a network, per unit of time, latency is the amount of time required to move a single byte between two points in a network. Latency is a combination of several kinds of delays incurred in the processing of network data. Latency values can range from less than a millisecond for local area networks up to tens or hundreds of milliseconds for cross-country network connections. To understand how latency affects performance, consider the following scenario: a client requests a new image, the server generates a new image, then sends the new image to the client. This communication pattern requires two stages of communication, one from the client to server, and the other from the server back to the client. If assuming an infinite bandwidth (network speed), then the maximum potential frame rate for a given amount of latency, L, measured in milliseconds is $1000/2L$. When $L = 50$ ms, then the maximum possible frame rate is 10 frames per second, regardless of the capacity of the underlying network or the amount of data moving between server and client. The performance of some pipeline partitionings are more sensitive to latency and capacity than others.

3.3 Send-Images Partitioning

In a send-images partitioning, all processing needed to compute a final image is performed on a server, then the resulting image data is transmitted to a client. Over the years, there have been several different approaches to implement this partitioning strategy.

The Virtual Network Computing (VNC) [25] system uses a client–server model for providing remote visualization and display services to virtually any application using a client–server model of operation: the custom client viewer on the local machine intercepts events and sends them to the VNC server running at the remote location; the server detects updates to the screen generated by either the windowing system or the graphics/visualization application, then packages up those updated regions of pixels; and, finally, sends them to the client for display, using one of a different number of user-configurable compression and encoding strategies. In its original design and implementation, VNC didn't support applications that used hardware-acceleration for rendering images. Paul et al. 2008 [23] describe a scalable rendering architecture, which is the subject of 3.9, that combines with VNC and overcomes this limitation for scalable rendering and image delivery.

OpenGL Vizserver [13] and VirtualGL [9] use a client–server model for the remote delivery of hardware-accelerated rendering of live-running applications. OpenGL Vizserver provides a remote image delivery of the contents of an application's OpenGL rendering window. VirtualGL uses two different communication channels: one for encoding and transmission of the OpenGL's window pixels, and another for Xlib traffic. In a sense, VirtualGL's implementation falls partly into the hybrid category, since it sends both images and the "draw commands" formed by the Xlib command stream.

Some visualization applications, such as VisIt [17] and ParaView [15], support a send-images mode of operation, where a scalable visualization and rendering back end provides images to a remote client. In these approaches, a remote client component requests a new frame; the server-side component responds by executing the local part of the visualization pipeline to produce and transmit a new image. Interestingly, VisIt supports a dynamic pipeline configuration model, where it can select between send-images and send-geometry paritionings at runtime. (For a detailed discussion, see 3.10.)

Chen et al. 2006 [6] describe a remote visualization architecture and implementation, MBender, that allows for the exploration of precomputed visualization results in a way that gives the appearance of semi-constrained interactive visualization. Their approach is akin to that of QuickTime VR [7] (QTVR) "object movies," which allow navigation through a 3D array of images. In contrast, QuickTime "panorama movies" consist of images taken or rendered from a fixed point where the view varies across azimuthal angles; object movie images typically represent viewpoints from different "latitude and longitude" positions, but with a common "look at" point, and with zoom-in and zoom-out options. Whereas QTVR supports navigation through up to three dimensions of images, Chen's MBender architecture supports navigation through N-dimensional image arrays, which is better suited for scientific visualization. Each of the N dimensions represents a variation across a single visualization parameter: isocontouring level, slicing, temporal evolution, and so forth. In practice, this approach has a significant space requirement for the source images, which are requested via a standard web server by a custom Java-based client, and works best when N is relatively small.

The primary advantage of the send-images partitioning is that there is an upper bound on the amount of data that moves across the network. That upper bound is a function of image size, I_s, rather than the size of the data set, D_s, being visualized. Typically, when $D_s \gg I_s$, the send-images partitioning has favorable performance characteristics, when compared to the others.

Its primary disadvantage is related to its primary advantage: there is a minimum amount of data, per frame, that must move across the network. The combination of latency to produce the frame and the time required to move it over the network may be an impediment on interactive levels of performance. For example, if the user desires to achieve a 30 frame-per-second throughput rate, and each frame is 4MB in size,[1] then, assuming zero latency, the network must provide a minimum of 120MB/s of bandwidth. Some systems, such as VNC, implement optimizations—compression and sending only the portion(s) of the screen that changes—to reduce the size of per-frame pixel payload.

The other disadvantage of send-images is the potential impact of network latency on interactivity. As discussed earlier in 3.2, network latency will impose an upper bound on absolute frame rate. This upper bound may be sufficiently high on local area networks to support interactive visualization when using the send-images approach, but may be too low on wide area networks. For example, achieving 10 frames per second is possible only on networks having less than 100 ms of round-trip latency: for $1000/2L \geq 10$ fps, then $L \leq 50$ ms.

3.4 Send-Data Partitioning

The send-data partitioning aims to move scientific data from server to client for visualization processing and rendering. The scientific data may be the "source data," prior to any processing, or it may be source data that has undergone some sort of processing, such as noise-reduction filtering, or a computation of a derived field. In the former case, the visualization pipeline, which resides entirely on the user/client machine, will perform "remote reads" to obtain data. In the latter case, some portion of the visualization pipeline may reside on the server, but the data, moved between server and client, consists of the same type and size of data, as if the client were performing remote reads.

Optimizing this type of pipeline partitioning can take several different forms. One form is to optimize the use of distributed network data caches and replicas, so a request for data goes to a "nearby" rather than "distant" source [1, 29], as well as to leverage high performance protocols that are more suitable than a TCP for bulk data transfers [1]. Other optimizations leverage data subsetting, filtering, and progressive transmission from remote sources to reduce the amount of data payload crossing the network [10, 22]. Some systems, like the Distributed Parallel Storage System (DPSS), provide a scalable, high performance, distributed-parallel data storage system that can be optimized for data access patterns and the characteristics of the underlying network [30]. The ViSUS system (see Chap. 19) implements several optimizations, including streaming (Chap. 10) and multiresolution (Chap. 8) send-data techniques.

Another approach is to use alternative representations for the data being sent from server to client. Ma et al. 2002 [19] describe an approach used for the transmission of particle-based data sets in which either particles, or server-

[1] 1024^2 pixels, each of which consists of RGBα tuples, one byte per color component, and no compression.

computed particle density fields, or both, are moved between server and client. The idea is that, in regions of high particle density, it may make more sense to render such regions using volume rendering, rather than using point-based primitives: in high density regions, point-based primitives would likely result in too much visual clutter. In their implementation, a user defines a transfer function for creating two derived data sets: one is a collection of particles that lie in "low density" regions, the other is a computed 3D density field that is intended to be smaller in size than the original particle field in "high density" regions. Then, the server component transmits both types of data sets to the client for rendering.

In practice, the send-data approach may prove optimal when two conditions hold true: (1) the size of the source data is relatively small, and (2) interactivity is a priority. However, as the size of scientific data grows, it is increasingly impractical to move full-resolution source data to the user's machine for processing, since the size of data may exceed the capacity of the user's local machine, and moving large amounts of data over the network may be cost-prohibitive.

3.5 Send-Geometry Partitioning

In the send-geometry partitioning, the payload, moving between the server and client, is "drawable" content. For visualization, a class of visualization algorithms, often referred to as "mappers," will transform scientific data, be it mesh-based or unstructured, and produce renderable geometry as an output [26]. In a send-geometry partitioning, the server component runs data I/O and visualization algorithms, producing renderable geometry, then transmits this payload to the client for rendering. One way to optimize this path is to send only those geometric primitives that lie within a view frustum. Such optimizations have proven useful in network-based walkthroughs of large and complex data [8]. Frustum culling, when combined with occlusion culling and level-of-detail selection at the geometric model level, can result in reduced transmission payload [16]. Both of those approaches require the server to have awareness of the client-side state, namely viewing parameters.

The send-geometry approach is included as part of several production applications, research applications, and libraries. The applications VisIt (Chap. 16), Ensight (Chap. 21), and ParaView (Chap. 18) all implement a form of send-geometry (as well as a form of send-images). Unmodified OpenGL and X11-based applications may be configured, through an environment variable, to transmit Xlib [21] or GLX [14] protocol "draw commands," respectively, from an application running on a server to a remote client.

The Chromium library [12] is a "drop-in" replacement for OpenGL that provides the ability to route an OpenGL command stream from an application to other machines for processing. A typical Chromium use scenario is for driving tiled displays, where each display tile is connected to a separate machine. Chromium will intercept the application's OpenGL command stream, and, via a user-configurable "stream processing" infrastructure, will route OpenGL commands to the appropriate machine for subsequent rendering. The application emitting OpenGL calls may be either a serial or a parallel application. Chromium provides constructs for an application to perform draw-time synchronization operations like barriers. Chromium has several other user-configurable "stream processing units" that also provide a form of send-images partitioning. Chromium serves as the basis for Chromium Renderserver, which is a high performance system for remote and distributed visualization (see 3.9). Chromium includes several types of processing capabilities, important for parallel rendering operations—render-time synchronization and barriers—to support sort-first parallel, hardware-accelerated volume rendering, complete with level-of-detail-based model switching to accelerate rendering [2].

One disadvantage of the send-geometry approach is the potential size of the renderable geometry payload. In some circumstances, the size of this payload may exceed the size of the original data set, or it may be so large as to exceed the capacity of the client to hold in memory all at once for rendering. Streaming approaches (Chap. 10) are one mechanism for accommodating rendering data sets too large for a client's memory, yet the relatively slower network connection may be a more significant barrier.

The primary advantage of the send-geometry approach is that, once the geometry content is resident in the client's memory, the client may be capable of very high rendering frame rates. This approach may be the best when: (1) the geometry payload fits entirely within the client memory, and (2) interactive client-side rendering rates are a priority.

3.6 Hybrid and Adaptive Approaches

Hybrid pipeline partitioning approaches are those that do not fall strictly into one of the above categories. Adaptive approaches are those where the application will alter its partitioning strategy at runtime to achieve better performance in light of changing condition, such as slower network speeds or a shift in the relative amount of altering between data, geometry, image data, moved between server and client.

The Visapult system [4, 3, 27], described in more detail later in 3.8, employs an architecture that leverages the concept of *co-rendering*, which is best thought of as a hybrid approach. In this approach, the server performs partial rendering of the source data set, and then transmits these partial rendering results to the client. The data sent by the server is not strictly geometry nor images: it is direct volume rendering (images) of subsets of the source data set. Once on the client, these partial renderings are further rendered in an interactive fashion to the user, who can interactively change the viewpoint, etc.

This approach, which hides latency from the user, strikes a trade-off between interactivity and data load, although the pipeline partitioning is static. If $O(n^3)$ volume data is input to the server, then the server moves $O(n^2)$ image-based data to the client. The trade-off is that the client has full interactive rendering capabilities of a reduced-sized version of the source data: it is able to render at full hardware-accelerated rates, but with known accuracy/fidelity limitations.

The VisIt visualization application (Chap. 16) uses an adaptive approach to decide when to use send-images vs. send-geometry. This adaptive approach is described in more detail later in 3.10.

3.7 Which Pipeline Partitioning Works the Best?

As part of an article examining how the Grid will affect the design and use of remote and distributed visualization systems, Shalf and Bethel 2003 [28] present a study that aims to determine which pipeline partitioning works the best. The main idea in the study is to set up three different partitionings of a common visualization algorithm and use various combinations of computational platforms to run different stages of the pipeline. They measure absolute end-to-end execution time for each pipeline partition as well as the time required for each individual component in each different pipeline configuration.

Figure 3.2 shows three potential ways to partition the visualization pipeline and shows the components of the visualization pipeline used to create an image of an isosurface. The links between the blocks indicate data flow. The *Desktop Only* pipeline places the isosurface and rendering components on the desktop machine. The *Cluster Isosurface* pipeline places a distributed-memory implementation of the isosurface component on the cluster, and transfers triangles over the network to a serial rendering component on the desktop. The *Cluster Render* pipeline performs isosurface extraction and rendering on the cluster, then transfers images to the desktop for display. The colors used in Figure 3.3a correspond to the groupings of components shown in Figure 3.2: dark gray for the Desktop Only pipeline, black for the Cluster Isosurface pipeline, and light gray for the Cluster Render pipeline.

For this study, the following conditions, optimizations, and simplifying assumptions are in effect:

- All geometry produced by the isosurface component are triangle strips. Each incremental triangle of the strip is represented with a single vertex, which consumes twenty-four bytes: six four-byte floats containing vertex and normal data.
- All graphics hardware is capable of rendering 50M triangles/second. An 8-node system, each node equipped with identical graphics hardware, can render 400M triangles/second.

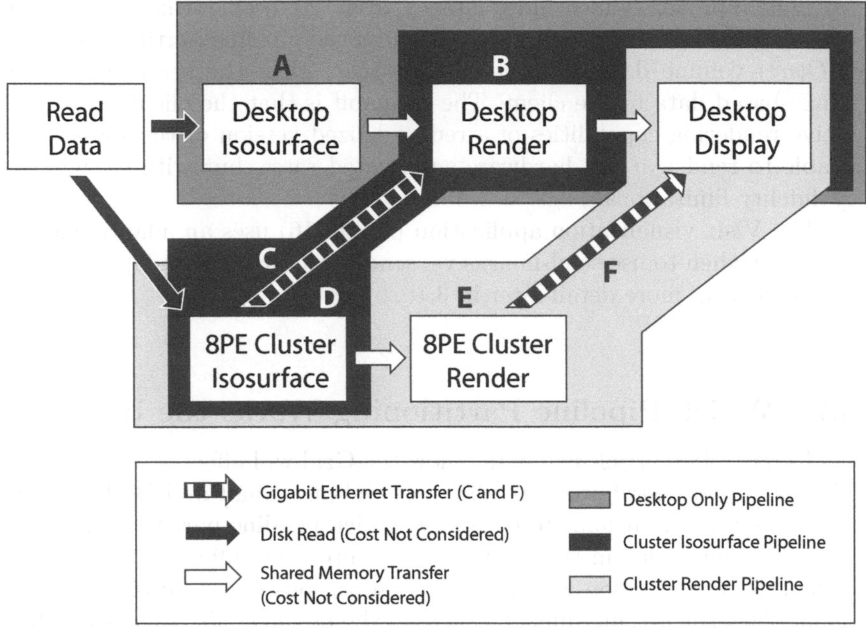

FIGURE 3.2: Three potential partitionings of a remote and distributed visualization pipeline performance experiment showing execution components and data flow paths. Colored arrows indicate each of the potential data flow paths. Image courtesy of E. Wes Bethel and John Shalf (LBNL).

- The image transfer assumes a 24-bit, high definition, 1920×1080 pixels, CIF framebuffer.
- Interconnects use a 1 GB network with perfect performance.
- The performance model does not consider the cost of data reads, the cost of scatter-gather operations, or the cost of displaying cluster-rendered images on the desktop.

Whereas Figure 3.3a shows the absolute runtime of each pipeline for varying numbers of triangles, Figure 3.3b presents the relative runtime of all components for varying numbers of triangles. In Figure 3.3b, the absolute runtime of all components is summed, totaling 100%, and the size of each colored segment in each vertical bar shows the relative time consumed by that particular component. Each component's execution time is normalized by the sum of all component execution times, so one may quickly determine which components, or network transfers, will dominate the execution time. Note, some components are used in more than one pipeline configuration.

The vertical bars in Figure 3.3b are color-coded by component: those in dark gray, cross-hatched and solid, are desktop-resident, and show the cost of computing isosurfaces on the desktop (A) and rendering isosurfaces on the

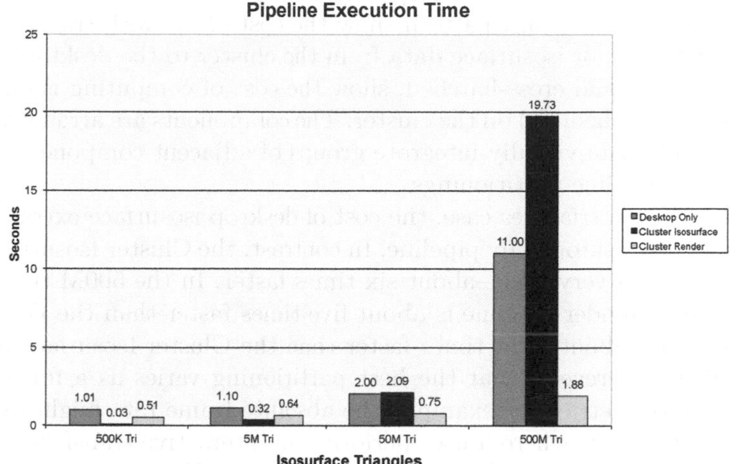

(a) Absolute runtime for each pipeline. *Desktop Only* performance is the sum of components A and B. Cluster Isosurface performance is the sum of components B, C and D. Cluster Render performance is the sum of components D, E and F.

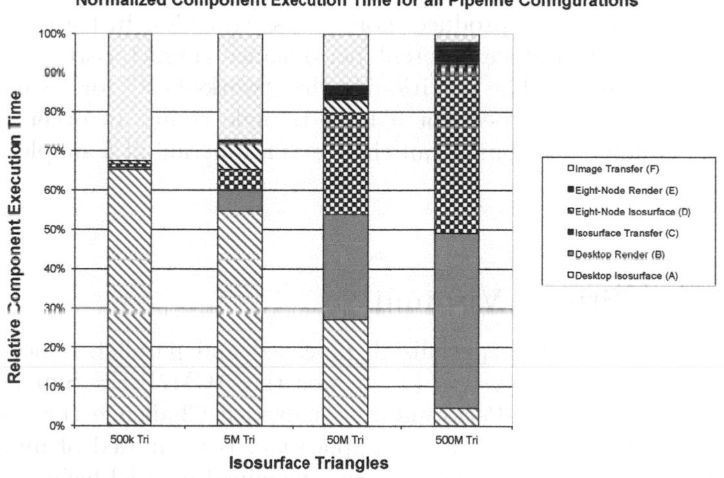

(b) Relative performance of execution components in the three pipelines. The desktop-only pipeline is the sum of $A+B$; the cluster isosurface pipeline is the sum of $B+C+D$; the cluster render pipeline is the sum of $D+E+F$.

FIGURE 3.3: The three potential partitionings have markedly different performance characteristics, depending on many factors, including some that are dependent upon the data set being visualized. Images courtesy of John Shalf and E. Wes Bethel (LBNL).

desktop (B) in both the desktop-only and cluster-isosurface configurations. Those in a checkerboard pattern show the cost of network transfers of either image data (F), or isosurface data from the cluster to the desktop (C). Those in black, solid and cross-hatched, show the cost of computing isosurfaces (D) and rendering them (E) on the cluster. The components are arranged vertically so the reader can visually integrate groups of adjacent components into their respective pipeline partitionings.

In the 500K triangles case, the cost of desktop isosurface extraction dominates in the Desktop Only pipeline. In contrast, the Cluster Isosurface pipeline would perform very well—about six times faster. In the 500M triangles case, the Cluster Render pipeline is about five times faster than the Desktop Only pipeline, and about eight times faster than the Cluster Isosurface pipeline.

This study reveals that the best partitioning varies as a function of the performance metric. For example, the absolute frame rate might be the most important metric, where a user performs an interactive transformation of 3D geometry produced by the isocontouring stage. The partitioning needed to achieve a maximum frame rate will vary according to the rendering load and rendering capacity of pipeline components.

Surprisingly, the best partitioning can also be a function of a combination of the visualization technique and the underlying data set. The authors' example uses isocontouring as the visualization technique and changes in the isocontouring level will produce more or less triangles. In turn, this varying triangle load will produce different performance characteristics of any given pipeline partitioning. The partitioning that "works best" for a small triangle count may not be the best for a large triangle count. In other words, the optimal pipeline partitioning can change as a function of a simple parameter change.

3.8 Case Study: Visapult

Visapult is a highly specialized, pipelined and parallel, remote and distributed, visualization system [4]. It won the ACM/IEEE Supercomputing Conference series High Performance Bandwidth Challenge three years in a row (2000–2002). Visapult, as an application, is composed of multiple software components that are executed in a pipelined-parallel fashion over wide-area networks. Its architecture is specially constructed to hide latency over networks and to achieve ultra-high performance over wide-area networks.

The Visapult system uses a multistage pipeline partitioning. In an end-to-end view of the system, from bytes on disk or in simulation memory, to pixels on screen, there are two separate pipeline stages. One is a send-geometry partitioning, the other is a send-data partitioning.

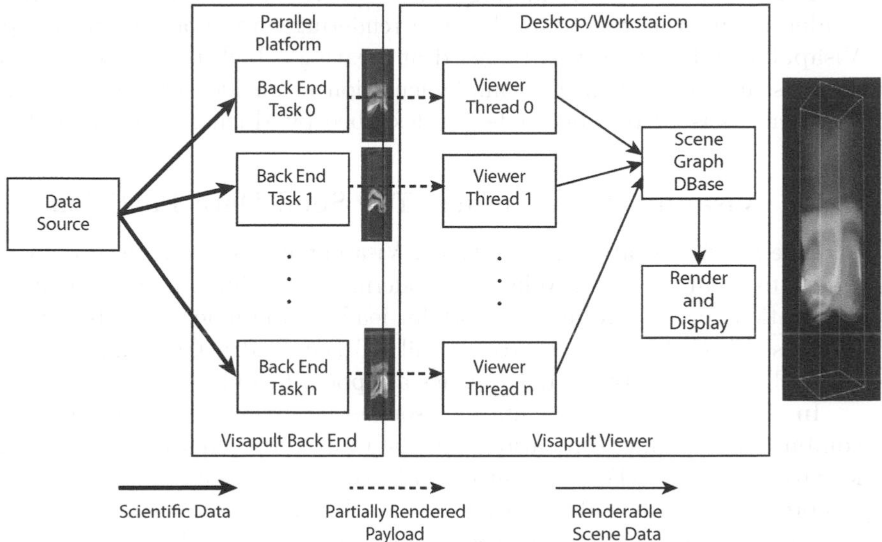

FIGURE 3.4: Visapult's remote and distributed visualization architecture. Image courtesy of E. Wes Bethel (LBNL).

3.8.1 Visapult Architecture: The Send-Geometry Partition

The original Visapult architecture, described by Bethel et al. 2000 [4], shown in Figure 3.4, consists of two primary components. One component is a parallel *back end*, responsible for loading scientific data and performing, among other activities, "partial" volume rendering of the data set. The volume rendering consists of applying a user-defined transfer function to the scalar data, producing an RGBα volume, then performing axis-aligned compositing of these volume subsets, producing semi-transparent textures. In a typical configuration, each data block in the domain decomposition will result in six volume rendered textures: one texture for each of the six principal axis viewing directions. As a result, if the Visapult back end loads $O(n^3)$ data, it produces $O(n^2)$ output in the form of textures.

Then, the back end transmits these "partially rendered" volume subsets to the *viewer*, shown in Figure 3.4 as `Partially Rendered Payload`, where the subsets are stored as textures in a high performance scene graph system in the viewer. The viewer, via the scene graph system, renders these semi-transparent textures on top of proxy geometry in the correct back-to-front order at interactive rates via hardware-acceleration. Inside the viewer, the scene graph system switches between each of the six source textures for each data block depending upon camera orientation to present the viewer with the best fidelity rendering.

This particular "co-rendering" idea—where the server performs partial rendering and the viewer finishes the rendering—was not new to Visapult. Visapult's architecture was targeted at creating a high performance, remote and distributed visualization implementation of an idea called image-based rendering assisted volume rendering described by Mueller et al. 1999 [20].

3.8.2 Visapult Architecture: The Send-Data Partition

Like all visualization applications, Visapult needs a source of data from which to create images. Whereas modern, production-quality visualization applications provide robust support for loading a number of well-defined file formats, Visapult's data source for all SC Bandwidth Challenge runs were remotely located network data caches as opposed to files.

In the SC 2000 implementation, source data consisted of output from a combustion modeling simulation. The data was stored on a Distributed Parallel Storage System (DPSS), which can be thought of as a high-speed, parallel remote block-oriented data cache [30]. In this implementation, the Visapult back end invoked DPSS routines, similar in concept to POSIX `fread` calls that, in turn, loaded blocks of raw scientific data in parallel, from a remote source and relied on the underlying infrastructure, the DPSS client library, to efficiently move data over the network.

In an effort to make an even better use of the underlying network, Shalf and Bethel, 2003 [27], extended Visapult to make use of a UDP-based "connectionless" protocol. They connected to a freely-running simulation, which provided a data source. This change resulted in the ability for the Visapult back end to achieve unprecedented levels of network utilization, close to 100% of the theoretical line rate, for sustained periods of time.

The TCP-based approach can be thought of as a process of "load a timestep's worth of data, then render it." Additionally, because the underlying protocol is TCP-based, there was no data loss between remote source and the Visapult back end.

Going to the UDP-based model required rethinking both the network protocol and the Visapult back end architecture. In the TCP approach, the Visapult back end "requests" data, which is a "pull" model. In the UDP approach, though, the data source streams out data packets as quickly as possible, and the Visapult back end must receive and process these data packets as quickly as possible. This approach is a "push" model. There is no notion of timestep or frame boundary in this push model; the Visapult back end has no way of knowing when all the data packets, for a particular timestep, are in memory. After all, some of the packets may be lost as UDP does not guarantee packet delivery. See Bethel and Shalf, 2005, [3] for more design change details and see Shalf and Bethel, 2003, [27] for the UDP packet payload design.

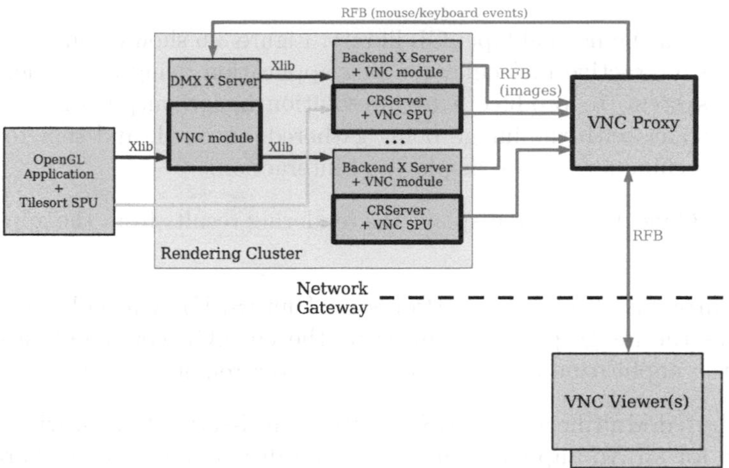

FIGURE 3.5: CRRS system components for a two-tile DMX display wall configuration. Lines indicate primary direction of data flow. System components outlined with a thick line are new elements from this work to implement CRRS; other components outlined with a thin line existed in one form or another prior to the CRRS work. Image source: Paul et al. 2008 [23].

3.9 Case Study: Chromium Renderserver

Paul et al. 2008 [23] describe Chromium Renderserver (CRRS), which is a software infrastructure that provides the ability for one or more users to run and view image output from unmodified, interactive OpenGL and X11 applications on a remote, parallel computational platform, equipped with graphics hardware-accelerators, via industry-standard Layer 7 network protocols and client viewers.

Like Visapult, CRRS has a multi-stage pipeline partitioning that uses both send-geometry and send-images approaches. Figure 3.5 shows a high-level architectural diagram of a two-node CRRS system. A fully operational CRRS system consists of six different software components, which are built around VNC's remote framebuffer (RFB) protocol. The RFB protocol is a Layer 7 network protocol, where image data and various types of control commands are encoded and transmitted over a TCP connection, between producer and consumer processing components. The motivation for using RFB is because it is well understood, and there exist (VNC) viewers for nearly all the current platforms. One of the CRRS design goals is to allow an unmodified VNC viewer application to be used as the display client in a CRRS application.

The CRRS general components are:

- The application is any graphics or visualization program that uses OpenGL and/or Xlib for rendering. Applications need no modifications to run on CRRS, but they must link with the Chromium *faker* library rather than the normal OpenGL library. Figure 3.6 shows visual output from an interactive molecular docking application being run under the CRRS system, to produce a high-resolution image output on a 3×2 tiled display, with the image being gathered, encoded, and sent to another machine for remote display and interaction.

- The VNC viewer, which displays the rendering results from the application.

- The Chromium VNC Stream Processing Unit (SPU), which obtains and encodes the image pixels produced by the OpenGL command, stream from the application and sends the pixels to a remote viewer.

- Distributed Multihead X (DMX), which is an X-server that provides the ability for an Xlib application to run on a distributed-memory parallel cluster.

- The VNC Proxy is a specialized VNC Server that takes encoded image input from VNC Servers and VNC SPUs, running on each of the parallel rendering nodes and transmits the encoded image data to the remote client(s). The VNC Proxy solves the problem of synchronizing the rendering results of the asynchronous Xlib and OpenGL command streams.[2]

- The VNC Server X Extension is present on the X server at each parallel rendering node. This component harvests, encodes and transmits the portion of the framebuffer modified by the Xlib command stream.

In a CRRS system, the send-geometry partitioning exists between the potentially parallel visualization application—which loads scientific data, performs visualization processing, and emits OpenGL draw commands—and a parallel machine, where those draw commands are routed by the Chromium Tilesort Stream Processing Unit [12]. In this part of the system, the visualization or graphics application will *push* draw commands over the wire to the rendering nodes.

There are actually two stages of send-images partitioning in the CRRS pipeline. The first exists between each of the nodes where parallel rendering occurs and the VNC Proxy component. The second exists between the remote viewer (a VNC viewer) and the centrally located VNC Proxy component. Whereas the send-geometry partitioning operates under a *push model*, the send-images partitioning operates under a *pull* model. In this mode of operation, the remote VNC Viewer will request an image update from the VNC Proxy. In turn, the VNC Proxy requests RFB-based image updates from each of the rendering nodes, which, in effect, is running a VNC server in the form of the VNC Chromium Stream Processing Unit.

[2]See `http://vncproxy.sourceforge.net` for more information about the VNC Proxy and CRRS setup and operation.

FIGURE 3.6: An unmodified molecular docking application, run in parallel, on a distributed-memory system using CRRS. Here, the cluster is configured for a 3 × 2 tiled display setup. The monitor for the "remote user machine" in this image is the one on the right. While this example has a "remote user" and a "central facility" connected via a local area network, the model typically used is one where the remote user is connected to the central facility via a low-bandwidth, high-latency link. Here, the image shows the complete 4800 × 2400 full-resolution image from the application running on the parallel, tiled system, appearing in a VNC viewer window on the right. Through the VNC viewer, the user interacts with the application to change the 3D view and orientation, as well as various menus on the application to select rendering modes. As is typically the case, this application uses Xlib-based calls for GUI elements (menus, etc.) and OpenGL for high performance model rendering. Image source: Paul et al. 2008 [23].

The authors of this study examined the end-to-end performance of several different types of optimizations, listed below. Collectively, all these optimizations, when enabled, produce a 20–25% performance improvement on all network types—local area network and three types of wide area networks. The optimizations fall into two broad classes: those that reduce end-to-end system latency and those that reduce the amount of payload moving between systems.

- *RFB caching* improves performance by maintaining a cache of RFB Update messages that contain encoded image data. The VNC Proxy responds to RFB Update Requests by sending cached responses, rather than encoding, possibly multiple times, the contents of its internal VFB. This optimization helps to reduce end-to-end latency.

- *Bounding box tracking.* Here, only the portions of the scene rendered by

OpenGL that have changed are rendered and transmitted to the remote client. This optimization helps to reduce the amount of RFB image payload.

- *Double buffering.* By maintaining a double-buffered VFB in the VNC SPU, two operations can occur simultaneously—application rendering and image encoding/transmission. This optimization helps to reduce end-to-end latency.

- *Frame synchronization.* While not strictly an optimization, this feature is needed to synchronize parallel rendering streams and to prevent frame dropping (spoiling), when images are rendered more quickly than they can be delivered to the remote client.

Another interesting element of this study was the examination of image compression algorithm combinations and their end-to-end performance, when used over various types of networks. The VNC system supports several types of image encoding/compression methods, the use of which is a user-selectable parameter. Some forms of compression are lossless; they produce a relatively lower levels of compression, but they execute relatively quicker. Other forms of compression are lossy; they produce relatively greater degrees of compression and execute relatively slower. The study identifies a trade-off in which lossless compression algorithms are better suited, from a performance perspective, on local area networks (LAN), which have a lower latency and higher bandwidth. In this setting, the runtime of the compression algorithm became a bottleneck. In contrast, on wide area networks (WAN), the extra runtime required for performing lossy compression was outweighed by the performance gain, resulting from sending less image payload data over a slower, higher latency network.

The two-stage CRRS design, which integrates send-geometry and send-images, combined with using the industry-standard RFB protocol, resulted in a system that was highly flexible, scalable, and that provided the ability to support a potentially large number of visualization and graphics applications accessible to remote users via a common, industry-standard viewer.

3.10 Case Study: VisIt and Dynamic Pipeline Reconfiguration

The VisIt parallel visualization application (see Chap. 16) supports both send-images and send-geometry pipeline partitionings, as well as a dynamic, runtime change from one partitioning to another, depending on several factors. This section describes how each of these partitionings work, along with the methodology VisIt uses for deciding to change from one pipeline partitioning to another.

3.10.1 How VisIt Manages Pipeline Partitioning

VisIt uses a construct, known as an `avtPlot` object, to manage pipeline execution and different forms of pipeline partitioning. An `avtPlot` object exists, in part, in both VisIt's server and client components. The two parts manage pipeline partitioning and also optimize the amount of computation used to update a visual output (plot), as inputs change. For example, some inputs impact only how geometry is rendered. Other inputs impact the geometry itself. The `avtPlot` object differentiates between these update operations and ensures that the appropriate parts of the pipeline are executed, only when inputs effecting them change.

An `avtPlot` object knows which parts of a pipeline require execution, as inputs change. This knowledge, then, affects how VisIt manages pipeline partitioning for any given plot. In a send-geometry partitioning, an input change that affects only the rendering step does not require any interaction with the server because all the computation can and will be handled on the client. On the other hand, when VisIt uses a send-images partitioning to produce and deliver the very same plot, then the same change in inputs will, indeed, involve interaction with the server. Nonetheless, even in this circumstance, the server will execute only those portions of the pipeline necessary to re-render the plot.

3.10.2 Send-Geometry Partitioning

The send-geometry mode of pipeline partitioning is typically most appropriate when visualizing small input data sets, where the results of visualization algorithms, such as renderable geometry, can fit entirely in the client's memory. In VisIt, the definition of "small" is determined by a user-specifiable parameter, indicating a maximum size threshold for the amount of geometry data produced by the server and sent to the client. This is known as the Scalable Rendering Threshold (SRT).

In this partitioning, the client can vary the visualization of a plot in a variety of ways, without requiring any further exchanges of data with the server. This includes variations in viewing parameters, such as pans, zooms and rotations, changes in transfer function, transparency, glyph types used for glyphed plots, and so forth. In addition, when visualizing a time-series and sufficient memory is available, the client can cache geometry for each timestep. This cache thereby enables the user to quickly animate the series, as well as vary the visualization in these ways, without the need for interactions with the server.

If, during the course of a run, the amount of geometry produced by visualization algorithms increases, due perhaps to processing larger or more complex data sets, it may exceed the client's ability to store and process in its entirety. Either the client will have to accept and render geometry in a streaming fashion (see Chap. 10), or the server will have to take on the responsibility for rendering. VisIt employs the latter approach. This is VisIt's send-images mode of operation.

3.10.3 Send-Images Partitioning

The send-images mode of pipeline partitioning is known as the *scalable rendering* mode in VisIt. It is typically used for large-scale data where the amount of geometry to be displayed exceeds the SRT. In the send-images partitioning, VisIt's server does all the work to compute the geometry to be visualized and then renders the final image. After, it sends this image to the client for a final display to the user. In parallel, each processor renders a portion of the geometry to the relevant pixels into a local z-buffered image. Results from different processors' local image are z-buffer-composited using a sort-last approach (see 4.2) via a user-defined MPI reduction operator, optimized for compositing potentially large images (e.g., $16K^2$ pixels). The server then sends the final image, with or without the associated z-buffer, to the client for display to the user. The z-buffer will be included if the client window into which the results will be displayed is also displaying results from other server processes. Use of the z-buffer in this fashion is what enables VisIt to properly handle multiple servers, each potentially employing a different pipeline partitioning but displaying to the user, via the same integrated visual output.

3.10.4 Automatic Pipeline Partitioning Selection

VisIt offers the user the ability to decide: to always or never use a send-images partitioning for execution; or to use an automatic mechanism for switching between send-geometry and send-images.

For the automatic selection mode, VisIt uses heuristics to decide which pipeline partitioning is best for a given situation. The heuristic relies on having an estimate of the geometry's volume data that would overwhelm the client either because it will require too much memory to store or because it will take too long to render. This estimate is governed by the SRT user-settable threshold. It is a count of polygons of which VisIt estimates the client can reasonably accommodate. VisIt does not measure this value explicitly, but instead relies on the user's judgment regarding the performance characteristics of the machine where the client is run, and then sets this threshold accordingly. By default, the SRT is set at two million polygons.

For each window, in the client employing a send-geometry partitioning, the server keeps track of the total polygon count of all plots being displayed there. When the count exceeds the SRT, VisIt automatically switches all the plots in that window to a send-images partitioning. When and if the polygon count drops below the SRT, VisIt will then automatically switch all plots to a send-geometry partitioning. There is some hysteresis included in this process to prevent oscillations in partitioning when the polygon count hovers right around the SRT.

For a parallel server, changes in pipeline partition are complicated somewhat, by the fact that the SRT can be exceeded when results from only some of the processors have been accumulated. Therefore, VisIt may have to do a reversal midway through a presumed send-geometry pipeline execution and restart it as a send-images partitioning.

3.11 Conclusion

RDV architectures are those in which the components of the visualization processing pipeline are distributed across multiple computational platforms. The differences between different RDV architectures tend to center around how the visualization pipeline is decomposed across distributed platforms. Some architectures move images from one machine to another, others move renderable geometry between machines, while others move raw scientific data between machines. It is an open question which form of partitioning works best. There is no one single solution that is optimal across all combinations of use scenario, specific data set and visualization/processing techniques, and computational and networking infrastructure.

There are a number of different RDV implementations, ranging from single applications, like VisIt, that have some form of RDV processing capability, to more general purpose solutions, like the Chromium Renderserver, which provide general purpose RDV capabilities along with scalable rendering, for many different applications. Interestingly, the notion of *in situ* processing, which is the subject of Chapter 9, includes a form of distributed visualization processing as one potential implementation path.

References

[1] Micah Beck, Terry Moore, and James S. Plank. An End-to-end Approach to Globally Scalable Network Storage. In *SIGCOMM '02: Proceedings of the 2002 Conference on Applications, Technologies, Architectures, and Protocols for Computer Communications*, pages 339–346, New York, NY, USA, 2002. ACM Press.

[2] E. Wes Bethel, Greg Humphreys, Brian Paul, and J. Dean Brederson. Sort-First, Distributed Memory Parallel Visualization and Rendering. In *Proceedings of the 2003 IEEE Symposium on Parallel and Large-Data Visualization and Graphics*, pages 41–50, Seattle, WA, USA, October 2003.

[3] E. Wes Bethel and John Shalf. Consuming Network Bandwidth with Visapult. In Chuck Hansen and Chris Johnson, editors, *The Visualization Handbook*, pages 569–589. Elsevier, 2005. LBNL-52171.

[4] E. Wes Bethel, Brian Tierney, Jason Lee, Dan Gunter, and Stephen Lau. Using High-Speed WANs and Network Data Caches to Enable Remote and Distributed Visualization. In *Supercomputing '00: Proceedings of the 2000 ACM/IEEE Conference on Supercomputing (CDROM)*, Dallas, Texas, USA, 2000. LBNL-45365.

[5] Ian Bowman, John Shalf, Kwan-Liu Ma, and E. Wes Bethel. Performance Modeling for 3D Visualization in a Heterogeneous Computing Environment. Technical report, Lawrence Berkeley National Laboratory, Berkeley, CA, USA, 94720, 2004. LBNL-56977.

[6] Jerry Chen, Ilmi Yoon, and E. Wes Bethel. Interactive, Internet Delivery of Visualization via Structured, Prerendered Multiresolution Imagery. *IEEE Transactions in Visualization and Computer Graphics*, 14(2):302–312, 2008. LBNL-62252.

[7] Shenchang Eric Chen. QuickTime VR—An Image-Based Approach to Virtual Environment Navigation. In *Proceedings of the 22nd Annual Conference on Computer Graphics and Interactive Techniques*, SIGGRAPH '95, pages 29–38, Los Angeles, CA, USA, 1995.

[8] Daniel Cohen-Or and Eyal Zadicario. Visibility Streaming for Network-based Walkthroughs. In *Proceedings of Graphics Interface*, pages 1–7, 1998.

[9] D. R. Commander. VirtualGL. `http://www.virtualgl.org`, 2011.

[10] H. Hege, A. Hutanu, R. Kähler, A. Merzky, T. Radke, E. Seidel, and B. Ullmer. Progressive Retrieval and Hierarchical Visualization of Large

Remote Data. In *Proceedings of the Workshop on Adaptive Grid Middleware*, pages 60–72, September 2003.

[11] Hans-Christian Hege, André Merzky, and Stefan Zachow. Distributed Visualizaton with OpenGL VizServer: Practical Experiences. ZIB Preprint 00-31, 2001.

[12] Greg Humphreys, Mike Houston, Ren Ng, Randall Frank, Sean Ahern, Peter D. Kirchner, and James T. Klosowski. Chromium: A Stream-Processing Framework for Interactive Rendering on Clusters. In *SIGGRAPH '02: Proceedings of the 29th Annual Conference on Computer Graphics and Interactive Techniques*, pages 693–702, San Antonio, TX, USA, 2002.

[13] Ken Jones and Jenn McGee. OpenGL Vizserver User's Guide Version 3.5. Technical report, Silicon Graphics Inc., Mountain View, CA, USA, 2005. Document Number 007-4245-014.

[14] Mark J. Kilgard. *OpenGL programming for the X Window System*. OpenGL Series. Addison-Wesley Developers Press, 1996.

[15] Kitware, Inc. and Jim Ahrens. ParaView: Parallel Visualization Application. `http://www.paraview.org/`.

[16] Dieter Kranzlmüller, Gerhard Kurka, Paul Heinzlreiter, and Jens Volkert. Optimizations in the Grid Visualization Kernel. In *IEEE Parallel and Distributed Processing Symposium (CDROM)*, pages 129–135, 2002.

[17] Lawrence Livermore National Laboratory. VisIt: Visualize It Parallel Visualization Application. `http://www.llnl.gov/visit/`.

[18] Eric J. Luke and Charles D. Hansen. Semotus Visum: A Flexible Remote Visualization Framework. In *VIS '02: Proceedings of the Conference on Visualization '02*, pages 61–68, Boston, MA, USA, 2002.

[19] Kwan-Liu Ma, Greg Schussman, Brett Wilson, Kwok Ko, Ji Qiang, and Robert Ryne. Advanced Visualization Technology for Terascale Particle Accelerator Simulations. In *Proceedings of Supercomputing 2002 Conference*, pages 19–30, Baltimore, MD, USA, November 2002.

[20] Klaus Mueller, Naeem Shareef, Jian Huang, and Roger Crawfis. IBR Assisted Volume Rendering. In *Proceedings of IEEE Visualization, Late Breaking Hot Topics*, pages 5–8, October 1999.

[21] A. Nye. *Xlib Programming Manual: For Version 11 of the X Window System*. Definitive guides to the X Window System. O'Reilly & Associates, 1992.

[22] Valerio Pascucci and Randall J. Frank. Global Static Indexing for Real-time Exploration of Very Large Regular Grids. In *Supercomputing '01: Proceedings of the 2001 ACM/IEEE Conference on Supercomputing (CDROM)*, Denver, CO, USA, 2001.

[23] Brian Paul, Sean Ahern, E. Wes Bethel, Eric Brugger, Rich Cook, Jamison Daniel, Ken Lewis, Jens Owen, and Dale Southard. Chromium Renderserver: Scalable and Open Remote Rendering Infrastructure. *IEEE Transactions on Visualization and Computer Graphics*, 14(3):627–639, May/June 2008. LBNL-63693.

[24] Steffen Prohaska, Andrei Hutanu, Ralf Kahler, and Hans-Christian Hege. Interactive Exploration of Large Remote Micro-CT Scans. In *VIS '04: Proceedings of the Conference on Visualization '04*, pages 345–352, Austin, TX, USA, 2004.

[25] Tristan Richardson, Quentin Stafford-Fraser, Kenneth R. Wood, and Andy Hopper. Virtual Network Computing. *IEEE Internet Computing*, 2(1):33–38, 1998.

[26] Will Schroeder, Kenneth M. Martin, and William E. Lorensen. *The Visualization Toolkit (2nd ed.): An Object-Oriented Approach to 3D Graphics*. Prentice-Hall, Upper Saddle River, NJ, USA, 1998.

[27] John Shalf and E. Wes Bethel. Cactus and Visapult: A Case Study of Ultra-High Performance Distributed Visualization Using Connectionless Protocols. *IEEE Computer Graphics and Applications*, 23(2):51–59, March/April 2003. LBNL-51564.

[28] John Shalf and E. Wes Bethel. How the Grid Will Affect the Architecture of Future Visualization Systems. *IEEE Computer Graphics and Applications*, 23(2):6–9, May/June 2003.

[29] Mohammad Shorfuzzaman, Peter Graham, and Rasit Eskicioglu. Adaptive Popularity-Driven Replica Placement in Hierarchical Data Grids. *Journal of Supercomputing*, 51(3):374–392, 2010.

[30] Brian Tierney, Jason Lee, Brian Crowley, Mason Holding, Jeremy Hylton, and Fred L. Drake. A Network-Aware Distributed Storage Cache for Data Intensive Environments. In *Proceedings of IEEE High Performance Distributed Computing Conference (HPDC-8)*, pages 185–193, Redondo Beach, CA, USA, 1999.

Chapter 4

Rendering

Charles Hansen
University of Utah

E. Wes Bethel
Lawrence Berkeley National Laboratory

Thiago Ize
University of Utah

Carson Brownlee
University of Utah

4.1 Introduction 49
4.2 Rendering Taxonomy 50
4.3 Rendering Geometry 52
4.4 Volume Rendering 53
4.5 Real-Time Ray Tracer for Visualization on a Cluster 56
4.5.1 Load Balancing 57
4.5.2 Display Process 57
4.5.3 Distributed Cache Ray Tracing 58
4.5.3.1 DC BVH 59
4.5.3.2 DC Primitives 60
4.5.4 Results 61
4.5.5 Maximum Frame Rate 62
4.6 Conclusion 66
References 67

4.1 Introduction

The process of creating an image from the data, as the last step before display, in the visualization pipeline is called *rendering*; this is a key aspect of all high performance visualizations. As shown in Figure 1.1, the rendering process takes data resulting from visualization algorithms and generates an image for display. This figure highlights that data is typically filtered, transformed, subsetted and then mapped to renderable geometry before rendering takes place. There are two forms of rendering typically used in visualization, based on the underlying mapped data: rendering of geometry generated through visualization mapping algorithms (see Fig. 1.1) and direct rendering from the data. For *geometric rendering*, typical high performance visualization packages use the OpenGL library, which converts the geometry generated by visualization mapping algorithms to colored pixels through rasterization [36]. Another method for generating images from geometry is ray tracing, where rays from the viewpoint through the image pixels, are intersected with geometry to form the colored pixels [35]. *Direct rendering* does not require an intermediate mapping of data to geometry, but rather, it directly generates the colored pixels through a mapping that involves sampling the data and for each sample, mapping the data to renderable quantities (color and transparency). Direct volume rendering is a common technique for rendering scalar fields directly into images.

The rest of this chapter will discuss a rendering taxonomy that is widely used in high performance visualization systems. Geometric rendering is then presented with examples of both rasterization and ray tracing solutions. Direct volume rendering is introduced. And finally, an example of a geometric rendering system is discussed, using ray tracing on a commodity cluster.

4.2 Rendering Taxonomy

In 1994, Molnar et al. described a taxonomy of parallel rasterization graphics hardware that has become the basis for most parallel implementations—both hardware and software based—and is widely used in high performance visualization systems [25]. While the taxonomy describes different forms of graphics hardware, the generalization of the taxonomy provides the basis for most software-based and GPU-based parallel rendering systems for both rasterization rendering and direct volume rendering.

The taxonomy describes three methods for the parallelization of graphics hardware performing rasterization, based on when the assignment to the image space (sometimes called screen-space) takes place. If one considers the rasterization process of consisting of two basic steps, geometry processing and rasterization (the generation of pixels from the geometry), then the sort (the assignment of data, geometry, or pixels), to image space can occur at three points.

As shown in the left of Figure 4.1, geometric primitives can be assigned to screen-space before geometry processing takes place. This is called sort-first. Since geometry processing involves the transformation of the geometry to image space, this requires *a priori* information about the geometry and the mapping to image space. Such *a priori* information can be obtained from previously rendered frames, heuristics, or by simply replicating the data among all parallel processes. The image space is divided into multiple tiles, each of which can be processed in parallel. Since the geometry is assigned to the appropriate image space tile, geometry processing, followed by rasterization, generates the final colored pixels of the image. This process can be conducted in parallel by each of the processors responsible for a particular image tile. The advantage of sort-first methods is that image compositing is not required, since subregions of the final image are defined uniquely. The disadvantage is that, without data replication or heuristics, the assignment of data to the appropriate image space subregion is difficult. Sort-first methods typically leverage the frame-to-frame coherency, which is often readily available in interactive applications.

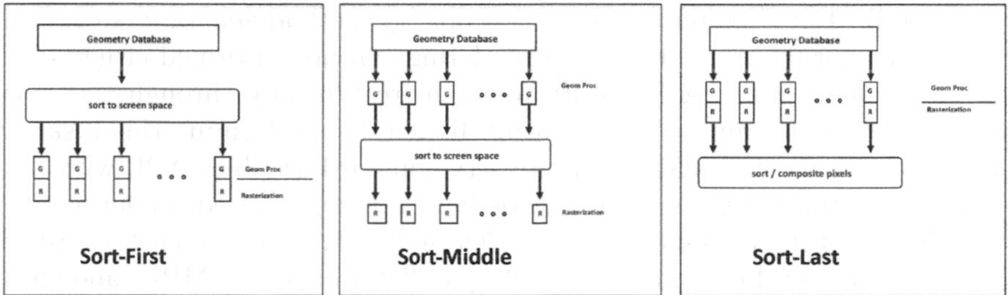

FIGURE 4.1: The rendering taxonomy based on Molnar's description. Left most is sort-first, the center is sort-middle, while the right most is sort-last.

Sort-middle architectures perform the sort to image space, after geometry processing, but before rasterization. As in sort-first, the image is divided into tiles and each rasterization processor is responsible for a particular image tile. Geometry is processed in parallel by the multiple geometry processors that obtain geometry in some manner, such as round-robin distribution. One of the steps of geometry processing is the transformation of the geometry to image space. The image space geometry is sorted and sent to the appropriate rasterization processor responsible for the image space partition covered by the processed geometry. While common in graphics hardware, high performance visualization systems typically use either sort-last or sort-first methodology for parallel rendering.

The right-most image in Figure 4.1 shows the sort-last technique. In this method, geometry is distributed to geometry processors, in some manner, such as round-robin distribution. Each processor in parallel transforms the geometry to image space and then, rasterizes the geometry to a local image. The rasterizer generates the image pixels. Note that depending on the geometry, the pixels may cover an entire image, or more typically, a small portion of the image. After all the geometry has been processed and rasterized, the resulting partial images are combined, through compositing, as described in Chapter 5. The advantage of the sort-last method is that renderable entities, geometry, or data can be partitioned among the parallel processors and the final image is constructed through parallel compositing. The disadvantage is that the compositing step can dominate the image generation time. Sort-last methods are the most common found in high performance visualization and are part of the VisIt and ParaView distributions.

4.3 Rendering Geometry

As described above, one method for visualization involves mapping the data to geometry, such as an isosurface or an intermediate representation, such as spheres, and then rendering the geometry into an image for display. As described in Chapter 3, the send-geometry method generates such geometry for rendering on a client. The rendering can be performed either serially, or in parallel and in either software or more typically through GPU hardware. The most common rasterization library for performing this task is the OpenGL library. OpenGL is an industry standard graphics API, with render implementations supporting GPU, or hardware-accelerated rendering. There is a long history of research in parallel and distributed geometry rasterization, focusing on both early massively parallel processors (MPP) and clusters of PCs.

Crockett and Orloff [9] describe a sort-middle parallel rendering system for message-passing machines. They developed a formal model to measure the performance and they were one of the first to use MPPs for parallel rendering. In comparison, Ortega et al. [30] introduce a data-parallel geometry rendering system using sort-last compositing. This implementation allowed integrated geometry extraction and rendering on the CM5 and was targeted for applications, which required extremely fast rendering for extremely large sets of polygons. The rendering toolkit enables the display of 3D shaded polygons directly from a parallel machine, avoiding the transmission of huge amounts of geometry data to a post-processing rendering system over the then slow networks. Krogh et al. [20] present a parallel rendering system for molecular dynamics applications, where the massive particle system is represented as spheres. The spheres are rendered with depth-enhanced sprites (screen aligned textures) and the results are composited with sort-last compositing. This system avoids rasterization by the use of sprites.

Correa et al. [7] present a technique for the out-of-core rendering of large models on cluster-based tiled displays, using sort-first. In their study, a hierarchical model is generated in a pre-process and at runtime, each node renders an image-tile in a multi-threaded manner, overlapping rendering, visibility computation, and disk operations. By using 16 nodes, they were able to match the performance of the, at the time, current high-end graphics workstations. Samanta et al. [34] introduce k-way replication for sort-first rendering where geometry is only replicated on k out of n nodes with k being much smaller than n. This small replication factor avoids the full data replication, typically required for sort-first rendering. It also supports rendering large-scale geometry, where the geometry can be larger than the memory capacity of an individual node. It simultaneously reduces communication overheads by leveraging frame-to-frame coherence, but allowing a dynamic, view-dependent partitioning of the data. The study found that parallel rendering efficiencies, achieved with small replication factors were, similar to the ones measured with full replication. VTK and VisIt use the Mesa OpenGL library, which implements the OpenGL API in software, to enable parallel rendering through sort-last compositing (see Chap. 17 for more information). Rendering, with either VisIt or Paraview, scales well on GPU clusters using hardware rendering and sort-last compositing, but they are bounded by the composite time for the image display. When using Mesa, the rendering is not scalable or interactive, due to the rasterization speed; though, replacing rasterization with ray tracing can improve the scalability of software rendering [4].

An alternative to rasterization for high performance visualization is to render images using ray tracing. Ray tracing refers to tracing rays from the viewpoint, through pixels, and intersecting the rays with the geometry to be rendered. There have been several parallel ray tracing implementations for high performance visualization. Parker et al. [31] implement a shared-memory ray-tracer, RTRT, which performs interactive isosurface generation and rendering on large shared-memory computers, such as the SGI Origin. This proves to be effective for large data sets, as the data and geometry are accessible by all processors. Bigler et al. [2] demonstrate the effectiveness of this system on large-scale scientific data. They extend RTRT to use various methods in order to visualize the particle data, from simulations using the material point method, as spheres. They also describe two methods for augmenting the visualization using silhouette edges and advanced illumination, such as ambient occlusion. In 4.5, there are details about how to accelerate ray tracing for large amounts of geometry on distributed-memory platforms, like commodity clusters.

4.4 Volume Rendering

Direct volume rendering methods generate images of a 3D volumetric data set, without explicitly extracting geometry from the data [21, 12]. Conceptually, volume rendering proceeds by casting rays from the viewpoint through pixel centers as shown in Figure 4.2. These rays are sampled where they intersect the data volume. Volume rendering uses an optical model to map sampled data values to optical properties, such as color and opacity [24]. During rendering, optical properties are accumulated along each viewing ray to form an image of the data, as shown in Figure 4.3. Although the data set is interpreted as a continuous function in space, for practical purposes it is represented as a discrete 3D scalar field. The samples typically are trilinearly interpolated data values from the 3D scalar field. On a GPU, the trilinear interpolation is performed in hardware and, therefore, is quite fast (see Chap. 11). Samples along a ray can be determined analytically, as in ray marching [35], or can be generated through proxy geometry on a GPU [14].

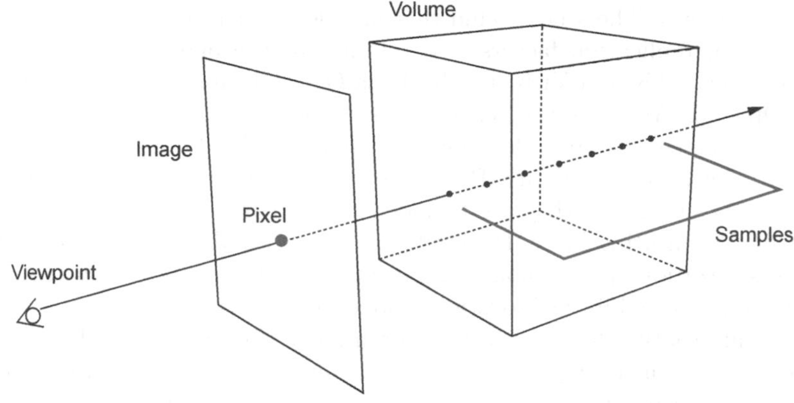

FIGURE 4.2: Ray casting for volume rendering.

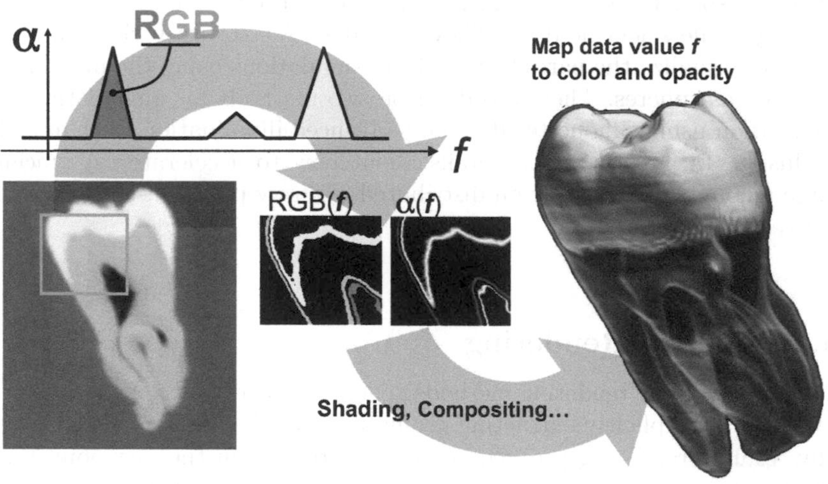

FIGURE 4.3: Volume rendering overview. The volume is sampled, samples are assigned optical properties (RGB, α). The different properties are shaded and composited to form the final image. Image courtesy of Charles Hansen (University of Utah).

In the quest towards interactivity and towards addressing the challenges posed by growing data size and complexity, there has been a great deal of work over the years on parallel volume visualization (see Kaufman and Mueller [17] for an overview). Volume rendering is easily parallelized, though care must be taken to achieve effective results. Typically, the 3D scalar field is partitioned into sub-volumes, which are assigned to the parallel nodes. Volume rendering can take place asynchronously, within the parallel nodes. Also, volume rendering synchronizes compositing between the parallel nodes, forming the final image (see Chap. 5).

The TREX system [19] is a parallel volume rendering application for shared-memory platforms that uses object-parallel data domain decomposition and texture-based, hardware-accelerated rendering, followed by a parallel, software-based composition phase with image space partitioning. TREX can map different portions of its pipeline to different system components on the SGI Origin; those mappings are intended to achieve optimal performance at each algorithmic stage and to achieve minimal inter-stage communication costs.

Muraki et al. [28] describe a custom hardware-based compositing system linking commodity GPUs used for volume rendering. They implemented a cluster composed of two 8-node systems that are linked by multiple networks. While the CPUs are used for volume processing, the GPUs are used for volume rendering and, custom compositing hardware forms the final image in a sort-last manner. Müller et al. [27] implement a ray casting volume renderer with object-order partitioning on a small 8-node GPU cluster, using programmable shaders. This system sustains frame rates in the single digits for data sets, as large as 1260^3 with an image size of 1024×768 pixels. More recently, Moloney et al. [26] implement a GLSL, texture-based volume renderer that runs on a 32-node GPU system. This work takes advantage of the sort-first architecture to accelerate certain types of rendering, like occlusion culling (see Chap. 11).

Peterka et al. [32] discuss end-to-end performance of parallel volume rendering on an IBM Blue Gene distributed-memory parallel architecture. They explore the system performance in terms of rendering, compositing and disk I/O. The system employs sort-last compositing using direct-send compositing (see Chap. 5). Their system was useful for very large data sets that could not be accommodated by GPU clusters and could produce frame times on the order of a few seconds for such data.

Yu et al. [40] describe an *in situ* parallel volume rendering implementation. The data partitioning is used directly from the domain decomposition for the corresponding simulation. Rendering combines both rasterization for geometry, representing particles, and volume rendering of the associated scalar field. Sort-last compositing, using 2-3 swap, was employed to form the final image. Their system tightly linked the visualization with the simulation and scaled to 15,360 cores.

Childs et al. [6] present a parallel volume rendering scheme for massive data sets (with one hundred million unstructured elements and a 3000^3 rectilinear data set). Their approach parallelizes over both input data elements and output pixels, which is demonstrated to scale well on up to 400 processors.

A detailed description of a hybrid approach to parallel volume rendering, which makes use of a design pattern common in many parallel volume rendering applications that uses a mixture of both object- and pixel-level parallelism [23, 37, 22, 1] is discussed in Chapter 12. The design employs an object-order partitioning that distributes source data blocks to processors where they are then rendered using ray casting [21, 33, 12, 38]. Within a processor, an image space decomposition is used, similar to Nieh and Levoy [29], to allow multiple rendering threads to cooperatively generate partial images that are later combined via compositing into a final image [12, 21, 38]. This hybrid design approach, which uses a blend of object- and pixel-level parallelism, has proven successful in achieving scalability and tackling large data sizes.

4.5 Real-Time Ray Tracer for Visualization on a Cluster

While the core ray tracing algorithm might be embarrassingly parallel, scaling a ray tracer to render millions of pixels at real-time frame rates on a cluster remains challenging. It is even more so if the individual nodes do not have enough memory to contain all the data and associated ray tracing acceleration structures (such as octrees or bounding volume hierarchies). The ability to render high-resolution images at interactive or real-time rates is important when visualizing data sets that contain information at the subpixel level. Since most commodity monitors are now capable of displaying at least a two megapixel HD resolution of 1920×1080, and higher end models can display up to twice that many pixels, it is important to make use of all those pixels when visualizing a data set. Current distributed ray tracing systems are not able to approach interactive rates for such pixel counts, regardless of how many compute nodes are used. Either lower resolutions are required for real-time rates, or less fluid frame rates are used in order to scale to larger image sizes. This section describes a distributed ray tracing system developed by Ize et al. that can scale, on an InfiniBand cluster, to real-time rates of slightly over 100fps, at full HD resolution, or 50–60fps at 4 megapixels with massive polygonal models [15]. The system uses sort-first parallelism, where the image is divided into tiles, but rather than pre-sort and assign geometry to nodes, the geometry is dynamically cached to the appropriate node taking advantage of frame-to-frame coherence.

Such a system can also handle massive out-of-core models that cannot reside inside the physical memory of any individual compute node by implementing a read-only distributed shared memory ray tracer [8] which is essentially a distributed cache (DC). The DC supports the use of any desired ray tracing shading models, such as shadows, transparent surfaces, ambient occlusion and even full path tracing. More advanced shading models than simple ray casting or rasterization allow for more productive and useful visualizations [13].

Ize et al. implemented their distributed ray tracer on the Manta Interactive Ray Tracer, a state-of-the-art ray tracer capable of scaling at real-time frame rates to hundreds of cores on a shared-memory machine [3]. They faced the challenge of ensuring that the distributed Manta implementation was able to scale to many nodes while also ensuring real-time frame rates.

The two main challenges of a real-time distributed ray tracing system are load balancing and ensuring the display is not a bottleneck. The real-time distributed ray tracer uses a master-slave configuration where a single process, the display, receives pixel results from render processes running on the other nodes. Another process, the load balancer, handles assigning tasks to the individual render nodes.

MPI does not guarantee fairness amongst threads in a process and the MPI library, in their implementation, forces all threads to go through the same critical section. Because of this, the display and load balancer were run as separate processes. Since they do not communicate with each other, nor do they share any significant state, this partitioning can be done without penalty or code complexity. Furthermore, since the load balancer has minimal communication, instead of running on a dedicated node, it can run alongside the display process without noticeably impacting overall performance.

4.5.1 Load Balancing

Manta uses a shared-memory dynamic load balancing work queue, where each thread is statically given a predetermined large tile of work to consume and then, when it needs more work, it progressively requests smaller tiles until there is no more work left. Ize et al. extended Manta's shared-memory load balancer to distributed memory, using a master dynamic load balancer with a work queue comprised of large tiles, which are given to each node (the first assignment is done statically and is always the same) and then each node has its own work queue where it distributes sub-tiles to each render thread. Inside each node, the standard Manta shared-memory load balancer distributes work amongst the threads. Each thread starts with a few statically assigned ray packets to render and then takes more ray packets from the node's shared work queue until no more work is available, at which point that thread requests another tile of work from the master load balancer for the entire node to consume. This approach effectively provides a two-level load balancer ensures that work is balanced both at the node level and at the thread level. Since the top level load balancer only needs to keep the work queues of the nodes full, instead of the queues for each individual thread, communication is kept low on the top level load balancer, which allows the system to scale to many nodes and cores.

4.5.2 Display Process

One process, the display, is dedicated to receiving pixels from the render nodes and placing those pixels into the final image. The display process shares a dedicated node with the load balancer process, which only uses a single thread and has infrequent communication. The display has one thread, which only receives the pixels from the render nodes into a buffer. The other threads in the display then take those pixels and copy them into the relevant parts of the final image. At first, Ize et al. used only a single thread to do both the receiving and copying to the final image, but they found that the maximum frame rates were significantly lower than what the InfiniBand network should be capable of. Surprisingly, it turned out that merely copying data into the local memory was introducing a bottleneck; and, for this reason, they employed several cores to do the copying.

Since InfiniBand packets are normally 2KB, any messages smaller than this 2KB will still consume 2KB of the network bandwidth. Therefore, it is necessary to send a full packet of pixels to maximize frame rate. If, for instance, one sends only a single seven-byte pixel at a time, it would actually require sending an effective $2048 \times 1920 \times 1080 = 3.96$GB of data, taking about two seconds over the high speed network, which is clearly too slow. Since ray packets contain 64 pixels, combining 13 ray packets into a single message offers the best performance, since it uses close to three full InfiniBand packets.

4.5.3 Distributed Cache Ray Tracing

The distributed cache infrastructure is similar to that of DeMarle et al. [11]. Data is distributed among nodes in an interleaved pattern. A templated direct-mapped cache is used for data that does not exist locally. Data is accessed at a block granularity of almost 8KB, with a little bit of space left for packet/MPI overhead, which results in each block containing 254 32-byte Bounding Volume Hierarchy (BVH) nodes or 226 36-byte triangles. Since multiple threads can share the cache, each cache line is controlled by a mutex so that multiple threads can simultaneously read from a cache line. In order to replace the data in a cache line, a thread must have exclusive access to that cache line so that when it replaces the cache element, it does not modify data that is being used by other threads. Remote reads are performed using a passive `MPI_Get` operation, which should, in turn, use an InfiniBand RDMA read to efficiently read the memory from the target node without any involvement of the target CPU. This approach allows for very fast remote reads, which do not impact performance on the target node and that scale to many threads and MPI processes [16].

DeMarle et al. assign block k to a node number $k \bmod numNodes$, so that blocks are interleaved across a distributed memory. If a node owns block k, it will then place it in location $k/numNodes$ of its resident memory array [10]; Ize et al. follow this convention. However, if a node does not own block k, the node places a copy of the block in its k mod *cacheSize* cache line. Ize et al. found this to be an inefficient mapping, because it does not make full utilization of the cache, as it does not factor in that some of the data might already reside in the node's resident memory. For instance, suppose, there are 2 nodes and the cache size is also 2, then node 0 will never be able to make use of cache line 0. When k mod $2 = 0$, then the owner of the data is $k/2 = 0$, which means that node 0 already has that data in its resident memory. A more efficient mapping that avoids the double counting is:

$$\left(k - \left\lfloor \frac{k + (numNodes - myRank)}{numNodes} \right\rfloor\right) \bmod cachSize.$$

Ray casting the Richtmyer–Meshkov data set (Fig. 4.4) with 60 nodes and more efficient mapping gives speedups of 1.16×, 1.31×, 1.48×, 1.46×, and 1.31× over the mapping of DeMarle et al. The respective cache sizes are 1/32, 1/8, 1/4, 1/2, and 1/1 of total memory, respectively.

4.5.3.1 DC BVH

Since the DC manager groups BVH nodes into blocks, Ize et al. reorder the memory locations of the BVH nodes so that the nodes in a block are spatially coherent in memory. Note, they are not reordering the actual BVH tree topology. They accomplish this coherency for a block size of B, BVH nodes by writing the nodes to memory according to a breadth first traversal of the first B nodes, thus creating a subtree that is coherent in memory. They then stop the breadth first traversal and instead recursively repeat that process for each of the B leaves of the newly formed subtree. If the subtree being created ends up with less than B leaves, it continues to the next subtree without introducing any gaps in the memory layout. It is therefore possible for a block to contain multiple subtrees. However, since the blocks are written according to a blocked depth-first traversal, the subsequent subtree will still often be spatially near the previous subtree.

In order to minimize memory usage, Manta's BVH nodes only contain a single child pointer rather than two, with the other child's memory location being adjacent to the first child. Because of this, the algorithm above recurses on each pair of child leaves, so that the two children stay adjacent in memory, rather than recursing on each of the B leaves.

Thus, the spatially coherent blocks contain mostly complete subtrees of B nodes so that when a block is fetched, one can usually expect to make $\log B$ traversals before a new block must be fetched from the DC manager. For 254 node blocks, this is about eight traversal steps for which the DC-BVH traversal performance should be roughly on par with the regular BVH traversal. Eight traversal steps are ensured by keeping track of the current block and not releasing that block (or re-fetching it) until either the traversal leaves the block or enters a new block. When the traversal enters a new block it must release the previously held block in order to prevent a deadlock condition where one of the following blocks requires the same cache line as the currently held block. Note, all multiple threads can safely share access to a block and thread stalling, while waiting for a block to be released, will only occur if one thread needs to use the cache line for a different block than is currently being held.

Since the root of the tree will be traversed by every thread in all nodes, rather than risk this data being evicted and then having to stall while the data becomes available again, the corresponding spatially coherent block is replicated across all nodes. Replication requires only an extra 8KB of data per node and ensures that those first 8 traversal steps are always fast because they only need to access resident memory.

FIGURE 4.4: On 60 nodes, the system can ray cast (left image) a 316M triangle isosurface from timestep 273 of a Richtmyer–Meshkov (RM) instability calculation, in HD resolution, at 101fps, with replicated data, 21,326MB of triangles and acceleration structures, and at 16fps if the DC is used to store only 2,666MB per node. Using one shadow ray or 36 ambient occlusion rays per pixel (right image) the system can achieve 4.76fps and 1.90fps, respectively. Image source: Ize et al. [15].

4.5.3.2 DC Primitives

While sharing vertices using a mesh will often halve the memory requirements, shared vertices do not map well to the DC since shared vertices require fetching a block from the DC manager to find the triangle and then once the vertex indices are known, one to three more fetches for the vertices. Thus, a miss results in between a 2×–4× slowdown for misses. Doubling the storage requirements is comparable in cost to halving the cache size, and halving the cache size empirically introduces less than a 2× performance penalty. Although DeMarle et al. used a mesh structure, Ize et al. did not. While it might appear that one could create a sub-mesh within each block so that less memory is used and only a single block need be fetched, due to a fixed block size, there will be wasted empty space inside each block. Attempting to place variable numbers of triangles in each block, so that empty space is reduced, it will also not work since then, it will not be possible to compute which block key corresponds to which triangle.

Ize et al. reorder the triangles into blocks by performing an in-order traversal of the BVH and outputting triangles into an array as they are encountered. This results in spatially coherent blocks that also match the traversal pattern of the BVH so that, if two leaf nodes share a recent ancestor, they are also spatially coherent and will likely have their primitives residing in the same block.

4.5.4 Results

Ize et al. use a 64 node cluster where each node contains two 4-core Xeon X5550s running at 2.67GHz, with 24GB of memory, and a 4× DDR InfiniBand interconnect between the nodes. All scenes are rendered at an HD resolution of 1920 × 1080 pixels. The data sets consist of a 316M triangle isosurface, computed from one timestep of a Richtmyer–Meshkov (RM) instability calculation (see Fig. 4.4) and the 259M triangle Boeing 777 CAD model (see Fig. 4.5). While the RM data set can be rendered using volume ray casting, this polygonal representation is used as an example of a massive polygonal model, where large parts of the model can be seen from one view. The Boeing data set consists of an almost complete CAD model for the entire aircraft and, unless it were made transparent, has significant occlusion so that regardless of the view, only a fraction of the scene can be viewed at any given time.

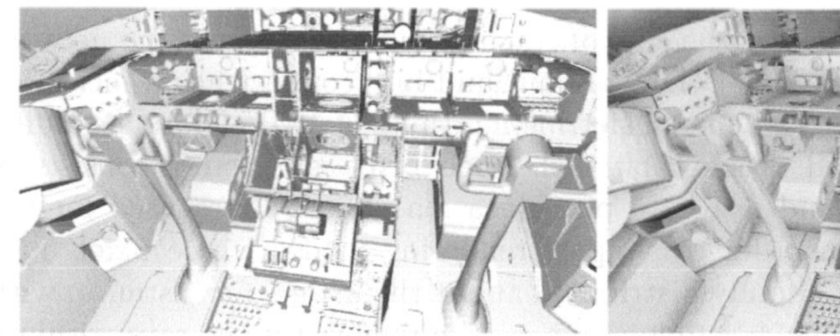
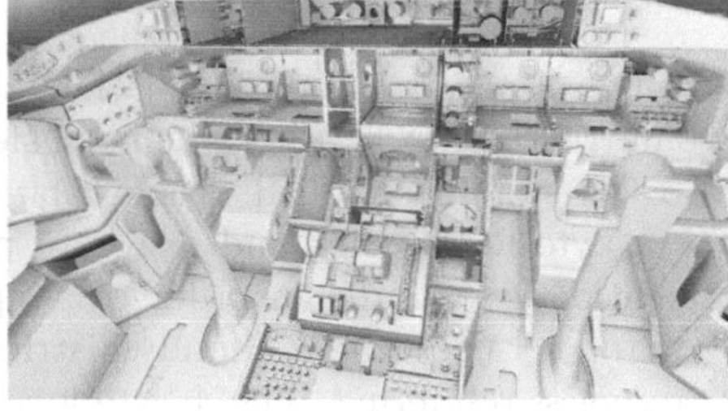

FIGURE 4.5: On 60 nodes, the 259M triangle Boeing data set using HD resolution can be ray cast (left image) at 96fps if all 15,637MB of triangle and acceleration structure data are replicated on each node and at 77fps if DC is used to store only 1,955MB of data and cache per node. Using 36 ambient occlusion rays per pixel (right image) the system achieves 1.46fps. Image source: Ize et al. [15].

In the study, scaling was reported where both data sets were rendered using two to sixty nodes, with one node used for display and load balancing, and the remaining nodes used for rendering. Simple ray casting was used for both models and ambient occlusion with 36 samples per shading point for the Boeing data set and a similar ambient occlusion, but with additional hard shadows for the RM data set.

Figure 4.6 shows how the system scales with increasing numbers of nodes—when ray casting both scenes—using replicated data across each node so that the standard BVH acceleration structure is used without any DC overhead. Then, a cache plus resident set size, $1/N$, of the total memory is used by the data set and acceleration structure. Assuming the nodes have enough memory, data replication allows the system to achieve near linear scaling to real-time rates and then begins to plateau as the rendering rate approaches 100fps. Since 1/1 has a cache large enough to contain all the data, no cache misses ever occur. As the cache is decreased in size, the rendering speed of the Boeing data set is not significantly impacted; indicating that the system was able to keep the working set fully in cache. The RM data set, on the other hand, has a larger working set, since more of it is in the view frustum, which causes cache misses to occur, and noticeably affects performance.

The total amount of memory required in order to keep data in-core depends on the cache size and the resident set size. On a single render node, all the data must be resident. For the RM data set, the resident set is 21GB and the cache is 0GB. Figure 4.7 shows that with more render nodes the size of the resident set becomes progressively smaller so that the cache size quickly becomes the limiting factor as to how much data the system can handle. Since the resident set decreases as more nodes are added, cache size can similarly be increased so that the node's capabilities are used to the fullest.

4.5.5 Maximum Frame Rate

Frame rate is limited by the cost of transferring pixels across the network to the display process. The 4×DDR InfiniBand interconnect has a measured bandwidth of 1868MB/s when transferring multiples of 2KB of data according to the system supplied `ib_read_bw` tool. Each pixel sent across consists of a 3 byte RGB color and a 4 byte pixel location. The 4 bytes used for the location is required in order to support rendering modes where pixels in a ray packet could be randomly distributed about the image, for instance, with frameless rendering. The 4 byte location can be removed if the rays within a ray packet form a rectangular tile, in which case it is only necessary to store the coordinates of the rendered rectangle over the entire ray packet instead of 4 bytes per ray. Further improvements might be obtained by compressing the pixels so that even less data needs to be transmitted. The time required by the display process to receive all the pixels from the other nodes is at best $\frac{1920*1080*7B}{1868MB/s} = 7.41$ms for a full HD image, which is an upper limit of 135fps. Since $13 * 64 = 832$ pixels are sent at a time, which occupies 95% of the InfiniBand packet, one would expect the implementation to achieve at best 127fps.

To see how close the system can approach the maximum frame rate by testing the system overhead, an HD image is rendered from a small model with the camera pointing away from the model, producing a blank screen and requiring only a minimal amount of ray tracing. To verify that the load balancer is not significantly competing for resources, the system is tested with the load balancer on its own dedicated node, and then, with the load bal-

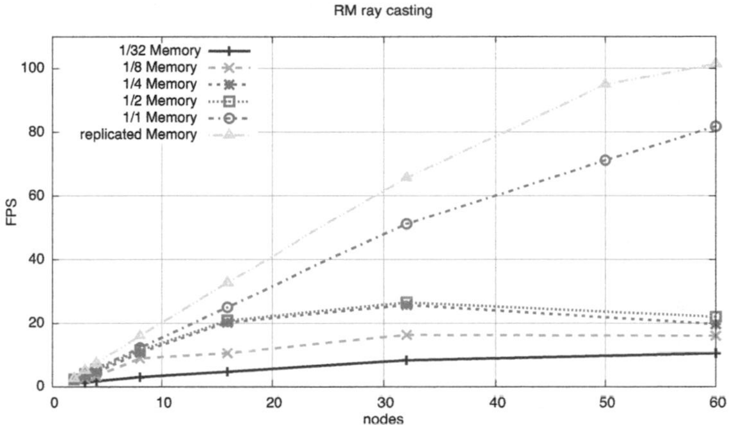

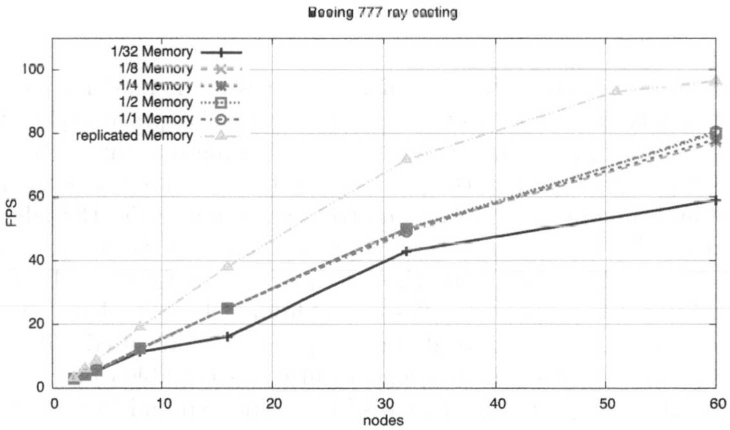

FIGURE 4.6: Frame rate when varying cache size and the number of nodes in the ray casted RM and Boeing data sets. The replicated data has almost perfect scaling until the display becomes network bound. Performance scales very well with the Boeing data set and DC, even with a small cache, because the working set is a fraction of the overall model. The scaling with the RM data set and DC performs well until about 20–25fps is reached. Image source: Ize et al. [15].

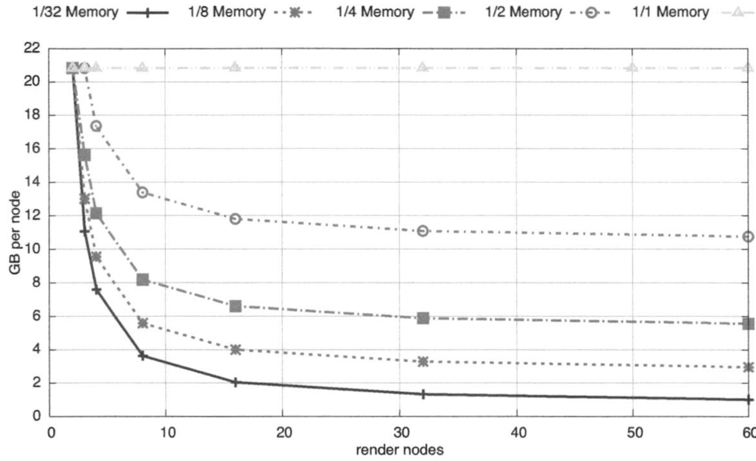

FIGURE 4.7: Total memory used per node for the RM data set. Image source: Ize et al. [15].

ancer sharing the node with the display process. The render nodes thus have a limited amount of work to perform and the display process quickly becomes the bottleneck. Figure 4.8 shows that when using one thread to receive the pixels and another to copy the pixels from the receive buffer to the image, the maximum frame rate is 55fps, no matter how many render threads and nodes being used. Using two threads to copy the pixels to the image results in up to a 1.6× speedup, three copy threads improves performance by up to 2.3×, but additional copy threads offer no additional benefit, demonstrating that copying is the bottleneck until three copy threads are used, after which the receive thread becomes the bottleneck since it is not able to receive the pixels fast enough to keep the copy threads busy. When around 18 threads are being used, be they on a few nodes or many nodes, the system can obtain a frame rate of 127fps. This is exactly the expected maximum of 127fps given by the amount of time it takes to transmit all the pixel data across the InfiniBand interconnect. More render threads result in lower performance due to the MPI implementation not being able to keep up with the large volume of communication. In order to achieve the results required, the MPI implementation must be tuned to use more RDMA buffers and turn off shared receive queues (SRQ); otherwise, the system can still achieve the same maximum frame rate of about 127fps with 17 render cores, but after that point, adding more cores causes performance to drop off quickly, with 384 render cores (48 render nodes) being 2× slower. However, this is a moot point since faster frame rates will offer no tangible benefit.

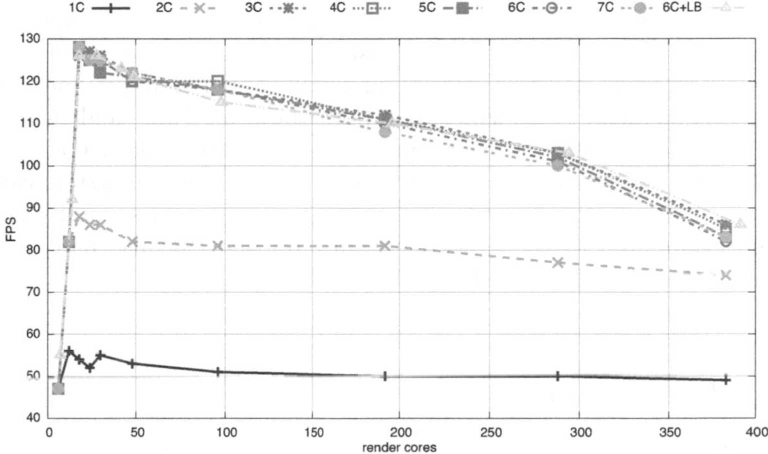

FIGURE 4.8: Display process scaling: frame rate using varying numbers of render cores to render a trivial scene when using one copy thread (1C) to seven copy threads (7C), and when using six copy threads with the load balancer process on the same node (6C+LB). Note that performance is not enhanced beyond three copy threads. Image source: Ize et al. [15].

Modern graphics cards can produce four megapixel images at 60fps. As this image size is roughly twice the HD image size, in our system the maximum frame rate would halve to about 60fps. Higher resolutions than 4 megapixels are usually achieved with a display wall consisting of a cluster of nodes driving multiple screens. In this case, the maximum frame rate will be given not by the time to transmit an entire image, but by the time it takes for a single display node to receive its share of the image. Assuming each node renders 4 megapixels, and the load balancing and rendering continue to scale, the frame rate will thus stay at 60fps, regardless of the resolution of the display wall.

Since three copy threads are able to keep up with the receiving thread, and the load balancer process is also running on the same node, there are three unused cores on the tested platform. If data is replicated across the nodes then these three cores can be used for a render process. This render process will also benefit from being able to use the higher-speed shared memory for its

MPI communication with the display and load balancer instead of the slower InfiniBand. However, if DC is required, then it will not be possible to run any render processes on the same node since those render processes will be competing with the display and load balancer for scarce network bandwidth and this will much more quickly saturate the network port and result in much lower maximum frame rates.

With modern hardware and software, the described system can ray trace massive models at real-time frame rates on a cluster and even show interactive to real-time rates when rendering distributed geometry using a small cache. The system is one to two orders of magnitude faster than previous cluster ray tracing implementations, which used both slower hardware and algorithms [11, 39], or had equivalent hardware but did not scale to as many nodes or to high frame rates [5]. Compared to compositing approaches, the system can achieve about a 4× improvement in the maximum frame rate for same size non-empty images compared to the state of the art [18] and can also handle advanced shading effects for improved visualization.

4.6 Conclusion

Parallel rendering methods for generating images from visualizations are an important area of research. In this chapter, a general framework for parallel rendering was presented and applied to both geometry rendering and volume rendering. In the future, as HPV moves into the exascale regime, parallel rendering methods will likely become more important as *in situ* methods require parallel rendering and the send-image method of parallel display will scale better than the send-geometry method. It is anticipated that GPUs will become integrated into compute nodes, which offer another avenue for parallel rendering in HPV.

References

[1] C. Bajaj, I. Ihm, G. Joo, and S. Park. Parallel Ray Casting of Visibly Human on Distributed Memory Architectures. In *VisSym '99 Joint EUROGRAPHICS-IEEE TVCG Symposium on Visualization*, pages 269–276, 1999.

[2] James Bigler, James Guilkey, Christiaan Gribble, Charles Hansen, and Steven Parker. A Case Study: Visualizing Material Point Method Data. In *EUROVIS the Eurographics /IEEE VGTC Symposium on Visualization*, pages 299–306. EuroGraphics, 2006.

[3] James Bigler, Abe Stephens, and Steven G. Parker. Design for Parallel Interactive Ray Tracing Systems. In *Proceedings of the 2006 IEEE Symposium on Interactive Ray Tracing*, pages 187–196, 2006.

[4] Carson Brownlee, John Patchett, Li-Ta Lo, David DeMarle, Christopher Mitchell, James Ahrens, and Charles Hansen. A Study of Ray Tracing Large-Scale Scientific Data in Parallel Visualization Applications. In *Proceedings of the Eurographics Workshop on Parallel Graphics and Visualization*, EGPGV '12, pages 51–60. Eurographics Association, 2012.

[5] Brian Budge, Tony Bernardin, Jeff A. Stuart, Shubhabrata Sengupta, Kenneth I. Joy, and John D. Owens. Out-of-Core Data Management for Path Tracing on Hybrid Resources. *Computer Graphics Forum*, 28(2):385–396, 2009.

[6] Hank Childs, Mark A. Duchaineau, and Kwan-Liu Ma. A Scalable, Hybrid Scheme for Volume Rendering Massive Data Sets. In *Eurographics Symposium on Parallel Graphics and Visualization*, pages 153–162, May 2006.

[7] Wagner T. Corrêa, James T. Klosowski, and Cláudio T. Silva. Out-of-Core Sort-First Parallel Rendering for Cluster-Based Tiled Displays. In *Proceedings of the 4th Eurographics Workshop on Parallel Graphics and Visualization*, EGPGV '02, pages 89–96. Eurographics Association, 2002.

[8] Brian Corrie and Paul Mackerras. Parallel Volume Rendering and Data Coherence. In *Proceedings of the 1993 Symposium on Parallel Rendering*, PRS '93, pages 23–26. ACM, 1993.

[9] Thomas W. Crockett and Tobias Orloff. A MIMD Rendering Algorithm for Distributed Memory Architectures. In *Proceedings of the 1993 Symposium on Parallel Rendering*, PRS '93, pages 35–42. ACM, 1993.

[10] David E. DeMarle. Ice Network Library. `http://www.cs.utah.edu/~demarle/software/`, 2004.

[11] David E. DeMarle, Christiaan Gribble, Solomon Boulos, and Steven Parker. Memory Sharing for Interactive Ray Tracing on Clusters. *Parallel Computing*, 31:221–242, 2005.

[12] Robert A. Drebin, Loren Carpenter, and Pat Hanrahan. Volume Rendering. *SIGGRAPH Computer Graphics*, 22(4):65–74, 1988.

[13] Christiaan P. Gribble and Steven G. Parker. Enhancing Interactive Particle Visualization with Advanced Shading Models. In *Proceedings of the 3rd Symposium on Applied Perception in Graphics and Visualization*, pages 111–118, 2006.

[14] Milan Ikits, Joe Kniss, Aaron Lefohn, and Charles Hansen. *Chapter 39, Volume Rendering Techniques*, pages 667–692. Addison Wesley, 2004.

[15] Thiago Ize, Carson Brownlee, and Charles D. Hansen. Real-Time Ray Tracer for Visualizing Massive Models on a Cluster. In *Proceedings of the 2011 Eurographics Symposium on Parallel Graphics and Visualization*, pages 61–69, 2011.

[16] W. Jiang, J. Liu, H.W. Jin, D.K. Panda, D. Buntinas, R. Thakur, and W.D. Gropp. Efficient Implementation of MPI-2 Passive One-Sided Communication on InfiniBand Clusters. *Recent Advances in Parallel Virtual Machine and Message Passing Interface, Lecture Notes in Computer Science*, 2131:450–457, 2004.

[17] Arie Kaufman and Klaus Mueller. Overview of Volume Rendering. In Charles D. Hansen and Christopher R. Johnson, editors, *The Visualization Handbook*, pages 127–174. Elsevier, 2005.

[18] W. Kendall, T. Peterka, J. Huang, H.W. Shen, and R. Ross. Accelerating and Benchmarking Radix-K Image Compositing at Large Scale. In *Proceedings Eurographics Symposium on Parallel Graphics and Visualization*, pages 101–110, 2010.

[19] Joe Kniss, Patrick McCormick, Allen McPherson, James Ahrens, Jamie Painter, Alan Keahey, and Charles Hansen. Interactive Texture-Based Volume Rendering for Large Data Sets. *IEEE Computer Graphics and Applications*, 21(4), July/August 2001.

[20] Michael Krogh, James Painter, and Charles Hansen. Parallel Sphere Rendering. *Parallel Computing*, 23(7):961–974, July 1997.

[21] Marc Levoy. Display of Surfaces from Volume Data. *IEEE Computer Graphics and Applications*, 8(3):29–37, May 1988.

[22] Kwan-Liu Ma. Parallel Volume Ray-Casting for Unstructured-Grid Data on Distributed-Memory Architectures. In *PRS '95: Proceedings of the IEEE Symposium on Parallel Rendering*, pages 23–30. ACM, 1995.

[23] Kwan-Liu Ma, James S. Painter, Charles D. Hansen, and Michael F. Krogh. A Data Distributed, Parallel Algorithm for Ray-Traced Volume Rendering. In *Proceedings of the 1993 Parallel Rendering Symposium*, pages 15–22. ACM Press, October 1993.

[24] Nelson Max. Optical Models for Direct Volume Rendering. *IEEE Transactions on Visualization and Computer Graphics*, 1(2):99–108, June 1995.

[25] Steven Molnar, Michael Cox, David Ellsworth, and Henry Fuchs. A Sorting Classification of Parallel Rendering. *IEEE Computer Graphics and Applications*, 14:23–32, 1994.

[26] Brendan Moloney, Marco Ament, Daniel Weiskopf, and Torsten Moller. Sort-First Parallel Volume Rendering. *IEEE Transactions on Visualization and Computer Graphics*, 17(8):1164–1177, 2011.

[27] C. Müller, M. Strengert, and T. Ertl. Optimized Volume Raycasting for Graphics-Hardware-Based Cluster Systems. In *Proceedings of Eurographics Parallel Graphics and Visualization*, pages 59–66, 2006.

[28] Shigeru Muraki, Masato Ogata, Eric Lum, Xuezhen Liu, and Kwan-Liu Ma. VG Cluster: A Low-Cost Solution for Large-Scale Volume Visualization. In *Proceedings of NICOGRAPH International Conference*, pages 31–34, May 2002.

[29] Jason Nieh and Marc Levoy. Volume Rendering on Scalable Shared-Memory MIMD Architectures. In *Proceedings of the 1992 Workshop on Volume Visualization*, pages 17–24. ACM SIGGRAPH, October 1992.

[30] Frank Ortega, Charles D. Hansen, and James Ahrens. Fast Data Parallel Polygon Rendering. In *Proceedings of the 1993 ACM/IEEE Conference on Supercomputing*, Supercomputing '93, pages 709–718. ACM, 1993.

[31] Steven Parker, Peter Shirley, Yarden Livnat, Charles Hansen, and Peter-Pike Sloan. Interactive Ray Tracing for Isosurface Rendering. In *Proceedings of the Conference on Visualization '98*, VIS '98, pages 233–238. IEEE Computer Society Press, 1998.

[32] Tom Peterka, Hongfeng Yu, Robert Ross, and Kwan-Liu Ma. Parallel Volume Rendering on the IBM Blue Gene/P. In *Proceedings of Eurographics Parallel Graphics and Visualization Symposium (EGPGV 2008)*, pages 73–80, April 2008.

[33] Paolo Sabella. A Rendering Algorithm for Visualizing 3D Scalar Fields. *SIGGRAPH Computer Graphics*, 22(4):51–58, 1988.

[34] Rudrajit Samanta, Thomas Funkhouser, and Kai Li. Parallel Rendering with K-Way Replication. In *Proceedings of the IEEE 2001 Symposium on Parallel and Large-Data Visualization and Graphics*, PVG '01, pages 75–84. IEEE Press, 2001.

[35] Peter Shirley, Michael Ashikhmin, Michael Gleicher, Stephen Marschner, Erik Reinhard, Kelvin Sung, William Thompson, and Peter Willemsen. *Fundamentals of Computer Graphics, 2nd Ed.* A. K. Peters, Ltd., 2005.

[36] Dave Shreiner. *OpenGL Programming Guide 7th Ed.* Addison-Wesley, 2009.

[37] R. Tiwari and T. L. Huntsberger. A Distributed Memory Algorithm for Volume Rendering. In *Scalable High Performance Computing Conference*, pages 247–251, May 1994.

[38] Craig Upson and Michael Keeler. V-buffer: Visible Volume Rendering. In *SIGGRAPH '88: Proceedings of the 15th Annual Conference on Computer Graphics and Interactive Techniques*, pages 59–64. ACM, 1988.

[39] Ingo Wald, Philipp Slusallek, and Carsten Benthin. Interactive Distributed Ray Tracing of Highly Complex Models. In *Proceedings of the 12th Eurographics Workshop on Rendering Techniques*, pages 274–285, 2001.

[40] Hongfeng Yu, Chaoli Wang, Ray W. Grout, Jacqueline H. Chen, and Kwan-Liu Ma. In-Situ Visualization for Large-Scale Combustion Simulations. *IEEE Computer Graphics and Applications*, 30(3):45–57, May/June 2010.

Chapter 5

Parallel Image Compositing Methods

Tom Peterka
Argonne National Laboratory

Kwan-Liu Ma
University of California at Davis

5.1 Introduction 72
5.2 Basic Concepts and Early Work in Compositing 72
  5.2.1 Definition of Image Composition 73
  5.2.2 Fundamental Image Composition Algorithms 74
  5.2.3 Image Compositing Hardware 77
5.3 Recent Advances 77
  5.3.1 2-3 Swap 77
  5.3.2 Radix-k 78
  5.3.3 Optimizations 80
5.4 Results 81
5.5 Discussion and Conclusion 82
  5.5.1 Conclusion 83
  5.5.2 Directions for Future Research 85
References 86

Image compositing is a fundamental part of high performance visualization on large-scale parallel machines. Aside from reading a data set from storage, compositing is the most expensive part of the parallel rendering pipeline because it requires communication among a large number of processes. On a modern supercomputer, compositing may generate literally hundreds of thousands of messages. Thus, developing compositing algorithms that scale with growing machine size is crucial. Such algorithms have enabled, for example, wall-size images that are tens of megapixels in resolution to be composited at interactive frame rates from all of the nodes of some of the world's largest supercomputers and visualization clusters. First, this chapter discusses a history of the classic parallel image compositing algorithms: direct-send and binary-swap. From there, the discussion moves to optimizations that have been proposed over the years, from scheduling to compression and load balancing. Advanced compositing on modern supercomputing architectures, however, is the main focus of this chapter, and in particular, 2-3 swap and radix-k for petascale HPC machines.

5.1 Introduction

The motivation for studying image compositing is the same as in most of this book: data sets consisting of trillions of grid points and thousands of timesteps are being produced by machines with hundreds of thousands of cores. Machine architectures are growing in size and in complexity, with multi-dimensional topologies and specialized hardware such as smart network adaptors, direct-memory access, and graphics accelerators. Against this backdrop, scientists in high energy physics, climate, and other domains are demanding more of visualization: real-time, multivariate, time-varying methods that maximize data locality and minimize data movement.

Image compositing is the final step in sort-last parallel rendering (see 4.2). In sort-last parallel rendering, each processor generates a finished image of its subset of data and these images must be combined into one final result.

Legacy image compositing algorithms were invented for much smaller systems. The history of the direct-send and binary-swap algorithms, began in the mid-1990s. Optimization features, such as compression, identification of active pixels, and scheduling, further improved performance. In the early 2000s, production compositing libraries implementing many of these features appeared, some of which are still used today.

New advances in the last five years feature more architecture awareness of HPC systems. The late 2000s yielded compositing algorithms with higher degrees of concurrency, scalability, and flexibility in the form of 2-3 swap and radix-k algorithms. The early 2010s continued optimization at a very large scale and also continued the implementation of the latest innovations in production. This chapter highlights some results of these recent advances with both theoretical and actual performance, and it concludes with future directions in highly parallel image compositing.

5.2 Basic Concepts and Early Work in Compositing

The previous chapter classified parallel rendering according to when rasterized images are sorted [17]: sort-first, sort-middle, and sort-last. One way to understand the difference in these methods is to identify what is distributed and what is replicated among the processes. The term *process* is used to designate a task executed in parallel with other processes, where processes each have separate memory address spaces and communicate via passing messages.

In sort-first rendering, the image pixels are usually distributed, and the data set is typically replicated. HPC applications generate data sets many times larger than the memory capacity of a single node, so sort-first rendering

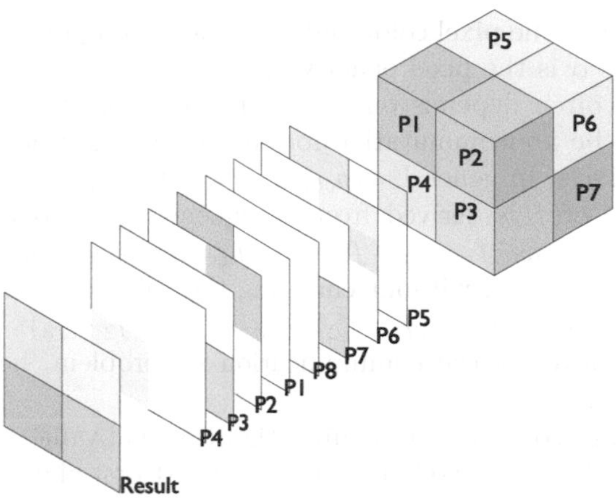

FIGURE 5.1: Sort-last image compositing.

is not frequently encountered in high performance visualization. Sort-middle rendering attempts to combine the best of sort-first and sort-last with both image pixels and data voxels distributed among processes. This is difficult to implement and even harder to scale, because the portion of the data that a process needs to maintain depends on the viewpoint and requires redistribution when the view frustum changes.

In sort-last rendering, data are distributed while pixels are replicated, and each process is responsible for rendering a complete image of its data subset. Of course, sort-last rendering comes at a cost, and that price is paid in compositing, when the pixels at each process must be combined into one final image. The basic idea behind sort-last rendering is illustrated in Figure 5.1. The data set is partitioned among processes; rendering occurs locally at each process, and the resulting images are composited at the end.

5.2.1 Definition of Image Composition

Image composition is the reduction of several images into one. It involves two subproblems: (1) communicating two or more images to one location where the reduction can occur, and (2) the computation of the reduction itself. The formal definition of image composition, therefore, is the definition of each of these subproblems, beginning with the reduction operator.

Reduction is a binary operation, where one of the inputs is the current value of the pixel p_{old}, and the other is an incoming value p_{new}. p_{new} is accumulated with p_{old} to produce an updated value: $p = f(p_{old}, p_{new})$. The exact definition of f depends on the application. For example, f can select the pixel with the closer depth value or the greater intensity, or it can be a combination of the two pixel values. A common example is the *over* operator [23], a linear combination of p_{old} and p_{new}, where $p = (1.0 - \alpha_{old}) * p_{new} + p_{old}$.

Here, p represents the pixel color and the opacity components, each computed separately, and α is the pixel opacity only.

Because f often depends on the depth order of blended input pixels, f is assumed to be noncommutative for the general case of compositing semi-transparent pixels. In general, $f(p_{old}, p_{new}) \neq f(p_{new}, p_{old})$, and moreover, if the final value of f is derived from a sequence (functional composition) of individual operations, $f = f_1 \circ f_2 \circ ... \circ f_n$, then the ordering of f_1 through f_n cannot be permuted without changing the value of f. Image composition is associative, however, so $(f_1 \circ f_2) \circ f_3 = f_1 \circ (f_2 \circ f_3)$. The remainder of this chapter addresses the communication subproblem, beginning with the problem definition.

A valid image composition requires that the final value of a pixel is derived from values of the same pixel on all processes. At first glance, this may appear to be an all-to-all communication pattern, with p^2 messages sent and received among p number of processes. Because f is accumulated, via an individual f_i, as shown above, however, each process can contribute to the final value f without communicating directly with every other process. Hence, a simple linear communication pattern with $p - 1$ messages suffices.

We can often do better by parallelizing this sequence with tree and pipeline communication patterns that trade fewer than $p - 1$ communication steps (or *rounds*) for a greater total number of messages, where each round involves multiple messages. The goal is for these multiple messages per round to be independent and for the network to support their concurrent transmission, thereby reducing the total communication cost to $O(\log p)$.

While the final image is often gathered to a single process, this is not strictly necessary, and a distributed result is not only acceptable, but often desirable. In message-passing parlance, this is an example of a noncommutative, reduce-scatter collective. Many algorithms for such collectives appear in literature [4, 3, 31, 2, 30, 25, 8, 5, 12, 13]. The algorithms in this chapter contain optimizations to the generic collective algorithm. These optimizations are possible due to the characteristics of the compositing algorithm.

5.2.2 Fundamental Image Composition Algorithms

Stompel et al. [28] surveyed methods for sort-last compositing, and Cavin et al. [6] analyzed the relative theoretical performance of these methods. These overviews show that compositing algorithms usually fall into one of three categories: direct-send, tree, and parallel pipeline. Pipeline methods are seldom found in practice and are not covered here.

In direct-send, each of the p processes takes ownership of $1/p$ of the final image, and all other processes send this subset of their images to the owner [10, 21]. Figure 5.2 shows an example of four processes executing a direct-send. The outermost rectangles represent each of the original images prior to compositing. The smaller highlighted rectangles represent the final image portions owned by each process after compositing. The arrows repre-

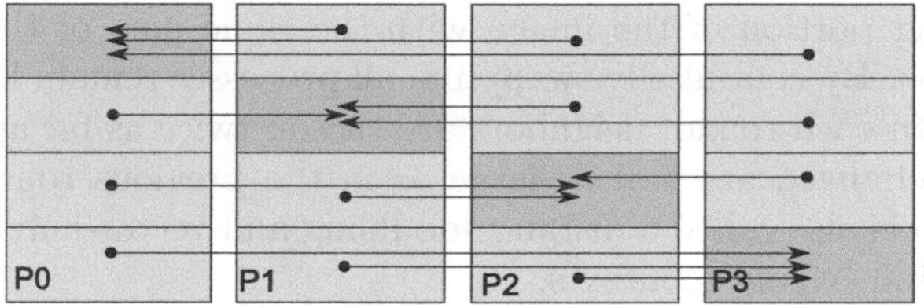

FIGURE 5.2: Example of the direct-send algorithm for four processes.

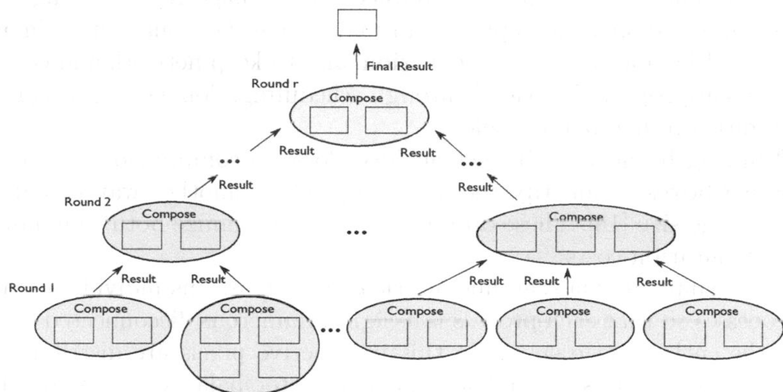

FIGURE 5.3: Tree-based compositing.

sent transmission of image pieces from sender to receiver. If the process, P_0, is the owner of the top 1/4 of the image, then P_1, P_2, and P_3 each send the top 1/4 of their image to P_0 to composite. The same thing happens at the other processes for the other three quarters of the image.

When p is large, direct-send can congest the network with many simultaneous messages. One remedy for relieving network contention is to use a tree for compositing. Figure 5.3 illustrates a hypothetical tree with a variety of group sizes. Tree methods mete out the work in levels of the tree, called rounds, and in each round, images are exchanged between a small number of processes. While the direct-send algorithm tries to do as much work as quickly as possible by generating the maximum number of messages in a single round, tree methods strive for a measured approach that generates fewer simultaneous messages over more rounds. Trees can be designed to span a broad range of this spectrum, but first let's consider the arguably best-known binary tree compositing algorithm, binary-swap.

Ordinary tree compositing, as Figure 5.3 shows, causes many of the processes that were busy in early rounds to go idle as execution proceeds to later rounds. Eventually, at the root of the tree, one process is performing the entire composition sequentially. To solve this bottleneck and keep all processes busy, Ma et al. [15] introduced the binary-swap algorithm. Figure 5.4 shows a binary-swap example of four processes and two rounds. Each process composes the incoming portion of the image with the same part of the image that it already owns. By continually swapping, all processes remain busy throughout all rounds. In each round, neighbors are chosen twice as far apart, and image portions exchanged are half as large as in the previous round. This is why binary-swap is also called a distance-doubling and vector-halving communication algorithm in some contexts.

Since the mid-1990s when direct-send and binary-swap were first published, numerous variations and optimizations to the basic algorithms have appeared. The basic ideas are to reduce active image regions using the spatial locality and sparseness present in many scientific visualization images, to better load balance after such reduction, and to keep network and computing resources appropriately loaded through scheduling. Some examples of each of these ideas are highlighted below.

Run-length encoding images achieves lossless compression [1], and using bounding boxes to identify the nonzero pixels is another way to reduce the active image size [15]. These optimizations can minimize both communication and computation costs.

Load balancing via scan line interleaving [29] assigns individual scan lines to processes so that each process is assigned numerous disconnected scan lines from the entire image space. In this way, active pixels are distributed more evenly among processes, and workload is better balanced. The drawback is that the resulting image must be rearranged once it is composited, which can be expensive for large images.

The SLIC [28] algorithm combines direct-send with active-pixel encoding, scan-line interleaving, and scheduling of operations. Spans of compositing tasks are assigned to processes in an interleaved fashion. Another way to schedule processes is to assign them different tasks. This is the approach taken by Ramakrishnan et al. [26], who scheduled some processes to perform rendering while others were assigned to compositing. The authors presented an optimal linear-time algorithm to compute this schedule.

Image compositing has also been combined with parallel rendering for tiled displays. The IceT library, (see Chap. 17) from Moreland et al. [19], performs sort-last rendering on a per-tile basis. Within each display tile, the processors that contributed image content to that tile perform either direct-

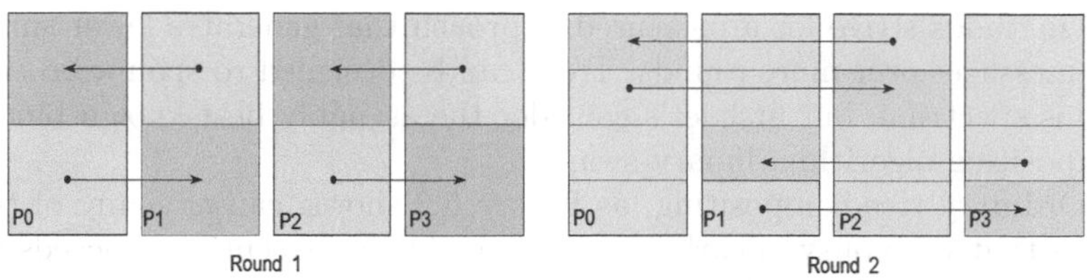

FIGURE 5.4: Example of the binary-swap algorithm for four processes and two rounds.

send or binary-swap compositing. Although the tile feature of IceT is not used much in practice, IceT has become a production-quality library that offers a robust suite of image compositing algorithms to scientific visualization tools. Both ParaView (see Chap. 18 and [32]) and VisIt (see Chap. 16 and [9]) use the IceT library for image compositing.

5.2.3 Image Compositing Hardware

While this chapter primarily studies the evolution of software compositing algorithms, it is worth noting that hardware solutions to the image compositing problem exist as well. Some of these have been made commercially available on smaller clusters, but as the interconnects and graphics hardware on visualization and HPC machines have improved over time, it has become more cost-effective to use these general-purpose machines for parallel rendering and image compositing rather than purchasing dedicated hardware for these tasks.

Sepia [16] is one example of a parallel rendering system that included PCI-connected FPGA boards for image composition and display. Lightning-2 [27] is a hardware system that received images from the DVI outputs of graphics cards, composited the images, and mapped them to sections of a large tiled display. Muraki et al. [20] described an eight-node Linux cluster equipped with dedicated volume rendering and image composition hardware. In a more recent system, the availability of programmable network processing units accelerated image compositing across 512 rendering nodes [24].

5.3 Recent Advances

Although the classic image composition algorithms and optimizations have been used for the past fifteen years, new processor and interconnect advances such as direct memory access and multi-dimensional network topology present new opportunities to improve the state of the art in image compositing. The next few sections explain the relationship between direct-send and binary-swap through a tree-based representation and how it is used to develop more general algorithms that combine both techniques. Two recent algorithms, 2-3 swap and radix-k, are presented as examples of more general communication patterns that can exploit new hardware.

5.3.1 2-3 Swap

Yu et al. [33] extended binary-swap compositing to nonpower-of-two numbers of processors with an algorithm they called 2-3 swap. One goal of this algorithm is to combine the flexibility of direct-send with the scalability of binary-swap, and the authors accomplished this by recognizing that direct-send and binary-swap are related and can be combined into a single algorithm.

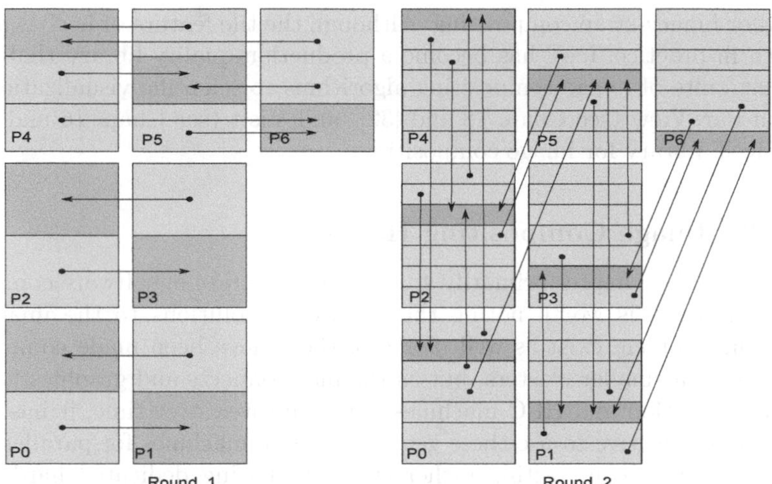

FIGURE 5.5: Example of the 2-3 swap algorithm for seven processes in two rounds. Group size is two or three in the first round, and between two and five in the second round.

Any natural number greater than one can be expressed as the sum of twos and threes, and this property is used to construct the first-round group assignment in the 2-3 swap algorithm. The algorithm proceeds to execute a sequence of rounds with group sizes that are between two and five. Each round can have multiple group sizes present within the same round. The number of rounds, r, is equal to the base of $\log p$, where p is the number of processes and need not be a power of two, as is the case for the binary-swap algorithm.

An example of 2-3 swap using seven processes is shown in Figure 5.5. In the first round, shown on the left, processes form groups in either twos or threes, as the name 2-3 swap suggests, and execute a direct-send within each group. (Direct-send is the same as binary-swap when the group size is two.) In the second round, shown on the right, the image pieces are simply divided into a direct-send assignment, with each process owning $1/p$, or $1/7$ in this example, of the image. By assigning *which* $1/7$ each process owns, however, Yu et al. proved that the maximum number of processes in a group is five, avoiding the contention in the ordinary direct-send. Indeed, Figure 5.5 shows that process P_5 receives messages from four other processes, while all other processes receive messages from only two or three other processes.

5.3.2 Radix-k

The next logical step in combining and generalizing direct-send and binary-swap is to allow more combinations of rounds and group sizes. To see how this is done, let k_i represent the number of processes in a communication

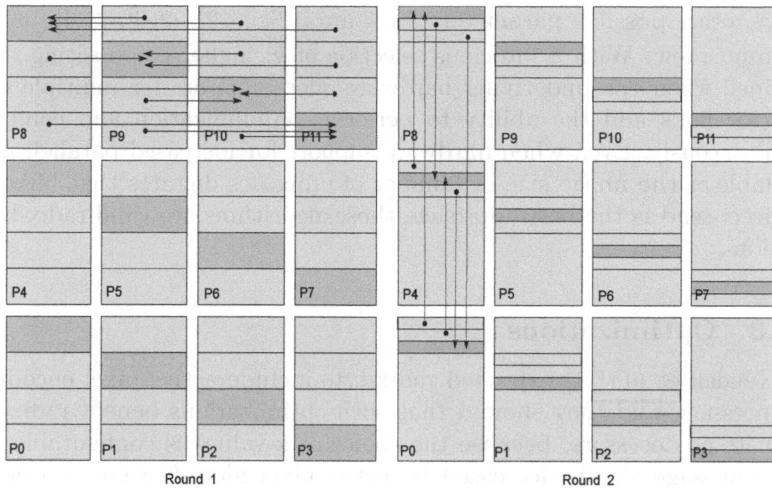

FIGURE 5.6: Example of the radix-k algorithm for twelve processes, factored into two rounds of $\vec{k} = [4, 3]$.

group in round i. The k-values for all rounds can be written as the vector $\vec{k} = [k_1, k_2, ..., k_r]$, where the number of rounds is denoted by r. Within each group, a direct-send is performed. The total number of processes is p.

More than a convenient notation, this terminology makes clear the relationship among the previous algorithms. Direct-send is now defined as $r = 1$ and $\vec{k} = [p]$; and binary-swap is defined as $r = \log p$ and $\vec{k} = [2, 2, 2, ...]$. Just as 2-3 swap was an incremental step in combining direct-send and binary-swap by allowing k-values that are either 2 or 3, it is natural to ask whether other combinations of r and $\vec{k}$ are possible. The radix-k algorithm [22] answers this question by allowing any factorization of p into $\prod_{i=1}^{r} k_i = p$. In radix-k, all groups in round i are the same size, k_i.

Figure 5.6 shows an example of radix-k for $p = 12$ and $\vec{k} = [4, 3]$. The processes are drawn in a 4×3 rectangular layout to identify the rounds and groups clearly. In this example, the rows on the left side form groups in the first round, while the columns on the right side form second-round groups. A convenient way to think about forming groups in each round is to envision the process space as an r-dimensional virtual lattice, where the size in each dimension is the k-value for that round. This is the convention followed in Figure 5.6 for two rounds drawn in two dimensions.

The outermost rectangles in the figure represent the image held by each process at the start of the algorithm. During the round i, the current image piece is further divided into k_i pieces, such that the image pieces grow smaller with each round. The image pieces are shown as highlighted boxes in each round.

Selecting different parameters can lead to many options; for the example above, other possible parameters for $\vec{k}$ are [12], [6,2], [2,6], [3,4], or [2,2,3], to name a few. With a judicious selection of $\vec{k}$, higher compositing rates are attained when the underlying hardware offers support for multiple communication links and the ability to perform communication and computation simultaneously. Even when hardware support for increased parallelism is not available or the image size or number of processes dictates that binary-swap or direct-send is the best approach, those algorithms are valid radix-k configurations.

5.3.3 Optimizations

Kendall et al. [11] extended radix-k to include active-pixel encoding and compression, and they showed that such optimizations benefit radix-k more than its predecessors, because the choice of k-values is configurable. Hence, when message size is decreased by active-pixel identification and encoding, k-values can be increased and performance can be further enhanced. Their implementation encodes nonempty pixel regions, based on the bounding box information, into two separate buffers: one for alternating counts of empty and nonempty pixels, and the other for the actual pixel values. This way, new subsets of the image can be taken by reassigning pointers rather than copying pixels, and images remain encoded throughout all of the compositing rounds. Non-overlapping regions are copied from one round to the next without performing the blending operation.

A set of empirical tests was performed to determine a table of target k-values for different image sizes, number of processes, and architectures at both the Argonne and Oak Ridge Leadership Computing Facilities. The platforms tested were IBM Blue Gene/P, Cray XT, and two graphics clusters. With this table, radix-k can look up the closest entry for a given image size, system size, and architecture, and automatically factor the number of processes into k-values as close to the target value as possible.

Moreland et al. [18] deployed and evaluated optimizations in a production framework, rather than in isolated tests: IceT serves as both this test and production framework. The advantages of this approach are that the tests represent real workloads and improvements are ready for use in a production environment sooner. These improvements include minimizing pixel copying through compositing order and scanline interleaving, and a new telescoping algorithm for the non-power-of-two number of processes that can further improve radix-k. A final advantage of using IceT for these improvements is that IceT provides unified and reproducible benchmarks that other researchers can repeat.

One of the improvements was devising a compositing order that minimizes pixel copying. The usual, accumulative order causes nonoverlapping pixels to be copied up to $k_i - 1$ times in round i, whereas tree methods only incur $\log k_i$ copy operations. Pixel copying can further be reduced while using a novel image interlacing algorithm. Rather than interleaving partitions according to

TABLE 5.1: Theoretical lower bounds for compositing algorithms.

Algorithm	Latency	Bandwidth	Computation
direct-send	$\alpha p/k$	$n\beta(p-1)/p$	$n\gamma(p-1)/p$
binary-swap	$\alpha \log_2 p$	$n\beta(p-1)/p$	$n\gamma(p-1)/p$
radix-k	$\alpha \log_k p$	$n\beta(p-1)/p$	$n\gamma(p-1)/p$

scanlines, as in [29], Moreland et al. realized that a desired property of the interleaving is that processes retain contiguous pixel regions after compositing, so as to not require further rearranging. The van der Corput sequence is one example of such an ordering.

While radix-k natively handles nonpower-of-two number of processes, it does not always do so gracefully. Some process counts, especially those with factorizations containing large prime numbers, can exhibit pathological performance. Moreland et al. [18] also devised a telescoping algorithm for nonpowers-of-two that continually looks for largest subgroups that are powers of two, and performs radix-k within each subgroup. Subgroups are further composited together afterwards.

5.4 Results

The theoretical communication and computation costs of direct-send, binary-swap, and radix-k are compared in Table 5.1 using the cost model in Chan et al. [7]. This model assumes p processors in a distributed-memory parallel architecture and the original image size has n total pixels. The communication cost is $\alpha + n\beta$, where α is the latency per message and β is the transmission time per data item (reciprocal of the link bandwidth). The computation time to compose one pixel is γ, making the total time to transmit and reduce a message consisting of n data elements $\alpha + n\beta + n\gamma$. The model further assumes a fully connected network, where k messages can occupy the network without link contention and no overlap between communication and computation.

Some of the theoretical assumptions are not true in practice, however, as calculating the relative cost of communication algorithms is simplified under these conditions. In actual implementations, radix-k group sizes vary between 8 and 128, depending on the architecture and optimizations. There is also an overlap between communication and computation on modern HPC systems that radix-k uses to further boost performance.

Nonetheless, Table 5.1 shows that, in theory, radix-k should perform as well or better than other algorithms—a theory confirmed in Figure 5.7. Figure 5.7 shows a test of the original algorithm in Peterka et al. [22] for a variety of process counts from 32 to 34,816 on the *Intrepid* Blue Gene/P supercomputer at Argonne National Laboratory. These graphs compare binary-swap and radix-k using k-values of 8 whenever possible, and include no other optimizations, such as compression. In the left graph, process count increases by 32 at each data point, up to 1024 processes, while the right graph is a continuation of the test by adding one Blue Gene rack (1,024 nodes) at each data point. On average, radix-k is approximately 20–40% faster than binary-swap.

The speedup increases when optimizations such as active-pixel encoding are enabled. While the target k-value for Intrepid was 8 in the original algorithm of Peterka et al. [22], Figure 5.8 shows that with optimizations, target k-values can be as high as 128. The reason is that smaller message sizes produced by active-pixel encoding and compression allow more messages to be injected into the system without contention. The resulting performance is shown in the strong scaling study of Figure 5.9. With the optimizations of Kendall et al. [11], performance is up to five times better than binary-swap with the same optimizations. These tests were conducted while parallel volume rendering the core-collapse supernova simulation output shown in Figure 5.10. Scientists visualize such simulations in order to understand the physics of thermonuclear and hydrodynamic instabilities in the death of some of the most massive stars in the universe.

5.5 Discussion and Conclusion

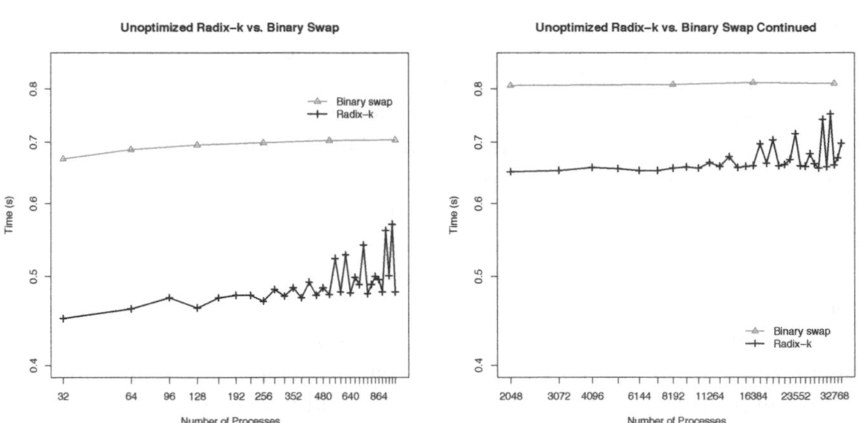

FIGURE 5.7: Performance comparing binary-swap with radix-k for an image size of eight megapixels. On the left, process counts from 32 to 1,024 in increments of 32. On the right, the same test continued at larger scale in increments of 1,024 processes. No optimizations, such as compression or active-pixel encoding, were applied to either algorithm for this test.

Table of target k-values for Intrepid BG/P and Jaguar XT5

Processes \ Megapixels	Intrepid 4	Intrepid 8	Intrepid 16	Intrepid 32	Jaguar 4	Jaguar 8	Jaguar 16	Jaguar 32
8	8	8	8	8	4	4	4	4
16	16	16	16	16	4	4	4	4
32	32	32	32	32	16	8	16	16
64	64	64	64	64	16	16	16	16
128	64	128	128	128	8	8	8	8
256	64	128	128	128	16	8	8	8
512	64	128	128	128	16	32	8	8
1 K	64	128	128	128	64	32	32	8
2 K	32	128	128	128	8	32	32	32
4 K	32	32	32	32	8	16	32	64
8 K	32	32	32	32	4	8	32	64
16 K	32	32	32	32	4	32	32	8
32 K	32	32	32	32	16	16	64	8

FIGURE 5.8: Target k-values for two different machines are shown. Optimizations such as active pixel encoding enable the use of higher k-values than before. In the original algorithm, k = 8 was used, but the tables above show that with active pixel encoding, k values as high as 128 are optimal, depending on the machine.

5.5.1 Conclusion

The radix-k algorithm allows the number of message partners per round and the number of rounds to be adjusted to maximize performance for a given

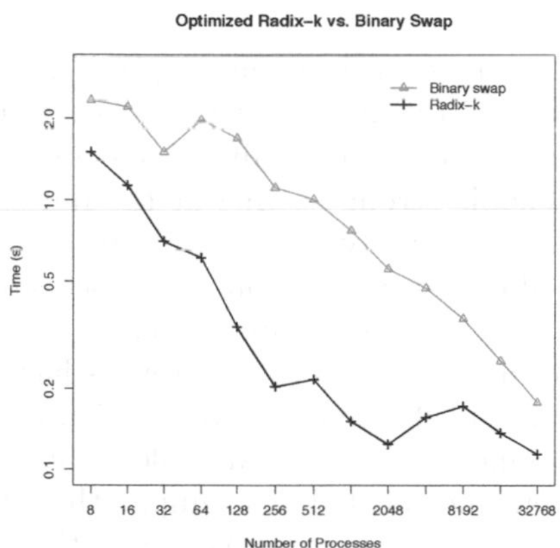

FIGURE 5.9: Performance comparing optimized binary-swap with radix-k shows overall improvement in volume rendering tests of core-collapse supernovae simulation output.

FIGURE 5.10: Core-collapse supernova volume rendered in parallel and composited using the radix-k algorithm.

problem size and architecture. Algorithms such as radix-k and 2-3 swap build on the previous contributions of binary-swap and direct-send, by generalizing and unifying two algorithms that previously were considered separate. Now, binary-swap and direct-send are just two points in an entire parameter space of configurations.

Optimizations to radix-k, such as active-pixel encoding, enabled the use of higher k-values, up to 128 in practice, and overall performance can be improved over binary-swap by up to five times for identical optimizations. Such improvements enable strong scaling at the full scale of HPC machines such as IBM Blue Gene and Cray XT, compared to legacy algorithms whose performance bottomed-out at a few thousand processes.

It is now possible to compose large-size images interactively as well; subsecond compositing of images up to 64 megapixels has been demonstrated on actual volume rendering of data produced by scientific computational simulations. These improvements are now available in IceT for production use in tools such as ParaView and VisIt. For example, the latest version of IceT in ParaView enabled image compositing, up to seven times faster and the ability to scale to 36,000 processes, which was not possible with previous versions of the library.

Of course, these improvements are predicated on an underlying architecture that can support additional message concurrency. Today, HPC networks have multiple data paths and DMA access and thus, can benefit from highly parallel algorithms, like radix-k. Because it is reasonable to expect more scientific visualizations to execute on supercomputers in the future, making efficient use of these architectures for compositing will be a critical part of the high performance visualization pipeline.

Modern compositing algorithms have demonstrated a tangible impact for high performance visualization and computational science. They allow scientists to render higher resolution images, at larger system scales, faster than before. Together, all of these benefits can result in an increased understanding of scientific data, and they will be absolutely essential going forward toward exascale computing.

5.5.2 Directions for Future Research

Highly parallel image compositing offers several avenues for continued exploration. Load balance in both computation and communication can be further improved and combined with fine-grained delineation of active pixels. Other potential solutions to load balancing may be found in time-varying parallel rendering, where the additional time dimension poses new scheduling opportunities.

New architectures are creating the need for continued study. Network topologies such as higher-dimensional torus networks are being investigated, as are clouds and distributed collections of heterogeneous processing elements [14]. With the ubiquity of multi-core processors, a natural next step is to parallelize the computing of the compositing operator across several pixels. Hybrid parallelism that combines message passing with shared-memory threading is being studied in numerous algorithms; image compositing can also potentially benefit from the hybrid programming models in Chapter 12.

Algorithms such as radix-k require empirical experiments to find appropriate k-values for a particular architecture though there exists a trend in communication algorithms toward self-tuning. Hence, another area for further study is automating parameter selection, a process sometimes referred to as auto-tuning. These ideas form the basis for Chapter 14.

References

[1] James Ahrens and James Painter. Efficient Sort-Last Rendering Using Compression-Based Image Compositing. In *Proceedings of the 2nd Eurographics Workshop on Parallel Graphics and Visualization*, pages 145–151, 1998.

[2] Mike Barnett, Satya Gupta, David G. Payne, Lance Shuler, Robert Geijn, and Jerrell Watts. Interprocessor Collective Communication Library. In *Proceedings of the Scalable High Performance Computing Conference*, pages 357–364. IEEE Computer Society Press, 1994.

[3] Mike Barnett, David G. Payne, Robert A. van de Geijn, and Jerrell Watts. Broadcasting on Meshes with Wormhole Routing. *Journal of Parallel Distributed Computing*, 35(2):111–122, 1996.

[4] Massimo Bernaschi and Giulio Iannello. Collective Communication Operations: Experimental Results vs. Theory. *Concurrency*, 10(5):359–386, 1998.

[5] Jehoshua Bruck, Ching-Tien Ho, Shlomo Kipnis, and Derrick Weathersby. Efficient Algorithms for All-to-All Communications in Multi-Port Message-Passing Systems. In *SPAA '94: Proceedings of the 6th Annual ACM Symposium on Parallel Algorithms and Architectures*, pages 298–309. ACM, 1994.

[6] Xavier Cavin, Christophe Mion, and Alain Fibois. COTS Cluster-based Sort-last Rendering: Performance Evaluation and Pipelined Implementation. In *Proceedings of IEEE Visualization 2005*, pages 111–118, 2005.

[7] Ernie Chan, Marcel Heimlich, Avi Purkayastha, and Robert van de Geijn. Collective Communication: Theory, Practice, and Experience: Research Articles. *Concurrency and Computation: Practice and Experience*, 19(13):1749–1783, 2007.

[8] Ernie Chan, Robert van de Geijn, William Gropp, and Rajeev Thakur. Collective Communication on Architectures that Support Simultaneous Communication over Multiple Links. In *PPoPP '06: Proceedings of the 11th ACM SIGPLAN Symposium on Principles and Practice of Parallel Programming*, pages 2–11. ACM, 2006.

[9] H. R. Childs, E. S. Brugger, K. S. Bonnell, J. S. Meredith, M. C. Miller, B. J. Whitlock, and N. L. Max. A Contract Based System for Large Data Visualization. In *Proceedings of IEEE Visualization 2005*, pages 190–198, 2005.

[10] William M. Hsu. Segmented Ray Casting for Data Parallel Volume Rendering. In *Proceedings of 1993 Parallel Rendering Symposium*, pages 7–14, 1993.

[11] Wesley Kendall, Tom Peterka, Jian Huang, Han-Wei Shen, and Robert Ross. Accelerating and Benchmarking Radix-K Image Compositing at Large Scale. In *Proceedings of Eurographics Symposium on Parallel Graphics and Visualization EG PGV'10*, pages 101–110, 2010.

[12] Sameer Kumar, Gabor Dozsa, Gheorghe Almasi, Philip Heidelberger, Dong Chen, Mark E. Giampapa, Michael Blocksome, Ahmad Faraj, Jeff Parker, Joseph Ratterman, Brian Smith, and Charles J. Archer. The Deep Computing Messaging Framework: Generalized Scalable Message Passing on the Blue Gene/P Supercomputer. In *ICS '08: Proceedings of the 22nd Annual International Conference on Supercomputing*, pages 94–103. ACM, 2008.

[13] Sameer Kumar, Gabor Dozsa, Jeremy Berg, Bob Cernohous, Douglas Miller, Joseph Ratterman, Brian Smith, and Philip Heidelberger. Architecture of the Component Collective Messaging Interface. In *Euro PVM/MPI '08: Proceedings of the 15th Annual European PVM/MPI Users' Group Meeting*, pages 23–32. Springer, 2008.

[14] Kwan-Liu Ma and David M. Camp. High Performance Visualization of Time-Varying Volume Data over a Wide-Area Network. In *Proceedings of the 2000 ACM/IEEE Conference on Supercomputing (CDROM)*, Supercomputing '00. IEEE Computer Society, 2000.

[15] Kwan-Liu Ma, James S. Painter, Charles D. Hansen, and Michael F. Krogh. Parallel Volume Rendering Using Binary-Swap Compositing. *IEEE Computer Graphics and Applications*, 14(4):59–68, 1994.

[16] Laurent Moll, Mark Shand, and Alan Heirich. Sepia: Scalable 3D Compositing using PCI Pamette. In *Proceedings of the Seventh Annual IEEE Symposium on Field-Programmable Custom Computing Machines*, FCCM '99, pages 146–155. IEEE Computer Society, 1999.

[17] Steven Molnar, Michael Cox, David Ellsworth, and Henry Fuchs. A Sorting Classification of Parallel Rendering. *IEEE Computer Graphics and Applications*, 14(4):23–32, 1994.

[18] Ken Moreland, Wesley Kendall, Tom Peterka, and Jian Huang. An Image Compositing Solution at Scale. In *SC '11: Proceedings of the 2011 ACM/IEEE Conference on Supercomputing*, pages 1–10, 2011.

[19] Kenneth Moreland, Brian Wylie, and Constantine Pavlakos. Sort-Last Parallel Rendering for Viewing Extremely Large Data Sets on Tile Displays. In *PVG '01: Proceedings of the IEEE 2001 Symposium on Parallel*

and Large-Data Visualization and Graphics, pages 85–92. IEEE Press, 2001.

[20] Shigeru Muraki, Masato Ogata, Kwan-Liu Ma, Kenji Koshizuka, Kagenori Kajihara, Xuezhen Liu, Yasutada Nagano, and Kazuro Shimokawa. Next-Generation Visual Supercomputing using PC Clusters with Volume Graphics Hardware Devices. In *Proceedings of the 2001 ACM/IEEE Conference on Supercomputing (CDROM)*, Supercomputing '01, pages 51–51. ACM, 2001.

[21] Ulrich Neumann. Communication Costs for Parallel Volume-Rendering Algorithms. *IEEE Computer Graphics and Applications*, 14(4):49–58, 1994.

[22] Tom Peterka, David Goodell, Robert Ross, Han-Wei Shen, and Rajeev Thakur. A Configurable Algorithm for Parallel Image-Compositing Applications. In *SC '09: Proceedings of the 2009 ACM/IEEE Conference on Supercomputing*, pages 4:1–4:10, 2009.

[23] Thomas Porter and Tom Duff. Compositing Digital Images. In *Proceedings of 11th Annual Conference on Computer Graphics and Interactive Techniques*, pages 253–259, 1984.

[24] David Pugmire, Laura Monroe, Andrew DuBois, and David DuBois. NPU-Based Image Compositing in a Distributed Visualization System. *IEEE Transactions on Visualization and Computer Graphics*, 13(4):798–809, 2007.

[25] Rolf Rabenseifner. *New Optimized MPI Reduce Algorithm.* 2004. `http://www.hlrs.de/organization/par/services/models/mpi/myreduce.html`.

[26] C. R. Ramakrishnan and Claudio Silva. Optimal Processor Allocation for Sort-Last Compositing under BSP-Tree Ordering. In *SPIE Electronic Imaging, Visual Data Exploration and Analysis IV*, pages 182–192, 1999.

[27] Gordon Stoll, Matthew Eldridge, Dan Patterson, Art Webb, Steven Berman, Richard Levy, Chris Caywood, Milton Taveira, Stephen Hunt, and Pat Hanrahan. Lightning-2: a High-Performance Display Subsystem for PC Clusters. In *Proceedings of the 28th Annual Conference on Computer Graphics and Interactive Techniques*, SIGGRAPH '01, pages 141–148. ACM, 2001.

[28] Aleksander Stompel, Kwan-Liu Ma, Eric B. Lum, James Ahrens, and John Patchett. SLIC: Scheduled Linear Image Compositing for Parallel Volume Rendering. In *Proceedings of IEEE Symposium on Parallel and Large-Data Visualization and Graphics*, pages 33–40, 2003.

[29] Akira Takeuchi, Fumihiko Ino, and Kenichi Hagihara. An Improved Binary-Swap Compositing for Sort-Last Parallel Rendering on Distributed Memory Multiprocessors. *Parallel Computing*, 29(11-12):1745–1762, 2003.

[30] Rajeev Thakur, Rolf Rabenseifner, and William Gropp. Optimization of Collective Communication Operations in MPICH. *International Journal of High Performance Computing Applications*, 19:49–66, 2005.

[31] Jesper Larsson Traff, Andreas Ripke, Christian Siebert, Pavan Balaji, Rajeev Thakur, and William Gropp. A Simple, Pipelined Algorithm for Large, Irregular All-gather Problems. In *Proceedings of Euro PVM/MPI*, pages 84–93, 2008.

[32] Brian Wylie, Constantine Pavlakos, Vasily Lewis, and Kenneth Moreland. Scalable Rendering on PC Clusters. *IEEE Computer Graphics and Applications*, 21(4):62–69, 2001.

[33] Hongfeng Yu, Chaoli Wang, and Kwan-Liu Ma. Massively Parallel Volume Rendering Using 2-3 Swap Image Compositing. In *SC '08: Proceedings of the 2008 ACM/IEEE Conference on Supercomputing*, pages 1–11. IEEE Press, 2008.

Chapter 6

Parallel Integral Curves

David Pugmire

Oak Ridge National Laboratory

Tom Peterka

Argonne National Laboratory

Christoph Garth

University of Kaiserslautern

6.1	Introduction	91
6.2	Challenges to Parallelization	94
	6.2.1 Problem Classification	94
6.3	Approaches to Parallelization	95
	6.3.1 Test Data	96
	6.3.2 Parallelization Over Seed Points	97
	6.3.3 Parallelization Over Data	99
	6.3.4 A Hybrid Approach to Parallelization	99
	6.3.5 Algorithm Analysis	103
	6.3.6 Hybrid Data Structure and Communication Algorithm	106
6.4	Conclusion	111
	References	112

Understanding vector fields resulting from large scientific simulations is an important and often difficult task. Integral curves (*IC*s)—curves that are tangential to a vector field at each point—are a powerful visualization method in this context. The application of an integral curve-based visualization to a very large vector field data represents a significant challenge, due to the non-local and data-dependent nature of *IC* computation. The application requires a careful balancing of computational demands placed on I/O, memory, communication, and processors. This chapter reviews several different parallelization approaches, based on established parallelization paradigms (across particles and data blocks) and current advanced techniques for achieving a scalable, parallel performance on very large data sets.

6.1 Introduction

The visualization of vector fields is a challenging area of scientific visualization. For example, the analysis of fluid flow that governs natural phenomena on all scales, from the smallest (e.g., Rayleigh–Taylor mixing of fluids) to the largest (e.g., supernovae explosions), relies on visualization to elucidate the patterns exhibited by flows and the dynamical structures driving them (see Fig. 6.1 and 6.2). Processes typically depicted by vector fields—such as transport, circulation, and mixing—are prevalently non-local in nature. Due to this specific property, methods and techniques that were developed and have proven successful for scalar data visualization are not readily generalized for the study of vector fields. Hence, while it is technically feasible to apply such methods directly to derived scalar quantities, such as vector magnitude, the resulting visualizations often fall short in explaining the mechanisms underlying the scientific problem.

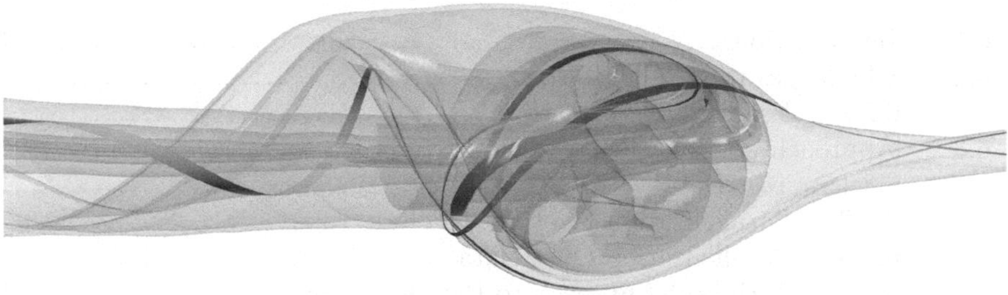

FIGURE 6.1: A stream surface visualizes recirculating flow in a vortex breakdown bubble. The surface, computed from high-resolution adaptive, unstructured flow simulation output, consists of several million triangles and was rendered using an adaptive approach for its transparency, designed to reveal the flow structure inside the bubble. Two stripes highlight the different trajectories taken by particles encountering the recirculation. Image courtesy of Christoph Garth (University of Kaiserslautern), data courtesy of M. Rütten (German Aerospace Center, Göttingen).

A large majority of visualization approaches for the visualization and analysis of vector fields are based on the study of integral curves (*IC*s). Naturally understood as trajectories of massless particles, such curves are ideal tools to study transport, mixing, and other similar processes. These *integration-based methods* make use of the intuitive interpretation of *IC*s as the trajectories of massless particles and were originally developed to reproduce physical flow visualization experiments based on small, neutrally buoyant particles. Mathematically, an *IC* is tangential to the vector field at every point along the curve and individual curves are determined by selecting an initial condition or *seed*

FIGURE 6.2: Streamlines showing outflows in a magnetic flow field computed by an Active Galactic Nuclei astrophysics simulation (FLASH code). Image courtesy of David Pugmire (ORNL), data courtesy Paul Sutter (UIUC).

point, the location from which the curve begins, and an *integration time* over which the virtual particle is traced.

*IC*s are applicable in many different settings. For example, the direct visualization of particles, and particle families and their trajectories, gives rise to such visualization primitives as streamlines, pathlines, streak lines, or integral surfaces (e.g., see Krishnan et al. [6]). To specifically study transport and mixing, *IC*s also offer an interesting change of perspective. Instead of considering the so-called *Eulerian* perspective that describes evolution of quantities at fixed locations in space, the *Lagrangian* view examines the evolution from the point of view of an observer attached to a particle moving with the vector field, offering a natural and intuitive description of transport and mixing processes. Consider, for example, the case of combustion: fuel burns while it is advected by a surrounding flow; hence, the burning process and its governing equations are primarily Lagrangian in nature.

Further, Lagrangian methods concentrate on deriving structural analysis from the Lagrangian perspective. For example, *Finite-Time Lyapunov Exponents* (FTLE) empirically determine exponential separation rates among neighboring particles from a dense set of *IC*s covering a domain of interest [7, 5]. From *ridges* in FTLE, that is locally maximal lines and surfaces, one can identify so-called *Lagrangian coherent structures* (LCS) that approximate hyperbolic material lines and represent the dynamic skeleton of a vector field that drives its structure and evolution. For an overview of modern flow visualization techniques, see McLoughlin et al. [8].

Computationally, *IC*s are approximated using numerical integration schemes (cf. [4]). These schemes construct a curve in successive pieces: starting from the seed point, the vector field is sampled in the vicinity of the current integration point, and an additional curve sequence is determined and appended to the existing curve. This is repeated until the curve has reached its desired length or leaves the vector field domain. To propagate an *IC* through a region of a given vector requires access to the source data. If the source data can remain in the main memory, it will be much faster than a situation where the source data must be paged into main memory from a secondary location. In this setting, the non-local nature of particle advection implies that the path taken by a particle largely determines which blocks of data must be loaded. This information is *a priori* unknown and depends on the vector field itself. Thus, general parallelization of *IC* computation is a difficult problem for large data sets on distributed-memory architectures.

6.2 Challenges to Parallelization

This chapter aims to both qualify and quantify the performance of three different parallelization strategies for *IC* computation. Before providing a discussion of particular parallelization algorithms in 6.3, this section will first present challenges that are particular to the parallelization of *IC* computation on distributed-memory systems.

6.2.1 Problem Classification

The parallel *IC* problem is complex and challenging. To design an experimental methodology that provides robust coverage of different aspects of algorithmic behavior, some of which is data set dependent, the following factors must be taken into account, which influence parallelization strategy and performance test design.

Data Set Size. If the data set is *small*, in the sense that it fits entirely into the memory footprint of each task, then it makes the most sense to distribute the *IC* computation workload. On the other hand, if the data set is *larger* than will fit in each task's memory, then more complex approaches are required that involve data distribution and possibly computational distribution.

Seed Set Size. If the problem at hand requires only the computation of a thousand streamlines, parallel computation takes secondary precedence to optimal data I/O and distribution. In this case, the corresponding seed set is *small*. Such small seed sets are typically encountered in interactive exploration scenarios, where relatively few *IC*s are interactively seeded by a user. A *large* seed set may consist of many thousands of seed points. Here, it will be desirable to distribute the *IC* computational workload, yet, the data distribution schemes need to allow for efficient parallel *IC* computation.

Seed Set Distribution. In the case where seed points are located densely within the spatial and temporal domain of a defined vector field, it is likely that the ICs will traverse a relatively small amount of the overall data. On the other hand, for some applications, like streamline statistics, a *sparse* seed point set covers the entire vector field evenly. This distribution results in ICs traversing the entire data set. Hence, the seed set distribution is a significant factor in determining if the performance stands to gain the most from parallel computation, data distribution, or both.

Vector Field Complexity. Depending on the choice of seed points, the structure of a vector field can have a strong influence on which parts of the data need to be taken into account in the IC computation process. Critical points, or invariant manifolds, of a strongly attracting nature draw streamlines towards them. The resulting ICs seeded in or traversing their vicinity will remain closely localized. In the opposite case, a nearly uniform vector field requires ICs to pass through large parts of the data. This dependency of IC computation on the underlying vector field is both counterintuitive and hard to identify without conducting a prior analysis to determine the field structure, as is done in Yu et al.'s study [13], for example. While such analysis can be useful for specific problems, this chapter's contents consider a more general setting, where the burden and cost of pre-processing is not considered.

6.3 Approaches to Parallelization

IC calculation places demands on almost all components of a computational system, including memory, processing, communication, and I/O. Because of the complex nature of vector fields, seeding scenarios, and the types of analyses, as outlined in 6.2, there is no single scalable algorithm suitable for all situations.

Generally, algorithms can be parallelized in two primary ways. The first way is parallelization across the seed points, where seeds are assigned to processors, and data blocks are loaded as needed. The second way is parallelization across the data blocks, where processors are assigned a set of data blocks, and particles are communicated to the processor that owns the data block. In choosing between these two axes of parallelization, the different assignments of particles and data blocks to processors will place different demands on the computing, memory, communication, and I/O subsystems of a cluster.

These two classes of algorithms have a tendency to perform poorly due to a workload imbalance, either through an imbalance in particle to processor assignment, or through loading too many data blocks, which become I/O bound. Recent research blending these two approaches shows promising results.

The subsequent sections outline algorithms that parallelize with respect to particles (6.3.2) and data blocks (6.3.3), as well as a hybrid approach (6.3.4) and discuss the algorithm's performance characteristics. These strategies aim to keep a balanced workload across the set of processors and to design efficient data structures for handling IC computation and communication.

In the approaches outlined below, the problem mesh is decomposed into

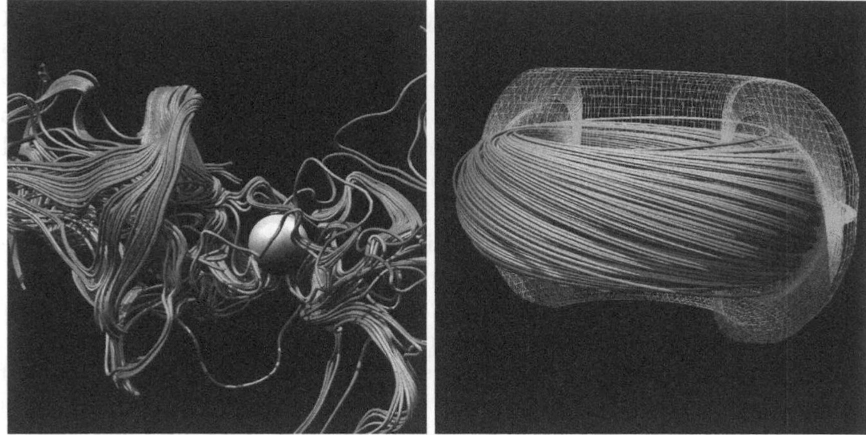

FIGURE 6.3: Data sets used for *IC* tests. Integral curves through the magnetic field of a GenASiS supernova simulation (left). Image courtesy of David Pugmire (ORNL), data courtesy of Anthony Mezzacappa (ORNL). Integral curves through the magnetic field of a NIMROD fusion simulation (right). Image courtesy of David Pugmire (ORNL), data courtesy of Scott Kruger (Tech-X Corporation).

a number of spatially disjoint data blocks. Each block may or may not have ghost cells for connectivity purposes. Each block has a timestep associated with it. Thus, two blocks that occupy the same space at different times are considered independent.

6.3.1 Test Data

To cover a wide range of potential problem characteristics, different seeding scenarios and different data sets were used in the performance experiments that follow. Although the techniques here are readily applicable to any mesh type and decomposition scheme, the data sets discussed here are multi-block and rectilinear. The simplest type of data representation was intentionally chosen, in order to exclude additional performance complexities that arise with more complex mesh types.

Astrophysics. The astrophysics test data is the magnetic field from a GenASiS code simulation of a core-collapse supernova. The search for the explosion mechanism of core-collapse supernovae and the computation of the nucleosynthesis in these stellar explosions is one of the most important and challenging problems in computational nuclear astrophysics. Understanding the magnetic field around the core is a fundamental component of this research, and *IC* analysis has proven to be an effective technique. The test data was computed by GenASiS, a multiphysics code being developed for the simulation of astrophysical systems involving nuclear matter [2, 3]. GenASiS

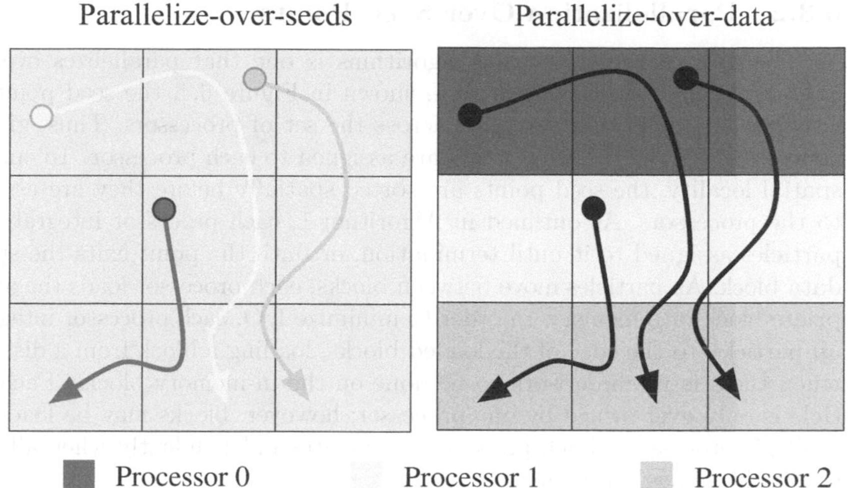

FIGURE 6.4: Parallelization over seeds versus parallelization over data algorithms.

computes the magnetic field at each cell face. For the purposes of this study, a cell-centered vector was computed by differencing the values at the faces in the X, Y, and Z directions. Node-centered vectors are computed by averaging adjacent cells to each node. To see how this algorithm would perform on very large data sets, the magnetic field was upsampled onto a total 512 blocks with 1 million cells per block. The dense seed set corresponds to seed points placed randomly in a small box around the collapsing core, whereas the sparse seed test places streamline seed points randomly throughout the entire data set domain.

Fusion. The second data set is from a simulation of magnetically confined fusion in a tokamak device. The development of magnetic confinement fusion, which may be a future source for low-cost power, is an important area of research. Physicists are particularly interested in using magnetic fields to confine the burning plasma in a toroidal-shaped device, known as a tokamak. To achieve stable plasma equilibrium, the field lines of these magnetic fields need to travel around the torus in a helical fashion. By using streamlines, the scientist can see and better understand the magnetic fields. The simulation was performed using the NIMROD code [12]. This data set has the unusual property that most streamlines are approximately closed and they repeatedly traverse the torus-shaped vector field domain. This behavior stresses the data cache.

6.3.2 Parallelization Over Seed Points

The first of the traditional algorithms is one that parallelizes over the axis of particles. In this algorithm, shown in Figure 6.4, the seed points, or particles, are uniformly assigned across the set of processors. Thus, given n processors, $1/n$ of the seed points are assigned to each processor. To enhance spatial locality, the seed points are sorted spatially before they are assigned to the processors. As outlined in Algorithm 1, each processor integrates the particles assigned to it until termination, or until the point exits the spatial data block. As particles move between blocks, each processor loads the appropriate block into memory. In order to minimize I/O, each processor integrates all particles to the edge of the loaded blocks, loading a block from a disk only when there is no more work to be done on the in-memory blocks. Each particle is only ever owned by one processor; however, blocks may be loaded by multiple processors. Each processor terminates independently when all of its streamlines have terminated.

In order to manage the handling of blocks, this algorithm makes use of the data block caching in a least-recently used (LRU) fashion. If there is insufficient memory for additional blocks when a new block must be loaded, the block that is purged is that which was least recently used. Clearly, having more main memory available decreases the need for I/O operations and increases the performance of this algorithm.

Another method for the handling of data blocks was introduced by Camp et al. 2011 [1], where an extended hierarchy of memory layers is used for data block management. When using a local disk, either SSD or hard drive, careful management of data block movement shows an increase in overall performance by about 2×.

Algorithm 1 Parallelization Over Seed Points Algorithm

```
while not done do
    activeParticles = particles that reside in a loaded data block
    inactiveParticles = particles that do not reside in a loaded data block
    while activeParicles not empty do
        advect particle
    end while
    if inactiveParticles then
        Load a dataset from inactiveParticles
    else
        done = true
    end if
end while
```

Advantages: This algorithm is straightforward to implement; it requires no communication and is ideal if I/O requirements are known to be minimal.

For example, if the entire mesh will fit into the main memory of a single process, or if the flow is known to be spatially local, the I/O will be minimal. In such cases, parallelization is trivial, and parallel scaling is expected to be nearly ideal. Also, the risk of processor imbalance is minimized because of the uniform assignment of seeds to processors.

Disadvantages: Because data blocks are loaded on demand, it is possible for I/O to dominate this algorithm. For example, in a vector field with circular flow and a data block cache that does not have large enough blocks, data will be continuously loaded and purged in the LRU cache. Additionally, workload imbalance is possible if the computational requirements for particles differ significantly. For example, if some particles are advected significantly farther than others, the processors assigned to these particles will be doing all the work, while the others sit idle.

6.3.3 Parallelization Over Data

The second of the traditional algorithms is one that parallelizes over the axis of data blocks. In this algorithm, shown in Figure 6.4, the data blocks are statically assigned to the processors. Thus, given n processors, $1/n$ of the data blocks are assigned to each processor. Each particle is integrated until it terminates or leaves the blocks owned by the processor. As a particle crosses block boundaries, it is communicated to the processor that owns the data block. A globally maintained active particle count is maintained so that all processors may monitor how many particles have yet to terminate. Once the count goes to zero, all processors terminate.

Advantages: This algorithm performs the minimal amount of I/O, where each data block is loaded once, and only once into memory. As I/O operations are orders of magnitude more expensive than computation, this is a significant advantage. This algorithm is well-suited to situations where the vector field is known to be uniform and the seed points are known to be sparse.

Disadvantages: Because of the spatial parallelization, this algorithm is very sensitive to seeding and vector field complexity. If the seeding is dense, only the processors where the seeds lie spatially will do any work. All other processors will be idle. Finally, in cases where the vector field contains critical points, or invariant manifolds, points will be drawn to these spatial regions, increasing the workload of some processors, while decreasing the work of the rest.

6.3.4 A Hybrid Approach to Parallelization

As outlined previously, parallelization over seeds or data blocks are subject to workload imbalance. Because of the complex nature of vector field analysis, it is often very difficult or impossible to know *a priori* which strategy

Algorithm 2 Parallelization Over Data Algorithm

```
activeParticleCount = N
while activeParticleCount > 0 do
    activeParticles = particles that reside in a loaded data block
    inactiveParticles = particles that do not reside in a loaded data block
    while activeParicles not empty do
        advect particle
        if particle terminates then
            activeParticleCount = activeParticleCount - 1
            Communicate activeParticleCount update to all
        end if
        if particle moves to another data block then
            Send particle to process that owns data block
        end if
    end while
    Receive incoming particles
    Receive incoming particle count updates
end while
```

would be most efficient. In general, where smooth flow and complex flow exist within the same data set, it is impossible to achieve perfectly scalable parallel performance using the traditional algorithms.

Research has shown that the most effective solution is a hybrid approach, where parallelization is performed over both seed points, and data blocks. In hybrid approaches, the resource requirements are monitored and the reallocation of resources can be done dynamically, based on the nature of a particular vector field and a seeding scenario. By reallocating as needed, resources can be dynamically used in the most efficient manner possible.

The algorithm introduced in Pugmire et al. [11] is a hybrid between parallelization across particles and data blocks. This hybrid approach dynamically partitions the workload of both particles and data blocks to processors in an attempt to load balance on-the-fly, based on the processor workloads and the nature of the vector field. It attempts to keep all processors busy while also minimizing I/O by choosing either to communicate particles to other processors or to have processors load duplicate data blocks based on heuristics.

Since detailed knowledge of flow is often unpredictable or unknown, the hybrid algorithm was designed to adapt during the computation to concentrate resources where they are needed, distributing particles where needed, and/or duplicating blocks when needed. Workload monitoring and balancing achieves parallelization across data blocks and particles simultaneously, and the algorithm is able to adapt to the challenges posed by the varying the characteristics of integration-based problems.

In the hybrid algorithm, the processors are divided up into sets of work groups. Each work group consists of a single master process, and a group of slave processes. Each master is responsible for monitoring the workload among the group of slaves, and making work assignments that optimize resource utilization. The master makes initial assignments to the slaves based on the initial seed point placement. As work progresses, the master monitors the length of each slave's work queue and the blocks that are loaded. The master reassigns particle computation to balance both slave overload and slave starvation. When the master determines that all streamlines have terminated, it instructs all slaves to terminate.

For scalable performance, work groups coordinate between themselves and work can be moved from one work group to another, as needed. This allows for work groups to focus resources on different portions of the problem, as needed.

The design of the slave process is outlined in Algorithm 3. Each slave continuously advances particles that reside in data blocks that are loaded. Similarly, to the parallelize over seed points algorithm, the data blocks are held in an LRU cache to the extent permitted by the memory. When the slave cannot advance any more particles or is out of work, it sends a status message to the master and waits for further instruction. In order to hide latency, the slave sends the status message right before it advances to the last particle.

Algorithm 3 Slave Process Algorithm

```
while not done do
    if new command from master then
        process command
    end if
    while active particles do
        process particles
    end while
    if state changed then
        send state to master
    end if
end while
```

The master process is significantly more complex, and outlined at a high level in Algorithm 4. The master process is responsible for maintaining information on each slave and managing their workloads. At regular intervals, each slave sends a status message to the master so that it maintains accurate global information. This status message includes: the set of particles owned by each slave, which data blocks those particles currently intersect, which data blocks are currently loaded into memory on that slave, and how many particles are currently being processed. The master enforces load balancing by monitoring the workload of the group of slaves, and deciding dynamically, how work should be shared. This is done using particle reassignment, and redundantly loading data blocks. All communication is performed using non-blocking send and receive commands.

Algorithm 4 Master Process Algorithm

```
get initial particles
while not done do
    if New status from any slave then
        command ← most efficient next action
        Send command to appropriate slave(s)
    end if
    if New status from another master then
        command ← most efficient next action
        Send command to appropriate slave(s)
    end if
    if Work group status changed then
        Send status to other master processes
    end if
    if All work groups done then
        done = true
    end if
end while
```

The master process uses a set of four rules to manage the work in each group. The rules are:

1. *Assign loaded block.* The master sends a particle to a slave that has the required data block already loaded into memory. The slave will add this particle to the active particle list and process the particle.

2. *Assign unloaded block.* The master sends a particle to a slave that does not have the required data block loaded into memory. The slave will load the data block into memory, add the particle to its active particle list, and process the particle. The master process uses this rule to assign the initial work to slaves when particle advection begins.

3. *Load block.* The master sends a message to a slave instructing it to load a data block required by one or more of the assigned particles. When the slave receives this message, the data block is loaded, and the particles requiring this data block are moved into the active particle list.

4. *Send particle.* The master sends a message to a slave instructing it to send particles to another slave. When the slave receives this command, the particle is sent to the receiving slave. The receiving slave will then place this particle in the active particle list.

The master in each work group monitors the load of the set of slaves. Using a set of heuristics [11], the master will use the four commands above to dynamically balance work across the set of slaves. The commands allow particle, and data block assignments for each processor to be changed over the course of the integration calculation.

6.3.5 Algorithm Analysis

The efficiency of the three algorithms are compared by applying each algorithm to several representative data sets and seed point distribution scenarios and by measuring various aspects of performance. Because each algorithm parallelizes differently over the particles and blocks, it is insightful to analyze the performance of the total runtime and also other key metrics that are impacted directly by the parallelization strategy choices.

One overall metric of the analysis is the total runtime or wall-clock time of the algorithm. This metric includes the total time-to-solution of each algorithm, including IC computation, I/O, and communication. Although, this metric alone is not sufficiently fine-grained enough to capture or convey performance characteristics of an algorithm on a given data set with a given initial seed set. The analysis of communication, I/O, and block management gives a more fine-grained insight into the performance efficiency.

Communication is a difficult metric to report and analyze since all communication in these algorithms are asynchronous. However, the time required to post *send* and *receive* operations and associated communication management measures the impact of parallelization that involves communication.

To measure the impact of I/O upon parallelization, time spent reading blocks from a disk is recorded, as well as the number of times blocks are loaded and purged. Because not all the blocks will fit into memory, an LRU cache, with a user-defined upper bound, is implemented to handle block purging. A ratio measures the efficiency of this aspect of the algorithm, which is defined by the *block efficiency*, E. The block's efficiency is the difference between number of blocks loaded and the number of blocks purged, over the number of blocks loaded.

$$E = \frac{B_L - B_P}{B_L}. \tag{6.1}$$

For the astrophysics data set, the wall-time graph in Figure 6.5a demonstrates the relative time performance of the three algorithms, with the hybrid parallelization algorithm demonstrating a better performance than either seed parallelization or data parallelization, for both spatially sparse and dense initial seed points. However, even at 512 processors, the difference between hybrid and data parallelization for the spatially sparse seed point set is only a factor of 3.8, so, in order to understand the parallelization, the analysis must include other metrics.

An examination of total time spent in I/O, as shown in Figure 6.5b, is particularly informative. In this graph, the hybrid algorithm performs very close to the ideal, as exemplified by the data parallelization algorithm. Though seed parallelization performs closely to the hybrid approach from a time point of view, it spends an order of magnitude more time in I/O for both seed point initial conditions.

Next, the block efficiency (see Eq. 6.1), is shown in Figure 6.5c for all three algorithms. The data parallelization performs ideally, loading each block once

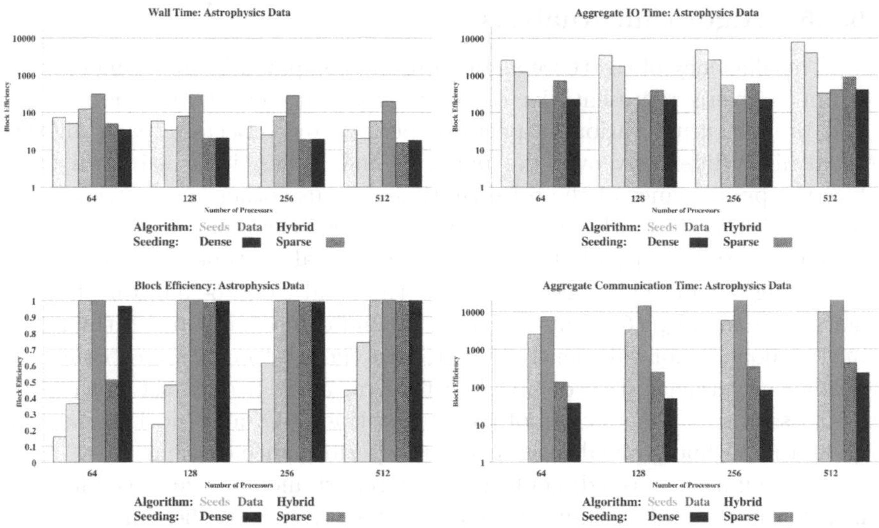

FIGURE 6.5: Graphs showing performance metrics for parallelization over seeds, parallelization over data and hybrid parallelization algorithms on the astrophysics data set. There were 20,000 particles used with both sparse and dense seeding scenarios. Image source: Pugmire et al., 2009 [11].

and never purging. The seed parallelization performs least efficiently, as blocks are loaded and reloaded many times. The performance of the hybrid algorithm is close to ideal for both sparse and dense seeding, which highlights the ability to dynamically direct computational resources, as needed.

It is also useful to consider the time spent in communication, shown in Figure 6.5d. As the seed parallelization algorithm does no communication, times are only shown for data and hybrid parallelization. For sparse seeding, data parallelization performs approximately 20 times more communication than the hybrid algorithm as *IC*s are sent between the processors that own the blocks. This trend remains even as the number of processors is scaled up. For the dense initial condition, the separation increases by another order of magnitude, as data parallelization performs between 165 and 340 times more communication as the processor count increases. This is because the ratio of blocks needed to total blocks decreases and large numbers of *IC*s communicate to the processors that own the blocks.

For the fusion data set, the wall-time graph in Figure 6.6a demonstrates the relative time performance of the three algorithms. In the fusion data set, the nature of the vector field, which rotates within the toroidal containment core, leads to some interesting performance results. Because of this, regardless of seed placement, the *IC*s tend to uniformly fill the interior of the torus. The data parallelization and hybrid algorithms perform almost identically for

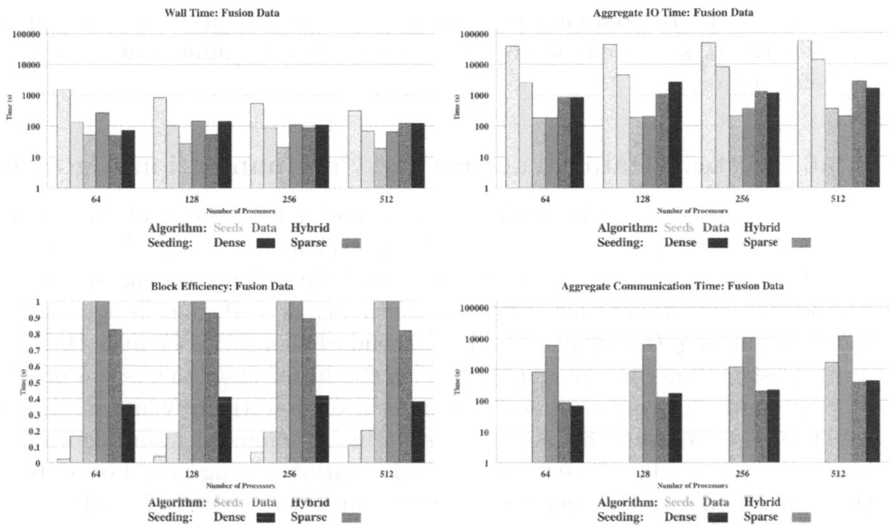

FIGURE 6.6: Graphs showing performance metrics for parallelization over seeds, parallelization over data and hybrid algorithms on the fusion data set. There were 20,000 particles used with both sparse and dense seeding scenarios. Image source: Pugmire et al., 2009 [11].

both seeding conditions. The seed parallelization algorithm performs poorly for spatially sparse seed points, but very competitively for a dense seed points.

In the case of seed parallelization with a dense seeding, good performance is obtained because the working set of active blocks fits inside memory and few blocks must be purged to advance the *IC*s around the toroidal core. An analysis of wall-clock time does not clearly indicate a dominant algorithm between data and hybrid parallelization, so further analysis is warranted. But it is encouraging that the hybrid algorithm can adapt to both initial conditions.

The graph of total I/O time is shown in Figure 6.6b. In both seeding scenarios, as expected, seed parallelization performs more I/O; however, it does not have the communication costs and latency of the other two algorithms and so, for the case of the dense seeding, seed parallelization is able to overcome the I/O penalty to show good performance overall.

The graph of block efficiency is shown in Figure 6.6c. It is interesting to note that the block efficiency of the hybrid algorithm is less than in the astrophysics case study. However, overall performance is still very strong. This indicates that for this particular data set, the best overall performance is found when more blocks are replicated across the set of resources.

The graph of total communication time is shown in Figure 6.6c. For a dense seeding, communication is very high for the data parallelization algorithm. Since the *IC*s tend to be concentrated in an isolated region of the torus, many ICs must be communicated to the block owning processors. For sparse seeding, the ICs are more uniformly distributed and communication costs are therefore lower.

6.3.6 Hybrid Data Structure and Communication Algorithm

Large seed sets and large time-varying vector fields pose additional challenges for parallel IC algorithms. The first challenge is the decomposition of 4D time-varying flow data, in space and time. This section discusses a hybrid data structure that combines both 3D and 4D blocks for computing time-varying ICs efficiently. The second challenge arises in both steady and transient flows but it is magnified in the latter case: this is the problem of performing nearest-neighbor communication iteratively while minimizing synchronization points during the execution. A communication algorithm is introduced, then, for the exchange of information among neighboring blocks that allows tuning the amount of synchronization desired. Together, these techniques are used to compute 1/4 million steady-state and time-varying ICs on vector fields over 1/2 terabyte in size [10].

Peterka et al. [9] introduced a novel 3D/4D hybrid data structure, which is shown in Figure 6.7. It is called "hybrid" because it allows a 4D space–time to be viewed as both a unified 4D structure and a 3D space × 1D time structure. The top row of Figure 6.7 illustrates this idea for individual blocks, and the bottom row shows the same idea for neighborhoods of blocks.

The same data structure is used for both steady-state and time-varying flows, by considering steady-state as a single timestep of the time-varying general case. After all, a particle is a 4D (x, y, z, t) entity. In the left column of Figure 6.7, a block is also a 4D entity, with both minimum and maximum extents in all four dimensions. Each 4D block sits in the center of a 4D neighborhood, surrounded on all sides and corners by other 4D blocks. A single neighborhood consists of 3^4 blocks: the central block and adjacent blocks are in both directions in the x, y, z, and t dimensions.

If the data structure were strictly 4D, and not hybrid, the center and right columns of Figure 6.7 would be unnecessary; but then, all timesteps of a time-varying flow field would be needed at the same time, in order to compute an IC. Modern CFD simulations can produce hundreds or even thousands of timesteps, and the requirement to simultaneously load the entire 4D data set into a main memory would exceed the memory capacity of even the largest HPC systems. At any given time in the particle advection, however, it is only necessary to load data blocks that contain vector fields at the current time and perhaps one timestep in the future. On the other hand, loading multiple timesteps simultaneously often results in a more efficient data access pattern from storage systems and reduced I/O time. Thus, the ability to step through time in configurable-sized chunks, or time windows, results in a flexible and robust algorithm that can run on various architectures with different memory capacities.

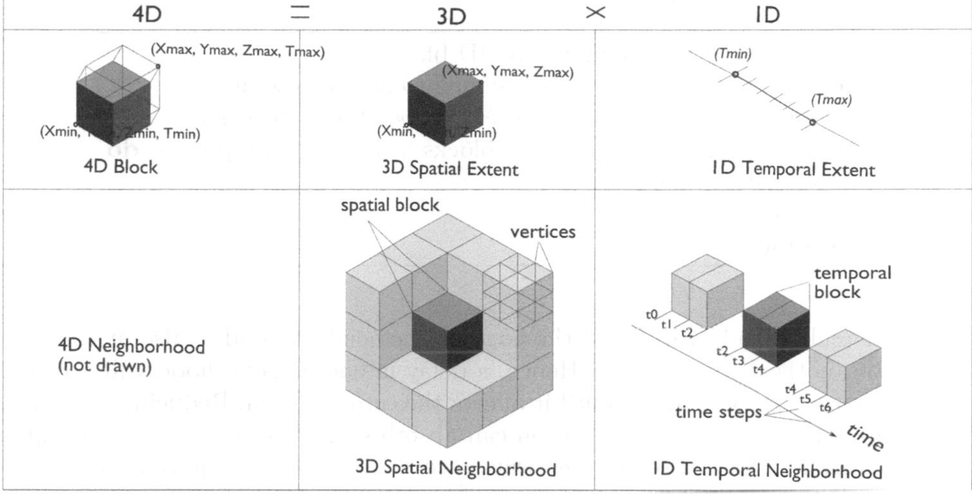

FIGURE 6.7: The data structure is a hybrid of 3D and 4D time–space decomposition. Blocks and neighborhoods of blocks are actually 4D objects. At times, such as when iterating over a sliding time window (or time block) consisting of one or more timesteps, it is more convenient to separate 4D into 3D space and 1D time. Image courtesy of Tom Peterka (ANL).

Even though blocks and neighborhoods are internally unified in space–time, the hybrid data structure enables the user to decompose space–time into the product of space × time. The lower right corner of Figure 6.7 shows the decomposition for a temporal block consisting of a number of individual timesteps. Varying the size of a temporal block—the number of timesteps contained in it—creates an adjustable sliding time window that is also a convenient way to trade the amount of in-core parallelism with out-of-core serialization. Algorithm 5 shows this idea in pseudocode. A decomposition of the domain into 4D blocks is computed given user-provided parameters, specifying the number of blocks in each dimension. The outermost loop then iterates over 1D temporal blocks, while the work in the inner loop is done in the context of 4D spatio-temporal blocks. The hybrid data structure just described, enables this type of combined in-core/out-of-core execution, and the ability to configure block sizes in all four dimensions is how Algorithm 5 can be flexibly tuned to various problem and machine sizes.

All parallel *IC* algorithms for large data and seed sets have one thing in common: the inevitability of interprocess communication. Whether exchanging data blocks or particles, nearest-neighbor communication is unavoidable and limits performance and scalability. Hence, developing an efficient nearest-neighbor communication algorithm is crucial. The difficulty of this communication stems from the fact that neighborhoods, or groups in which communication occurs, overlap. In other words, blocks are members of more than one

Algorithm 5 Main Loop

```
decompose entire domain into 4D blocks
for all 1D temporal blocks assigned to my process do
    read corresponding 4D spatio-temporal data blocks into memory
    for all 4D spatio-temporal blocks assigned to my process do
        advect particles
    end for
end for
```

neighborhood: a block at the edge of one neighborhood is also at the center of another, and so forth. Hence, a delay in one neighborhood will propagate to another, and so on, until it affects the entire system. Reducing the amount of synchronous communication can absorb some of these delays and improve overall performance and one way to relax such timing requirements is to use nonblocking message passing.

In message-passing systems like MPI, nonblocking communication can be confusing. Users often have the misconception that communication automatically happens in the background, but in fact, messages are not guaranteed to make progress without periodic testing of their status. After nonblocking communication is initiated, control flow is returned to the caller and hopefully, some communication occurs while the caller executes other code, but to ensure this happens, the caller must periodically check on the communication and perhaps wait for it to complete.

The question then becomes how frequently one should check back on nonblocking messages, and how long to wait during each return visit. An efficient communication algorithm that answers these questions was introduced by Peterka et al. [10] and is outlined in Algorithm 6. The algorithm works under the assumption that nearest-neighbor communication occurs in the context of alternating rounds between the advection and communication of particles. In each round, new communication is posted and previously posted messages are

Algorithm 6 Asynchronous Communication Algorithm

```
for all processes in my neighborhood do
    pack message of block IDs and particle counts
    post nonblocking send
    pack message of particles
    post nonblocking send
    post nonblocking receive for IDs and counts
end for
wait for enough IDs and counts to arrive
for all IDs and counts that arrived do
    post blocking receive for particles
end for
```

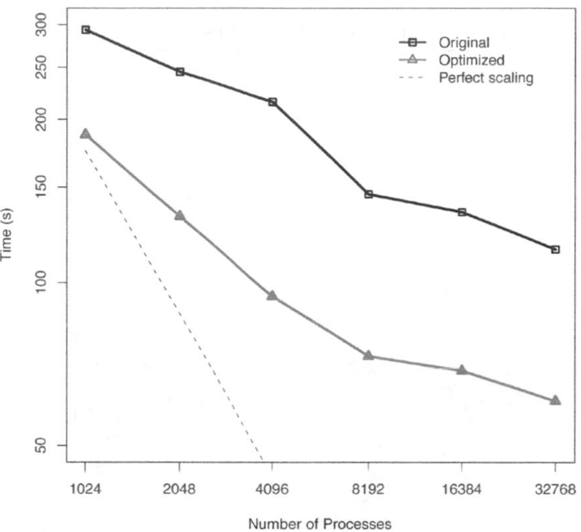

FIGURE 6.8: The cumulative effect on end-to-end execution time of a hybrid data structure and an adjustable communication algorithm is shown for a benchmark test of tracing 256K particles in a vector field that is 2048 × 2048 × 2048.

checked for progress. Each communication consists of two messages: a header message containing block identification and particle counts, and a payload message containing the actual particle positions.

The algorithm takes an input parameter that controls the fraction of previously posted messages for which to wait in each round. In this way, the desired amount of synchrony/asynchrony can be adjusted and allows the "dialing down" of synchronization to the minimum needed to make progress. In practice, waiting for only 10% of the pending messages to arrive in each round is the best setting. This way, each iteration makes a guaranteed minimum amount of communication progress, without imposing excessive synchronization. Reducing communication synchronization accelerates the overall particle advection performance and is an important technique for communicating across large-scale machines where global code synchronization becomes more costly as the number of processes increases.

The collective effect of these improvements is a 2× speedup in overall execution time, compared to earlier algorithms. This improvement is demonstrated in Figure 6.8 with a benchmark test of tracing 1/4 million particles in a steady state thermal hydraulics flow field that is 2048^3 in size. Peterka et al. [10] benchmarked this latest fully-optimized algorithm using scientific data sets from problems in thermal hydraulics, hydrodynamics (Rayleigh–Taylor instability), and combustion in a cross-flow. Table 6.1 shows the results on

TABLE 6.1: Parallel *IC* performance benchmarks.

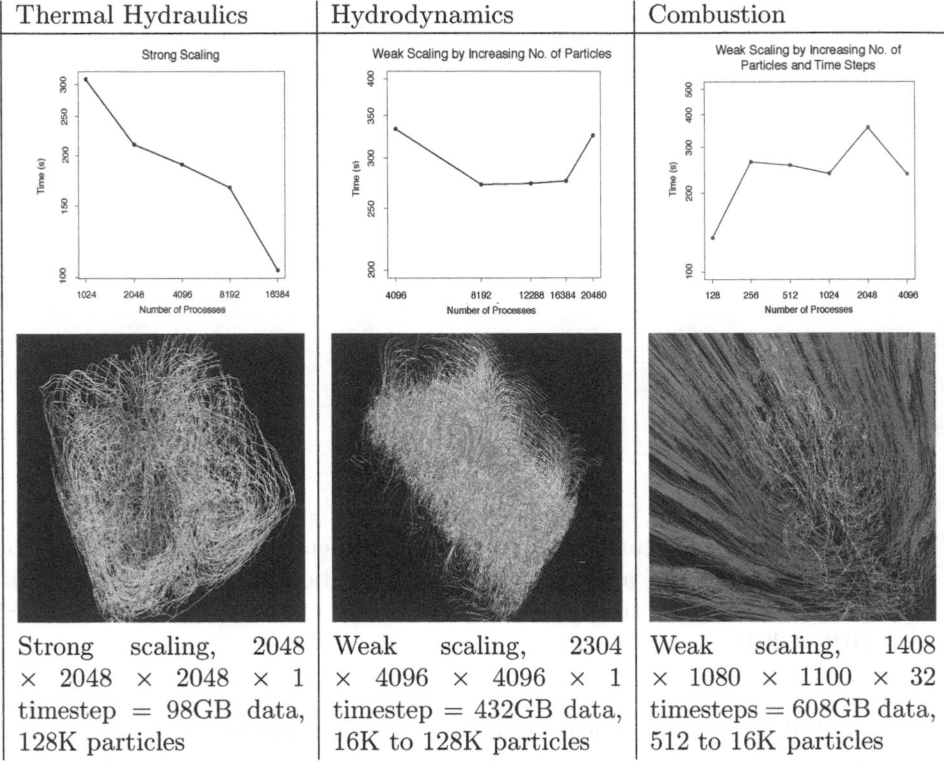

Thermal Hydraulics	Hydrodynamics	Combustion
Strong scaling, 2048 × 2048 × 2048 × 1 timestep = 98GB data, 128K particles	Weak scaling, 2304 × 4096 × 4096 × 1 timestep = 432GB data, 16K to 128K particles	Weak scaling, 1408 × 1080 × 1100 × 32 timesteps = 608GB data, 512 to 16K particles

both strong and weak scaling tests of steady and unsteady flow fields. The left-hand column of Table 6.1 shows good strong scaling (where the problem size remains constant) for the steady-state case, seen by the steep downward slope of the scaling curve. The center column of Table 6.1 also shows good weak scaling (where the problem size increases with process count), with a flat overall shape of the curve. The right-hand column of Table 6.1 shows a weak scaling curve that slopes upward for the time-varying case and it demonstrates that even after optimizing the communication, the I/O time to read each time block from storage remains a bottleneck.

Advances in hybrid data structures and efficient communication algorithms can enable scientists to trace particles of time-varying flows during their simulations of CFD vector fields at a concurrency that is comparable with their parallel computations. Such highly parallel algorithms are particularly important for peta- and exascale computing, where more analyses will be needed to execute at simulation runtime, in order to avoid data movement and avail full-resolution data at every time step. Access to high-frequency data that are only available *in situ* is necessary for accurate *IC* computation; typically, simulation checkpoints are not saved frequently enough for accurate time-varying flow visualization.

Besides the need to visualize large data sizes, the ability to trace a large number of particles is also a valuable asset. While hundreds of thousands or millions of particles may be too many for human viewing, a very dense field of streamlines or pathlines may be necessary for accurate follow-on analysis. The accuracy of techniques such as identifying Lagrangian coherent structures and querying geometric features of field lines relies on a dense field of particle traces.

6.4 Conclusion

IC methods are a foundational technique that have proven to be a valuable tool in the analysis and understanding of complex flow with scientific simulations. These techniques provide a powerful framework for a variety of specific analyses, including streamlines, pathlines, streak lines, stream surfaces, as well as advanced Lagrangian techniques, such as Finite Time Lyapunov Exponents (FTLE), and Lagrangian Cohrent Structures.

Due to the complexities of *IC* calculation in a parallel, distributed memory setting, as outlined in 6.2, it is clear that there is no optimal algorithm suitable for all types of vector fields, seeding scenarios, and data set sizes. The complexities of parallel *IC* computations are likely to stress the entire system, including compute, memory, communication, and I/O. This chapter outlined several different algorithms that showed a good scalability on large computing resources, on very large data sets, that were built on the standard algorithms for parallelization across both seeds and data. These include the dynamic workload monitoring of both particles and data blocks and the rebalancing of work among the available processors, efficient data structures, and techniques for maximizing the efficiency of communication between processors.

References

[1] David Camp, Hank Childs, Amit Chourasia, Christoph Garth, and Kenneth I. Joy. Evaluating the Benefits of An Extended Memory Hierarchy for Parallel Streamline Algorithms. In *Proceedings of the IEEE Symposium on Large-Scale Data Analysis and Visualization (LDAV)*, pages 57–64. IEEE Press, October 2011.

[2] C. Cardall, A. Razoumov, E. Endeve, E. Lentz, and A. Mezzacappa. Toward Five-Dimensional Core-Collapse Supernova Simulations. *Journal of Physics: Conference Series*, 16:390–394, 2005.

[3] E. Endeve, C. Y. Cardall, R. D. Budiardja, and A. Mezzacappa. Generation of Strong Magnetic Fields in Axisymmetry by the Stationary Accretion Shock Instability. *ArXiv e-prints*, November 2008.

[4] E. Hairer, S. P. Nørsett, and G. Wanner. *Solving Ordinary Differential Equations I, Second Edition*, volume 8 of *Springer Series in Comput. Mathematics*. Springer-Verlag, 1993.

[5] G Haller. Lagrangian Structures and the Rate of Strain in a Partition of Two-Dimensional Turbulence. *Physics of Fluids*, 13(11):3365, 2001.

[6] Hari Krishnan, Christoph Garth, and Kenneth I. Joy. Time and Streak Surfaces for Flow Visualization in Large Time-Varying Data Sets. 15(6):1267–1274, October 2009.

[7] Francois Lekien, Shawn C. Shadden, and Jerrold E. Marsden. Lagrangian Coherent Structures in N-Dimensional Systems. 48(6):065404, 2007.

[8] Tony McLoughlin, Robert S. Laramee, Ronald Peikert, Frits H. Post, and Min Chen. Over Two Decades of Integration-Based, Geometric Flow Visualization. In *EuroGraphics 2009 - State of the Art Reports*, pages 73–92. Eurographics Association, April 2009.

[9] Tom Peterka, Wesley Kendall, David Goodell, Boonthanome Nouanesengsey, Han-Wei Shen, Jian Huang, Kenneth Moreland, Rajeev Thakur, and Robert Ross. Performance of Communication Patterns for Extreme-Scale Analysis and Visualization. *Journal of Physics, Conference Series*, pages 194–198, July 2010.

[10] Tom Peterka, Robert Ross, Boonthanome Nouanesengsey, Teng-Yok Lee, Han-Wei Shen, Wesley Kendall, and Jian Huang. A Study of Parallel Particle Tracing for Steady-State and Time-Varying Flow Fields. In *Proceedings of IPDPS 11*, pages 577–588, Anchorage AK, 2011.

[11] Dave Pugmire, Hank Childs, Christoph Garth, Sean Ahern, and Gunther H. Weber. Scalable Computation of Streamlines on Very Large Datasets. In *Proceedings of Supercomputing (SC09)*, Portland, OR, USA, November 2009.

[12] C.R. Sovinec, A.H. Glasser, T.A. Gianakon, D.C. Barnes, R.A. Nebel, S.E. Kruger, S.J. Plimpton, A. Tarditi, M.S. Chu, and the NIMROD Team. Nonlinear Magnetohydrodynamics with High-order Finite Elements. *Journal of Computational Physics*, 195:355, 2004.

[13] Hongfeng Yu, Chaoli Wang, and Kwan-Liu Ma. Parallel Hierarchical Visualization of Large Time-Varying 3D Vector Fields. In *Proceedings of Supercomputing (SC07)*, 2007.

Part II

Advanced Processing Techniques

Chapter 7

Query-Driven Visualization and Analysis

Oliver Rübel

Lawrence Berkeley National Laboratory

E. Wes Bethel

Lawrence Berkeley National Laboratory

Prabhat

Lawrence Berkeley National Laboratory

Kesheng Wu

Lawrence Berkeley National Laboratory

7.1	Introduction	118
7.2	Data Subsetting and Performance	119
	7.2.1 Bitmap Indexing	120
	7.2.2 Data Interfaces	122
7.3	Formulating Multivariate Queries	124
	7.3.1 Parallel Coordinates Multivariate Query Interface	125
	7.3.2 Segmenting Query Results	127
7.4	Applications of Query-Driven Visualization	129
	7.4.1 Applications in Forensic Cybersecurity	129
	7.4.2 Applications in High Energy Physics	132
	7.4.2.1 Linear Particle Accelerator	133
	7.4.2.2 Laser Plasma Particle Accelerator	135
7.5	Conclusion	139
	References	140

This chapter focuses on an approach to high performance visualization and analysis, termed *query-driven visualization and analysis* (QDV). QDV aims to reduce the amount of data that needs to be processed by the visualization, analysis, and rendering pipelines. The goal of the data reduction process is to separate out data that is "scientifically interesting" and to focus visualization, analysis, and rendering on that interesting subset. The premise is that for any given visualization or analysis task, the data subset of interest is much smaller than the larger, complete data set. This strategy—extracting smaller data subsets of interest and focusing of the visualization processing on these subsets—is complementary to the approach of increasing the capacity of the visualization, analysis, and rendering pipelines through parallelism, which is the subject of many other chapters in this book. This chapter discusses the fundamental concepts in QDV, their relationship to different stages in the visualization and analysis pipelines, and presents QDV's application to problems in diverse areas, ranging from forensic cybersecurity to high energy physics.

7.1 Introduction

Query-driven visualization and analysis (QDV) addresses the big-data analysis challenge by focusing on data that is "scientifically interesting" and, hence, reducing the amount of data that needs to be processed by the visualization, analysis, and rendering pipelines. The data reduction is typically based on range queries, which are used to constrain the data variables of interest. The term QDV was coined by Stockinger et al. [31, 30] to describe the combination of high performance index/query methods with visual data exploration methods. Using this approach, the complexity of the visualization, analysis and rendering is reduced to $O(k)$, with k being the number of objects retrieved by the query. The premise of QDV, then, is that for any given visualization or analysis task, the data subset of interest, k, is much smaller than the larger, complete data set. QDV methods are among a small subset of techniques that are able to address both large *and* highly complex data.

In a typical QDV task, the user begins the analysis by forming definitions for data subsets of interest. This characterization is done through the concept of range queries, that is, the user defines a set of constraints for variables of interest. For example, in the context of a combustion simulation, a scientist may only be interested in regions of a particular temperature t and pressure p, such that (1000F$< t <$ 1500F) AND ($p <$ 800Pa). QDV uses such range queries to filter data before it is passed to the subsequent data processing pipelines, thereby, focusing the visualization and analysis exclusively on the data that is meaningful to the user. In practice, a user may not have a precise notion of which parts of the data are of the most interest. Rather than specifying the exact data subset(s) of interest, a user may begin the analysis by excluding large data portions that are known to be of no interest and/or specifying regions that are known to contain the data of interest. Through a process of iterative refinement of queries and analysis of query results, the user is able to gain further insight into the data and identify and specify more precisely the data subset(s) of interest.

QDV inherently relies on two main technologies: fast methods for evaluating data queries; and efficient methods for specification, validation, visualization, and analysis of query results. In the context of QDV of scientific data, approaches based on bitmap indexing have shown to be very effective for accelerating multivariate data queries, which will be the focus of 7.2. Then, 7.3 discusses different methods used for specification, validation, and visualization of data queries, with a particular focus on parallel coordinates as an effective interface for formulating multivariate data queries (see 7.3.1). To facilitate the query-based data analysis process, the QDV approach has also been combined with automated analysis methods to suggest further query refinements, as well as to automate the definition of queries and extraction of features of interest. As an example, 7.3.2 discusses various methods for the segmentation and labeling of query components. To illustrate the effectiveness and wide applicability of QDV methods, two case studies are investigated: the applications of QDV to network traffic data in 7.4.1, and particle accelerator simulation data in 7.4.2. This chapter concludes with an evaluation of the QDV approach and a discussion of future challenges in 7.5.

7.2 Data Subsetting and Performance

Efficient methods for data subsetting are fundamental to the concept of QDV. The database community has developed a variety of indexing methods to accelerate data searches. Many popular database systems use variations of the B-tree [2]. The B-tree was designed for transaction-type applications, exemplified by interactions between a bank and its customer. Interactions with transactional data are characterized by the frequent modification of data records, one record at a time, and retrieval of a relatively small number of data records, such as the look-up of a customer's banking information. In contrast, scientific data is typically not modified after creation, but is processed in a read-only fashion. Furthermore, scientific search operations typically retrieve many more data records than typical transaction-type queries, and the number of records may also vary considerably more. For example, when studying the ignition process in a combustion data set, a query that isolates regions of high temperature may initially only retrieve very few data records; however, as the combustion process progresses from an initial spark into a large flame, the same query may result in the retrieval of a sizable fraction of the data. Scientific search operations also often involve a large number of data dimensions. For example, when analyzing the combustion process, a scientist may be interested in regions that exhibit high temperature, high pressure, and contain a high/low concentration of different chemical species. For these types of versatile, high-dimensional, read-only query operations, the bitmap index is a more appropriate indexing structure [28], which is the subject of 7.2.1.

There is a pressing need for efficient data interfaces that make indexing technology accessible to the scientific data processing pipelines. Across scientific applications, array-based data models are most commonly used for data storage. However, common array-based scientific data formats, such as HDF5 [32] and NetCDF [33], do not support semantic indexing methods. A number of array-based research databases, such as SciDB [6] or MonetDB [4], provide advanced index/query technology, are optimized for scientific data. However, scientific data is commonly stored outside of a database system. In practice, it is often too expensive to make copies of extremely large scientific data sets for the purpose of indexing, and the conversion of data to different file formats may break downstream data processing pipelines. Query-based analyses, hence, require efficient data interfaces that make semantic indexing methods accessible within state-of-the-art scientific data formats (see 7.2.2).

The first section, 7.2.1, discusses bitmap indexing in detail. Then, in 7.2.2, FastQuery [9] is introduced as an example index and query system for integrating bitmap indexing technology with state-of-the-art scientific data formats, including extensions of bitmap indexing to the processing of queries on large distributed-memory platforms.

7.2.1 Bitmap Indexing

A bitmap index stores the bulk of its information in a series of bit sequences known as bitmaps. Figure 7.1 shows a small example with one variable named **I**, with a value range of 0, 1, 2, or 3. For each of the four possible values, a separate bitmap is created to record which rows, or records, contain the corresponding value. For example, the value 0 only appears in row 1. The bitmap representing the value 0, hence, contains a 1, as the first bit, while the remaining bits are 0. This is the most basic form of a bitmap index [22].

A scientific data set often involves many variables, and a query may involve an arbitrary combination of the variables. When the B-Tree index [10] or a multidimensional index [13] is used, the most efficient way to answer a query is to create an index with the variables involved in the query. This approach requires a different index for each combination of variables. Because the number of possible combinations is large, a large amount of disk space is needed to store the different indices. These combinations are necessary because the results from these tree-based indices cannot be easily combined to produce the final answer. In contrast, when using a bitmap index to answer a query, a bitmap is produced to represent the answer. A multivariate query can be answered by resolving the condition on each variable separately, using the bitmap index for that variable, and then combining all the intermediate answers with bitwise logical operations to produce the final answer. In this case, the total size of all bitmap indices grows linearly with the number of variables in the data. Typically, the bitwise logical operations are well-supported by computer hardware. Therefore, the bitmap indices can answer queries efficiently.

The variable **I** shown in Figure 7.1 has only four possible values. In general, for a variable with C distinct values, the basic bitmap index needs C bitmaps of N bits each, where N is the number of rows (or records) in the data set [8]. If C is the cardinality of the variable, then in a worst-case scenario, $C = N$, and the corresponding bitmap index has N^2 bits. Even for a moderate size data set, N^2 bits can easily fill up all disks of a large computer system. Ideally,

RID	I	Bitmap Index			
		=0	=1	=2	=3
1	0	1	0	0	0
2	1	0	1	0	0
3	2	0	0	1	0
4	3	0	0	0	1
5	3	0	0	0	1
6	3	0	0	0	1
7	3	0	0	0	1
8	1	0	1	0	0
		b_1	b_2	b_3	b_4

FIGURE 7.1: A sample bitmap index where RID is the record ID and **I** is the integer attribute with values in the range of 0 to 3.

the index size should be proportional to N, instead of N^2. The techniques for controlling the bitmap index sizes roughly fall in three categories: compression, encoding, and binning.

Compression: In principle, any lossless compression method can be used to compress bitmap indices. In practice, the most efficient bitmap compression methods are based on run-length encoding, as they support bitwise logical operations on the compressed data without decompressing the bitmaps [1, 37]. Working directly with the compressed bitmaps reduces the time needed for reading the bitmaps from disk to memory, as well as the amount of time to perform the necessary computations. Furthermore, using Word-Aligned Hybrid (WAH) compression, it was shown that the time needed to evaluate a query is, in the worst case, a linear function of the number of records h that satisfy the query [38]. Therefore, the WAH compressed bitmap index has an optimal computational complexity of $O(h)$. In the context of QDV, this means that the computational complexity of the analysis no longer depends on the total of number of data records, N, but only on the number of records, h, that define the feature(s) of interest defined by the query.

Encoding: The basic bitmap encoding used in Figure 7.1 is also called *equality-encoding*, as it shows the best performance for *equality queries*, such as temperature, t = 100F. Other well-known bitmap encoding methods include the *range encoding* [7] and the *interval encoding* [8]. They are better suited for other query types, such as (35.8Pa $< p <$ 56.7Pa). There are also a number of different ways of composing more complex bitmap indices by using combinations of the equality, range, and interval encoding [40].

Binning: The basic bitmap index works well for low-cardinality attributes, such as "gender," "types of cars sold per month," or "airplane models produced by Airbus and Boeing." However, for scientific data, the variables often have a large value-range, while different values are hardly ever repeated. These types of scientific data are referred to as high-cardinality data. A bitmap index for a high-cardinality variable typically requires many bitmaps. One can reduce the number of bitmaps needed by binning the data. However, the corresponding index would not be able to answer some queries accurately. The cost of resolving the query accurately could be quite high [25]. Additional data structures might be needed to answer the queries with predictable performance [42].

The bitmaps inside the bitmap indices also offer a way to count the number of records satisfying certain conditions quickly. This feature can be used to quickly compute *conditional histograms.* For example, in addition to directly counting from the bitmaps, it is also possible to count the rows in each histogram bucket by reading the relevant raw data, or use a combination of bitmaps and the raw data. Stockinger et al. [29] described a set of algorithms for efficient parallel computation of conditional histograms. The ability to quickly compute conditional histograms, in turn, accelerates many histogram-based visualization methods, such as histogram-based parallel coordinates described later in 7.3.1. Figure 7.2a illustrates the serial performance for computing conditional histograms, using bitmap indexing by way of the FastBit [36] software.

Histogram-based analysis methods—such as density-based segmentation and feature detection methods—require the ability to evaluate bin-queries efficiently [26]. A *bin-query* extracts the data associated with a set of histogram bins and is comprised of a series of queries, for example, in the 3D case, of the form $[(x_i \leq x < x_{i+1}) AND (y_i \leq y < y_{i+1}) AND (z_i \leq z < z_{i+1})]$, where i indicates the index of a selected bin and x, y, and z refer to the dimensions of the histogram. Instead of evaluating complex bin-queries explicitly, one can use bitmaps—one for each nonzero histogram bin—to efficiently store the inverse mapping from histogram-bins to the data. Figure 7.2b illustrates the performance advantage of this approach for evaluating bin-queries. In practice, the overhead for computing the per bin bitmaps depends on the number of nonzero bins, but is, in general, moderate.

A number of the above described methods have been implemented in an open-source software called FastBit [36]. In the context of QDV, FastBit is used to process range queries and equality queries, as well as to compute conditional histograms and bin-queries. FastBit has also been integrated with the parallel visualization system VisIt, described later in Chapter 16, making FastBit-based QDV capabilities available to the user community.

7.2.2 Data Interfaces

In order to make effective use of semantic indexing methods to accelerate data subselection, advanced data interfaces are needed that make index and query methods accessible within state-of-the-art scientific data formats. Such interfaces should ideally have the following characteristics. First, they enable access to a large range of scientific data formats. Second, they avoid costly data copy and file conversion operations for indexing purposes. Third, they

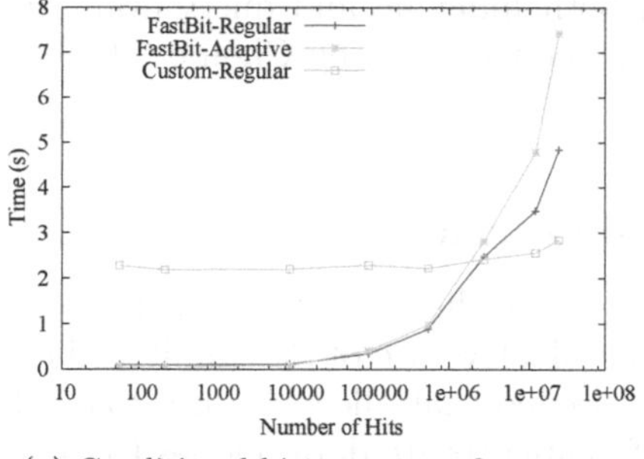

(a) Conditional histogram performance.

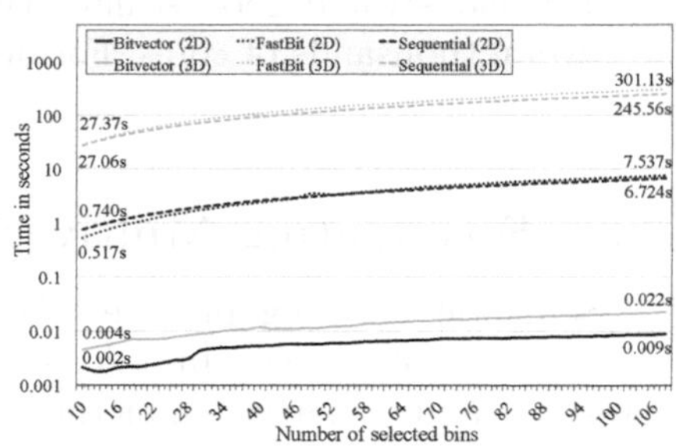

(b) Bin-query performance.

FIGURE 7.2: On the left, (a) shows timings for serial computation of regularly and adaptively binned 2D histograms on a 3D ($\approx 90 \times 10^6$ particles) particle data set using bitmap indexing and a baseline sequential scan method. Image source: Rübel et al. 2008 [27]. Timings for serial evaluation of 3D bin-queries are shown in (b), using a 2D ($\approx 2.4 \times 10^6$ particles) and 3D ($\approx 90 \times 10^6$ particles) particle data set. Bin queries are evaluated using: (1) per-bin bitvectors returned by FastBit, (2) FastBit queries, and (3) a baseline sequential scan method. Image source: Rübel et al., 2010 [26].

scale to massive data sets. Fourth, they are capable of performing indexing and query operations on large, distributed, multicore platforms. For example, Gosink et al. [16] described HDF5-FastQuery, a library that integrates serial index/query operations using FastBit with HDF5.

Recently, Chou et al. introduced FastQuery [9], which integrates parallel-capable index/query operations using FastBit—including bitmap index computation, storage, and data subset selection—with a variety of array-based data formats. FastQuery provides a simple array-based I/O interface to the data. Access to the data is then performed via data format specific readers that implement the FastQuery I/O interface. The system is, in this way, agnostic to the underlying data format and can be easily extended to support new data formats. FastQuery also uses FastBit for index/query operations. To enable parallel index/query operations, FastQuery uses the concept of subarrays. Similar to Fortran and other programming languages, subarrays are specified using the general form of *lower : upper : stride*. Usage of subarrays provides added flexibility in the data analysis, but more importantly—since subarrays are, by definition, smaller than the complete data set—index/query times can be greatly reduced compared to approaches that are constrained to processing the entire data set. This subarray feature, furthermore, enables FastQuery to divide the data into chunks during index creation, chunks that can be processed in parallel in a distributed-memory environment. Evaluation of data queries is then parallelized in a similar fashion. In addition to subarrays, FastQuery also supports parallelism across files and data variables. Chou et al. demonstrated good scalability of both indexing and query evaluation, to several thousands of cores. For details, see the work by Chou et al. [9].

7.3 Formulating Multivariate Queries

Semantic indexing provides the user with the ability to quickly locate data subsets of interest. In order to make effective use of this ability, efficient interfaces and visualization methods are needed that allow the user to quickly identify data portions of interest, specify multivariate data queries to extract the relevant data, and validate query results. As mentioned earlier, a QDV-based analysis is typically performed in a process of iterative refinement of queries and analysis of query results. To effectively support such an iterative workflow, the query interface and visualization should provide the user with feedback on possible strategies to refine and improve the query specification, and be efficient to provide the user with fast, in-time feedback about query-results, in particular, within the context of large data.

Scientific visualization is very effective for the analysis of physical phenomena and plays an important role in the context of QDV for the validation of query results. Highlighting query results in scientific visualizations provides an effective means for the analysis of spatial structures and distributions of the selected data portions. However, scientific visualization methods are limited with respect to the visualization of high-dimensional variable space in that only a limited number of data dimensions can be visualized at once. Scientific visualizations, hence, play only a limited role as interfaces for formulating multivariate queries.

On the other hand, information visualization methods—such as scatterplot matrix and parallel coordinate plots—are very effective for the visualization and exploration of high-dimensional variable spaces and the analysis of relationships between different data dimensions. In the context of QDV, information visualizations, therefore, play a key role as interfaces for the specification of complex, multidimensional queries and the validation of query results.

Both scientific and information visualization play an important role in QDV. In the context of QDV, multiple scientific and information visualization views—each highlighting different aspects of the data—are, therefore, often linked to highlight the same data subsets (queries) in a well-defined manner to facilitate effective coordination between the views. In literature, this design pattern is often referred to as brushing and linking [34]. Using multiple views allows the user to analyze different data aspects without being overwhelmed by the high dimensionality of the data.

To ease validation and refinement of data queries, automated analysis methods may be used for the post-processing of query results to, for example, segment and label the distinct spatial components of a query. Information derived through the post-processing of query results provides important means to enhance the visualization, to help suggest further query refinements, and to automate the definition of queries to extract features of interest.

The next sections describe the use of parallel coordinates as an effective interface to formulate high-dimensional data queries (see 7.3.1). Afterwards, the post-processing of query results are discussed, enhancing the QDV-based analysis through the use of automated methods for the segmentation of query results and methods for investigation of the importance of variables to the query solution and their interactions (see 7.3.2).

7.3.1 Parallel Coordinates Multivariate Query Interface

Parallel coordinates are a common information visualization technique [18]. Each data variable is represented by a vertical axis in the plot (see Fig. 7.3). A parallel coordinates plot is constructed by drawing a polyline connecting the points where a data record's variable values intersect each axis.

Parallel coordinates provide a very effective interface for defining multi-dimensional range queries. Using sliders attached to each parallel axis of the plot, a user defines range thresholds in each displayed data dimension. Selection is performed iteratively by defining and refining thresholds one axis at a time. By rendering the user-selected data subset (the focus view) in front of the parallel coordinates plot created from the entire data set—or a subset of the data defined using a previous query—(the context view), the user receives immediate feedback about general properties of the selection (see Fig. 7.7). Data outliers stand out visually as single or small groups of lines diverging from the main data trends. Data trends appear as dense groups of lines (or bright colored bins, in the case of histogram-based parallel coordinates). A comparison of the focus and context view helps to convey understanding about similarities and differences between the two views. Analysis of data queries defined in parallel coordinates, using additional linked visualizations—such as physical views of the data—then provides additional information about the structure of a query. These various forms of visual query feedback help to validate and refine query-based selections. Figure 7.7 exemplifies the use of parallel coordinates for the interactive query-driven analysis of laser plasma particle acceleration data discussed in more detail later in 7.4.2.2.

Parallel coordinates also support iterative, multiresolution exploration of data at multiple time-scales. For example, initially a user may view data on a weekly scale. After selecting a week(s) of interest, this initial selection may be used as context view. By scaling the parallel axis to show only the data ranges covered by the context selection—also called dimensional scaling—the selection can be viewed and refined in greater detail at daily resolution.

In practice, parallel coordinates have disadvantages when applied to very large data. In the traditional approach, the parallel coordinates plot is rendered by drawing a polyline for each data record. When there are relatively few data records, this approach is reasonable and produces legible results. But when applied to large data sets, the plot can be quickly cluttered and difficult

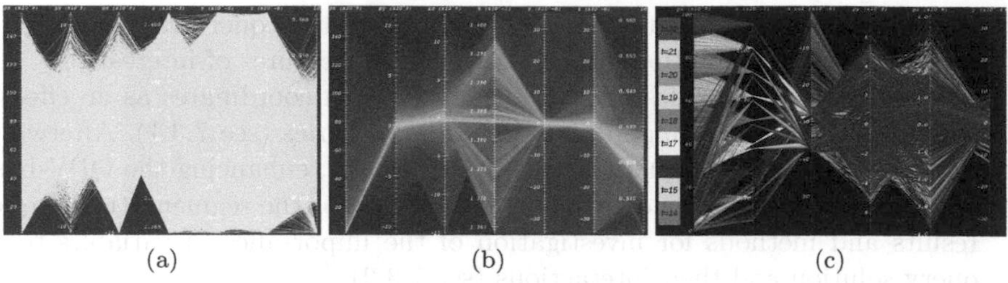

(a) (b) (c)

FIGURE 7.3: The left images, (a) and (b), show a comparison of two different parallel coordinate renderings of a particle data set consisting of 256,463 data records and 7 variables using: (a) traditional line-based parallel coordinates and (b) high-resolution, histogram-based parallel coordinates with 700 bins per data dimension. The histogram-based rendering reveals many more details when displaying large numbers of data records. Image (c) shows the temporal histogram-based parallel coordinates of two particle beams in a laser-plasma accelerator data set, at timesteps $t = [14, 22]$. Color is used to indicate the discrete timesteps. The two different beams can be readily identified in x (second axis). Differences in the acceleration can be clearly seen in the momentum in the x direction, px (first axis). Image source: Rübel et. al, 2008 [27].

to interpret, as is the case in Figure 7.3a. Worst of all, the computational and rendering complexity of parallel coordinates is proportional to the size of the data set. As data sizes grow, these problems quickly become intractable.

Histogram-based parallel coordinates [27, 21] are an efficient, density-based method for computing and rendering parallel coordinates plots. Rather than viewing the parallel coordinates plot as a collection of polylines (one polyline per data record), one can discretize the relationship of all data records between pairs of parallel axes using 2D histograms.

Based on the 2D histograms—one per neighboring axes pair—rendering proceeds by drawing one quadrilateral per nonempty bin, where each quadrilateral connects two data ranges between the neighboring axes. Quadrilateral color and opacity is a function of histogram bin magnitude, so more densely populated regions are visually differentiated from regions with lower density. This approach has a significant advantage for large data applications: the rendering complexity no longer depends on the size of the original data, but only on the resolution of the underlying histograms. The histograms, for the context view, need to be computed only once. The calculation of the focus view histograms happens in response to user changes to query ranges and are accelerated by FastBit. Previous studies examine the scalability characteristics of those algorithms on large supercomputers when applied to very large scientific data sets [27, 29, 3].

Figure 7.3b shows how more information in data is revealed through a combination of visualization and rendering principles and techniques. For example, color brightness or transparency can convey the number of records per bin, and rendering order takes into account region density so that important regions are not hidden by occlusion. Different colors could be assigned to different timesteps, producing a temporal parallel coordinates plot (see Fig. 7.3c). This approach is useful in helping to reveal multivariate trends and outliers across large, time-varying data sets.

7.3.2 Segmenting Query Results

A query describes a binary classification of the data, based on whether a record satisfies the query condition(s). However, typically, the feature(s) of interest to the end users are not individual data records, but regions of space defined by connected components of a query—such as ignition kernels or flame fronts. Besides physical components of a query, a feature of interest may also be defined by groups of records (clusters) in high-dimensional variable space. Combining QDV with methods for the segmentation and classification of query results can facilitate the analysis of large data sets by supporting the identification of subfeatures of a query, by suggesting strategies for the refinement of queries, or by automating the definition of complex queries for automatic feature detection.

A common approach for enhancing the QDV analysis process consists of identifying spatially connected components in the query results. This has been accomplished in the past using a technique known as connected component labeling [35, 39]. Information from connected component labels provides the means to enhance the visualization—as shown in Figure 7.4b—and enables further quantitative analysis. For example, statistical analysis of the number and distributions in size or volume of physical components of query results can provide valuable information about the state and evolution of dynamic physical processes, such as a flame [5].

Connected component analysis for QDV has a wide range of applications. Stockinger et al. [31] applied this approach to combustion simulation data. In this context, a researcher might be interested in finding ignition kernels, or regions of extinction. On the other hand, when studying the stability of magnetic confinement for fusion, a researcher might be interested in regions with high electric potential because of their association with zonal flows, critical to the stability of the magnetic confinement—as shown in Figure 7.4a [41].

In practice, connected component analysis is mainly useful for data with known topology, in particular, data defined on regular meshes. For scattered data—such as particle data—where no connectivity is given, density-based clustering approaches may provide a useful alternative to group selected particles based on their spatial distribution. One of the main limitations of a connected component analysis, in the context of QDV, is that the labeling of components by itself does not yield any direct feedback about possible strategies for the refinement of data queries.

To address this problem, Gosink et al. [17] proposed the use of multivariate statistics to support the exploration of the solution space of data queries. To

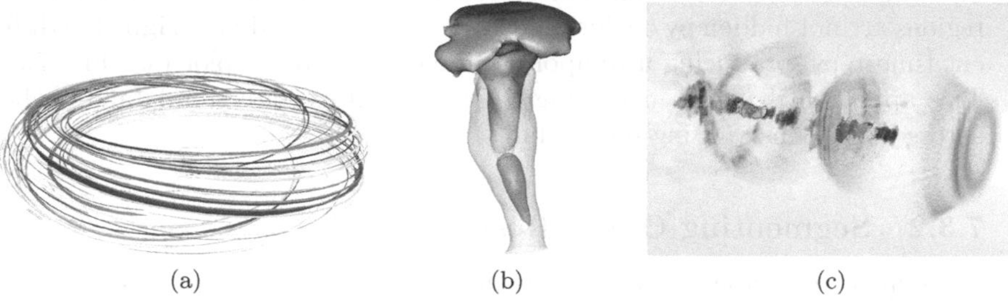

(a) (b) (c)

FIGURE 7.4: Applications of segmentation of query results. Image (a) displays the magnetic confinement fusion visualization showing regions of high magnetic potential colored by their connected component label. Image source: Wu et. al, 2011 [41]. Image (b) shows the query selecting the eye of a hurricane. Multivariate statistics-based segmentation reveals three distinct regions in which the query's joint distribution is dominated by the influence of pressure (blue), velocity (green), and temperature (red). Image source: Gosink et. al, 2011 [17]. Image (c) is the volume rendering of the plasma density (gray), illustrating the wake of the laser in a plasma-based accelerator. The data set contains $\approx 229 \times 10^6$ particles per timestep. The particles of the two main beams, automatically detected by the query-based analysis, are shown colored by their momentum in acceleration direction (px). Image source: Rübel et. el, 2009 [26].

analyze the structure of a query result, they used kernel density estimation to compute the joint and univariate probability distributions for the multivariate solution space of a query. Visual exploration of the joint density function—for example, using isosurfaces in physical space—helps users with the visual identification of regions in which the combined behavior of the queried variables is statistically more important to the inquiry. Based on the univariate distribution functions, the query solution is then segmented into different subregions in which the distribution of different variables is more important in defining the queries' solution. Figure 7.4b shows an example of the segmentation of a query defining the eye of a hurricane. Segmentation of the query solution based on the univariate density functions reveals three regions in which the query's joint distribution is dominated by the influence of pressure (blue), velocity (green), and temperature (red). The comparison of a query's univariate distributions to the corresponding distributions, restricted to the segmented regions, can help identify parameters for further query-refinement.

As illustrated by the above example, understanding the structure of a query defined by the trends and interactions of variables within the query's solution, can provide important insight to help with the refinement of queries. To this end, Gosink et al. [15] proposed the use of univariate cumulative distribution functions restricted to the solution space of the query to identify principle level sets for all variables that are deemed the most relevant to the query's solution. Additional derived scalar fields, describing the local pairwise correlation between two variables in physical space are then used to identify a pairwise variable interactions. Visualization of these correlation fields, in conjunction with principle isosurfaces, then enables the identification of statistically important interactions and trends between any three variables.

The manual, query-driven exploration of extremely large data sets enables the user to build and test new hypotheses. However, manual exploration is often a time-consuming and complex process and is, therefore, not well-suited for the processing of large collections of scientific data. Combining the QDV concept with methods to automate the definition of advanced queries to extract specific features of interest helps to support the analysis of large data collection. For example, in the context of plasma-based particle accelerators, many analyses rely on the ability to accurately define the subset of particles that form the main particle beams of interest. To this effect, 7.4.2 describes an efficient query-based algorithm for the automatic detection of particle beams [26]. Combining the automatic query-based beam analysis with advanced visualization supports an efficient analysis of complex plasma-based accelerator simulations, as shown in Figure 7.4c.

7.4 Applications of Query-Driven Visualization

QDV is among the small subset of techniques that can address both large and highly complex data. QDV is an efficient, feature-focused analysis approach that has a wide range of applications. The case studies that follow all make use of supercomputing platforms to perform QDV and analysis on data sets of unprecedented size. One theme across all these studies is that QDV reduces visualization and analysis processing time from hours or days, to seconds or minutes.

The study results presented in 7.4.1 aim to accelerate analysis within the context of forensic cybersecurity [3, 29]. Two high energy physics case studies are presented in 7.4.2, both of which are computational experiments aimed at designing next-generation particle accelerators, such as: a free-electron laser (7.4.2.1) and a plasma-wakefield accelerator (7.4.2.2).

7.4.1 Applications in Forensic Cybersecurity

Modern forensic analytics applications, like network traffic analysis, consist of hypothesis testing, knowledge discovery, and data mining on very large data sets. One key strategy to reduce the time-to-solution is to be able to quickly focus analysis on the subset of data relevant for a given analysis. This case study, presented in earlier work [3, 29], uses a combination of supercomputing platforms for parallel processing, high performance indexing for fast searches, and user interface technologies for specifying queries and examining query results.

The problem here is to find and analyze a distributed network scan attack buried in one year's worth of network flow data, which consists of 2.5 billion records. In a distributed scan, multiple distributed hosts systematically probe for vulnerabilities on ports of a set of target hosts. The traditional approach, consisting of command line scripts and using ASCII files, could have required many weeks of processing time. The approach presented below reduces this time to minutes.

Data: In this experiment, the data set consists of network connection data for a period of 42 weeks—a total of 2.5 billion records—acquired from a Bro system [23] running at a large supercomputing facility. The data set contains for each record, the "standard" set of connection variables, such as source and destination IP addresses, ports, connection duration, etc. The data is stored in flat files, using an uncompressed binary format for a total size of about 281GB. The IP addresses are split into four octets, $A : B : C : D$, to improve query performance. For instance, IPS_A refers to the class A octet of the source IP address. The FastBit bitmap indices used to accelerate data queries require a total of $\approx$ 78.6GB of space.

Interactive Query Interface: The study's authors showed a histogram-based visualization and query interface to facilitate the exploration of network traffic data. After loading the metadata from a file, the system invokes FastBit to generate coarse temporal resolution histograms containing, for example, counts of variables over the entire temporal range at a weekly resolution. The user then refines the display by drilling into a narrower temporal range and at a finer temporal resolution. The result of such a query is another histogram, which is listed as a new variable in the user interface. A more complex query may be formed by the "cross-product" of histograms of arbitrary numbers of data variables, and may be further qualified with arbitrary, user-specified conditions.

One advantage of such a histogram-centric approach is that it allows for an effective iterative refinement of queries. This iterative, multiresolution approach, which consists of an analyst posing different filtering criteria to examine data at different temporal scales and resolutions, enables effective context and focus changes. Rather than trying to visually analyze 42 weeks worth of data at one-second resolution all at once, the user can identify higher-level features first and then analyze those features and their subfeatures in greater detail.

Another advantage is performance. The traditional methodology in network traffic analysis consists of running command-line scripts on logfiles. Due to the nature of how FastBit stores and operates on bitmap indices, it can compute conditional histograms much more quickly than a traditional sequential scan. This idea was the subject of several different performance experiments [3, 29].

Network Traffic Analysis: The objective of this analysis example is to ex-

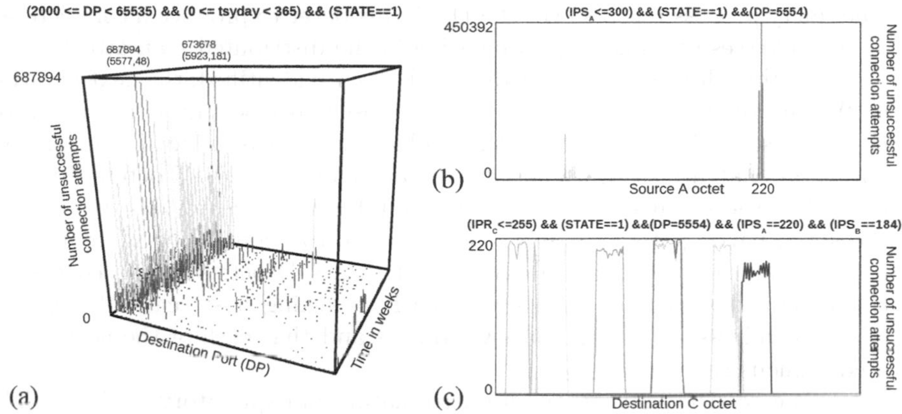

FIGURE 7.5: Histograms showing the number of unsuccessful connection attempts (with radiation excluded): (a) on ports 2000 to 65535 over a 42-week period, indicating high levels of activity on port 5554 during the 7th week; (b) per source A octet during the 7th week on port 5554, indicating suspicious activity from IPs with a 220 A octet; (c) per destination C octet scanned by seven suspicious source hosts (color), indicating a clear scanning pattern. Image source: Bethel et al., 2006 [3].

amine 42 weeks worth of network data to investigate an initial intrusion detection system (IDS) alert indicating a large number of scanning attempts on TCP port 5554, which is indicative of a so-called "Sasser worm." The first step, then, is to search the complete data set to identify the ports for which large numbers of unsuccessful connection attempts have been recorded by the Bro system. Figure 7.5a shows the counts of unsuccessful connection attempts on all ports over the entire time range at weekly temporal resolution. The chart reveals a high degree of suspicious activity in the seventh week, on destination port 5554.

Now that the suspicious activity has been confirmed and localized, the next goal is to identify the addresses of the host(s) responsible for the unsuccessful connection attempts. The splitting of IP addresses into four octets ($A : B : C : D$) enables an iterative search to determine the A, then B, C, and finally, D octet addresses of the attacking hosts. The first query then asks for the number of unsuccessful connection attempts on port 5554 during the seventh week over each address within the Class A octet. As can be seen in Figure 7.5b, the most suspicious activity originates from host(s), having IP addresses with a 220 A octet. Further refinements of the query, then reveal the B octet (IP=220 : 184 : $x : x$) and a list of C octets ($IPS_C = \{26, 31, 47, 74, 117, 220, 232\}$) from which most of the suspicious activity originates, indicating that a series of hosts on these seven IPS_C network segments may be involved in a distributed scan.

The query process repeats over the D address octet to produce a complete set of IP addresses of all hosts participating in the distributed scan attack.

Once the addresses for the attacking hosts are identified, the next question is "What are the access patterns of these host addresses through destination IP addresses?" In other words, the problem is to determine the set of host addresses that are being attacked. The destination Class C octets, scanned by each of the seven source hosts—shown in Figure 7.5c—reveals that each of the seven participating hosts is sending traffic to about 21 or 22 contiguous Class C addresses. Further analysis of the class D octets in the context of the identified C octets, reveals that each attacking host scans through all Class D addresses for each Class C address and that each of the hosts scans a contiguous range of IPR_C's.

This case study reveals the iterative nature of exploratory analytics. The first inquiry examined all data at a coarse temporal resolution. Subsequent inquiries became more focused, looking at smaller and smaller subsets of the data. One of the previous studies estimated the time required for executing a single query to be approximately 43 hours, using traditional shell-based scripts, compared to about five seconds, using the QDV approach. Given that this case study encompasses many queries, one conclusion is that advanced exploratory analytics on very large data sets may simply not be feasible using traditional approaches.

7.4.2 Applications in High Energy Physics

Particle accelerators are among the most important and versatile tools in scientific discovery. They are essential to a wealth of advances in material science, chemistry, bioscience, particle physics, and nuclear physics. Accelerators also have important applications to the environment, energy, and national security. Example applications of particle accelerators include studying bacteria for bioremediation, exploring materials for solar cells, and developing accelerator-based systems to inspect cargo for nuclear contraband. Accelerator-based systems have also been proposed for accelerator-driven fission energy production and as a means to transmute nuclear waste.

Accelerators have a direct impact on the quality of people's lives through applications in medicine, such as the production of medical radioisotopes, pharmaceutical drug design and discovery, and through the thousands of accelerator-based irradiation therapy procedures that occur daily at U.S. hospitals. Given the importance of particle accelerators, it is essential that the most advanced high performance computing tools be brought to bear on accelerator R&D, and on the design, commissioning, and operation of future accelerator facilities. The following sections describe the application of QDV to the analysis of output from numerical simulations that model the behavior of electron linear particle accelerators (linacs) in 7.4.2.1 and laser plasma particle accelerators (LPA) in 7.4.2.2.

Different variations of particle-in-cell (PIC) based simulations are used to model both linacs, as well as LPA experiments [24, 20]. In PIC simulations, collections of charged particles, such as electrons or protons, are modeled as computational macro-particles that can be located anywhere in the computational domain. The electromagnetic field is spatially discretized, often using a regular mesh. At each simulation step, the particles are moved under the electromagnetic forces obtained through interpolation from the fields. The current carried by the moving particles is then deposited onto the simulation grid to update the electric and magnetic fields. The data sets produced by accelerator simulations are extremely large, heterogeneous (both particles and fields), multivariate and often of highly varying spatio-temporal resolution.

The analytics process is similar to that of the previous network traffic analysis example (see 7.4.1): a scientist begins the investigation with a very large data set and formulates a set of questions, then iteratively extracts subsets of data for inspection and analysis. The pace at which knowledge is discovered is a function of the rate at which this entire process can occur. The combination of QDV and supercomputers can reduce this duty cycle from hours to seconds, thereby helping to accelerate the discovery of scientific knowledge.

7.4.2.1 Linear Particle Accelerator

The linac case study applies QDV to the analysis of data from large-scale, high-resolution simulations of beam dynamics in linacs for a proposed next-generation x ray, free electron laser (FEL) at LBNL [11]. PIC-based simulations on this type of accelerator require large numbers of macroparticles ($> 10^8$) to control the numerical macroparticle shot noise and to avoid overestimation of the microbunching instability, and they produce particle data sets that are massive in size [24]. The authors, Chou et al. [9], focus on applying QDV to study beam diagnostics of the transverse halo and beam core.

Data: The source data in this study was generated by the IMPACT-T simulation code [24], and consists of 720 timesteps with ≈ 1 billion particles per timestep. The 50TB data set contains nine variables describing the physical particle location and momentum. The variables x and y are the transverse location of particles. The longitudinal particle location, t, is in a relative coordinate system—that is, t identifies the particle's arrival time at the physical location z.

Transverse Halo: The transverse halo of a particle beam is a low-density portion of the beam usually defined as those particles beyond some specified radius or transverse amplitude in physical space, shown in Figure 7.6a. Controlling the beam halo is critical because the particles of the halo have the potential to reach very large transverse amplitudes, eventually striking the beam pipe and causing radioactivation of accelerator components and possibly component damage. Also, the beam halo is often the limiting factor in increasing the beam intensity to support scientific experiments. To enable a detailed analysis of this phenomenon—to determine the origin of the halo

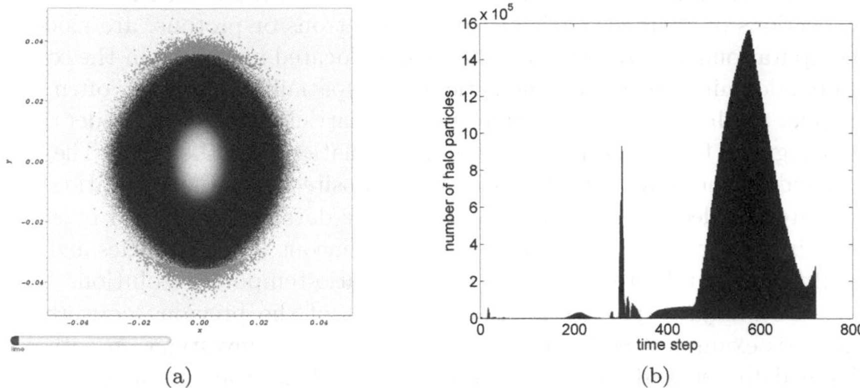

FIGURE 7.6: (a) Particle density plot (gray) and selected halo-query particles (red) for timestep 20 of the simulation. (b) Plot showing the number of halo particles per timestep in a 50TB electron linear particle accelerator data set. Image source: Chou et al., 2011 [9].

particles and to help develop halo mitigation strategies—physicists need to be able to efficiently identify and extract the halo particles.

The authors identify halo particles using the query $r > 4\sigma_r$. The derived quantity $r = \sqrt[2]{x^2 + y^2}$, which is separately computed and indexed, describes the transverse radial particle location. The transverse halo threshold is given by $\sigma_r = \sqrt[2]{\sigma_x^2 + \sigma_y^2}$, where σ_x and σ_y denote the root mean square (RMS) beam sizes in x and y, respectively. Using r for identification of halo particles is based on the assumption of an idealized circular beam cross section. To ensure that the query adapts more closely to the transverse shape of the beam, one may relax the assumption of a circular beam cross section to an elliptical beam cross section, by scaling the x and y coordinates independently by the corresponding RMS beam size, σ_x and σ_y using, for example, a query of the form $r_s^2(x,y) > 16$, with $r_s^2(x,y) = (\frac{x}{\sigma_x})^2 + (\frac{y}{\sigma_y})^2$.

Figure 7.6b shows the number of halo particles per timestep identified by the halo query. It shows large variations in the number of halo particles, while in particular the larger number of halo particles at later timesteps are indicative of a possible problem. In this case, the halo particles and observations of an increase in the maximum particle amplitude were found to be due to a mismatch in the beam as it traveled from one section of the accelerator to the next. These type of query-based diagnostics provide accelerator designers with evidence that further improvement of the design may be possible, and also provides quantitative information that is useful for optimizing the design to reduce halo formation and beam interception with the beam pipe, which will ultimately improve accelerator performance.

Core: Creating beams with good longitudinal beam quality, which is defined in terms of the longitudinal RMS emittance, and maintaining that beam quality during the acceleration process, is critical in certain types of accelerators involving intense electron beams. Frequently, the longitudinal "head" and "tail" of an electron bunch will be very different from its "center" region, also known as the longitudinal "core." Focusing on the longitudinal core of a particle bunch eliminates strong variations at the head and tail, which would otherwise distort the analysis results.

The authors define the core of the beam using the query $\bar{t} - 200\delta \leq t \leq \bar{t} + 200\delta$, with $\bar{t}$ being the longitudinal bunch centroid. The parameter $\delta = \frac{10^{-9}}{0.036728}$ has been provided by domain scientists and corresponds to $\approx$ 1nm; or more precisely, the time it takes an electron to travel 1nm. To identify within-core variations in particle density, uncorrelated energy spread, and transverse emittance, the core is typically further subdivided into slices for analysis purposes. In order to divide the core into, say, 400, 1nm wide slices, one could evaluate, at each timestep, multiple queries of the form $t_{min}(i) \leq t \leq t_{max}(i)$ with $t_{min} = \bar{t} - (i - 201)\delta$, $t_{max}(i) = \bar{t} - (i - 200)\delta$ and $i \in [1, 400]$. Such an analysis of all 720 timesteps of the example data set would require 288,000 queries, further highlighting the need for efficient query methods.

Performance: The authors used the FastQuery [9] system for the parallel processing of index and query operations on the NERSC Cray XE6 supercomputing system Hopper.[1] The pre-processing to build the indexes for all timesteps of the 50TB data set required about 2 hours. Using bitmap indexing, the evaluation of the halo and core query required only 12 seconds, using 3000 cores, while practically reasonable times of around 20 seconds can be achieved using only approximately 500 compute cores. The combination of visual data exploration and efficient data management in the QDV concept enables, in this way, repeated, complex, large-scale query-based analysis of massive data sets, which would otherwise not be practical, with respect to both time as well as computational cost.

7.4.2.2 Laser Plasma Particle Accelerator

Plasma-based particle accelerators (LPAs) [12] use a short ($\leq$ 100fs), ultra-high intensity ($\geq 10^{18} W/cm^2$) laser pulse to drive waves in a plasma. Electrons in a hydrogen plasma are displaced by the radiation pressure of the laser pulse, while the heavier ions remain stationary. This displacement of the electrons in combination with the space-charge restoring force of the ions, drives a wave (wake) in the plasma. Similar to a surfer riding a wave, electrons that become trapped in the plasma wave are accelerated by the wave to high energy levels. LPAs can achieve electric and magnetic fields thousands of times stronger than conventional accelerators, enabling the acceleration of particles to high

[1]System specification: #compute nodes $\cong$ 6,500, #cores/node=24, memory/node=32GB, filesystem=Lustre, peak I/O=25GB/s. For details see `http://www.nersc.gov/nusers/systems/hopper2/`.

energy levels within very short distances (centimeters to meters). Researchers at the LOASIS[2] program have demonstrated high-quality electron beams at 0.1 to 1 GeV using millimeter to centimeter long plasmas [14, 19].

One central challenge in the analysis of large LPA simulation data arises from the fact that, while large numbers of particles are required for an accurate simulation, only a small fraction of the particles are accelerated to high energies and subsequently form particle features of interest. During the course of a simulation, multiple high-energy particle beams may form and additional particle bunches may appear in secondary periods of the plasma wave. In this context, scientists are particularly interested in the analysis of the main particle beam. This section discusses query-driven analysis of LPA simulations used to extract, analyze, and compare high-energy particle beams.

Data: This case study uses data sets produced by VORPAL [20], an electromagnetic PIC-based simulation code. Accurate modeling of LPAs is computationally expensive, in particular, due to large differences in scale, e.g., length of the plasma versus the laser wavelength. To save computational resources and storage space, VORPAL employs a moving-window simulation approach. In this method, only a window around the laser pulse is simulated at each timestep, and it is moved along at the speed of light, as the laser propagates through the plasma. The output data contains particle position (x, y, z), particle momentum (px, py, pz), particle weight (wt), and particle identifier (id). The data is indexed using FastBit to accelerate the evaluation of range queries, id-based equality queries, and the computation of conditional histograms.

Interactive Query-Driven Data Exploration: To gain a deeper understanding of the acceleration process, physicists need to address a number of complex questions, such as: Which particles become accelerated? How are particles trapped and accelerated by the plasma wave? How are the beams of highly accelerated particles formed? And, how do the particle beams evolve over time? The case study tackles these questions via an interactive, query-driven visual exploration approach for histogram-based parallel coordinates (see 7.3.1) and high performance scientific visualization [27].

To identify those particles that were accelerated, the authors, who included an accelerator physicist, performed an initial threshold selection in px at a late timestep of the simulation. This initial selection restricts the analysis to a small set of particles with energies above the base acceleration of the plasma wave. Figure 7.7a shows an example of a parallel coordinates plot of such a selection (gray).

Based on the results of the first query, the selection was further refined to select the main particle beams of interest. The authors first increase the px threshold to extract the particles of highest energy. This initial refinement

[2]Lasers, Optical Accelerator Systems Integrated Studies (LOASIS) Program at LBNL; `http://loasis.lbl.gov/`

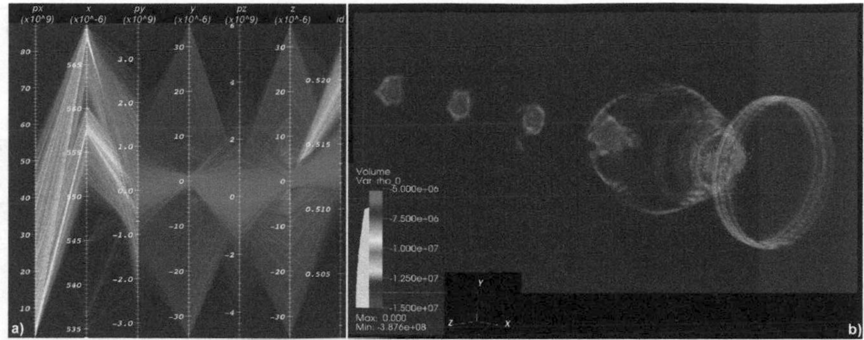

FIGURE 7.7: QDV of large 3D plasma-based particle acceleration data containing $\approx 90 \times 10^6$ particles per timestep. On the left (a), parallel coordinates of timestep $t = 12$ showing: (1) all particles above the base acceleration level of the plasma wave (Query: $px > 2 \times 10^9$) (gray) and (2) a set of particles that form a compact beam in the first wake period following the laser pulse (Query: $(px > 4.856 \times 10^{10})AND(x > 5.649 \times 10^{-4})$)(red). On the right (b), volume rendering of the plasma density illustrating the 3D structure of the plasma wave. The selected beam particles are shown in addition in red. Image source: Rübel et al., 2008 [27].

often results in the selection of multiple beam-like features trapped in different periods of the plasma wave.

As illustrated in Figure 7.7a (red lines), to separate the different particle beams and to extract the main particle beam, the selection is then often further refined through range queries in the longitudinal coordinate x and transverse coordinates y and z. In the selection process, parallel coordinates provide interactive feedback about the structure of the selection, allowing for fast identification of outliers and different substructures of the selection. High performance scientific visualization methods are then used to validate and further analyze the selected particles Figure 7.7b.

The next step is to trace, or follow, and analyze the high-energy particles through time. Particle tracing is used to detect the point in time when the beam reaches its peak energy and to assess the quality of the beam. Tracing the beam particles further back in time, to the point at which the particles enter the simulation window, supports analysis of the injection, beam formation, and beam evolution processes. Figure 7.3c shows an example in which parallel coordinates are used to compare the acceleration behavior of two particle beams. Based on the information from different individual timesteps, a user may refine a query, to select, for example, beam substructures that are visible at different discrete timesteps. Rübel et al. [27] demonstrated that using FastBit-accelerated equality queries, the time needed for tracing beam particles can be reduced from many hours to less than a second.

Automatic Query-Based Beam Detection: Interactive query-driven visual

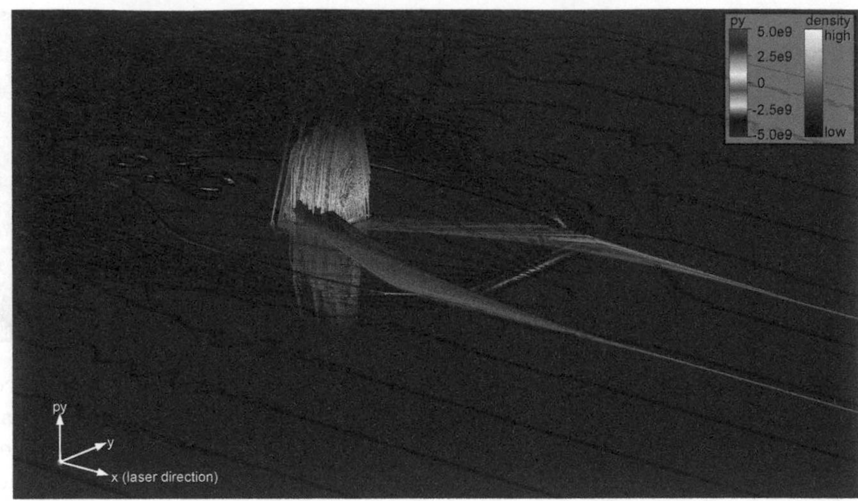

FIGURE 7.8: Visualization of the relative traces of a particle beam in a laser plasma accelerator. The traces show the motion of the beam particles relative to the laser pulse. The xy-plane shows isocontours of the particle density at the timepoint when the beam reaches its peak energy, illustrating the location of the beam within the plasma wave. The positive vertical axis and color of the particle traces show the particle momentum in the transverse direction, py. The image shows particles being injected from the sides and oscillating while being accelerated. Image source: Rübel et al., 2009 [26].

data exploration is effective in that it provides great flexibility and supports detailed data analysis, allowing scientist to define and validate new hypotheses about the data and the physical phenomena being modeled. However, manual detection and extraction of the acceleration features of interest, such as particle beams, is time consuming. To support the analysis of large collections of accelerator simulations, efficient methods are needed for automatic detection and extraction of particle beams.

Rübel et al. [26] described an efficient query-based algorithm for automatic detection of particle beams. At a single timestep, the analysis proceeds as follows. First, the algorithm computes a 3D conditional histogram of (x, y, px) space, restricted to high-energy particles selected by the query $px > 1e10$. Using a two-stage segmentation process, the algorithm first identifies the region in the physical space (x, y), containing the highest energy particles. The so-defined spatial selection is then refined using a 3D region-growing-based segmentation centered around the highest-density histogram bin. Histogram-based bin-queries (see 7.2.1) are then used to extract the particles belonging to the so-identified feature. The identified per-timestep features are then traced over time and refined through additional per-timestep segmentations.

Finally, the particles of interest are traced over time using a series of

ID-based equality queries, and additional particle-to-feature distance fields are computed. In this query-based analysis process, the computation of conditional histograms, the evaluation of histogram-based bin queries, and the identifier-based particle tracing are accelerated through FastBit.

Even for a large 3D data set (approximately 620GB) consisting of 26 timesteps with approximately 230×10^6 particles per timestep, the analysis requires only about 3 minutes in serial, which is already much less than what a single histogram-based bin-query would require without using FastBit. As shown in Figure 7.2b, the analysis has to evaluate tens of bin queries, 52 threshold and equality queries. Combining the query-based beam analysis with advanced visualization supports automated analysis and comparison of particle beams in complex LPA simulations (see Fig. 7.4c). Figure 7.8 shows as an example a visualization of the injection and acceleration process of an automatically detected particle beam.

7.5 Conclusion

Query-driven visualization and analysis is a blend of technologies that limit visualization and analysis processing to the subset of data that is "interesting." The premise is that, in any given investigation, only a small fraction of a large data set is of interest.

QDV is useful for fast visual exploration of extremely large data sets. It has been shown to be useful to accelerate and automate complex feature detection tasks, in particular, when combined with other analysis methods, such as unsupervised learning. As a general concept, QDV is very flexible and can be easily integrated with a large range of other methods. Advanced index/query systems (e.g., FastQuery) and integration of these data interfaces with advanced visualization systems (such as VisIt, Chap. 16), make QDV-based analysis much more accessible to the scientific community.

QDV is among the small subset of techniques that can address visualization and analysis of large and highly complex data. The case studies in this chapter, taken from forensic cybersecurity and high-energy physics applications, make use of supercomputing platforms to perform QDV and analysis on data sets of unprecedented sizes. Other recent work in QDV, not discussed here, applies the concepts to diverse application areas, like computational finance, climate modeling and analysis, magnetically confined fusion, and astrophysics. One theme across all of these studies is that QDV reduces visualization and analysis processing time from hours or days to seconds or minutes.

References

[1] G. Antoshenkov. Byte-aligned Bitmap Compression. Technical report, Oracle Corp., 1994. U.S. Patent number 5,363,098.

[2] Rudolf Bayer and Edward McCreight. Organization and Maintenance of Large Ordered Indexes. *Acta Informatica*, 3:173–189, 1972.

[3] E. Wes Bethel, Scott Campbell, Eli Dart, Kurt Stockinger, and Kesheng Wu. Accelerating Network Traffic Analysis Using Query-Driven Visualization. In *Proceedings of 2006 IEEE Symposium on Visual Analytics Science and Technology*, pages 115–122. IEEE Computer Society Press, October 2006. LBNL-59891.

[4] Peter A. Boncz, Marcin Zukowski, and Niels Nes. MonetDB/X100: Hyper-Pipelining Query Execution. In *Proceedings of the Biennial Conference on Innovative Data Systems Research (CIDR)*, pages 225–237, January 2005.

[5] Peer-Timo Bremer, Gunther H. Weber, Valerio Pascucci, Marcus S. Day, and John B. Bell. Analyzing and Tracking Burning Structures in Lean Premixed Hydrogen Flames. *IEEE Transactions on Visualization and Computer Graphics*, 16(2):248–260, Mar/Apr 2010. LBNL-2276E.

[6] Paul G. Brown. Overview of sciDB: Large Scale Array Storage, Processing and Analysis. In *Proceedings of the 2010 ACM SIGMOD International Conference on Management of Data*, SIGMOD '10, pages 963–968, Indianapolis, IN, USA, 2010.

[7] C. Y. Chan and Y. E. Ioannidis. Bitmap Index Design and Evaluation. In *Proceedings of the 1998 ACM SIGMOD International Conference on Management of Data*, SIGMOD '98, pages 355–366, New York, NY, USA, June 1998. ACM.

[8] C. Y. Chan and Y. E. Ioannidis. An Efficient Bitmap Encoding Scheme for Selection Queries. In *SIGMOD*, Philadelphia, PA, USA, June 1999. ACM Press.

[9] Jerry Chou, Mark Howison, Brian Austin, Kesheng Wu, Ji Qiang, E. Wes Bethel, Arie Shoshani, Oliver Rübel, Prabhat, and Rob D. Ryne. Parallel Index and Query for Large Scale Data Analysis. In *Proceedings of 2011 International Conference for High Performance Computing, Networking, Storage and Analysis*, SC '11, pages 30:1–30:11, Seattle, WA, USA, November 2011.

[10] Douglas Comer. The Ubiquitous B-Tree. *Computing Surveys*, 11(2):121–137, 1979.

[11] J. N. Corlett, K. M. Baptiste, J. M. Byrd, P. Denes, R. J. Donahue, L. R. Doolittle, R. W. Falcone, D. Filippetto, J. Kirz, D. Li, H. A. Padmore, C. F. Papadopoulos, G. C. Pappas, G. Penn, M. Placidi, S. Prestemon, J. Qiang, A. Ratti, M. W. Reinsch, F. Sannibale, D. Schlueter, R. W. Schoenlein, J. W. Staples, T. Vecchione, M. Venturini, R. P. Wells, R. B. Wilcox, J. S. Wurtele, A. E. Charman, E. Kur, and A. Zholents. A Next Generation Light Source Facility at LBNL. In *Proceedings of PAC 2011*, New York, NY, USA, April 2011.

[12] E. Esarey, C. B. Schroeder, and W. P. Leemans. Physics of Laser-Driven Plasma-Based Electron Accelerators. *Reviews of Modern Physics*, 81:1229–1285, 2009.

[13] V. Gaede and O. Günther. Multidimension Access Methods. *ACM Computing Surveys*, 30(2):170–231, 1998.

[14] C. G. R. Geddes, Cs. Toth, J. van Tilborg, E. Esarey, C. B. Schroeder, D. Bruhwiler, C. Nieter, J. Cary, and W. P. Leemans. High-Quality Electron Beams from a Laser Wakefield Accelerator Using Plasma-Channel Guiding. *Nature*, 438:538–541, 2004. LBNL-55732.

[15] Luke Gosink, John C. Anderson, E. Wes Bethel, and Kenneth I. Joy. Variable Interactions in Query-Driven Visualization. *IEEE Transactions on Visualization and Computer Graphics (Proceedings of Visualization 2007)*, 13(6):1400–1407, November/December 2007. LBNL-63524.

[16] Luke Gosink, John Shalf, Kurt Stockinger, Kesheng Wu, and E. Wes Bethel. HDF5-FastQuery: Accelerating Complex Queries on HDF Datasets using Fast Bitmap Indices. In *Proceedings of the 18th International Conference on Scientific and Statistical Database Management*, pages 149–158. IEEE Computer Society Press, July 2006. LBNL-59602.

[17] Luke J. Gosink, Christoph Garth, John C. Anderson, E. Wes Bethel, and Kenneth I. Joy. An Application of Multivariate Statistical Analysis for Query-Driven Visualization. *IEEE Transactions on Visualization and Computer Graphics*, 17(3):264–275, 2011. LBNL-3536E.

[18] Alfred Inselberg. *Parallel Coordinates Visual Multidimensional Geometry and Its Applications.* Springer-Verlag, Secaucus, NJ, USA, 2008.

[19] W. P. Leemans, B. Nagler, A. J. Gonsalves, Cs. Toth, K. Nakamura, C. G. R. Geddes, E. Esarey, C. B. Schroeder, and S. M. Hooker. GeV Electron Beams from a Centimetre-Scale Accelerator. *Nature Physics*, 2:696–699, 2006.

[20] C. Nieter and J. R. Cary. VORPAL: A versatile plasma simulation code. *Journal of Computational Physics*, 196(2):448–473, 2004.

[21] Matej Novotný and Helwig Hauser. Outlier-Preserving Focus+Context Visualization in Parallel Coordinates. *IEEE Transactions on Visualization and Computer Graphics*, 12(5):893–900, 2006.

[22] P. O'Neil. Model 204 Architecture and Performance. In *Proceedings of the 2nd International Workshop on High Performance Transaction Systems*, pages 40–59, Asilomar, CA, USA, September 1987. Springer-Verlag.

[23] Vern Paxson. Bro: A System for Detecting Network Intruders in Real-Time. *Computer Networks*, 31(23-24):2435–2463, 1999.

[24] J. Qiang, R. D. Ryne, M. Venturini, and A. A. Zholents. High Resolution Simulation of Beam Dynamics in Electron Linacs for X-Ray Free Electron Lasers. *Physical Review*, 12(10):100702–1–100702–11, 2009.

[25] Doron Rotem, Kurt Stockinger, and Kesheng Wu. Optimizing I/O Costs of Multi-dimensional Queries Using Bitmap Indices. In *Proceedings of the 16th International Conference on Database and Expert Systems Applications*, DEXA'05, pages 220–229, 2005.

[26] Oliver Rübel, Cameron G. R. Geddes, Estelle Cormier-Michel, Kesheng Wu, Prabhat, Gunther H. Weber, Daniela M. Ushizima, Peter Messmer, Hans Hagen, Bernd Hamann, and E. Wes Bethel. Automatic Beam Path Analysis of Laser Wakefield Particle Acceleration Data. *IOP Computational Science & Discovery*, 2(1):015005, November 2009. LBNL-2734E.

[27] Oliver Rübel, Prabhat, Kesheng Wu, Hank Childs, Jeremy Meredith, Cameron G. R. Geddes, Estelle Cormier-Michel, Sean Ahern, Gunther H. Weber, Peter Messmer, Hans Hagen, Bernd Hamann, and E. Wes Bethel. High Performance Multivariate Visual Data Exploration for Extemely Large Data. In *Supercomputing 2008 (SC08)*, Austin, Texas, USA, November 2008. LBNL-716E.

[28] Arie Shoshani and Doron Rotem, editors. *Scientific Data Management: Challenges, Technology, and Deployment.* Chapman & Hall/CRC Press, Boca Raton, FL, USA, 2010.

[29] Kurt Stockinger, E. Wes Bethel, Scott Campbell, Eli Dart, and Kesheng Wu. Detecting Distributed Scans Using High-Performance Query-Driven Visualization. In *SC '06: Proceedings of the 2006 ACM/IEEE Conference on High Performance Computing, Networking, Storage and Analysis.* IEEE Computer Society Press, November 2006. LBNL-60053.

[30] Kurt Stockinger, John Shalf, E. Wes Bethel, and Kesheng Wu. DEX: Increasing the Capability of Scientific Data Analysis Pipelines by Using Efficient Bitmap Indices to Accelerate Scientific Visualization. In *Proceedings of Scientific and Statistical Database Management Conference (SSDBM)*, pages 35–44, Santa Barbara, CA, USA, June 2005. LBNL-57203.

[31] Kurt Stockinger, John Shalf, Kesheng Wu, and E. Wes Bethel. Query-Driven Visualization of Large Data Sets. In *Proceedings of IEEE Visualization 2005*, pages 167–174. IEEE Computer Society Press, October 2005. LBNL-57511.

[32] The HDF Group. HDF5 User Guide. `http://hdf.ncsa.uiuc.edu/HDF5/doc/H5.user.html`, 2010.

[33] Unidata. The NetCDF Users' Guide. `http://www.unidata.ucar.edu/software/netcdf/docs/netcdf/`, 2010.

[34] Michelle Q. Wang Baldonado, Allison Woodruff, and Allan Kuchinsky. Guidelines for Using Multiple Views in Information Visualization. In *Proceedings of the Working Conference on Advanced Visual Interfaces*, AVI '00, pages 110–119, New York, NY, USA, 2000. ACM.

[35] K. Wu, W. Koegler, J. Chen, and A. Shoshani. Using Bitmap Index for Interactive Exploration of Large Datasets. In *Proceedings of the 15th International Conference on Scientific and Statistical Database Management*, SSDBM '03, pages 65–74. IEEE Computer Society Press., July 2003.

[36] Kesheng Wu, Sean Ahern, E. Wes Bethel, Jacqueline Chen, Hank Childs, Estelle Cormier-Michel, Cameron Geddes, Junmin Gu, Hans Hagen, Bernd Hamann, Wendy Koegler, Jerome Lauret, Jeremy Meredith, Peter Messmer, Ekow Otoo, Victor Perevoztchikov, Arthur Poskanzer, Prabhat, Oliver Rübel, Arie Shoshani, Alexander Sim, Kurt Stockinger, Gunther Weber, and Wei-Ming Zhang. FastBit: Interactively Searching Massive Data. In *SciDAC 2009*, 2009. LBNL-2164E.

[37] Kesheng Wu, Ekow Otoo, and Arie Shoshani. On the Performance of Bitmap Indices for High Cardinality Attributes. In *Proceedings of the Thirtieth International Conference on Very large Data Bases—Volume 30*, VLDB '04, pages 24–35. VLDB Endowment, 2004.

[38] Kesheng Wu, Ekow Otoo, and Arie Shoshani. Optimizing Bitmap Indices with Efficient Compression. *ACM Transactions on Database Systems*, 31:1–38, 2006.

[39] Kesheng Wu, Ekow Otoo, and Kenji Suzuki. Optimizing Two-Pass Connected-Component Labeling Algorithms. *Pattern Analysis & Applications*, 12(2):117–135, 2009. `http://www.springerlink.com/index/B67258V347158263.pdf`.

[40] Kesheng Wu, Arie Shoshani, and Kurt Stockinger. Analyses of Multi-Level and Multi-Component Compressed Bitmap Indexes. *ACM Transactions on Database Systems*, 35(1):1–52, 2010. `http://doi.acm.org/10.1145/1670243.1670245`.

[41] Kesheng Wu, Rishi R Sinha, Chad Jones, Stephane Ethier, Scott Klasky, Kwan-Liu Ma, Arie Shoshani, and Marianne Winslett. Finding Regions of Interest on Toroidal Meshes. *Computational Science & Discovery*, 4(1):015003, 2011.

[42] Kesheng Wu, Kurt Stockinger, and Arie Shosani. Breaking the Curse of Cardinality on Bitmap Indexes. In *Proceedings of the 20th International Conference on Scientific and Statistical Database Management*, SSDBM '08, pages 348–365. Springer-Verlag, 2008.

Chapter 8

Progressive Data Access for Regular Grids

John Clyne

Computational Information Systems Laboratory, National Center for Atmospheric Research

8.1 Introduction 145
8.2 Preliminaries 146
8.3 Z-Order Curves 147
8.3.1 Constructing the Curve 149
8.3.2 Progressive Access 149
8.4 Wavelets 151
8.4.1 Linear Decomposition 153
8.4.2 Scaling and Wavelet Functions 154
8.4.3 Wavelets and Filter Banks 156
8.4.4 Compression 158
8.4.5 Boundary Handling 159
8.4.6 Multiple Dimensions 164
8.4.7 Implementation Considerations 164
8.4.7.1 Blocking 165
8.4.7.2 Wavelet Choice 165
8.4.7.3 Coefficient Addressing 166
8.4.8 A Hybrid Approach 166
8.4.9 Volume Rendering Example 167
8.5 Further Reading 167
References 169

This chapter presents three progressive refinement methods for data sampled on a regular grid. Two of the methods are based on multiresolution: the grid may be coarsened or refined as needed by dyadic factors. The third is based on the energy compaction properties of the discrete wavelet transform, which enables the sparse representation of signals. In all cases, the objective is to afford the end user the ability to make trade-offs between fidelity and speed, in response to the available computing resources, when visualizing or analyzing large data.

8.1 Introduction

Fueled by decades of exponential increases in microprocessor performance, computational scientists in a diverse set of disciplines have enjoyed unprecedented supercomputing capabilities. However, with the increase in computing power, comes more sophisticated and realistic computing models and an invariable increase in resolution of the discrete grids used to solve the equations of state. A direct result of the increase in grid resolution is the profuse amount of stored data. Unfortunately, the ability to generate data has not been matched by our ability to consume it. Whereas microprocessor floating point performance, combined with communication interconnect bandwidth and latencies, largely determines the scale by which numerical simulations are run, it is often primary storage capacities and I/O bandwidths that constrain access to data during visualization and analysis. These latter technologies, and I/O bandwidths, in particular, have not kept pace with the performance advancements of CPUs, GPUs, or high performance communication fabrics. For many visualization and analysis workflows, there is a bottleneck caused by the rate at which data is delivered to the computational components of the analysis pipeline.

This chapter discusses methods used to reduce the volume of data that must be processed in order to support a meaningful analysis. Ideally, the user should be able to trade-off data fidelity for increased interactivity. In many applications, aggressive data reduction may have a negligible impact, or no impact whatsoever, on the resulting analyses. The specific target workflow is highly interactive, quick-look exploratory visualization enabled by coarsened approximations of the original data, followed by a less interactive validation of results using the refined or original data, as needed. This model and these methods, which are referred to as *progressive data access* (PDA), are similar to those employed by the ubiquitous GoogleEarth™—coarsened imagery is transmitted and displayed when the viewpoint is far away, and continuously refined as the user zooms in on a region of interest.

The PDA approaches in this chapter are all intended to minimize the volume of data accessed from secondary storage, whether the storage is a locally attached disk, or a remote, network-attached service. Additionally, the following properties of the data model are also deemed important: (1) the ability to quickly access coarsened approximations of the full data domain; (2) the ability to quickly extract subsets of the full domain at a higher quality; (3) lossless reconstruction of the original data; and (4) minimal storage overhead.

8.2 Preliminaries

A trivial approach to supporting PDA for regular grids is the generalization of 2D texture MIP mapping to higher dimensions. A MIP map, also referred to in the literature as an *image pyramid*, is a precalculated hierarchy of approximations created by successively coarsening a 2D image, typically reducing the sampling along each dimension by a factor of 2, with each pass. Without loss of generality, if assuming a 3D grid of dimension N^3, a series of approximations is created of the dimensions $(\frac{N}{2})^3$, $(\frac{N}{4})^3$, $(\frac{N}{8})^3$, and so on. An unfortunate consequence of this simplistic tack, though, is the additional storage required by each approximation. For a d-dimensional hypercube, the fractional storage increase is given by

$$\sum_{j=1}^{J} \frac{1}{2^{dj}},$$

where J is the number of approximations in the hierarchy—an ostensibly small multiplier, but one that may be unacceptable when N is large and multiple variables and timesteps are involved.

In the following sections, there are more sophisticated methods than MIP mapping to support PDA. The overall goal is to improve upon MIP mapping by eliminating, or at least reducing, the additional storage required. The notion of multiple resolutions—based on a hierarchical decomposition of the grid sampling space—plays a role in all of the methods presented. The convention adopted throughout this chapter is that the level in a multiresolution hierarchy is given by j, with j_{min} referring to the coarsest member of the hierarchy, and j_{max} referring to the finest member. The total number of members is $j_{max} - j_{min} + 1$. In general, $j_{min} = 0$, but not always. A distinction is always made when j_{min} is not zero.

8.3 Z-Order Curves

A space-filling curve is one that passes through all points of a 2D square, or more generally, all points of a d-dimensional hypercube. In the context of a regular grid of dimension N^d, where $N = 2^n$, a space-filling curve visits all vertices of the mesh exactly once, providing a mapping between 1D and higher dimensional space. Here, we will focus our attention on the 2D and 3D cases. The Morton space-filling curve, also referred to as the *Z-order* curve due to its shape (see Fig. 8.1), visits all vertices in the order of a depth-first traversal of a quadtree (or octree in 3D) [8]. Figure 8.1d shows the complete Z-order curve traversal, starting at the grid point with coordinates $(0, 0)$, of a 16×16 2D mesh. When a third dimension is added, plane-adjacent pairs of Z shapes are connected by a diagonal.

Informally, the property of Z-order curves that makes them appealing is that, in contrast to a conventional row or column major order traversal of grid vertices, the Z-order curve preserves the locality of data points; points whose integer offsets are near each other along the Z-order curve are geometrically close to each other, as well. Moreover, the grid points that are members of a

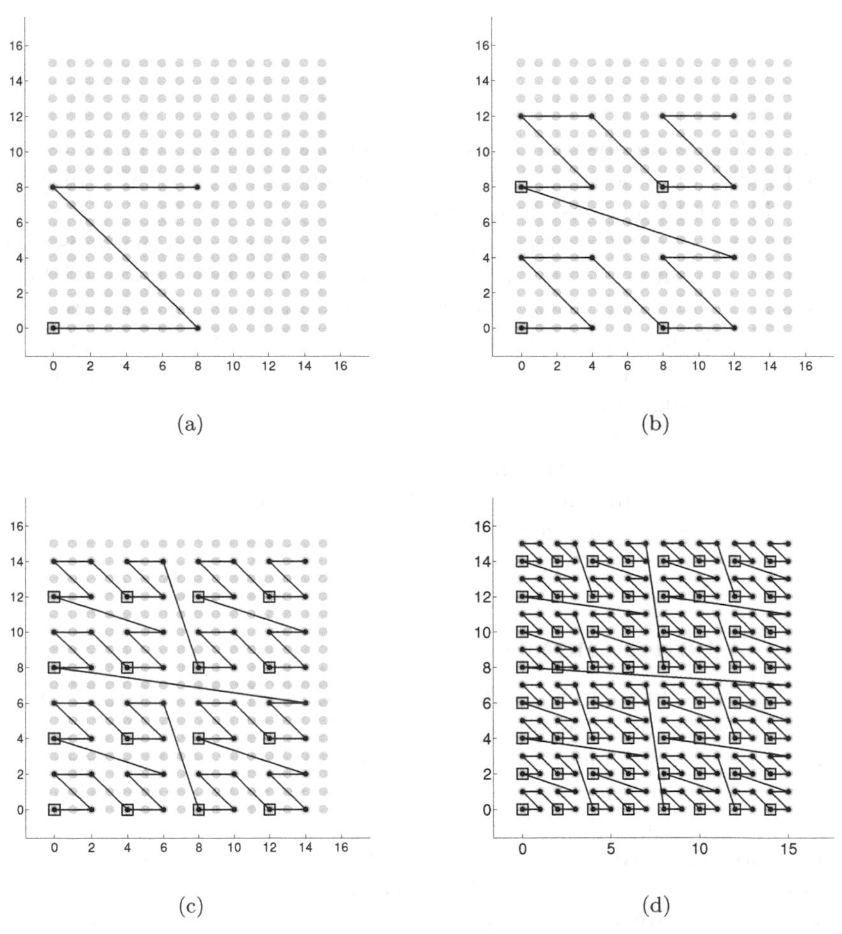

FIGURE 8.1: The Z-order, or Morton, space-filling curve maps multi-dimensional data into one dimension while preserving the locality of neighboring points.

particular level of the quadtree (octree) hierarchy are identified in a straightforward manner.

If, instead of using a unit stride, one chooses a suitable non-unit value to traverse the Z-order curve, $s = 2^{d(j_{max}-j)}$, then, all points can be visited at any level j of the tree, while still preserving the locality of points. The maximum number of levels in the hierarchy defined by the Z-order curve is given by $log_2(N)+1$, thus, $0 \leq j \leq j_{max} = log_2(N)$. The coarsened grid corresponding to $j = 0$, in the Z-order curve decomposition, contains a single point located at the origin of the original grid. Figures 8.1a–d show the resulting curves when $s = 64$, 16, 4, and 1, respectively. Not shown is the coarsest level: $j = 0$.

8.3.1 Constructing the Curve

The construction of the 1D Z-order curve from the the d-dimensional hypercube is accomplished by interleaving the bits of the binary representation of each grid point index $(i_0, \ldots i_{d-1})$ to form an integer offset, z, along the curve. If each dimension index, i_k is expressed in binary with n bits as $b_k^0 b_k^1 \ldots b_k^{n-1}$, then the 1D reference, z, containing nd bits, is constructed by interleaving the bits of each d-dimensional index starting with the slowest changing dimension. For example, in the 2D case with grid point addresses given by (i_0, i_1), i_1 varying slowest, z would be given by the string of $2n$ bits: $b_1^0 b_0^0 \ldots b_1^{n-1} b_0^{n-1}$. The curve offsets, z, are then sorted from smallest to largest to provide the depth-first traversal order. This establishes a mapping from a point's d-dimensional address, $(i_0, \ldots i_{d-1})$, to a point's offset along the Z-order curve, z.

8.3.2 Progressive Access

As noted, changing the Z-order curve stride, s, can restrict the visitation of points in the tree up to a level j. More formally, a hierarchy of ordered sets of grid point Z-order curve offsets can be constructed, such that:

$$S_0 \subset S_1 \subset \ldots \subset S_{j_{max}}$$

where

$$S_j = \{z_j | z_j = si, i \in \mathbb{Z}, 0 \leq si < 2^{dj_{jmax}}\},$$

$s = 2^{d(j_{max}-j)}$, and the cardinality $|S_j| = 2^{dj}$.

A progressive access storage layout can be affected simply by ordering the data on disk from S_0 to $S_{j_{max}}$, with the order of elements within each S_j given by z. Any approximation level in the hierarchy can then be accessed, reading only those points contained in S_j. An added benefit of preserving the Z-ordering within each S_j is the locality of points. The data are effectively partitioned into blocks, which can offer substantially improved I/O performance, over row (column) major ordering on block-based storage when subsets of the data are read.

There is one problem with the scheme just described: data replication.

Each set S_j is a subset of S_{j+1}. To avoid data replication a set differences is constructed, such that:

$$S'_j = S_j - S_{j-1}.$$

S'_j is then stored, instead of S_j. Reconstructing S_j from S'_j requires reading all S'_i, where $0 \leq i \leq j$, and forming the union $S_j = S'_0 \cup \ldots \cup S'_j$. In each of the Figures 8.1a–d, a circle enclosed inside a square is used to denote a grid point added through the union of the sets S'_j and S_{j-1}.

There are a few additional properties of the Z-order curve that are helpful to support the construction and access of S'_j.

The level in the hierarchy that a point with Z-order curve index z belongs to is given by:

$$j = \begin{cases} 0 & \text{if } z = 0, \\ j_{max} - \lfloor \frac{n}{d} \rfloor & \text{if } z > 0, \end{cases} \tag{8.1}$$

where 2^n is the largest power-of-two factor of z.

The cardinality of S'_j is

$$|S'_j| = \begin{cases} 1 & \text{if } j = 0, \\ 2^{dj} - 2^{d(j-1)} & \text{if } j > 0. \end{cases} \tag{8.2}$$

Finally, the z value from a point in the set S'_j can be reconstructed as:

$$z = (l + \lfloor \frac{l}{2^d - 1} \rfloor + 1)2^{d(j_{max} - j)}, \tag{8.3}$$

where l is the integer offset of the point within the set (recall set elements are ordered by z).

When reordering data on disk from row (column) major order to Z-order curve, Equation 8.1 allows the user to walk the Z-order curve and determine what point belongs to which set of S'_j. When reconstructing row (column) major order from Z-order curve data stored on disk, Equation 8.3 provides a point's z offset based on its position in S'_j.

There are a few other items worth noting. No floating point calculations are performed on the data in the scheme described above and the storage of the data is simply reordered in a manner that lends itself to progressive access and subregion extraction. Hence, the reconstructed data at the highest level of the hierarchy are bit-for-bit identical to the original data. In general, this will not be true of other progressive access models. A less desirable aspect of the approach is that data coarsening is performed through simple subsampling. The quality of the coarsened approximations will not be as high as when achieved by other methods.

8.4 Wavelets

In the preceding section, the Z-order curve offers a convenient way to create and access a multiresolution hierarchy composed of the nodes of a gridded hypercube. While the method allows for bit-for-bit identical reconstruction of the original grid and does not impose any additional storage overhead, the restriction to the grids with regular symmetry and power-of-two dimensions, or the crudeness of coarsened approximations created by sub-sampling, may prove limiting. In this section, both of these shortcomings are addressed, but they give up exact, bit-for-bit reconstruction.

Consider the following set of samples from a 1D signal:

$$c_3[n] = [1\ 3\ 3\ 5\ 9\ 7\ 7\ 5].$$

The eight samples can be approximated with pair-wise, unweighted averaging to produce signal at half the resolution:

$$c_2[\frac{n}{2}] = [2\ 4\ 8\ 6].$$

The samples in $c_2[\frac{n}{2}]$ are constructed by

$$c_{j-1}[i] = \frac{1}{2}(c_j[2i+1] + c_j[2i]). \tag{8.4}$$

Now consider the signal constructed with

$$d_{j-1}[i] = \frac{1}{2}(c_j[2i+1] - c_j[2i]). \tag{8.5}$$

So, for the $d_2[\frac{n}{2}]$ example:

$$d_2[\frac{n}{2}] = [1\ 1\ -1\ -1].$$

The signal $d_2[\frac{n}{2}]$, when combined with $c_2[\frac{n}{2}]$, provides a way to reconstruct the original $c_3[n]$:

$$\begin{aligned} c_j[2i] &= c_{j-1}[i] - d_{j-1}[i], \\ c_j[2i+1] &= c_{j-1}[i] + d_{j-1}[i]. \end{aligned} \tag{8.6}$$

The samples in c_j are referred to as *approximation* coefficients, and those in d_j, which represent the differences or the error introduced by averaging two neighboring data values, are referred to as *detail* coefficients. The calculations given by Equations 8.4 and 8.5 can be recursively applied to c_j until a single approximation coefficient is left—the average of all the samples in our original signal—and, in this example, seven detail coefficients:

$$c_0, d_{0\ldots2} = [5\ 2\ 1\ -1\ 1\ 1\ -1\ -1].$$

Thus, a linear transform has been applied to the original signal that allows

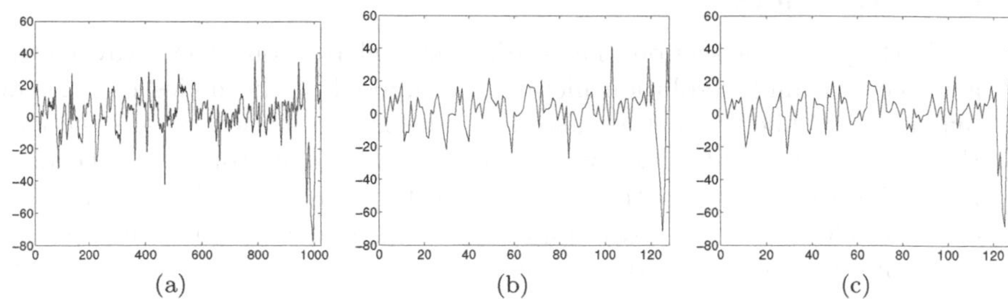

FIGURE 8.2: PDA based on multiresolution. The original signal containing 1024 samples (a), reduced to $1/8^{th}$ resolution using subsampling, analogous to the Z-order curve (b); and the Haar wavelet approximation coefficients (c).

a reconstruction of an approximation with 1, 2, 4, or 8 samples, but without any storage overhead. In the context of data read from disk, the coarsest approximation, c_0, can be retrieved by reading a single sample, the next approximation level, c_1, by reading one additional coefficient, d_0, and applying Equation 8.6, and the next, by reading two more coefficients, and so on. Multiple dimensions can be easily generalized by averaging neighboring points along each dimension to generate approximation coefficients, and generate detail coefficients by taking the difference between the approximation coefficients and each of the points used to find its average.

Mathematically, this is the 1D, unnormalized, Haar wavelet transform. The Haar wavelet transform, and other wavelet transforms that are discussed later, constructs a multiresolution representation of a signal. In one dimension, each multiresolution level j contains, on average, half the samples of the $j+1$ level approximation, with coarser level samples created by averaging finer level samples. Note, while the Haar transform exhibits what is known in the digital signal processing world as the property of *perfect reconstruction*, due to limited floating point precision the signal reconstructed from the c_j and d_j coefficients will not, as was the case with the Z-order curve, be an exact copy of the original signal. Moreover, even if the input signal consists entirely of the integer data the division by two in Equation 8.4 will result in floating point coefficients. However, integer-to-integer wavelet transforms do exist [3], but they are not discussed here.

Figure 8.2 shows a qualitative comparison of a 1D sampling of the x-component of vorticity obtained from a 1024^3 simulation by Taylor-Green turbulence [7]. The original signal is seen in Figure 8.2a. The signal generated by nearest neighbor sampling (analogous to Z-order curve approximation) at $1/8^{th}$ resolution is shown in Figure 8.2b, and at the same coarsened resolution, approximated with the Haar wavelet, as in Figure 8.2c.

Equations 8.4–8.6 provide everything that is needed to support a PDA scheme based purely on a multiresolution hierarchy, not suffering from the limitations of the Z-order curve, but giving up bit-for-bit identical reconstruction. One drawback of both the Z-order curve and the wavelet based multiresolution approach is that they fail to take advantage of any coherence in the data. Better approximations of a function can be achieved for a given number of coefficients, by further exploiting some of the properties of wavelets.

8.4.1 Linear Decomposition

Often it is advantageous to represent a signal or function f as a superposition:

$$f(t) = \sum_k a_k u_k(t) \qquad k \in \mathbb{Z}, a \in \mathbb{R}, \tag{8.7}$$

where a's are *expansion coefficients*, and $u_k(t)$'s are a set of functions of time, t, that form a basis for L^2—the space of square, integrable (finite energy) functions. If, for example, u_k's are complex exponentials, then the expansion is a Fourier series of f. More generally, if the basis functions, $u_k(t)$, are orthogonal, that is if

$$\langle u_k(t), u_l(t)\rangle = \int u_k(t)u_l(t)dt = 0 \quad l \neq k \tag{8.8}$$

then the coefficients a_k can be easily calculated by the inner product:

$$a_k = \langle f(t), u_k(t)\rangle = \int f(t)u_k(t)dt. \tag{8.9}$$

The wavelet expansions of signals are also superpositions of basis functions but they are represented as a two-parameter expression:

$$f(t) = \sum_k c_{j_{min},k}\phi_{j_{min},k}(t) + \sum_j\sum_k d_{j,k}\psi_{j,k}(t), \tag{8.10}$$

where $c_{j,k}$ and $d_{j,k}$ are again real-value coefficients. $\phi_{j,k}(t)$ and $\psi_{j,k}(t)$ are the *scaling* and *wavelet* basis functions, respectively. The parameterization of time (or space) is provided by k, while frequency or scale, not coincidentally, is parameterized by j.

Several properties of wavelets and the above wavelet expansion are applicable to all wavelets discussed in this section:

1. The wavelet expansion computes efficiently. For discrete $f(t)$, the upper bound on the parameter j is $log_2(N)$, where N is the number of samples in f. Similarly, for many wavelets both $\phi_{j,k}(t)$ and $\psi_{j,k}(t)$ have *compact support* and are zero-valued outside of a small interval of t. Thus, the expansion is computed with only a handful of multiplications and additions, making its complexity $O(N)$. The same is true for the wavelet transform used to compute $c_{j,k}$ and $d_{j,k}$. Compare this to $O(Nlog_2(N))$ complexity for the Fourier expansion.

2. Unlike the Fourier transform, the wavelet transform localizes in time (space) the frequency information of a signal. Most of the information or energy of a signal is represented by a small number of expansion coefficients (a property exploited later for progressive data access).

3. Finally, Equation 8.10 provides a multiresolution decomposition of f. The first term of the right-hand side of the equation describes a coarsened, low-resolution approximation of f, while the second term, for increasing j, provides finer and finer details—that represent the missing information not contained in the first term. Another view is that the first term contains the low frequency components of f, while the second term provides the high frequency components of f. Hence, the labels *approximation* and *detail* are applied to the coefficients, c_j and d_j, respectively.

8.4.2 Scaling and Wavelet Functions

Up to this point, scaling and wavelet functions have been discussed without formally defining them. Unlike the Fourier basis functions, the complex exponentials, the wavelet and scaling functions are an infinitely large family of functions. Here, the discussion will strictly pertain to normalized, orthogonal, or *orthonormal*, families of wavelets as defined by the relations:

$$\left.\begin{aligned} \langle \phi_{j,k}(t), \phi_{j,l}(t)\rangle &= \delta(k,l) \\ \langle \psi_{j,k}(t), \psi_{j,l}(t)\rangle &= \delta(k,l) \\ \langle \phi_{j,k}(t), \psi_{j,l}(t)\rangle &= 0 \end{aligned}\right\} \quad \text{for all } j, k, \text{ and } l, \tag{8.11}$$

where $\delta(k,l)$ is 1 for $k = l$, otherwise $\delta(k,l)$ is 0.

The multiresolution properties of the wavelet expansion arise from the scaling function and the ability to recursively construct this basis function from a canonical version of itself. That is

$$\phi(t) = \sum_n h_0[k]\sqrt{2}\phi(2t - n) \quad n, k \in \mathbb{Z}. \tag{8.12}$$

Equation 8.12 is a *dilation* equation; it shows how to construct the scaling function, $\phi(t)$, from the superposition of scaled, translated, and dilated copies of itself. The translation $(t - n)$ shifts $\phi(t)$ to the right. The multiplier $2t$ narrows (dilates) $\phi(t)$, and $\sqrt{2}$ scales the function. In other words, $\phi(t)$ is expressed as a linear expansion, similar in form to Equation 8.7, but by using itself as a basis! By convention, the expansion coefficients are relabeled from a_k to $h_0[k]$.

For the Haar basis function discussed in the beginning of 8.4, the canonical form of $\phi(t)$ is:

$$\phi(t) = \begin{cases} 1 & 0 \leq t < 1, \\ 0 & otherwise. \end{cases} \tag{8.13}$$

Thus, the Haar scaling function is a box of unit height over the interval $[0, 1)$. The expansion coefficients, $h_0[k]$, for the Haar scaling function are $h_0[0] = h_0[1] = 1/\sqrt{2}$.

Using shifted, scaled, and dilated versions of the scaling function $\phi(t)$ alone, we can construct a multiresolution hierarchy of $f(t)$. However, much like the first attempt with the Z-order curve, this leads to more replications. For these and other reasons that are beyond the scope of this text, the wavelet expansion includes a second basis function, $\psi(t)$, that provides a means to capture the information that is missing from the approximation based on $\phi(t)$.

Similar to the scaling functions, the wavelets are also defined by the canonical scaling function:

$$\psi(t) = \sum_n h_1[k]\sqrt{2}\phi(2t - n). \tag{8.14}$$

Here, h_1 are a different set of expansion coefficients from h_0. Without proof, the orthogonal wavelets h_0 and h_1 are related to each other by

$$h_1[n] = (-1)^n h_0(N - 1 - n), \tag{8.15}$$

where N is the support size of h_0.

Thus, for the Haar wavelet, the non-zero coefficients in Equation 8.14 are $h_1[0] = 1/\sqrt{2}$ and $h_1[1] = -1/\sqrt{2}$. The canonical Haar wavelet function is therefore:

$$\psi(t) = \begin{cases} 1 & 0 \leq t < 1/2, \\ -1 & 1/2 \leq t < 1, \\ 0 & otherwise. \end{cases} \tag{8.16}$$

A multitude of wavelet families exist, each with a variety of parameterizations. A reasonable question to ask is: If Haar wavelets possess all of the multiresolution properties required, why consider other wavelets? Again, an in-depth treatment is beyond the scope of this text, and the remarks will be limited to those properties related only to the goals of this discussion.

The tendency of wavelet transform expansion coefficients is to quickly become small in magnitude [5]. Many signals can be accurately approximated by only a small number of coefficients (see 8.4.4). The Haar wavelets, constructed from box functions, are the most efficient in representing signals that are piecewise-constant; spans of the signal that are constant-valued between dyadic points (multiples of powers of two) can be represented by approximation coefficients alone, and without any need for detail coefficients. While smoother signals, in fact, any signal in $L2$, can be represented by the Haar basis, the number of non-zero detail coefficients will, in general, be large. To represent signals sparsely, a basis is required whose properties, in some way, match those of the signal. One measure of the wavelet function's ability to match a signal is given by the number of vanishing moments, p, where,

$$\int x^k \psi(t)dt = 0 \quad \text{for } 0 \leq k < p, \tag{8.17}$$

which shows that ψ is orthogonal to any polynomial of degree $p - 1$. If f resembles a Taylor polynomial of degree $k < p$ over some interval then the detail coefficients of the wavelet transform of f will be small at fine scales 2^j. For this reason much research has gone into the design of wavelets that have support of minimum size for a given number of vanishing moments, p. Not surprisingly, within a family of wavelets, increasing vanishing moments comes at the cost of wider support.

8.4.3 Wavelets and Filter Banks

Now that the wavelets and wavelet transforms have been reviewed, this subsection explores the application of progressive data access from a digital filter bank point of view. The discussion of this point of view is necessary because the digital filter bank matches the discrete, gridded data better than the function expansion point of view. Now, instead of a continuous function $f(t)$, consider a discretely sampled, finite function, $x[n]$, $0 \leq n < N$, where N is the total number of samples. To compute the wavelet expansion coefficients of Equation 8.10 there is no need to actually evaluate the scaling or wavelet function. The normalized scaling and wavelet expansion coefficients of Equations 8.12 and 8.14—h_0 and h_1—are the only coefficients needed to compute $c_{j,k}$ and $d_{j,k}$. Without proof, the following relations are known as the *Discrete Wavelet Transform* (DWT):

$$c_{j,k} = \sum_n \tilde{h}_0[2k - n]c_{j+1}[n], \tag{8.18}$$

$$d_{j,k} = \sum_n \tilde{h}_1[2k - n]c_{j+1}[n], \tag{8.19}$$

where $\tilde{h}_0(n) = h_0(-n)$ and $\tilde{h}_1(n) = h_1(-n)$. These equations show that the approximation and detail coefficients for a level j are computed via a simple two channel filter bank, recursively applied in a process known as *analysis* (see Fig. 8.3a). Analysis is equivalent to convolving the input signal, c_{j+1}, with time-reversed copies of the filter coefficients, $h_0[-n]$ and $h_1[-n]$, and then downsampling the filtered signal by discarding every odd term. The analysis filter implemented by h_0 is a low-pass filter, and the one implemented by h_1 is a high-pass filter. Thus, the filters split the signal into its low frequency and high frequency components. The cascade of outputs formed by each iteration, or pass, of the filter pairs divides the frequency spectrum into a logarithmic sequence of bands. By downsampling the results of each convolution, the total number of non-zero coefficients is held constant through each pass.

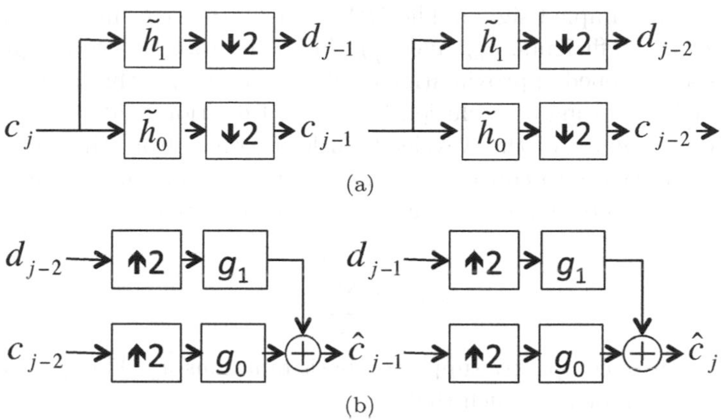

FIGURE 8.3: The *analysis* of the input signal, c_j, is performed by a cascade of convolutions, followed by downsampling steps in (a). Reconstruction of c_j is performed by a cascade of upsampling, followed by convolution steps in a process known as *synthesis* in (b).

A reconstructed signal, $\hat{c}_{j,k}$, is obtained with the *Inverse Discrete Wavelet Transform* (IDWT) in a process known as *synthesis*, where,

$$\hat{c}_{j,k} = \sum_n g_0[k-2n]c_{j-1}[n] + g_1[k-2n]d_{j-1}[n]. \tag{8.20}$$

Analogous to Equations 8.18 and 8.19, the IDWT is equivalent to upsampling c_j and d_j by the introduction of zeros in every other sample and summing the convolutions of c_j and d_j with g_0 and g_1, respectively, as illustrated in Figure 8.3b. To ensure that $\hat{c}_{j,k} = c_{j,k}$, the inverse transform must exhibit the property of *perfect reconstruction.* The reconstruction filters, g_0 and g_1, must be chosen accordingly. But how does one know that g_0 and g_1 even exist? Conveniently, for orthogonal filters the reconstruction (synthesis) filters do exist and are related to the analysis filters by:

$$\begin{aligned} g_0[n] &= h_0[n] = \tilde{h}_0[N-n], \\ g_1[n] &= h_1[n] = \tilde{h}_1[N-n]. \end{aligned} \tag{8.21}$$

An obvious question that arises is this: Given a signal $x[k]$, how are the initial $c_{j_{max},k}$ computed? Fortunately, if the samples of $x[k]$ are computed above the Nyquist rate, it can be assumed that $c_{j_{max},k} \approx x[k]$.

8.4.4 Compression

Now, all the machinery is in place to provide a multiresolution representation of a 1D sampled signal. The DWT transforms $x[n]$ into a wavelet space sequence of coefficients, $c_{j,k}$ and $d_{j,k}$. From the wavelet space representation there is a coarsened approximation of $x[n]$ given by $c_{j=0}$ that can progressively be refined by adding a resolution by way of Equation 8.20.

The real power of the wavelet transform comes from the aforementioned ability of wavelets to concentrate energy into a small number of coefficients. Assume a discrete signal, $x[n]$, expressed as the expansion

$$x[n] = \sum_{k=0}^{K-1} a_k u_k[n],$$

for some set of compactly supported basis functions u_k. The goal is to find an approximation of $x[n]$, such that,

$$\tilde{x}[n] = \sum_{k=0}^{\tilde{K}-1} \tilde{a}_k u_k[n],$$

where $\tilde{K} < K$. Moreover, for a given $\tilde{K}$, it is desired that the approximation $\tilde{x}[n]$ to be the best possible approximation for $x[n]$ by some error metric. One possibility is $||x[n] - \tilde{x}[n]||_p < \epsilon$ for some norm, p. For an orthonormal basis, for $p = 2$,

$$||x[n] - \tilde{x}[n]||_2^2 = \sum_{k=\tilde{K}}^{K-1} (a_{\pi(k)})^2, \tag{8.22}$$

where $\pi(k)$ is a permutation of $0 \ldots K-1$ such that

$$|a_{\pi(0)}| \geq \ldots \geq |a_{\pi(K-1)}|,$$

and $\tilde{x}$ uses the coefficients corresponding to the first $\tilde{K} - 1$ elements of $\pi(k)$. Formally:

$$\tilde{a}_k = \begin{cases} a_k & \text{if } \pi(k) < \tilde{K}, \\ 0 & \text{otherwise.} \end{cases}$$

Thus, the L^2 error of $\tilde{x}[n]$ is given by the sum of the square of the coefficients $a[k]$ that are not included in the expansion of $\tilde{x}[n]$. To minimize the L^2 error for a given $\tilde{K}$, a_k is sorted by decreasing the magnitude and only the $\tilde{K} - 1$ largest coefficients are included in the approximation. Alternatively, if a particular ϵ is wanted, Equation 8.22 allows the calculation of how many and which coefficients can be discarded to ensure $||x[n] - \tilde{x}[n]||_p < \epsilon$. Note, for any $a[k] = 0$ its exclusion introduces no error in the approximation.

Equation 8.22 sheds some light on the choice of wavelet. As discussed earlier (see Eq. 8.17), a wavelet possessing a higher number of vanishing moments will permit the compaction of more energy (information) into fewer expansion coefficients. Figure 8.4 qualitatively compares the compression of a signal by a factor of 8, by using wavelets with different numbers of vanishing moments. For ease of comparison, Figures 8.4a and 8.4b reproduce Figures 8.2a and 8.2b, respectively. Results using the Haar (Fig. 8.4c) and Daubechies D4 (Fig. 8.4d) wavelet are shown with one and two vanishing moments, respectively. Note the "blockiness" resulting from the box shape of the Haar wavelet. Also note that while the multiresolution approximation of Figure 8.4b may appear visually more appealing than Figure 8.4c, many extreme values in the original signal are lost, which is the result of the cascade of applications of the low-pass scaling filter.

Finally, unlike the pure multiresolution approaches based on Z-order curve and wavelets, an arbitrary number of approximations are now produced. In the extreme case, the approximation is refined one coefficient at a time.

8.4.5 Boundary Handling

Finite length signals present a couple of challenges that have been ignored until this point. Consider the application of the DWT to a signal $x[n]$ with a low-pass filter h of length $L = 4$. By Equation 8.18, the computation of $c[0]$ is given by

$$c[0] = \tilde{h}[3]x[-3] + \tilde{h}[2]x[-2] + \tilde{h}[1]x[-1] + \tilde{h}[0]x[0].$$

But, for a finite signal, the samples $x[-3]$, $x[-2]$, and $x[-1]$ do not exist! One simple solution is to extend $x[n]$ so that it is defined for $n < 0$ and $n \geq N$. There are a number of choices for the extension values, such as zero padding or repeating the boundary samples, each with its own set of trade-offs. In the general case, however, if the input signal, x, is extended, creating a new input signal, x_e of length $N_e > N$, the resulting output after filtering and downsampling no longer has exactly N non-zero and non-redundant approximations and detail coefficients. Moreover, perfect reconstruction of x_e requires all of the N_e coefficients. Due to downsampling in the case of the DWT, each iteration of the filter bank requires an extension of the incoming approximation coefficients. For a 1D signal, this overhead may not be significant, but later, when moving on to 2D and 3D grids, the additional coefficients can consume a substantial amount of space.

A filter bank that is *nonexpansive* is more preferable, as it preserves the number of input coefficients on output, and does not introduce discontinuities at the boundaries. The solution is the employment of symmetric filters combined with a symmetric signal extension. Only the case of signals of length $N = 2^n$, and odd length filters where the symmetry is about the center boundary coefficients, is considered here. For an extensive discussion on symmetric

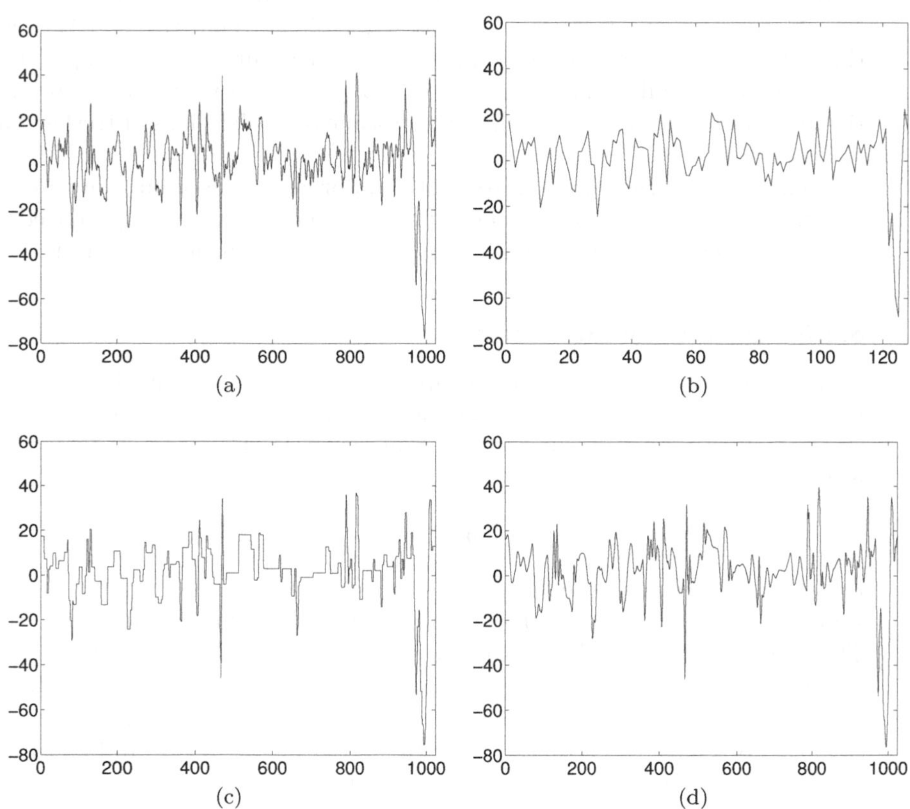

FIGURE 8.4: A test signal (a) with 1024 samples and a multiresolution approximation, (b), at $1/8^{th}$ resolution, reproduced from Figure 8.2. The test signal reconstructed using $1/8^{th}$ of the expansion coefficients with the largest magnitude from the Haar, (c), and Daubechies 4-tap wavelet, (d), respectively.

filters of even length, which introduce considerably more complexity, or handling $N \neq 2^n$ (see [1, 12]). For clarity of exposition, assume that the filter is *whole-sample* symmetric about $h[0]$. That is, $h[n] = h[-n]$, and $h[0]$ is not repeated. For example, if $L = 3$, the center point is $h[0]$, and $h[-1] = h[1]$. Moreover, the input signal must be made a whole-sample symmetric about the first $(n = 0)$ and last $(n = N - 1)$ samples. The motivation for this symmetry is straightforward: the output samples for $n < 0$ and $n \geq N$ are redundant and need not be explicitly stored. Consider the calculation in a general linear transform of the expansion coefficient $a[-1]$, by the convolution of x with $h[-n]$ for whole-sample symmetric x, and h, with $L = 3$:

$$a[-1] = h[0]x[-2] + h[1]x[-1] + h[2]x[0].$$

Due to symmetry, $h[n] = h[-n]$, and $x[n] = x[-n]$. Therefore:

$$a[1] = a[-1] = h[0]x[2] + h[1]x[1] + h[2]x[0],$$

and $a[-1]$ is redundant and need not be explicitly stored!

Things become a little more complicated with the DWT, which operates as a dual channel convolution filter, followed by downsampling. Here, centering the filter must be done carefully for each channel, such that symmetry is preserved after downsampling, and the total number of coefficients output by the two channels, equals the number of input coefficients. For even N, each channel outputs exactly $N/2$ samples.

Symmetric filters combined with a symmetric signal extension provide a straightforward mechanism for dealing with finite length signals and the DWT. Unfortunately, with the exception of the Haar wavelet, there are no orthogonal wavelets with compact support possessing both the property of symmetry and perfect reconstruction. The solution to this dilemma is the relaxation of the orthogonality requirement and the introduction of *biorthogonal* wavelets. For biorthogonal wavelets, the properties of Equation 8.11 no longer hold. Different *analysis* scaling and wavelet basis functions, $\tilde{\phi}(t)$ and $\tilde{\psi}(t)$, and *synthesis* scaling and wavelet basis, $\phi(t)$ and $\psi(t)$, must be introduced. Similarly, new analysis, $\tilde{h}_0$ and $\tilde{h}_1$, and synthesis filters, g_0 and g_1, will appear.

From a filter bank perspective, h_1 no longer relates to h_0 by a simple expression. However, the following cross relationship between synthesis and analysis filters hold:

$$\tilde{h}_0[n] = (-1)^n g_1(N - 1 - n), \quad g_0[n] = (-1)^n \tilde{h}_1(N - 1 - n), \qquad (8.23)$$

where N is the support size of the filter.

With the analysis filters no longer related to each other by Equation 8.15, the support of these respective filters need not be the same. The support sizes of $\tilde{h}_0$ and $\tilde{h}_1$ are then denoted as L_0 and L_1, respectively, leading to new analysis equations:

$$c_{j,k} = \sum_{n=-(L_0-1)/2}^{(L_0-1)/2} \tilde{h}_0[n]c_{j+1}[2k-n], \tag{8.24}$$

$$d_{j,k} = \sum_{n=-(L_1-1)/2}^{(L_1-1)/2} \tilde{h}_1[n]c_{j+1}[2k-n+1]. \tag{8.25}$$

Note that Equation 8.24 processes the even samples of c_{j+1}, while Equation 8.25 processes the odd samples—a necessity for nonexpansive output [12].

Because of the shift in the inputs to the analysis equation, a new synthesis equation is also necessary, which must shift the $d[n]$ coefficients:

$$\hat{c}_{j,k} = \sum_n g_0[k-2n]c_{j-1}[n] + g_1[k-2n-1]d_{j-1}[n]. \tag{8.26}$$

As already noted, the correct behavior of Equations 8.24 and 8.25 is predicated on treating the input signal, $c_{j+1}[n]$, as exhibiting whole-sample symmetry about the left and right boundary. Perfect reconstruction from Equation 8.26 requires a mixture of whole-sample and *half-sample* symmetry, where the point of symmetry lies halfway between two samples. Signals with the left boundary, half-sample symmetry are symmetric about the non-integer point $-\frac{1}{2}$, while right boundary, half-sample symmetric signals are symmetric about the point $N-\frac{1}{2}$—for the left boundary, $x[-1]=x[0]$, $x[-2]=x[1]$ and so on. For Equation 8.26, the left and right boundaries of $c[n]$ and $d[n]$ must be made whole-sample symmetric, respectively, while the right and left boundaries of $c[n]$ and $d[n]$ must be half-sample symmetric, respectively. Lastly, for both the symmetric analysis filters, $\tilde{h}_0[n]$ and $\tilde{h}_1$[n], and the symmetric synthesis filters, g_0[n] and g_1[n], the filters have a zero value for $n < (L-1)/2$ and $n > (L-1)/2$, where L is the support size of the filter.

Although symmetry is gained and the number of inputs and output are preserved, by giving up orthogonality, other important properties are lost. Most significantly, Equation 8.22 no longer holds, and the L^2 error between a signal x and its approximation $\tilde{x}$ can no longer be determined by the coefficients (not included in the construction of $\tilde{x}$). Nevertheless, the L^2 error, introduced in the reconstruction after discarding coefficients, is still minimized by discarding the coefficients of smallest magnitude.

Figure 8.5 shows plots of the Cohen-Daubechies-Feauveau (CDF) 9/7 biorthogonal synthesis and the analysis functions. Note the symmetry in all of the functions. Table 8.1 provides the filter coefficients for the CDF 5/3 and 9/7 normalized, biorthogonal wavelets. These filters are the foundation for the JPEG2000 image compression standard. The reader is cautioned that the naming of the CDF family of biorthogonal wavelets is inconsistent in the literature. Here, the naming convention based on the support size in the low-pass analysis and synthesis filters, respectively (e.g., CDF 9/7), is adopted. Other authors use a naming scheme based on the number of analysis and synthesis filter vanishing moments, respectively (e.g., bior4.4).

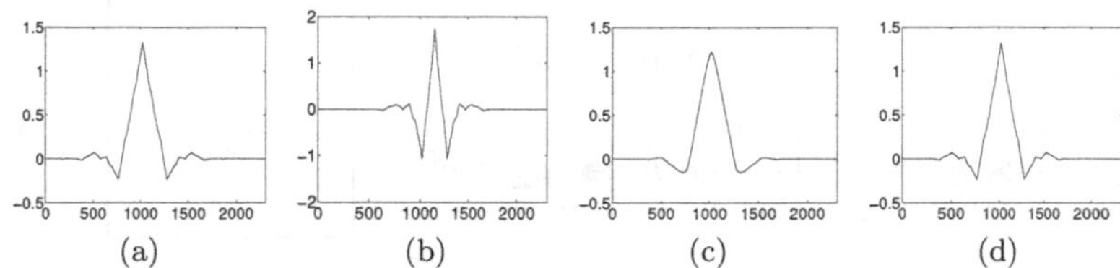

FIGURE 8.5: The CDF 9/7 biorthogonal wavelet and scaling functions: analysis scaling (a) and wavelet functions (b), and synthesis scaling (c) and wavelet functions (d).

TABLE 8.1: Biorthogonal wavelet filter coefficients for the CDF 5/3 (top) and 9/7 (bottom) wavelets

p	n	$h_0[n]$	$g_0[n]$
2	0	1.06066017177982	0.70710678118655
	$-1, 1$	0.35355339059327	0.35355339059327
	$-2, 2$	-0.17677669529663	0
4	0	0.852698679008894	0.788485616405583
	$-1, 1$	0.377402855612831	0.418092273221617
	$-2, 2$	-0.110624404418437	-0.0406894176091641
	$-3, 3$	-0.023849465019557	-0.0645388826286971
	$-4, 4$	0.037828455507264	0.0

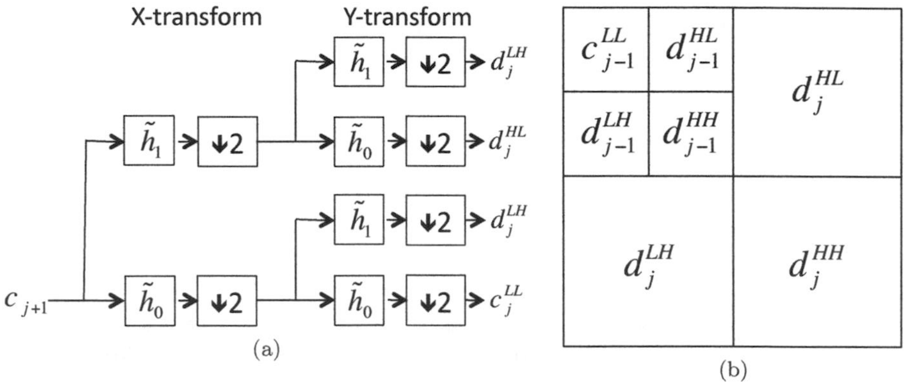

FIGURE 8.6: On the left, (a) shows a single pass of the 2D DWT resulting in one set of approximation coefficients, and three sets of detail coefficients. The right side of the figure (b) shows the resulting decomposition after two passes of the DWT. As each coefficient is the result of two filtering steps, one along each dimension, the superscripts in (a) and (b) indicate highpass, H, or lowpass, L, filtering.

8.4.6 Multiple Dimensions

Extending the 1D wavelet filter bank to multiple dimensions is straightforward. The 1D analysis filter is simply applied along each dimension is illustrated in the 2D example in Figure 8.6. Thus, for a $M \times N$ grid, a single pass of the 2D DWT yields, on average $\frac{MN}{4}$ approximation coefficients and $\frac{3MN}{4}$ detail coefficients. For 3D data, seven times as many detail coefficients as approximation coefficients are generated for each iteration of the DWT.

8.4.7 Implementation Considerations

There are two possible approaches used to construct a PDA model, based on the forward and inverse DWT. A multiresolution hierarchy can be constructed, just like with the Z-order space-filling curve, and by exposing the scale parameter, j, in Equation 8.10 the grid may be coarsened or refined by factors of 2^d. This approach is called *frequency truncation.* Each iteration of the analysis filters uses the normalized coefficients provided in Table 8.1, which scales the amplitude of the approximation coefficients by $\sqrt{2.0}$. If the approximation coefficients are used as an approximation of the original signal, this scaling should be undone by multiplication by $2^{-1/j}$, where j is the number of iterations of the analysis filter. Alternatively, the wavelet expansion representation's power can be exploited to concentrate most of a signal's information content into a small number of expansion coefficients. This approach is called *coefficient prioritization.* With either method, a number of practical implementation issues arise.

8.4.7.1 Blocking

The multidimensional DWT, discussed in 8.4.6, can be applied directly to a d-dimensional grid. However, for high-resolution grids, there are a number of reasons for considering decomposing the grid into a collection of smaller blocks and applying the DWT to each individual block. Computational efficiency is one consideration. Significantly improved performance may be achieved on cache-based microprocessors by operating on a collection of smaller blocks that better accommodate the memory hierarchy rather than one large, single volume. The second reason, and perhaps the more compelling one, is efficient region extraction. For many visualization workflows—such as drawing a cutting plane through a 3D volume or zooming in and volume rendering a small subregion of a larger volume—only a subset of the data is required. By decomposing the volume into blocks, it is possible to access (from storage) only those blocks that intersect the region of interest. The merits of data blocking with regard to I/O are discussed in *Hierarchical and Geometrical Methods in Scientific Visualization* [9].

The best choice of block size is somewhat application dependent. Given the dyadic nature of the wavelet, and the complexities of boundary handling discussed in 8.4.5, power-of-two dimensions are sensible. As each pass of the DWT reduces the number of approximation coefficients by half along each dimension, larger block sizes permit more passes of the DWT resulting in a deeper multiresolution hierarchy. A deeper hierarchy offers more refinement levels for a progressive access scheme based on frequency truncation. Similarly, if coefficient prioritization is employed, more coefficients present greater degrees of freedom for compression. However, if the block size is too large, the goal of performing efficient region extraction and computational performance may be defeated. As a reasonable compromise for 3D grids, the suggested block sizes are 64^3 or 128^3.

For data that is not block-aligned, padding of the boundary blocks is required. Care should be taken to use an extension strategy that does not introduce sharp discontinuities. Constant, linear extrapolation, or symmetric extension, generally produces reasonable results.

8.4.7.2 Wavelet Choice

The advantages of symmetric, biorthogonal wavelets have been discussed; and filter coefficients for two wavelets commonly used have been provided in image compression applications in 8.4.5. Other biorthogonal wavelets possessing more vanishing moments and, therefore, greater information compaction capability also exist at the cost of additional filter taps. It is worth noting that while filters with better energy compaction capabilities may yield sparser representations, the additional filter coefficients incur additional computational cost. Also, the wider the filter the greater the number of input coefficients required, which may limit the possible number of cascades of the DWT. For example, with a block of dimension 64^3 after three stages of the DWT, there are 8^3 approximation coefficients—too few for another pass with the nine-tap CDF9/7 filter, but sufficient for one more stage of the narrower, five-tap CDF5/3 filter.

8.4.7.3 Coefficient Addressing

An important attribute of regular grids is the implicit addressing of grid vertices. For example, each vertex in the grid can be addressed by an integer index, (i, j, k) in 3D, whose offsets can be implicitly determined by the serialized storage order. Therefore, only the sampled field value for each grid point needs to actually be stored. The implicit addressing of expansion coefficients is easily preserved during storage with frequency truncation. The coefficients are simply written in the order that they are output from each stage of the filter bank. With coefficient prioritization, however, the coefficients must be ordered by their information content. Their addresses are no longer implicit by their position in the output stream, and must be explicitly preserved. The number of bits required to uniquely address each of the N expansion coefficients is $log_2(N)$, which introduces a sizeable storage overhead.

Consider, however, binning the N expansion coefficients output by the DWT (which will generically refer to as $c[k]$) into a collection of P sets,

$$S_p = \{c_{\pi(i)} | C(p-1) \leq i < C(p), 0 \leq i < N-1\}, \qquad (8.27)$$

where $C(0) = 0$, $C(p \geq 1) = \sum_1^p |S_p|$, and, as before, $\pi(i)$ is a permutation of $0 \ldots N-1$ such that $|c_{\pi(0)}| \geq \ldots \geq |c_{\pi(N-1)}|$. Then, if within each set S_p, the expansion coefficients are stored in their relative order of output from the filter bank, the addresses of the coefficients need only be stored for sets $S_1 \ldots S_{P-1}$. The addresses of the elements of S_P can be inferred from the addresses of $S_1 \ldots S_{P-1}$. By keeping track of the ordered set of addresses not found in $S_1 \ldots S_{P-1}$, the lowest address not found in $S_1 \ldots S_{P-1}$ is the address of the first coefficient found in S_P and so on. If $|S_P|$ is sufficiently large, the storage savings can be considerable.

8.4.8 A Hybrid Approach

Frequency truncation and coefficient prioritization each have their strengths and weaknesses. Frequency truncation preserves the implicit ordering of grid vertices, thereby imposing minimum storage overhead, but it offers only a very coarse grain control over the the level of detail available; incrementing or decrementing the j scaling parameter by one changes the number of grid points by a factor 2^d. Furthermore, unlike coefficient prioritization, all coefficients are treated equally regardless of their contribution to the signal. On the other hand, though, coefficient prioritization allows for arbitrary control over level of detail, and generally offers better approximations for a given bit budget, but prioritization requires additional storage to keep track of coefficient addresses. Perhaps, a less obvious difference between the two schemes is that, by virtue of possessing fewer grid points, multiresolution approximations (frequency truncation) benefit all the resources of the visualization pipeline including: physical memory, floating point calculations, and I/O. Frequency truncation, on the other hand, only benefits the transmission of the data (e.g., access to secondary storage). Once a signal is reconstructed—albeit even if from fewer expansion coefficients than samples in the original signal—the number of samples in the reconstructed signal is the same as for the original signal. After the reconstruction is finished, there is no further realized benefit of the wavelet expansion.

The merits of both methods may readily be combined with little or no additional effort. By allowing control over both the j parameter in Equation 8.26, which controls the resolution, and (if the coefficients are binned by their information content) the p parameter, then, Equation 8.27 controls the expansion coefficients used in reconstruction.

8.4.9 Volume Rendering Example

Figure 8.7 shows at varying approximations a direct volume rendering of an enstrophy field derived from a subregion of a 1024^3 isotropic and homogeneous Taylor–Green (TG) incompressible turbulence simulation [7]. The original data is shown in Figure 8.7a. Approximations based on coefficient prioritization, using reduction factors of $\frac{1}{500}$, $\frac{1}{100}$, and $\frac{1}{10}$, are shown in Figures 8.7b–d, respectively. The data was decomposed into 128^3 blocks, and transformed with the CDF9/7 wavelet.

8.5 Further Reading

The presentation on Z-order curves and their application in progressive data access is based almost entirely on the work of Pascucci and Frank. We refer the reader to their papers for further details on the method, suggested implementation strategies, and a discussion of their results with real data sets and applications [10, 9].

Wavelets are a relatively new field in mathematics with a wealth of applications related to data analysis that go far beyond progressive data access. Here we have only scratched the surface on their theory and their capabilities. For the interested reader, an excellent introduction on wavelets and wavelet transforms may be found in the books by Burrus et al. [2] and Stollnitz et al. [11]. For a deeper understanding we recommend the authoritative books by Mallat [6], and Strang and Nguyen [12], and the seminal work by Daubechies [4].

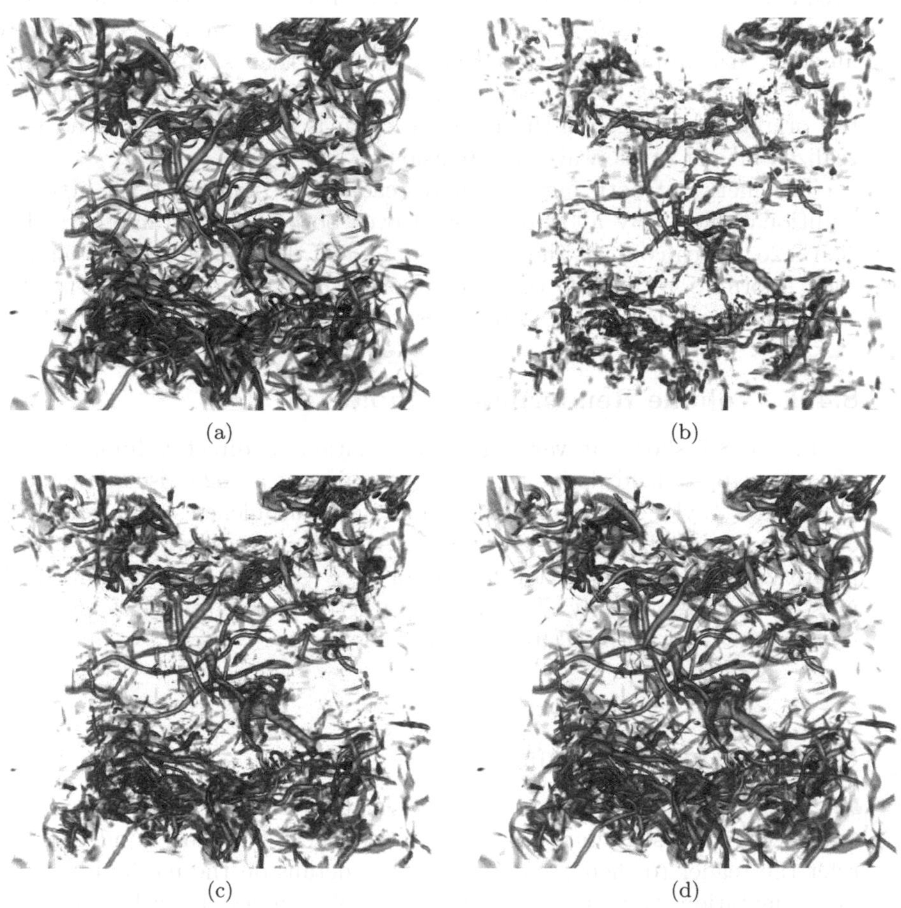

FIGURE 8.7: Direct volume rendering of an enstrophy field: original data (a), and shown in (b)–(d), respectively, are results after reduction of the number of expansion coefficients used in reconstruction by factors of $\frac{1}{500}$, $\frac{1}{100}$, and $\frac{1}{10}$.

References

[1] Christopher M. Brislawn. Classification of Nonexpansive Symmetric Extension Transforms for Multirate Filter Banks. *Applied and Computational Harmonic Analysis*, 3(4):337 – 357, 1996.

[2] Sidney C. Burrus, Ramesh A. Gopinath, and Haitao Guo. *Introduction to Wavelets and Wavelet Transforms: A Primer.* Prentice Hall, August 1997.

[3] A. R. Calderbank, Ingrid Daubechies, Wim Sweldens, and Boon-Lock Yeo. Wavelet Transforms that Map Integers to Integers. *Applied and Computational Harmonic Analysis*, 5(3):332–369, 1998.

[4] Ingrid Daubechies. *Ten Lectures on Wavelets.* Society for Industrial and Applied Mathematics, June 1992.

[5] D. L. Donoho. Unconditional Bases are Optimal Bases for Data Compression and for Statistical Estimation. *Applied and Computational Harmonic Analysis*, 1(1):100–115, 1993.

[6] Stephane Mallat. *A Wavelet Tour of Signal Processing, Third Edition: The Sparse Way.* Academic Press, December 2008.

[7] P. D. Mininni, A. Alexakis, and A. Pouquet. Large-Scale Flow Effects, Energy Transfer, and Self-Similarity on Turbulence. *Physical Review E*, 74:016303, Jul 2006.

[8] G. M. Morton. A Computer Oriented Geodetic Data Base and a New Technique in File Sequencing. Technical report, IBM Ltd., Ottawa, Ontario, Canada, 1966.

[9] V. Pascucci and R. J. Frank. Hierarchical Indexing for Out-of-Core Access to Multi-Resolution Data. In G.Farin, B. Hamann, and H. Hagen, editors, *Hierarchial and Geometrical Methods in Scentific Visualization*, pages 225–241. Springer-Verlag, Berlin, 2002. UCRL-JC-140581.

[10] Valerio Pascucci. Global Static Indexing for Real-time Exploration of Very Large Regular Grids. In *Proceedings of Supercomputing (SC01)*, 2001.

[11] Eric J. Stollnitz, Anthony D. Derose, and David H. Salesin. *Wavelets for Computer Graphics (The Morgan Kaufmann Series in Computer Graphics).* Morgan Kaufmann, January 1996.

[12] Gilbert Strang and T. Nguyen. *Wavelets and Filter Banks.* Wellesley-Cambridge Press, 1997.

Chapter 9

In Situ Processing

Hank Childs

Lawrence Berkeley National Laboratory

Kwan-Liu Ma, Hongfeng Yu

University of California, Davis and Sandia National Laboratories

Brad Whitlock, Jeremy Meredith, Jean Favre

Lawrence Livermore National Laboratory, Oak Ridge National Laboratory, and Swiss Center for Scientific Computing

Scott Klasky, Norbert Podhorszki

Oak Ridge National Laboratory

Karsten Schwan, Matthew Wolf

Georgia Institute of Technology, Atlanta

Manish Parashar

Rutgers University

Fan Zhang

Rutgers University

9.1 Introduction 172
9.2 Tailored Co-Processing at High Concurrency 174
9.3 Co-Processing With General Visualization Tools Via Adaptors 175
9.3.1 Adaptor Design 178
9.3.2 High Level Implementation Issues 178
9.3.3 In Practice 179
9.3.4 Co-Processing Performance 180
9.4 Concurrent Processing 183
9.4.1 Service Oriented Architecture for Data Management in HPC 183
9.4.2 The ADaptable I/O System, ADIOS 184
9.4.3 Data Staging for *In Situ* Processing 185
9.4.4 Exploratory Visualization with VisIt and Paraview Using ADIOS 186

9.5 *In Situ* Analytics Using Hybrid Staging 187
9.6 Data Exploration and *In Situ* Processing 189
9.6.1 *In Situ* Visualization by Proxy 190
9.6.2 *In Situ* Data Triage 191
9.7 Conclusion 193
References 194

Traditionally, visualization is done via post-processing: a simulation produces data, writes that data to disk, and then, later, a separate visualization program reads the data from disk and operates on it. *In situ* processing refers to a different approach: the data is processed while it is being produced by the simulation, allowing visualization to occur without involving disk storage. As recent supercomputing trends have simulations producing data at a much faster rate than I/O bandwidth, *in situ* processing will likely play a bigger and bigger role in visualizing data sets on the world's largest machines.

The push towards commonplace *in situ* processing has a benefit besides saving on I/O costs. Already, scientists must limit how much data they store for later processing, potentially limiting their discoveries and the value of their simulations. *In situ* processing, however, enables the processing of this unexplored data.

This chapter describes the different approaches for *in situ* processing and discusses their benefits and limitations.

9.1 Introduction

The terms used for *in situ* processing have not been used consistently throughout the community. In this book, *in situ* processing refers to a spectrum of processing techniques. The commonality between these techniques is that they enable visualization and analysis techniques without the significant—and increasing expensive—cost of I/O. On one end of the *in situ* spectrum, referred to in this book as *co-processing,* visualization routines are part of the simulation code, with direct access to the simulation's memory. On the other end of the *in situ* spectrum, referred to in this book as "concurrent processing," the visualization program runs separately on distinct resources, with data transferred from the simulation to the visualization program via the network. Hybrid methods combine these approaches: data is processed and reduced using direct access to the simulation's memory (i.e., co-processing) and then sent to a dedicated visualization resource for further processing (i.e., concurrent processing). Table 9.1 summarizes these methods.

For approaches that process data via direct access to the simulation's memory (i.e., co-processing and hybrid approaches), another important consideration is whether to use custom, tailored code, written for a specific simulation, or whether to leverage existing, richly featured visualization software. Tai-

Technique	Co-processing	Concurrent Processing	Hybrid
Aliases	Tightly coupled Synchronous	Loosely coupled Asynchronous Staging	None
Description	Vis routines have direct access to memory of simulation code	Vis runs on dedicated, concurrent resources and access data via network	Data is reduced via co-processing and sent to a concurrent resource.
Negative Aspects	Memory constraints. Large impact on simulation (crashes, performance).	Data movement costs. Requires separate resources.	Complex. Also shares negatives of other approaches.

TABLE 9.1: Summary of *in situ* processing techniques

lored code is most likely to be well-optimized for performance and memory requirements, for example by using a simulation's data structures without copying them into another form. However, tailored code is often not portable or re-usable, making it tied to a specific simulation code. Utilizing existing, richly featured visualization software is more likely to work well with many simulations. But, typically, the price of this generality may be increased due to the usage of resources, such as memory, network bandwidth, or CPU time.

This chapter discusses case studies for three *in situ* processing scenarios:

- a tailored, co-processing approach at high levels of concurrency, described in 9.2;
- a co-processing approach via an adaptor layer to a richly featured visualization tool, described in 9.3; and
- a concurrent approach in the context of the ADIOS system, described in 9.4.

Hybrid processing is an emerging idea, with little published work; its concepts and recent results are discussed in 9.5. Finally, 9.6 discusses how to use *in situ* processing when the visualizations and analyses to be performed are not known *a priori*.

9.2 Tailored Co-Processing at High Concurrency

Tailored co-processing is the most natural approach when implementing an *in situ* solution from scratch and focusing on integration with just one simulation code. The techniques and data structures are implemented with a specific target in mind, resulting in high efficiency in performance and memory overhead, since the simulation's data does not need to be copied. Not surprisingly, many of the largest scale examples of *in situ* processing, to date, have been done using tailored co-processing. Co-processing has been practiced by some researchers in the past with the objective to either monitor or steer simulations [10, 21, 17, 20]. Later results focused on speeding up the simulation by reducing I/O costs [32, 37].

Co-processing has many advantages:

- Accessing the data is very efficient, as the data is already in the primary memory. In contrast, post-processing accesses data via the file system and concurrent processing accesses data via the network.

- Visualization routines can access more data than is typically possible. Simulation codes limit the number of time slices they output for post-processing, since I/O is so expensive. But, with co-processing, the visualization routines can be applied to every time slice, and at a low cost. This prevents discoveries from being missed because the data could never be explored.

- The integrity of the data does need to be compromised. Some simulations produce so much data that the data must be reduced somehow, for example by subsampling; this is not necessary when co-processing.

Co-processing also has some disadvantages:

- Any resources, such as memory or network bandwidth, consumed by visualization routines will reduce what is available to the simulation. Further, existing visualization algorithms are usually not optimized to use the domain decomposition and data structures designed for the simulation code. To be used *in situ*, the algorithms may have to be reformulated so as not to duplicate data and incur excessive interprocessor communication for the visualization calculations.

- The visualization calculations should only take a small fraction of the overall simulation time. If they take longer, the visualization routines can impact the ability of the simulation to advance if they can not finish quickly enough. Although sophisticated visualization methods offer visually compelling results, they are often not acceptable for *in situ* visualization.

- If the visualization routines crash or corrupt memory, then the simulation will be affected—possibly crashing itself.

- The visualization routines must run at the same level of concurrency (in terms of numbers of nodes) as the simulation itself, which may require rethinking some visualization algorithm design and implementation.

In short, it is both challenging and beneficial to design scalable parallel *in situ* visualization algorithms. The resulting visualization should be cost-effective and highlight the best features of interest in the modeled phenomena without constantly acquiring global information about the spatial and temporal domains of the data. A good design must take into account the domain knowledge about the modeled phenomena and the simulation model. As such, a co-processing visualization solution should be developed as a collective effort between the visualization specialists and simulation scientists. Other design considerations for co-processing *in situ* visualization are discussed by Ma [18].

Finally, the recent results of Yu et al. [38] are briefly summarized here,to understand a real-world, tailored co-processing approach. They achieved highly scalable *in situ* visualization of a turbulent combustion simulation [6] on a Cray XT5 supercomputer, using up to 15,260 cores. Figure 9.1 shows a set of timing test results. This work was novel outside of its extreme scale, because it introduced an integrated solution for visualizing both volume data and particle data, allowing the scientists to examine complex particle-turbulence interaction, like the type shown in Figure 9.2. The mixed types of data required considerable coordination for data occurring along the boundaries [38] and highly streamlined routines, like the 2-3 swap compositing algorithm (see 5.3.1). Finally, their efforts had benefits beyond performance. Before employing *in situ* visualization, the combustion scientists subsampled the data to reduce both data movement and computational requirements for post-processing visualization. This work allowed them to operate on the original full-resolution data.

9.3 Co-Processing With General Visualization Tools Via Adaptors

As with tailored code, co-processing with a general visualization system incorporates visualization and analysis routines into the simulation code, allowing direct access to the simulation's data and compute resources. Whereas co-processing with tailored code integrates tightly-coupled visualization routines adapted to the simulation's own data structures, co-processing using general routines from fully featured visualization systems uses an adaptor layer to access simulation data. The adaptor layer is simply a set of routines that the simulation developer provides to expose a simulation's data structures in a manner that is compatible with the visualization system's code.

The complexity of adaptors can vary greatly. If the data structures of the simulation and the visualization system differ, then the adaptor's job is to copy and reorganize data. For example, a simulation may contain particle data, with spatial coordinates and other attributes for each particle. If the

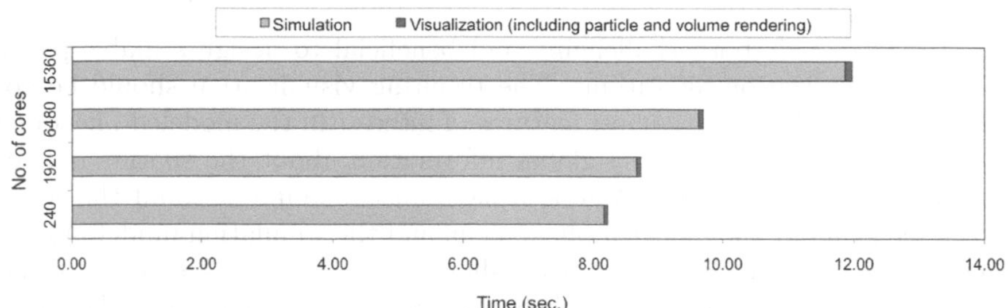

FIGURE 9.1: Timing results for both simulation and visualization using different numbers of cores. In this set of tests, the number of grid points on each core is fixed so using more cores provides higher resolution calculations. Visualization is computed every ten timesteps, which is at least 20–30 times more frequent than usual. As shown, the visualization cost looks negligible. The simulation time increases as more cores are used due to the communication required to exchange data after each iteration. The visualization time also grows as more cores are used because of the communication required for image compositing, but it grows at a much slower rate than the simulation time.

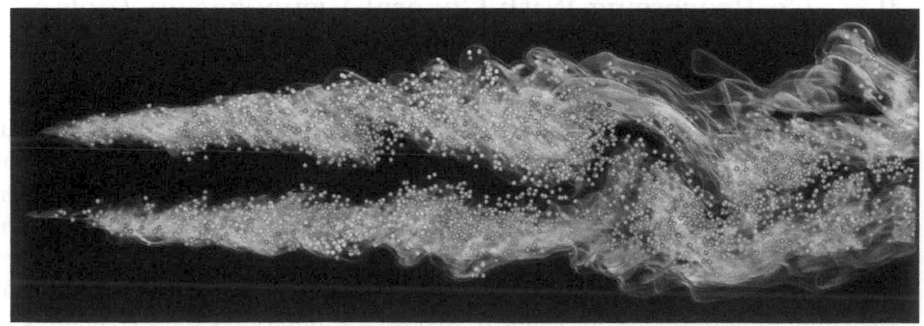

FIGURE 9.2: Visualization of the particles and CH_2O field.

simulation's data is organized as an array of structures, and the visualization system requires a structure of arrays, the simulation data must be reorganized to match that of the visualization code, thus consuming more resources. Conversely, adaptors for codes with very similar data structures may share pointers to the simulation's arrays with the visualization code, a "zero-copy" scenario.

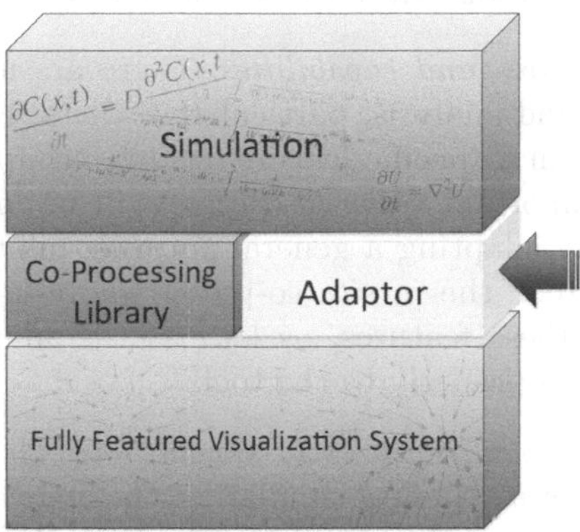

FIGURE 9.3: Diagram of the adaptor that connects fully featured visualization systems with arbitrary simulation codes.

Tailored code and general co-processing share many of the same strengths and weaknesses. Both approaches operate in the same address space as the simulation, potentially operating on data directly without conversion. Both approaches cause some extra code to be included in the simulation executable and both may generate additional data, subtracting from the memory available to the simulation. This last property poses problems on many supercomputers, as the trend is to have less and less memory available per core. Memory considerations are exacerbated for co-processing approaches since adaptors may copy data and general visualization tools might have higher memory usage requirements than tailored solutions. The two approaches may also differ in their performance. Tailored code is optimized for the simulation data structures, and so it may not match an optimal layout for already-written general visualization algorithms. But it may be possible with tailored code to optimize these algorithms for the simulation's native data layout.

On the other hand, co-processing with general visualization tools can often be easily enabled within a simulation through the addition of a few function calls and by adding an adaptor layer. An adaptor code is far less complex to create than the visualization routines themselves, often consisting of only tens of lines of source code. Consequently, adaptor code is cheaper to develop than tailored code and benefits from the use of existing well-exercised routines from fully featured visualization systems. Furthermore, once the data are exposed via the adaptor, they can be visualized in many ways using the general visualization system as opposed to a tailored approach, which might provide highly optimized routines for creating limited types of visualizations.

9.3.1 Adaptor Design

A successful adaptor layer connecting computational simulations to general purpose visualization tools will have several features inherent in its design:

- *Maximize features and capabilities.* There are numerous use cases for visualization and analysis, particularly *in situ*. Focusing an *in situ* implementation on a specific feature set, for example making movies, will mean that important other use cases will be missing, like interactive debugging. By adapting a general purpose, fully-featured visualization and analysis tool, the *in situ* co-processing tasks gain the capability to use many of these features, as long as the adaptor layer exposes the simulation data properly to the tool.
- *Minimize code modifications to simulations.* The less effort it takes to apply the adaptor to a new simulation, the more easily an *in situ* library can be adopted by a wide range of codes in the HPC community. General purpose tools are already capable of supporting the data model for a wide range of simulation applications, so the barrier for translating data structures should be minimal.
- *Minimize impact to simulation codes when in use.* Ideally, codes should be able to run the same simulation with or without *in situ* analysis. For example, the limited memory situation in distributed systems is exacerbated by co-processing via shared computational nodes, and zero-copy and similar features mitigate the negative impact.
- *Zero impact to simulation codes when not active.* Simulation codes should be able to build a single executable, with *in situ* support built-in, and suffer no detrimental side effects, such as reduced performance or additional dependencies on libraries. Without this feature, users must decide before running a simulation code—or worse, before compiling it—whether or not they wish to pay a penalty for the possible use of *in situ* capabilities, and may choose to forgo the option. Allowing users to start an *in situ* session on demand enables many visualization and analysis tasks, like code debugging, that the user could not have predicted beforehand.

9.3.2 High Level Implementation Issues

Co-processing is implemented successfully in the general visualization applications ParaView [9] and VisIt [35], two codes built on top of the Visualization Toolkit (VTK) [23]. The codes are similar in many ways beyond their mutual reliance on VTK, including their use of data flow networks to process data. As described in Chapter 2, data flow networks are composed of various interchangeable filters through which data flows from one end to the other with transformations occurring in the filters. In an *in situ* context, the output of the last filter in the data flow network is the product of the *in situ* operation. It may be a reduced data set, statistical results, or a rendered image. For *in situ* use, both applications use adaptor layers to expose simulation data to their visualization routines by ultimately creating VTK data sets that are facades to the real simulation data. ParaView's adaptor layer must be written in C++ and it directly creates VTK objects that will be passed into its data flow networks. VisIt adaptor layers may be written in C, Fortran, and Python and also create VTK data sets that can pass through VisIt's data flow networks.

The co-processing strategies used by ParaView and VisIt are somewhat different where the simulation interacts with the co-processing library. ParaView's co-processing library consists of a few top-level routines inserted into the simulation's main loop. The simulation data are transformed by the adaptor layer and passed into a co-processing routine, which constructs a data flow network based on an XML specification from ParaView. The results of the pipeline may be written to a disk or staged over the network to concurrent resources, where ParaView may be used for further analysis. VisIt's interaction with the simulation emphasizes a more exploratory style of interaction where, rather than executing fixed data flow networks, the VisIt client dynamically connects to the simulation and directs it to create any number of data flow networks. The end user can be involved with this process and decide which analyses to perform on-the-fly. Since VisIt is able to send plotting commands to the simulation, it integrates differently into the simulation's main loop than ParaView and its adaptor layer is executed only when necessary.

9.3.3 In Practice

This section explores *Libsim*, the *in situ* library in VisIt, in more detail. *Libsim* wraps all of VisIt's functionality in an interface that is accessible to simulations. In effect, it transforms the simulation into the server in its client-server design. *Libsim* has features consistent with the four design goals listed in 9.3.1: it connects to VisIt, a fully-featured visualization and analysis tool and can expose almost any type of data VisIt understands; it requires only a handful of lines of code to instrument a simulation; it minimizes data copies wherever possible; and it defers loading any heavyweight dependencies until *in situ* analysis tasks begin.

There are two interfaces in *Libsim*: a control interface that drives the VisIt server, and a data interface that hands data to the VisIt server upon request. The control interface is contained in a small, lightweight static library that is linked with the simulation code during compilation, and is literally a front-end to a second, heavier-weight runtime library that is pulled in only when *in situ* analysis begins at runtime. The separation of *Libsim* into two pieces prevents inflation of executable size, even after linking with *Libsim*, while still retaining the ability to access the full VisIt feature set when it is needed.

The control interface is capable of advertising itself to authorized VisIt clients, listening for incoming connections, initiating the connection back to the VisIt client, handling VisIt requests like plot creation and data queries, and letting the client know when the simulation has advanced and that new data is available.

Several aspects of *Libsim* result in benefits for both interacting and interfacing with simulations. The separation between the front-end library and the runtime library is one such aspect. In particular, modifications to simulation codes to support *Libsim* can be minimal, as the front-end library contains only approximately twenty simple control functions (most of which take no arguments), and only as many data functions as the simulation code wishes to expose. This separation also enables the front-end library to be written in pure C. As such, despite VisIt being written in C++, no C++ dependencies are introduced into the simulation by linking to the front-end library, which is critical since many simulations are written in C and Fortran. Additionally, by providing a C interface, the process of automatically generating bindings to other programming languages, like Python, is greatly simplified. The separation into front-end and runtime library components also means that the runtime library implementation is free to change as VisIt is upgraded, letting the simulation benefit from these changes automatically, without relinking to create a new executable. In addition, by deferring the heavyweight library loading to runtime, there is effectively zero overhead and performance impact on a simulation code linked with *Libsim* when the library is not in use.

Another beneficial aspect is the manner in which data are retrieved from the simulations. First, as soon as an *in situ* connection is established, the simulation is queried for metadata containing a list of meshes and fields that the simulation wishes to expose for analysis. The metadata comes from a function in the adaptor layer. Just as in a normal VisIt operation, this list is transmitted to the client, where users can generate plots and queries. Once the user creates a plot and VisIt starts executing the plot's data flow network, the other functions in the simulation's data adaptor layer are invoked to retrieve only the data needed for the calculation. VisIt's use of contracts (see 2.5.1) ensures that the minimal set of variables will be exposed by the adaptor layer, reducing any unnecessary data copying from converting variables that are not involved in a calculation. Whenever possible, simulation array duplication is avoided by including simulation arrays directly into the VTK objects, minimizing the impact on the performance and scaling capabilities of simulation codes.

9.3.4 Co-Processing Performance

An experiment performed by Whitlock, Favre, and Meredith [35] demonstrated that the adaptor approach performs and scales well. Their experiments were conducted on a 216 node visualization cluster with two, six-core 2.8GHz Intel Xeon 5660 processors and 96GB of memory per node. The test system utilizes an InfiniBand QDR high-speed interconnect and a Lustre parallel file system. They ran two types of tests to characterize the performance of this library. The first test determined the cost associated with introducing *Libsim* into the main loop of a prototype simulation. The second test investigated the performance of *in situ* visualization versus I/O in a real simulation code.

Libsim introduces additional overhead in the simulation's main loop: it listens for incoming connections and broadcasts *in situ* analysis requests to distributed tasks. Whitlock et al. wanted to measure the associated time overhead when VisIt was not connected to the simulation, so they added timing code to *Libsim*'s `updateplots` example program. They then ran the example program through 10K main loop iterations. The time spent was small: $2\mu s$ and $8\mu s$, respectively, for 512 core runs. The measurements remain consistent once VisIt connects and is not requesting data. Connecting to VisIt does increase the amount of memory used by the simulation as this connection requires loading the VisIt runtime libraries, which imposes a one-time cost of approximately one second.

A study by Childs et al. (see Chap. 13) demonstrated that I/O dominates VisIt's execution time on diverse supercomputer architectures and that VisIt's isocontouring performance at 8K up to 64K cores was, on average, over an order of magnitude faster than the I/O operations needed to obtain the data from disk. *In situ* visualization uses simulation arrays directly, of course, and usually allows VisIt's pipeline to execute in a fraction of the time required for I/O. Although tests using thousands of cores would better represent large supercomputers, *in situ*'s performance advantage over I/O even became apparent at modest core counts.

The study by Whitlock et al. [35] was designed to measure the impacts and benefits of *in situ* processing via general visualization tools and adaptors. They instrumented GADGET-2 [26], a distributed-memory parallel code for cosmological simulations of structure formation (see Figure 9.4). They ran GADGET-2 at three levels of concurrency: 32, 256, and 512 cores to measure *in situ* performance versus I/O performance. They isolated the time required for GADGET-2 to write its data to disk in two modes: one using collective I/O to a single file, and again in a mode where each task writes its own disk file. They separately recorded the timings of just the visualization/analysis processing in the VisIt pipeline, rendering a Pseudocolor plot of a scalar variable and saving a 2048 square pixel image to disk. They ran each of these tests using two sets of initial conditions for GADGET-2, generating particle sets with 16M and 100M particles.

In the larger test case where 100M particles were saved, GADGET-2 would

16 Million Particles			
	32 cores	256 cores	
I/O 1 file	2.76s	4.72s	
I/O N files	0.74s	0.31s	
VisIt pipeline	0.77s	0.34s	
100 Million Particles			
	32 cores	256 cores	512 cores
I/O 1 file	24.45s	26.7s	25.27s
I/O N files	0.69s	1.43s	2.29s
VisIt pipeline	1.70s	0.46s	0.64s

TABLE 9.2: Performance of visualization/analysis vs. I/O for 16M and 100M particles at different levels of concurrency.

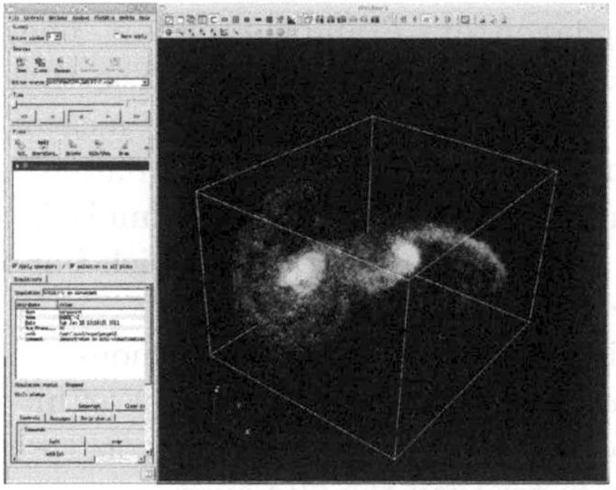

FIGURE 9.4: VisIt client connected to GADGET-2 instrumented with *Libsim*. Image source: Whitlock et. al [35].

generate a 2.8GB snapshot file. Since the amount of data was constant for each test run, they found that the timings for collective I/O are relatively consistent. Independent files are the fastest means for writing files with smaller core counts, though performance in this mode is inversely proportional to the core count, as the I/O subsystem can absorb only a limited number of simultaneous requests before its performance starts to degrade. By substituting a visualization operation, in this case a Pseudocolor plot, for writing a full 2.8GB GADGET-2 snapshot file, they were able to reduce the amount of data written to a single 12MB image file. From this performance data, they observed that, for the selected visualization operation, *in situ* processing is competitive with single-file I/O and exceeds collective I/O performance. For smaller core counts with large sets of particles, the work performed per core is higher, resulting in a longer runtime vs. single-file I/O. However, as the number of cores increases, the runtime of the visualization processing alone is far lower than that of either single-file I/O or collective I/O. The margin of performance is large enough that they could have generated several *in situ* visualizations in the time needed to write a full size snapshot file. This example demonstrates that *in situ* analysis can be an attractive use of extra compute cycles in upcoming exascale computers and a powerful acceleration technique for large scale computations.

9.4 Concurrent Processing

The concurrent *in situ* approach allows simulation and visualization software to run with minimal interaction, similar in spirit to non-*in situ* usage. The approach resembles a file-based approach, but uses the network for transferring data instead of storage. By keeping the simulation and visualization programs on separate resources, potential errors during processing do not affect the simulation itself. It also allows for scaling the simulation and the visualization code separately. Compared to co-processing, however, there is an extra cost: all data for visualization and analysis has to be transferred to separate processing nodes through the internal network. Fortunately, these costs can be partially ameliorated via smart data filtering [1] and/or compression [15].

In this section, a service-oriented approach to data processing, analysis and visualization is presented, in the context of the ADIOS I/O framework. The framework provides a single I/O API for the simulation to output data (as if writing to files), and the visualization tool to read input data (as if reading from files). It also provides services to transfer data from the output to the reader(s) and a plug-in-based architecture to allow new data processing routines for analysis or visualization steps. ADIOS can organize these steps into a workflow, including a scheduling that takes into account available resources.

9.4.1 Service Oriented Architecture for Data Management in HPC

The Service Oriented Architecture (SOA) is a methodology for designing software to create interoperable services. It was defined by Thomas Erl of SOA Systems [8]. Some of its requirements must be loosened, however, for performance or complexity reasons when considering data intensive HPC. Specifically, services need not be stateless or opaque to other services. Furthermore, costly service ideas are avoided, such as ontology-based service contracts, per-invocation based service selection, and utilization of an explicit service bus. The following SOA principles have been and are being followed when designing ADIOS for I/O and *in situ* data processing.

- *Abstract and reusable services.* The backbone of a flexible system is its ability to deal with multiple types of input data. For example, a compression service that simply treats data as a bucket of bits, instead of a semantically knowable data structure, will not be sufficient for the needs of extreme scale data management. A self-describing data format, including metadata for introspection, allows services to leverage semantic knowledge about the data to be processed.

- *Service contracts.* The SOA approach requires services to define an interface, or a contract, that defines how the service interacts with external entities. Given the vast range of domains and semantics that need to be supported for an HPC framework, a well understood read/write/query API forms the foundation of the ADIOS infrastructure. Computational scientists are already intimately familiar with the standard data input and output operations. Actions such as opening a file and writing a group of variables are very useful abstractions in providing a familiar yet dynamic and flexible interface to the scientists. In order to address the requirements of data format descriptions there is, however, one additional requirement for the user: the creation of metadata from which the service framework can obtain reasonable semantic information about the data structure.

- *Service location transparency.* While service location transparency is not part of the basic SOA definition, it is an important aspect of attaining the highest level of performance for data intensive computing. In particular, it is paramount that services can be flexibly placed within the data processing pipeline, to maximize the exploitation of available resources.

9.4.2 The ADaptable I/O System, ADIOS

ADIOS [16] has been developed with the goal of addressing the needs of users of leadership computing facilities (such as those at Oak Ridge and Argonne National Laboratories), as well as users of other large supercomputers. Many past efforts in providing advanced data management infrastructures have burdened scientists with complex APIs and forced them to make extensive changes to their applications. And, while providing adequate functionality, they do not usually perform acceptably at high concurrency. ADIOS has been designed and developed with constant interaction from developers of large-scale simulations from many domains, such as computer science, nuclear physics, combustion, and astrophysics. The primary focus of research has been performance for typical I/O scenarios while maintaining a simple API for a self-describing data format. The result is a framework that provides several methods for performing I/O, allowing users to choose the one best for their needs.

The interface to ADIOS is a simple read/write API. An application can describe the content and organization of an output file in an XML file, which can be created manually by application developers. The XML file contains variables with name, type and dimensionality, and also assigned attributes. Alternatively, the program itself can declare the variables by invoking a function. The I/O part of the program code can be generated from the XML automatically, if all of the routines for writing data are in a single location. This is not required, however, as traditional open, write, and close functions can also be used to create the output file. Additionally, the method for writing a group of variables can be chosen in the XML file. This means that if the currently used method is not performing well anymore, then porting the I/O part of the application to a new computer architecture or new I/O system requires only a small change in the XML file. For example, an I/O strategy where all processes write into a single file using MPI-IO, can be replaced with an aggregating strategy, where a small number of processes gather all data and write efficiently into separate files on a parallel file-system. Further, the XML definition of the content allows for an easy way to describe semantic information according to an application domain's predefined storage schema or domain knowledge. Moreover, it allows users to prescribe actions for visualization too, including which tools to use and operations to perform.

The buffering provided by ADIOS works with all of its I/O methods that output in the ADIOS-BP file format, which was designed for efficient utilization of parallel file systems. Buffering allows for using bulk data transfers, which is usually orders of magnitudes faster than writing small chunks of data. The BP format is the fastest format used in ADIOS, in most writing and reading scenarios. It is a self-describing format, where the file content can be discovered and queried from the information stored in the file itself. The format avoids the bottlenecks of reorganizing data from application memory to the file, which are common with other file formats, and also minimizes the contention of file access patterns from many processors to the limited number of file servers.

9.4.3 Data Staging for *In Situ* Processing

I/O bandwidth, relative to the processing power of current supercomputers, is often insufficient and this trend is predicted to only get worse. Already, storing data to files is often a significant bottleneck during simulation, even with the best I/O solutions. In response, ADIOS developers turned their research focus to *in situ* processing for as many data management tasks as possible before writing data to permanent storage. ADIOS had many advantages in transitioning their product for the *in situ* space, including the separation of API from the choice of I/O methods, the buffering-as-service provided to all methods, and the simplicity and completeness of the API to describe data sets.

There are many benefits to using a set of nodes as a staging area, where I/O can be placed before moving it to disk. First, data can be asynchronously and efficiently moved to staging nodes using Remote Data Memory Access (RDMA), as with DataStager [2]. Second, multiple, parallel applications can simultaneously read a multidimensional array, with arbitrary decomposition, as with DataSpaces [7]. And, third, services can apply data processing actions to data in the staging area and as it is being moved from the (parallel) writer to the (parallel) reader, as with PreDatA [40]. In short, combining data processing actions with I/O in a staging area has been shown to be flexible and efficient. Further, the approach minimizes power and I/O issues and is an active focal point for exascale research (see Chap. 15). Finally, the user experience is simplified: applications do not need to manage how and where each piece of data is stored, as each process observes and accesses all data sets by defining the geometries of the subsets it wants to write or read.

Virtual data spaces can be used for code coupling. In one example, they facilitated multiphysics simulations by combining different plasma fusion models. Although the models were written for exchanging data via the file system, virtual data spaces allowed them to move data asynchronously from memory to memory. The file-based coupling was changed to a faster memory-to-memory coupling by changing the name of the used method in their respective ADIOS configuration XML files. This service-oriented approach to data movement let the application developer teams keep the code of their physics models separate, avoiding the software development and maintenance issues of a combined code.

The abstractions of data access and transfer, which are performed asynchronously between tasks, enable task pipelining. This means the data can be processed without worrying about the placement of a task and how it will get the data it needs. However, effective description of the data is essential: a self-describing data format and API for reading and writing data are necessary for a task to understand the data it receives. For example, there is a need for semantically meaningful data chunks or portions, such as the data representing entire variables in simulation codes, and there also must be descriptions about those chunks available when processing actions are carried out. The abstractions of data access and transfer, and the effective data description enable a generic plug-in architecture, where both common and domain-specific analysis and visualization tasks can be created as self-contained codes with a well-defined input and output that behave as services and can be publicly released and reused by many applications.

9.4.4 Exploratory Visualization with VisIt and Paraview Using ADIOS

A client-server design where the server was parallelized and read data (in parallel) from files is described in 2.4. In this section, a similar configuration is considered, but the data is read from the network, not files. Once a simulation code opts to use the ADIOS API, ADIOS' plug-in architecture allows it to dynamically switch from a file-based I/O to a mode where data is sent directly to a staging server for processing, enabling exploratory visualization using concurrent *in situ*. VisIt (Chap. 16) and ParaView (Chap. 18) both have readers for this ADIOS read API, and support both file-based and *in situ*-based ADIOS usage. From the perspective of these tools, there is virtually no difference in obtaining data from the network instead of the disk. The only change needed was to handle timesteps coming separately from the simulation, instead of having access to all timesteps in a file.

An example of concurrent visualization is described here, namely, how output from the Pixie3D plasma fusion code [5] is visualized (see Fig. 9.5). This code's output is not directly used for visualization because its internal geometries are not useful when visualized. A separate code, Pixplot, is used to read in the output and, on a much smaller number of processes, transforms and writes the data and meshes in Cartesian coordinates. In the past, this post-processing step generated files, which the user of Pixie3D would study with VisIt in a post-processing fashion. The three components—the Pixie3D simulation, the Pixplot transformation and the VisIt visualization tool—represent a typical scenario for concurrent *in situ* visualization. Since each of these codes used ADIOS for writing and reading data for post-processing exploration, it was straightforward to modify them for concurrent visualization. Pixie3D required no change at all. Pixplot required its processing loop to change from opening one file and iterating over all timesteps to loop for an unknown amount of iterations while there is data coming from the simulation, and read a timestep from the staging area, including possibly waiting if the next timestep is not yet available. The same was true for VisIt's ADIOS reader, which had to be modified to deal with only a single timestep at a time.

Additionally, there is a sequential tailored code for Pixie3D data, Pixmon, which generates 2D slices from the 3D output of Pixie3D itself. These small images are stored on a disk, which is accessible by the ESiMon dashboard [29], so the user of the code can monitor the status of the simulation through a set of 2D snapshots, in close to real-time. For this reason, they used the DataSpaces staging service, which runs separately and can serve two readers of Pixie3D data: Pixplot and Pixmon. Pixplot's output is also stored in the distributed, shared space provided by this same staging server. Several timesteps are stored, so that VisIt or, more precisely, the user driving the VisIt client can explore a timestep long after it was generated by the simulation.

9.5 *In Situ* Analytics Using Hybrid Staging

To achieve better performance and efficiency, high performance computing (HPC) systems are increasingly based on node architectures with large core counts and deep memory hierarchies. For example, Oak Ridge's Titan system uses 16-core AMD Opteron nodes, and the forthcoming petascale supercomputers like Mira at Argonne and Sequoia at Lawrence Livermore will both employ the IBM BlueGene/Q 18-core nodes to exploit more on-chip parallelism. This architectural trend increases the performance gap between on-chip data sharing and off-chip data transfers. As a result, moving large volumes of data using the communication network fabric can significantly impact performance. Further, minimizing the amount of inter-application data

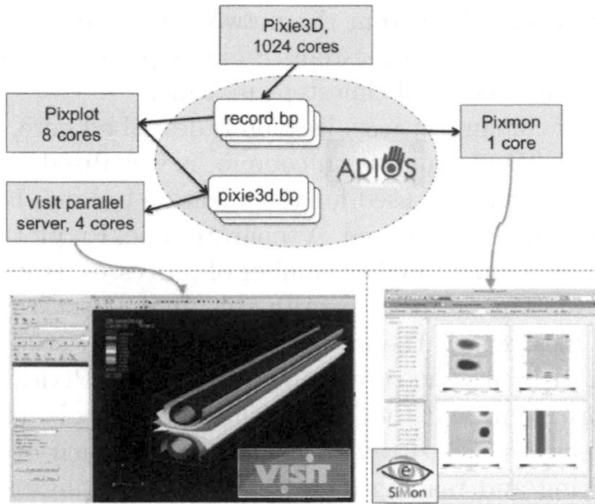

FIGURE 9.5: This example has four separate programs—the Pixie3D simulation, Pixplot, Pixmon, and VisIt—strung together as a workflow on the same supercomputer. A fifth program, the staging server (ADIOS), serves the read and write requests of these programs through its API.

exchanges across compute nodes and its consequent network use are critical to achieving high application performance and system efficiency.

Consider a pipeline that consists of more than one task for processing simulation data, for example, one that includes filtering, pre-processing, analytics, visualization, etc. Using an exclusively concurrent *in situ* solution may result in too much network traffic, while a purely co-processing setup may overtax the resources present on compute nodes. Furthermore, certain analytics processing tasks are inherently difficult to scale, such as those requiring frequent collective communications. But, to effectively utilize the potential of current and emerging HPC systems, coupled data-intensive visualization pipelines must exploit compute node-level data locality and core-level parallelism to the greatest extent possible. Each task should be schedulable for either co-processing or concurrent processing, as per each pipeline's processing characteristics. In other words, there must be flexibility in where each data processing action is performed, including on compute nodes, on staging nodes, or both [41, 19]. Achieving such flexibility is challenging, requiring locality-aware mapping of tasks from separate-yet-coupled applications onto the same cores and efficient support for coordination and data exchange between these coupled applications.

Hybrid *in situ* (illustrated in Figure 9.6) is a technique that combines dedicated cores for co-processing to reduce data movement with dedicated nodes for concurrent data processing. Its goal is to efficiently enable visualization and analysis of data by reducing data movement by moving computation closer to the data. Successful deployment of such a hybrid *in situ* framework can enable novel data processing scenarios, such as moving data customization and filtering operations to where the simulation data is being produced—transforming data as it moves from source to sink, supporting *in situ* coupling to increase intra-node data sharing, and dynamically deploying data processing plug-ins (binary and/or source code) for execution within the staging area.

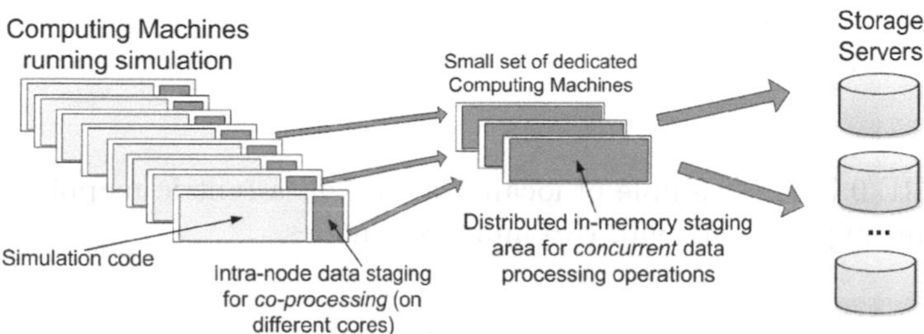

FIGURE 9.6: Illustration of hybrid data staging architecture for *in situ* co-processing and concurrent processing.

There have been several research efforts in understanding the efficacy of hybrid *in situ* processing. Early work on commodity networks established that source-based data filtering and compression can substantially reduce the volumes of data being moved and the associated movement overheads [4, 36]. Later work [1, 3] explored selective co-processing actions that move only the data that is needed from compute to staging nodes. Zheng et al. [41] and Parashar [19] then described the importance of such flexibility and the design of a stream processing system that provided location-flexible operations.

The Co-Located DataSpaces (CoDS) framework [39] uses a workflow-like approach to maximize the on-chip exchange of data for coupled visualization applications by building on DataSpaces [7] and ADIOS (see 9.4). The design of CoDS includes a distributed data sharing and task execution runtime. It employs data-centric task placement to map computations from the coupled application components onto cores so that a large portion of the data exchanges can be performed using the on-node shared memory (see Fig. 9.7). It also provides a shared space programming abstraction that supplements existing parallel programming models (e.g., message passing) with specialized one-sided asynchronous data access operators, and can be used to express coordination and data exchanges between the coupled applications. The framework builds a scalable, semantically specialized virtual, shared space that is distributed across cores on compute nodes, and provides simple abstractions for coordination, interaction, and data-exchange. The CoDS framework currently supports execution and data-centric task mapping for a workflow that is composed of data parallel application components with regular multidimensional data meshes and domain decompositions.

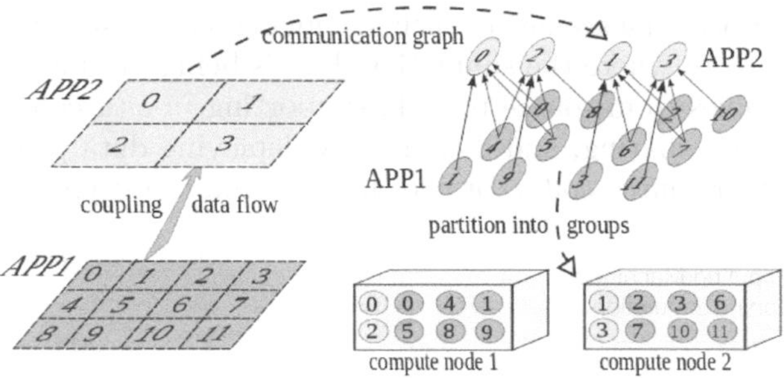

FIGURE 9.7: An example of locality-aware, data-centric mapping of two interacting applications onto two multi-core nodes.

9.6 Data Exploration and *In Situ* Processing

In situ visualization is most appropriate when end users know what they want to look for *a priori*. In this case, visualization and analysis may be performed as soon as the data is generated and the end user does not need to be involved. However, the case where the user does *not* know what they want to look for *a priori*—that is, where they want to explore the data—is more problematic, since the amount of time to explore the data is variable and may take hours or days. These time scales make stalling the simulation or keeping a copy of the data in memory unpalatable. Further, some types of exploration may require studying temporal data; keeping multiple time slices in memory almost certainly will exceed memory constraints.

An alternate approach is to transform (and hopefully reduce) the data *in situ* and then explore the data with traditional post-processing. Common approaches involve compression (either spatially or temporally), statistical sampling, data subsetting [28, 27, 12, 11, 22], or feature identification and tracking [24, 25, 14, 13].

The following subsections describe techniques specifically designed for leveraging *in situ* processing to enable exploration. An algorithm for reducing data while allowing users to modify the characteristics of a volume rendering is described in 9.6.1. An approach where data is compared to a reference simulation and only the most different regions are stored for post-processing is described in 9.6.2.

9.6.1 *In Situ* Visualization by Proxy

Typically, *in situ* visualization is used to generate an animation that captures features at high temporal resolution. This results in static images rendered based on a fixed set of rendering and visualization parameters and exploration is not possible. This section describes *visualization by proxy*, introduced by Tikhonova et al. in 2010 [30], an approach that enables exploration without requiring the full raw data and a supercomputer. The core idea is that,

Number of cores	240	1920	6480
Simulation time (sec)	8.7204	9.3393	9.5573
I/O time (sec)	9.4563	26.051	52.565
Volume rendering time (sec)	0.3817	0.6155	0.7359
RAF computing time (sec)	1.2775	1.3938	1.3973

TABLE 9.3: Timing results for visualization by proxy on a Cray XT5. Each core is assigned a region of $27 \times 40 \times 40$ grid points. Image resolution used is 1024×1024. The number of layers used for RAF is 16. The time for RAF is significantly less than the I/O time, demonstrating the advantage of an *in situ* approach.

instead of generating fully rendered results, proxy images are computed that essentially encode isosurface layers by taking into account accumulated attenuation in a view dependent fashion. The resulting proxy images (also called ray attention functions, or RAF, in [30]) enable exploration in the transfer function space. The storage requirement for proxy images corresponds to the number of layers chosen by the user. Ten to fifteen layers were found to be generally sufficient to capture enough details in the data. As a result, the proxy images are typically 10–15 times larger than a single regular image; however, the proxy images are substantially smaller than the size of a video that can offer the same level of exploration.

Tikhonova et al. extended their visualization by proxy technique into a parallel computing environment to support *in situ* visualization [31]. They showed that computing proxy images resembles volume rendering, and can be performed directly on the distributed data out of the simulation without replication. The calculation is as scalable as parallel volume rendering. Table 9.3 displays some timing results. All simulation data was stored as double-precision floating point values. The proxy images are saved in 32-bit precision. The timing for simulation, I/O, volume rendering, and RAF computing was measured for one timestep. I/O time is how long it took for the simulation to write data to a disk. We can see that volume rendering and RAF computing time accounts only for a small fraction of the total time. The authors noted that, although the fraction for calculating the proxy images is small, real-world scenarios would have the fraction be smaller still, since users would not visualize every timestep in practice.

9.6.2 *In Situ* Data Triage

Another strategy for using *in situ* techniques to reduce data for subsequent exploration is to conduct a data triage such that the data is stored in a way to effectively supports particular analysis and visualization tasks following the simulation. The basic approach is to reduce and organize the data accord-

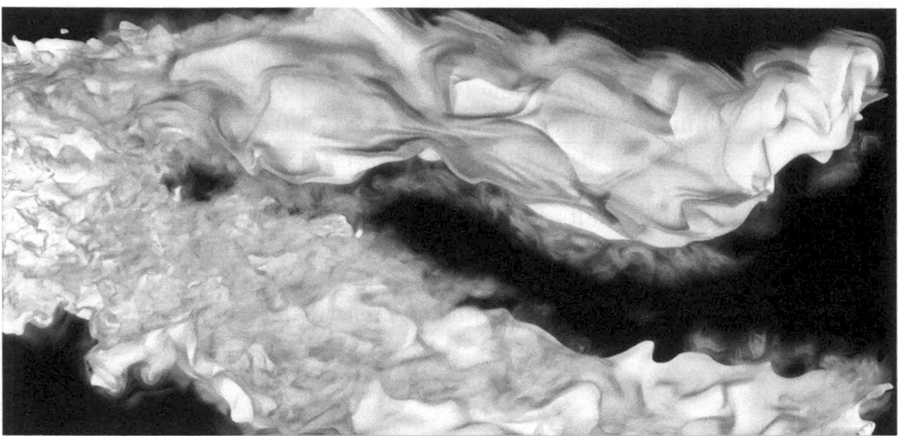

FIGURE 9.8: Visualization uncovering the interaction between the small turbulent eddies and the preheated layer of flame. Image source: Wang et al. [34].

ing to scientists' prior knowledge about certain properties of the simulation and features of interest. The resulting data should be compact and organized specifically to facilitate the offline visualization operations.

A few designs to support co-processing *in situ* data triage have shown opportunities for substantial data reduction and visualization of previously hidden features in the data. Wang et al. [33] computed the importance of each unit block of a flow field, based on how much information and the uniqueness of the information the block contains with respect to other blocks. They showed that the resulting importance field can be used to characterize the flow field, assist in feature extraction process, and guide compression and adaptive rendering. In another study, Wang et al. [34] computed the importance value of each voxel in volume data output from a turbulent combustion simulation according to its distance from the reference features defined by the simulation scientist. First, by factoring the important values into compression, they showed they can achieve impressive savings in storage space. Second, by mapping the importance values to optical properties in the rendering step, the resulting visualization, as shown in Figure 9.8, allows scientists to see the interaction of small turbulent eddies with the preheated layer of a turbulent flame, a region previously obscured by the multiscale nature of turbulent flow.

The effectiveness of this data triage approach relies on how much prior knowledge about the modeled phenomena and features of interest is available for the design of distance functions for importance measures. *Distance* may have a much broader meaning to consider than distance in spatial domain, temporal domain, and any particular high-dimensional space defined by the scientists, making the *in situ* data triage approach possibly applicable to many simulations.

9.7 Conclusion

In situ processing refers to a family of techniques that operate on simulation data as it is produced, as opposed to the traditional post-processing model where the simulation stores its data to disk and a visualization program subsequently reads it in. This technique is widely predicted to play a larger and larger role as supercomputers march towards the exascale era, as I/O is increasingly insufficient relative to compute power. Concurrent processing is the *in situ* form where separate resources are allocated for visualization and analysis and data is transferred over the network. Co-processing refers to the form where visualization algorithms have direct access to the memory of the simulation code. It can be done with tailored code that optimizes memory, network bandwidth and CPU time, or through an adaptor layer to a general visualization tool, which brings a wealth of techniques and can work with many simulation codes. Finally, hybrid techniques use elements of concurrent and co-processing to provide some of the advantages of concurrent processing while minimizing data movement. All of these techniques have been shown to be successful in real-world settings. Finally, *in situ* can prevent a potential loss of science; it creates an opportunity to process more data than is possible with post-processing.

In situ processing is not a panacea, however, as it is incongruent with explorative use cases, since they rarely have *a priori* knowledge. That said, recent work has looked at using *in situ* processing to extract key information that can later be post-processed. Although these techniques are typically domain-specific or algorithm-specific, they provide hope that exploration can still occur even in configurations where *in situ* processing is required.

References

[1] Hasan Abbasi, Greg Eisenhauer, Matthew Wolf, Karsten Schwan, and Scott Klasky. Just in Time: Adding Value to the IO Pipelines of High Performance Applications with JITStaging. In *Proceedings of Symposium on High Performance Distributed Computing (HPDC)*, pages 27–36, July 2011.

[2] Hasan Abbasi, Matthew Wolf, Greg Eisenhauer, Scott Klasky, Karsten Schwan, and Fang Zheng. DataStager: Scalable Data Staging Services for Petascale Applications. In *Proceedings of the 18th ACM International Symposium on High Performance Distributed Computing*, HPDC '09, pages 39–48, New York, NY, USA, 2009. ACM.

[3] V. Bhat, M. Parashar, and S. Klasky. Experiments with In-Transit Processing for Data Intensive Grid Workflows. In *8th IEEE International Conference on Grid Computing (Grid 2007)*, pages 193–200, 2007.

[4] Fabián E. Bustamante, Greg Eisenhauer, Patrick Widener, Karsten Schwan, and Calton Pu. Active Streams: An Approach to Adaptive Distributed Systems. In *Proceedings of the 8th Workshop on Hot Topics in Operating Systems (HotOS-VIII)*, page 163, Schoss Elmau, Germany, May 2001.

[5] L. Chacón, D. A. Knoll, and J. M. Finn. An Implicit, Nonlinear Reduced Resistive MHD Solver. *Journal of Computational Physics*, 178:15–36, May 2002.

[6] J. H. Chen, A. Choudhary, B. de Supinski, M. DeVries, E. R. Hawkes, S. Klasky, W. K. Liao, K.-L. Ma, J. Mellor-Crummey, N. Podhorszki, R. Sankaran, S. Shende, and C. S. Yoo. Terascale Direct Numerical Simulations of Turbulent Combustion Using S3D. *Computational Science and Discovery*, 2(015001), 2009.

[7] Ciprian Docan, Manish Parashar, and Scott Klasky. DataSpaces: An Interaction and Coordination Framework for Coupled Simulation Workflows. In *Proceedings of 19th International Symposium on High Performance and Distributed Computing (HPDC'10)*, June 2010.

[8] Thomas Erl. *Service-Oriented Architecture: Concepts, Technology, and Design.* Prentice Hall PTR, Upper Saddle River, NJ, USA, 2005.

[9] N. Fabian, K. Moreland, D. Thompson, A.C. Bauer, P. Marion, B. Geveci, M. Rasquin, and K.E. Jansen. The ParaView Coprocessing Library: A Scalable, General Purpose In Situ Visualization Library. In *IEEE Symposium on Large Data Analysis and Visualization (LDAV), 2011*, pages 89 –96, Oct. 2011.

[10] A. Globus. A Software Model for Visualization of Time Dependent 3-D Computational Fluid Dynamics Results. Technical Report RNR 92-031, NASA Ames Research Center, 1992.

[11] Luke Gosink, John C. Anderson, E. Wes Bethel, and Kenneth I. Joy. Variable Interactions in Query-Driven Visualization. *IEEE Transactions on Visualization and Computer Graphics (Proceedings of Visualization 2007)*, 13(6):1400–1407, October 2007.

[12] Luke J. Gosink, Christoph Garth, John C. Anderson, E. Wes Bethel, and Kenneth I. Joy. An Application of Multivariate Statistical Analysis for Query-Driven Visualization. *IEEE Transactions on Visualization and Computer Graphics*, 17(3):264–275, 2011.

[13] G. Ji and H.-W. Shen. Efficient Isosurface Tracking Using Precomputed Correspondence Table. In *Proceedings of Symposium on Visualization (VisSym) '04*, pages 283–292. Eurographics Association, 2004.

[14] G. Ji, H.-W. Shen, and R. Wegner. Volume Tracking Using Higher Dimensional Isocontouring. In *Proceedings of IEEE Visualization '03*, pages 209–216. IEEE Computer Society, 2003.

[15] Sriram Lakshminarasimhan, Neil Shah, Stéphane Ethier, Scott Klasky, Rob Latham, Rob Ross, and Nagiza F. Samatova. Compressing the Incompressible with ISABELA: In-situ Reduction of Spatio-temporal Data. In *Euro-Par (1)*, pages 366–379, 2011.

[16] J. Lofstead, Fang Zheng, S. Klasky, and K. Schwan. Adaptable, Metadata Rich IO Methods for Portable High Performance IO. In *IEEE International Symposium on Parallel Distributed Processing (IPDPS)*, pages 1–10, May 2009.

[17] Kwan-Liu Ma. Runtime Volume Visualization of Parallel CFD. In *Proceedings of International Parallel Computational Fluid Dynamics Conference (ParCFD)*, pages 307–314, 1995.

[18] Kwan-Liu Ma. In Situ Visualization at Extreme Scale: Challenges and Opportunities. *IEEE Computer Graphics and Applications*, 29(6):14–19, November/December 2009.

[19] Manish Parashar. Addressing the Petascale Data Challenge Using In-Situ Analytics. In *Proceedings of the 2nd International Workshop on Petascale Data Analytics: Challenges and Opportunities*, PDAC '11, pages 35–36, New York, NY, USA, 2011. ACM.

[20] S. G. Parker and C. R. Johnson. SCIRun: A Scientific Programming Environment for Computational Steering. In *Proceedings of ACM/IEEE Supercomputing Conference (SC)*, 1995.

[21] J. Rowlan, G. Lent, N. Gokhale, and S. Bradshaw. A Distributed, Parallel, Interactive Volume Rendering Package. In *Proceedings of IEEE Visualization Conference*, pages 21–30, October 1994.

[22] Oliver Rübel, Prabhat, Kesheng Wu, Hank Childs, Jeremy Meredith, Cameron G. R. Geddes, Estelle Cormier-Michel, Sean Ahern, Gunther H. Weber, Peter Messmer, Hans Hagen, Bernd Hamann, and E. Wes Bethel. High Performance Multivariate Visual Data Exploration for Extemely Large Data. In *Supercomputing 2008 (SC08)*, Austin, Texas, USA, November 2008.

[23] William J. Schroeder, Kenneth M. Martin, and William E. Lorensen. The Design and Implementation of an Object-Oriented Toolkit for 3D Graphics and Visualization. In *VIS '96: Proceedings of the 7th Conference on Visualization '96*, pages 93–100. IEEE Computer Society Press, 1996.

[24] D. Silver and X. Wang. Tracking and Visualizing Turbulent 3D Features. *IEEE Transactions on Visualization and Computer Graphics (TVCG)*, 3(2):129–141, 1997.

[25] D. Silver and X. Wang. Tracking Scalar Features in Unstructured Datasets. In *Proceedings of IEEE Visualization '98*, pages 79–86. IEEE Computer Society Press, 1998.

[26] V. Springel. The Cosmological Simulation Code GADGET-2. *Monthly Notices of the Royal Astronomical Society*, 364:1105–1134, December 2005.

[27] Kurt Stockinger, E. Wes Bethel, Scott Campbell, Eli Dart, and Kesheng Wu. Detecting Distributed Scans Using High-Performance Query-Driven Visualization. In *SC '06: Proceedings of the 2006 ACM/IEEE Conference on High Performance Computing, Networking, Storage and Analysis*. IEEE Computer Society Press, October 2006.

[28] Kurt Stockinger, John Shalf, Kesheng Wu, and E. Wes Bethel. Query-Driven Visualization of Large Data Sets. In *Proceedings of IEEE Visualization 2005*, pages 167–174. IEEE Computer Society Press, October 2005.

[29] R. Tchoua, S. Klasky, N. Podhorszki, B. Grimm, A. Khan, E. Santos, C. Silva, P. Mouallem, and M. Vouk. Collaborative Monitoring and Analysis for Simulation Scientists. In *International Symposium on Collaborative Technologies and Systems (CTS)*, pages 235–244, May 2010.

[30] A. Tikhonova, C. Correa, and Kwan-Liu Ma. Visualization by Proxy: A Novel Framework for Deferred Interaction with Volume Data. *IEEE Transactions on Visualization and Computer Graphics*, 16(6):1551–1559, 2010.

[31] A. Tikhonova, Hongfeng Yu, C. Correa, and Kwan-Liu Ma. A Preview and Exploratory Technique for Large Scale Scientific Simulations. In *Proceedings of the Eurographics Symposium on Parallel Graphics and Visualization*, pages 111–120, 2011.

[32] Tiankai Tu, Hongfeng Yu, Leonardo Ramirez-Guzman, Jacobo Bielak, Omar Ghattas, Kwan-Liu Ma, and David R. O'Hallaron. From Mesh Generation to Scientific Visualization: An End-to-End Approach to Parallel Supercomputing. In *Proceedings of the ACM/IEEE Supercomputing Conference (SC)*, 2006.

[33] Chaoli Wang, Hongfeng Yu, and Kwan-Liu Ma. Importance-Driven Time-Varying Data Visualization. *IEEE Transactions on Visualization and Computer Graphics*, 14(6):1547–1554, 2008.

[34] Chaoli Wang, Hongfeng Yu, and Kwan-Liu Ma. Application-Driven Compression for Visualizing Large-Scale Time-Varying Data. *IEEE Computer Graphics and Applications*, 30(1):59–69, January/February 2010.

[35] Brad Whitlock, Jean Favre, and Jeremy Meredith. Parallel In Situ Coupling of Simulation with a Fully Featured Visualization System. In *Proceedings of 11th Eurographics Symposium on Parallel Graphics and Visualization (EGPGV'11)*, pages 101–109, April 2011.

[36] Y. Wiseman and K. Schwan. Efficient End to End Data Exchange Using Configurable Compression. In *Proceedings of the 24th International Conference on Distributed Computing Systems*, ICDCS '04, pages 228–235, March 2004.

[37] Hongfeng Yu, Tiankai Tu, Jacobo Bielak, Omar Ghattas, Julio C. Lpez, Kwan liu Ma, David R. Ohallaron, Leonardo Ramirez-guzman, Nathan Stone, Ricardo Taborda-rios, and John Urbanic. Remote Runtime Steering of Integrated Terascale Simulation and Visualization. In *HPC Analytics Challenge, ACM/IEEE Supercomputing 2006 Conference (SC06)*, 2006.

[38] Hongfeng Yu, Chaoli Wang, Ray W. Grout, Jacqueline H. Chen, and Kwan-Liu Ma. In Situ Visualization of Large-Scale Combustion Simulations. *IEEE Computer Graphics and Applications*, 30(3):45–57, May/June 2010.

[39] Fan Zhang, Ciprian Docan, Manish Parashar, Scott Klasky, Nobert Podhorszki, and Hasan Abbasi. Enabling In-situ Execution of Coupled Scientific Workflow on Multi-core Platform. In *Proceedings 25th IEEE International Parallel and Distributed Processing Symposium (IPDPS'12)*, May 2012.

[40] Fang Zheng, H. Abbasi, C. Docan, J. Lofstead, Qing Liu, S. Klasky, M. Parashar, N. Podhorszki, K. Schwan, and M. Wolf. PreDatA—Preparatory Data Analytics on Peta-scale Machines. In *IEEE International Symposium on Parallel Distributed Processing (IPDPS)*, pages 1–12, Apr. 2010.

[41] Fang Zheng, Hasan Abbasi, Jianting Cao, Jai Dayal, Karsten Schwan, Matthew Wolf, Scott Klasky, and Norbert Podhorszki. In-situ I/O Processing: a Case for Location Flexibility. In *Proceedings of the Sixth Workshop on Parallel Data Storage*, PDSW '11, pages 37–42, 2011.

Chapter 10

Streaming and Out-of-Core Methods

David E. DeMarle
Kitware Inc.

Berk Geveci
Kitware Inc.

Jon Woodring
Los Alamos National Laboratory

Jim Ahrens
Los Alamos National Laboratory

10.1 External Memory Algorithms 200
10.2 Taxonomy of Streamed Visualization 202
10.3 Streamed Visualization Concepts 204
10.3.1 Data Structures 204
10.3.2 Repetition 205
10.3.3 Algorithms 205
10.3.4 Sparse Traversal 206
10.4 Survey of Current State of the Art 209
10.4.1 Rendering 209
10.4.2 Streamed Processing of Unstructured Data 210
10.4.3 General Purpose Systems 211
10.4.4 Asynchronous Systems 212
10.4.5 Lazy Evaluation 214
10.5 Conclusion 215
References 216

One path toward visualizing ever-larger data sets is to use external memory or out-of-core (OOC) algorithms. OOC algorithms are designed to process more data than can be resident in the main memory at any one time, and the OOC algorithms do this by optimizing external memory transfers.

Even if only within a pre-processing phase, OOC algorithms run through the entire data, processing some portion, and then freeing up space before moving on to the next portion. *Streaming* is the concept of piecemeal data processing and applying some computation kernel on each portion. In this way, managing the memory footprint directly addresses the large data challenge and it can also improve performance, even on small data because of the effect of forced locality of data references on the memory hierarchy [32, 9, 34, 40].

However, there are challenges to applying streaming in general purpose visualization. Some data structures are not well-suited for partitioning, and not every algorithm can work within the confines of a restricted portion of the data. Also, since a streaming formulation of an algorithm may involve many passes through the data where each pass can be time consuming, optimizations are essential for interactive data exploration. Fortunately, given that the data is to be processed in distinct portions, streaming facilitates skipping (culling) unimportant portions of the data, which can amount to a majority of the input.

This chapter first discusses OOC processing and streaming, in general. First, it considers the design parameters for applying the concept of streaming to scientific visualization. Second, it presents some observations on the potential optimizations and hazards of doing scientific visualization within a streaming framework. With an appreciation of these fundamentals, recent works are surveyed in which the streaming concept plays an important part of the implementation. The discussion will conclude by summarizing the state of the art and a conjecture about the future.

10.1 External Memory Algorithms

The most popular approach to large data visualization is data parallelism. In data parallelism, the system takes advantage of the aggregate memory of a set of processing elements by splitting up the data among all of them. Each processing element only computes the portion for which it is responsible. Chapters 16, 18, and 21 describe current scientific visualization systems that make use of data parallelism.

Data parallelism implicitly assumes that the data fits entirely in a large aggregate memory space, somewhere. Getting access to a large enough memory space can be expensive and otherwise impractical for many users. Streaming exchanges resources for time—instead of using ten machines to visualize a problem, one machine might visualize the same problem but take ten times longer. Streaming then allows all users to process larger data sets than they otherwise could, regardless of the size of the machine.

External memory algorithms are written with the assumption that the entire problem cannot be loaded all at once. These algorithms are described well in surveys by Vitter [35], Silva [31], and Lipsa [22]. Vitter first classified external memory algorithms into two groups: online problems and batch problems.

Online problems deal with large data by first pre-processing data and then making queries into an organized whole, out of which each query can be satisfied by touching only a small portion. The data structures and algorithms

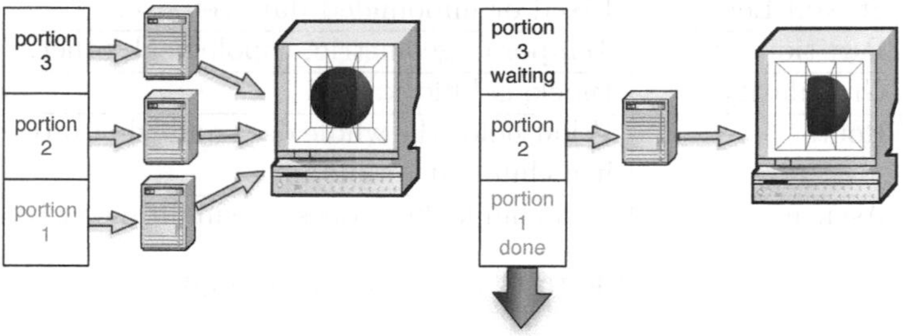

FIGURE 10.1: A comparison of data parallel visualization and streaming.

used are designed so that they never overflow the available RAM, and they require as few slow external memory accesses as possible by amortizing the cost of each access across many related data items. Pre-processing is very effective for visualization because typically only a small fraction of the total data contributes to a meaningful visualization [9]. For example, a pre-processing step can sort or index the data by data values to help accelerate thresholding and isocontouring operations.

Since scientific visualization presents the results to the user in a graphical format, the user can also take advantage of the numerous techniques developed for rendering. If data is spatially sorted into coherent regions, large areas can be skipped, and screen space appropriate details quickly rendered. iWalk [8], an interactive walkthrough system, does this by using an OOC sorting algorithm to build an octree on a disk containing scene geometry in the nodes. At runtime, visible nodes are determined and fetched (speculatively) by a thread that runs asynchronously from the rendering. Likewise, many level-of-detail systems [28, 13] compute summary information during the preprocessing stage; they later fetch that information from a disk and render approximations while the user interacts and whenever the full-resolution data covers too few pixels for full details to be seen. To be fully scalable, the algorithm must be end-to-end out-of-core, meaning that neither the runtime nor the pre-processing phases allow the data to exceed the available RAM.

Batch problems operate by iterating through the data one time or many times, touching only a small portion of the large data at any given time, and then accumulating or writing out new results as they are generated, until completion. These algorithms are most often called "streamed" processing in visualization. This helps to eliminate confusion with the more typical meaning of "batch," producing visualizations in unattended sessions.

In a survey of stream processing within the wider field of computer science, Stephens shows that the streaming concept has been used since at least the 1960s [33]. Within his framework of stream processing systems, data flow networks hold the greatest amount of interest, since they are the dominant

Stream Length	Fixed or unbounded data set size.
Partition Axes	Temporal, geometric, topological, other.
Granularity	Block, subblock, cell.
Route	What holds the whole data and what processes it a chunk at a time.
Asynchrony	Can chunks be processed simultaneously?

TABLE 10.1: Taxonomy of streaming.

architecture for constructing general purpose visualization programs today. This chapter primarily focuses on data flow visualization systems.

Stream processing has been used within visualization for decades as well [14, 32]. Law et al. [21] added a streaming framework to a general purpose scientific visualization library, the Visualization Toolkit (VTK) [30], and later to its sister toolkit, the Insight Toolkit (ITK) [15]. In their streaming framework, a series of pipeline passes allow filters to negotiate how a structured data set is split into memory size-limited portions called *pieces*. In the first pass the network determines the overall problem domain. In the next pass, the network prepares to visualize one specific piece. Next, the pipeline runs fully to produce a visualization of that piece. In this way, each piece is processed, in turn, by all filters along the visualization pipeline.

As illustrated in Figure 10.1, there is a strong relationship between streaming and data parallel processing. If the data can be processed in portions, many portions could be processed simultaneously. In later work, the same architecture is used as the basis for data parallel visualization within VTK [1]. The two techniques are in fact complementary and can be combined.

10.2 Taxonomy of Streamed Visualization

In general, it is assumed that streamed visualization is the processing and rendering of a sequence of data values, which in turn, will only keep a portion of the whole data set in memory at any given time. Under this assumption, streaming algorithms are parameterized by the contents of the stream, how the data is ordered, how long the stream is, and how exactly it is processed. Under this classification, a taxonomy of streaming is described within scientific visualization.

One axis of the taxonomy is whether the stream is of fixed or unbounded length. Consider the case of operating on an endless stream, as is the case in signal processing frameworks. An analogy within the visualization community is *in situ* visualization. As explained in Chapter 9 with *in situ* visualization, the visualization engine computes graphical results as data are generated or obtained. When the visualization program purges old values to keep memory usage bounded, this is a form of streaming.

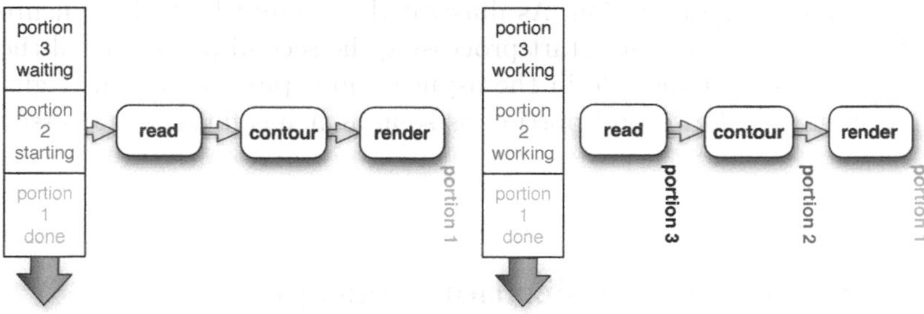

FIGURE 10.2: Synchronous (left) and asynchronous (right) streaming in a data flow visualization network.

The second axis of the taxonomy is which axes the data is divided along: spatial, temporal, topological, or some combination of these. Most time-varying simulations produce a new data set that covers the entire spatial domain at each computed timestep. It is almost always impractical to store all of the time slices in memory at once, so visualization of time-varying data is almost always streamed. Streaming through the timesteps works because many visualization algorithms require only one or a few timesteps to be loaded at once [24, 4]. Still, in many cases, a single timestep is too large to fit into the memory all at once, therefore, the data is partitioned in the spatial domain as well.

The third axis is the granularity by which the data is divided. The largest possible portion size is just smaller than the available memory of the machine and the smallest is the size of an individual data item. There are computational trade-offs though, with respect to the level of granularity [9, 37, 2]. Smaller portions make a better utilization of cache architectures and allow sparse data traversal to skip over more of the data that ultimately does not contribute to the visualization. Larger portion sizes do not force the visualization system to update as many times to cover the same space.

The fourth axis is the path that data takes as it is streamed. So far, the cases being discussed are the case of streaming data off a disk, through a visualization algorithm and onto the screen. However, streaming also includes the case in which data is too large for a local disk, and therefore, it is acquired from a server and processed locally, one portion at a time [9, 29, 20]. Likewise, streaming applies when data is streamed to and processed within a GPU.

The fifth axis of the taxonomy is the degree of asynchrony within the visualization system. Once data is partitioned, a system may allow different portions to be processed simultaneously. For example, in a client–server framework, the client might display some results while the server fetches additional portions for it to process. As another example, the modules within a visualization pipeline can operate as a staged parallel pipeline [32, 1, 34, 36] if there is space for inter-stage buffering. As illustrated in Figure 10.2, the synchronous pipeline, on the left, cannot start processing the second portion until the first is finished. The first module in the asynchronous pipeline, on the right, will start working on the second portion as soon as it has finished processing the first one.

10.3 Streamed Visualization Concepts

This section discusses some general aspects of applying the concept of streaming to scientific visualization. The discussion will be limited to traditional qualitative scientific visualization in which data sets are processed and rendered in an effort to gain qualitative insights into the problem at hand.

10.3.1 Data Structures

Ideally, streaming visualization must break scientific data apart into coherent regions of space and time. The different classes of input data that are to be visualized have properties that make them more or less well-suited for partitioning.

A completely structured data set has the property where the topology and geometry are implicit and the data values represent regularly separated samples. Structured data sets are efficiently stored as data value arrays. Streaming with structured data is relatively easy:

- The mapping between geometric location and a location within a data array is implicit and easily calculated.
- Portions containing spatially coherent regions (slices along the slowest changing direction) are straightforward [32, 34] without any preprocessing.
- With pre-processing, portions containing contiguous regions [9] and multiresolution representations [26] are also trivial to acquire.
- Strided sampling can produce lower resolution data [27, 3].

Unstructured data are not as well-suited to streaming because of their unconstrained topology and geometry. For example, in the context of the VTK library, to know the location of any given cell, the program must access the cell's vertices by walking the cell's vertex index list. The vertices are stored in their own array, and the vertices, for any cell will be arbitrarily scattered in that array. It is not trivial to create coherent portions since the cell and vertex arrays may not fit into the memory, and because geometric information is spread across that memory layout. Once partitioned, the fact that a portion's bounding box is not known until after the portion is entirely read can also be problematic.

Modern external memory and streaming formats for unstructured data solve the first problem by in-lining the vertex data with the cell data (also called normalized or triangle soup representations [6]). The cells and vertices are kept together in portions called *meta-cells* [5], or by intermixing the vertex and cell lists so that the vertices and the cells that reference them are kept very close together in the data stream [19]. The second problem is solved by keeping meta-information along with each portion.

A notable exception to the rule that unstructured data is hard to stream is the point set. It is an unstructured data type that consists only of a set of locations in space without any connections between them. The fact that one level of indirection (from cell list to point list) is removed greatly simplifies the handling of large point sets. There are also many algorithms that work on point sets that do not require spatially coherent regions. For those algorithms that do, standard OOC sorting algorithms [35] can be used in a pre-processing step to produce regions that do not overlap.

Composite data sets, which consist of sets of structured or unstructured data sets (i.e., blocks), can be straightforward to stream, as long as metadata describing each block is available. In this situation it is straightforward to process each block, in turn.

10.3.2 Repetition

In streaming visualization, the visualization pipeline has to support repetition to iterate through all of the portions that make up the whole data set. As Figure 10.3 illustrates, both push and pull pipeline models can support streaming. In the push model, the system reads a portion at a time. The act of reading each portion causes the pipeline to push it through toward the rendering module. In the pull model, it is the rendering module that requests each portion, in turn (see also 2.2.2).

Comparing the two, a push model is better suited to unbounded length streams since the availability of new data can simply trigger the next cycle. Push models also have less overhead in asynchronous streaming because downstream modules can begin as soon as results are ready [36]. On the other hand, push models are not as effective at sparse traversal because downstream modules can not be as selective about what the pipeline processes [23].

With either approach, streaming visualization systems must aggregate results over time as each new result is produced. For example, with a scan-converted rendering of opaque surfaces, the color and depth buffers are not cleared in between portions, so that each pixel ends up with the color of the nearest primitive out of the entire data set.

10.3.3 Algorithms

Some, but not all, visualization algorithms are inherently well-suited for streaming [32, 34, 11]. The defining factor is how much support an algorithm

push	pull

`create_pipeline`
`(source, ..., sink)`
`while (!endofstream)`
`source.readMore()`
	`create_pipeline`
	`(source, ..., sink)`
	`while (piece<#pieces)`
	`sink.render(piece++)`

FIGURE 10.3: In push and pull pipelines, the source or sink direct the pipeline to process the next portion.

requires: local support or global support? Localized algorithms, typified by fragment programs with pixel processing, are well-suited, because they analyze each point or cell independently, and are thus insensitive to data partitioning and traversal order. Many extraction algorithms fall into this category. Algorithms that require information from a local neighborhood of elements, for example differential operators, are less well-suited as both the primary element and its neighbors must be resident together. This is typically solved by using a ghost cell technique, where portions include a layer of cells belonging to their neighbors.

Algorithms that need global support must be recast into a local problem to work efficiently under a streaming paradigm [40]. Problems such as streamline integration, which can suffer from load imbalances in a data parallel solution, also tend to cause excessive passes through the data in a streaming solution. In a distributed-memory data parallel context, in which the data fits in the parallel machine's aggregate memory, the data needed by one processor to continue work is always present somewhere. An algorithm can resort to communicating between different processors to continue forward work progress. In a streaming context, data dependencies are more critical, since the data needed to continue work may not have been loaded yet, or it might have been processed and evicted from the memory to make space for current data. This problem may require additional passes to solve. The problem of interdependence between portions can sometimes be solved by carefully considering the order in which portions are processed, such as in the parallel streamed computation of ghost cells [16].

Rendering algorithms may be local or global in nature. Aggregating scan-converted or splat-rendered results, neglecting translucency, are handled simply, as described earlier. Volume rendering and translucent surface rendering is problematic for two reasons: first, as the interior of the data is visible, a larger fraction of it contributes to the image and fewer portions can be culled; and second, one must maintain a back-to-front (or front-to-back) traversal order [32]. Ray tracing and streamlines calculations are problematic because of their unstructured memory access characteristics (see 14.3 and Chap. 6).

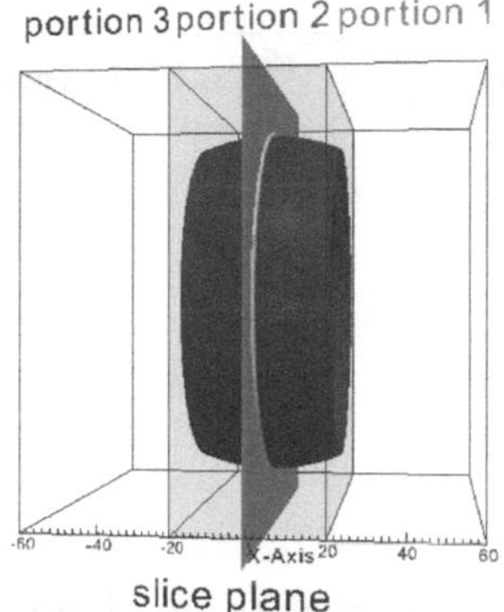

FIGURE 10.4: Culling unimportant data leaves only portion 2 when slicing at $X = 0$. Likewise, lazy evaluation could operate on only the important cells within the portion.

10.3.4 Sparse Traversal

In streamed visualization, where each pass through the data takes a significant amount of time and where many passes may be needed, achieving interactivity may require reducing the number of portions processed. Any region of the data that can be quickly determined as being unimportant to the final visualization should be culled to avoid time-consuming processing [23, 7, 2]. If a portion's extreme data ranges do not encompass a contour module's isovalue, the entire portion can be ignored by the whole pipeline. Similarly, as shown in Figure 10.4, a slice module could reject a portion by testing its bounding box against a slicing plane. These "lightweight," "metadata" operations can be performed quickly and are able to save time by skipping "heavyweight" module operations. At a finer granularity, expensive derived computations on cells within a passing portion may also be avoided.

Other lightweight computations, such as rendering optimization, can save time as well. For example, portions that lie entirely outside the view frustum, or that are occluded by other portions, should be skipped. In some situations, back-facing portions are assumed to be occluded, so back-facing portions can be rejected in the same way that back-facing polygons are rejected in rendering [28]. Level-of-detail techniques are also appropriate for reducing data and processing passes. With large data, there may be many more display primitives (polygons, voxels, etc.,) than there are pixels, so only a subset of those primitives actually contribute [29, 37, 3]. During interaction, responsiveness is more important than accuracy, so the level of detail can be further reduced. In practice, the appropriate level of detail can be determined by projecting the portion's bounding boxes into screen-space; low-resolution data can be obtained by downsampling either on-the-fly or in a pre-processing step.

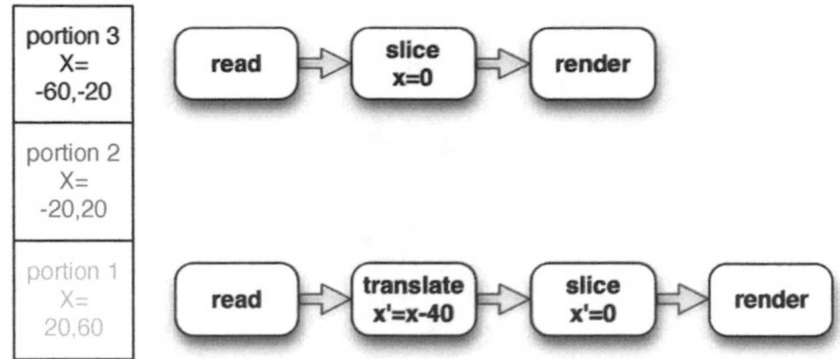

FIGURE 10.5: In order to be used to avoid processing portions, metadata must remain accurate without processing the data in the portion. The slice module in the pipeline on top can trust the metadata from the source and will ignore portions 1 and 3. The translate module in the pipeline on the bottom will alter the real data, and the slice module should ignore portions 2 and 3 instead.

Sparse traversal adds an additional requirement to have accurate metadata with the partitioned data. In a few cases, simulation codes store metadata along with the data. If metadata is not present, the data must be preprocessed, or, the metadata can be derived on the first pass through the data, during which sparse traversal will not be possible.

Once the metadata is obtained, it must be managed through the pipeline [10]. The pipeline avoids processing chunks based on the contents of their metadata. A portion that is off-screen will be ignored. However, pipeline modules manipulate data and thereby invalidate every portion's metadata. For example, a module that repositions data may move an offscreen portion back into view. Downstream modules can not cull anything then, unless upstream modules change the metadata before processing the data. Figure 10.5 demonstrates another example. In many cases the metadata can not be determined *a priori*. The system should fall back to marking a portion's metadata as "known to be clean" or "potentially dirty," and then cull portions based only on the clean information. Later, the dirty information can be updated and marked clean after data processing once it is known.

Lastly, the output of the visualization pipeline tends to be much smaller than the whole data set. Although the input data may be very large, the salient features within it tend to be small. Some salient features are: the part of the data within a particularly important region; slices, isocontours and other techniques that reduce dimensionality; descriptive statistical metrics; locations of extremal values; the presence or absence of particular events; etc. Because the features are small, it is often possible to cache results [23, 10] in the available memory, in order to avoid redundant processing while the user interacts with previously generated results. Figure 10.6 illustrates an example

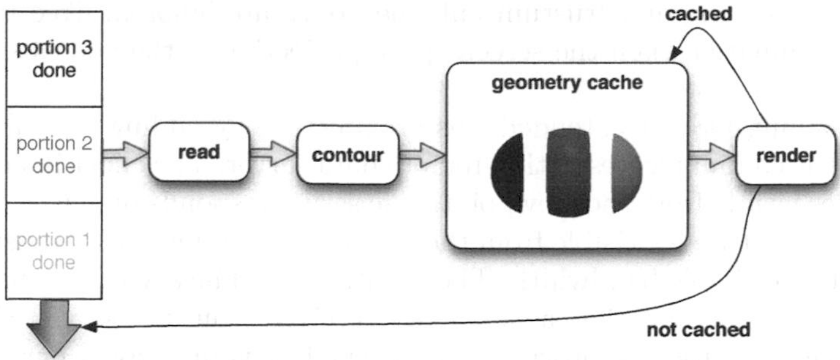

FIGURE 10.6: Assuming the cache can fit the results, the cache module will prevent upstream re-executions when the camera moves but, otherwise, the data remains unchanged.

in which placing a module that caches polygonal results before the rendering module in a streaming pull pipeline will prevent reprocessing of portions, during camera motion or other interactions that do not invalidate the cached results.

10.4 Survey of Current State of the Art

This section highlights recent works that use streaming for large data visualization, and categorize each in the terms of the streaming taxonomy of 10.2.

10.4.1 Rendering

Streaming has been used effectively in rendering large data for more than a decade. With high-resolution scanner technologies, there is a need to display very high-resolution triangle meshes, stored at remote data repositories, on local workstations. Besides the challenge of overcoming the available compute and memory resource limitations for rendering on a local workstation, the slow network pipe between the remote repository and the viewer's workstation make it infeasible to download these large meshes entirely before rendering them locally.

One system designed to overcome the problem of rendering such a large set of data is the streaming QSplat [29] system. This is a remote rendering extension of the QSplat [28] large model viewer. The original implementation of QSplat achieves interactive rendering speeds for massive triangle meshes, through point splat rendering, using a pre-processed bounding sphere hierarchy representation for the input mesh. Portions, called nodes, within the model are selected from a view-dependent, top-down traversal of the tree where back-facing and frustum cullable nodes are ignored. Tree traversal is depth terminated when the screen space projection of the node's splat size is small enough.

Streaming QSplat extended this architecture by adding a local rendering client that communicates with a remote data server. The client issues requests for nodes visible from the viewpoint. The server responds by sending only previously unsent nodes visible from the viewpoint and limits the transmission to match the available bandwidth. The client, meanwhile asynchronously renders the nodes it has and the caches nodes received from the server. Over time, the image resolution improves as the client builds up a local partial copy of the server's hierarchy to deeper and deeper resolution levels. Later systems improved on the coarse appearance of the splat rendered points by rendering triangles [20, 13] in a streamed, multiresolution view-dependent fashion.

These systems pre-process the data to use view-dependent rendering optimizations that cull portions from consideration and show data in only as much detail as is required. In terms of the streaming taxonomy, these systems operate on a fixed-size stream, partition the geometry (in a multiresolution hierarchy), and have a coarse granularity of pages of triangles. They stream data from disk to memory on the server and then over the network to the client, which uses asynchronous processing to render data while transferring other data and also, relies heavily on pre-processing. They are streaming techniques only in that they send small portions of data to the client so that it is not overwhelmed. Progressive transmission is very beneficial to users, as they receive early visual feedback and do not have to wait for the whole model to be transferred before interacting with it. On a spectrum between Vitter's online and batch classification of OOC techniques, QSplat and similar systems are closest to online.

10.4.2 Streamed Processing of Unstructured Data

Compared to structured data sets, unstructured data sets are harder to stream because of the loose coupling between the cells and vertices in memory, which makes it difficult to partition the data and keep a low-memory footprint. Unstructured mesh data is traditionally stored in an indexed format, consisting of a list of vertices and a separate list of triangles, tetrahedra, or other cell types that refer to locations in the vertex list that fully specify the cells' geometry. The potentially random indexing between the cell list and the vertex list is a challenge for many out-of-core algorithms, since cache locality is poor and page swaps frequently occur.

Isenberg and Lindstrom's streaming mesh [19] format advanced the state of the art by developing a cache friendly data representation and OOC construction procedure. Their construction procedure reorders the triangles in the mesh and intersperses vertex and cell information within it. The resulting mesh representation keeps vertices and the cells that use them nearby, and minimizes the lifetime that any element must remain in memory. This ordering reduces the memory footprint, which improves memory hierarchy caching performance and allows the mesh to be processed in a streaming fashion. Because of reduced memory pressure, their streaming mesh processing algorithms were found to outperform efficient OOC algorithms. For example, the authors were able to simplify a streamed version of the 323 million triangle/58MB St. Matthew model to one million triangles in 45 minutes using only 9MB of RAM. For comparison purposes, a pure external memory approach, which efficiently pages data from a disk, took 14 hours and 80MB of RAM.

This and later extensions of the technique for both meshes [17, 40] and point sets [25, 18] facilitated the construction of connected processing pipes. Here, independent processes can be chained together, each processing a strand of mesh data and immediately forwarding each transformed cell onward to the next streaming algorithm. In terms of the streaming taxonomy, these systems could potentially operate on an unlimited stream length if it arrived in a streaming mesh format. The systems partition over the mesh connectivity rather than the spatial axes and have a fine granularity. They stream data from external memory to the processing pipeline, in which streaming operators operate in asynchronous pipelines. These systems are less dependent on preprocessing than the rendering systems discussed in 10.4.1, and they are closer to the batch end of the OOC spectrum.

10.4.3 General Purpose Systems

There has been significant work in utilizing streaming in general purpose, large data visualization systems. The flexible nature of these systems, which accept a variety of input data formats and allow users to apply any number of different processing algorithms, make it difficult to apply streaming techniques consistently.

Law et al. [21, 1] add a streaming framework to the VTK library by managing the way the data flow pipeline executes. The authors insert a set of new pipeline passes into VTK's update mechanism in which filters ask for the processing of some small region within the whole of any atomic data type (called a piece). Applications that use VTK can stream data by iterating through the set of pieces to build up a result incrementally.

VisIt [7] implements streaming on a block or atomic data set level rather than on the piece level of regions within a block (see 2.5). Each processor in the visualization process streams through many blocks because the simulations are usually run on larger, parallel machines than the visualization engine. In parallel streamed visualization, processors obtain blocks on demand, which helps balance the load.

In VisIt, "contracts" allow metadata management over the pipeline. Contracts are a container where modules in the pipeline store arbitrary metadata, which allows them to agree upon how to process the data. Nonessential blocks can be culled from consideration because contracts can hold metadata, like geometric bounds and data array ranges. Likewise, processing modules that require global support and do collective communications put a flag in the contract that disables the streaming entirely.

VTK uses a similar metadata transferring mechanism to allow culling and prioritization [2] of pieces. An additional pipeline pass allows the computation of metadata about any given piece, before the piece itself is fully processed. This metadata computation pass is relatively fast and returns a number that represents the importance or priority of the specified piece. Applications use the priority to schedule the comparatively slow passes that process the real data. Figure 10.7 shows how the prioritization pass allows the basic pull pipeline streaming algorithm presented in Figure 10.3 to be optimized.

In a culling test [2], the authors found that only 117 of the 512 total pieces of a test model were processed. Culling garnered a 3.22× speedup over a non-streamed approach, which was smaller than the ideal (4.38×), because of the overhead of processing many pieces instead of one. However, prioritization also allows for the most important pieces to be processed first. Since pieces are prioritized by distance from the camera, the user's initial view is more informative, which helps the user to adjust parameters and interact with the data more quickly. In the same tests, the rendered image approached 75% of the pixel accuracy within the first 7.7% of the total 16.25 second render time.

To further enhance interactivity in remote visualization using VTK and ParaView, research has since focused on multiresolution streamed progressive visualization [3, 10]. Once again, the facility is implemented in a new pipeline pass. The bottom algorithm in Figure 10.7 shows how applications can use this pass to request pieces at various resolution levels. Pipeline sources respond to resolution requests by producing sampled versions of pieces. For example, structured data readers retrieve lower resolution data from pre-processed files or read raw data in a strided manner. Producing lower resolution pieces primarily reduces I/O time and also allows downstream modules to run faster. Multiresolution streaming improves response time for the whole spatial domain because of lower I/O and processing costs, and because it requires a smaller amount of network bandwidth between the remote server and local client. The same framework also serves to visualize adaptively sampled, large point sets [38] and wavelet-compressed, structured data sets [39].

In terms of the streaming taxonomy, these systems operate on a fixed-size stream and they partition the temporal and geometric domains independently at a coarse level of granularity. They stream data from external memory to the processing pipeline and also stream data, or images, from the server to a client. Both VisIt and ParaView are fully synchronous. These systems do not rely on pre-processing to prepare visualization data for streaming, but they take advantage of it when it is available.

10.4.4 Asynchronous Systems

Several systems perform independent operations simultaneously to operate more efficiently. This is called *functional parallelism* and stands in contrast to

Streaming with culling

```
while (piece<#pieces)
   if sink.prioritize(piece)>0
      sink.render(piece)
   piece++
```

Streaming with prioritization

```
while (piece<#pieces)
   priority = sink.prioritize(piece)
   priorityQueue.insert(piece,priority)
   piece++

while (!priorityQueue.Empty())
   piece, priority = priorityQueue.pop()
   if (priority > 0)
      sink.render(piece)
```

Streaming with multiresolution refinement

```
piece = 0 #entire domain
resolution = 0 #coarsest resolution
priority = 1.0
priorityQueue.insert(piece,resolution,priority)

while (!priorityQueue.Empty())
   piece, resolution, priority = priorityQueue.pop()
   if (priority > 0)
     sink.render(piece,resolution)
     children = split(piece,resolution)#*
     #Piece's children cover the same area as piece
     #but all have higher resolution.
     foreach child:
        childsPriority =
           sink.prioritize(child, childsResolution)
        priorityQueue.push(child, childsPriority,
                                 childsResolution)#**
```

FIGURE 10.7: Culling, prioritizing, and multiresolution streaming in a pull pipeline. The prioritize call initiates a pipeline pass that determines the importance of a piece. In practice, it is not necessary to split up every piece, and it is necessary to use two queues to ensure that the entire domain is drawn before resolution increases.

data parallelism wherein the same operations are performed simultaneously on different data items. Examples of functional parallelism include overlapping asynchronous requests for data with rendering, as occurs in out-of-core rendering systems [8, 29], and causing modules along a visualization pipeline to operate on different portions simultaneously [34]. This is called *pipeline parallelism.*

Vo et al. [36] added streamed, pipelined parallel processing to VTK by scheduling module execution in a push pipeline. They found that a push pipeline can have less overhead than a pull pipeline, because a pull pipeline must stall the entire pipeline to obtain an up-to-date result. In a push pipeline, the end-to-end delay is lower, because filters can process data portions as soon as they become ready. In general, asynchrony allows the pipeline to overcome serial bottlenecks. Ideally, an N element long pipeline will yield an N times speedup when run on N pipelined processors. In a test, Vo's asynchronous VTK approach achieved 77% parallel efficiency on eight threads where the standard VTK achieved only 41.8%.

10.4.5 Lazy Evaluation

Lazy evaluation is a variation on streamed visualization in which computations are aggregated and performed on demand so that unimportant data is not considered. For example, with the Demand Driven Visualizer [23, 24] system, when the user computes derived quantities for data values that lie on a slicing plane, only the data values on the plane are fed into the expression. No time is wasted in evaluating the expression for the majority of the values in the data, which do not happen to lie on the plane.

The Demand Driven Visualizer makes sparse traversal part of its core data access architecture. Paged modules and time dependent modules service requests for data in space and time by loading the requested pages from external memory on demand and managing the removal of unused pages transparently. Data accessors are layered so that arbitrarily derived expressions can be computed without requiring the expression to be evaluated over the entire domain. A cached accessor class ensures that results can be reused many times without an additional I/O, once they are computed initially. A set of visualization modules, built on top of the sparse memory access API, produce images parsimoniously.

Duke et al. [12, 11] implement a lazily evaluated streamed visualization system within Haskell, a functional language with built-in support for a lazy evaluation. The key insight is that demand-driven pipelines and streaming are naturally expressed within lazy evaluated functional languages. In an object-oriented language, visualization modules may be implemented as a class with methods to connect the modules together, to compute a result, and to communicate results across the module connections. Important optimizations, such as reusing completed results and working in a streaming fashion, add considerable code complexity. In a lazy evaluated functional language, a pipeline is expressed simply as: `render(contour(readfile()))`. The implementation of each function or processing module is restricted to the specific algorithm while the language handles the details of connecting modules together, evaluating and passing results between modules.

In the evaluation of their visualization system in Haskell, the result was a fine-grained streaming visualization system that scaled to large data sizes by minimizing memory consumption. In the comparison of isocontouring performance and VTK, the study results show that although isocontouring performance for small data suffered, with large enough data the streaming approach was faster, and it had a smaller memory footprint because both the input and output data sizes were constrained. A 128MB volume that produced a 199 million-vertex isosurface took 517 seconds in the functional approach and consumed 8.35MB of RAM, while the VTK implementation took 755 seconds and consumed 1.27GB of RAM because of the complex output surface.

In terms of the streaming taxonomy, these operate on a finite stream and partition, both in the geometric and temporal domains, at a fine level of granularity. They stream data from external memory to the processing pipeline asynchronously.

10.5 Conclusion

This chapter has given an overview of streaming, including hazards and a potential optimization. By surveying multiple works, this chapter recognizes the effectiveness of the streaming technique in a variety of settings. In entering the era of exascale computing, the streaming technique is likely to become more widespread. The generation of supercomputers that will come online soon will produce even larger data sets than today's generation, and they will have less random access memory (RAM) associated with and tied more tightly to each processing element. In this setting the streaming concepts of piecemeal processing, with parsimonious data traversal and remote delivery become even more appealing. One might envision a complete solution consisting of a true asynchronous data flow system, where the modules operate simultaneously on fine-grained data streams, which allow for arbitrary processing of arbitrary scientific data of an unlimited size.

References

[1] J. Ahrens, K. Brislawn, K. Martin, B. Geveci, C. Law, and M. Papka. Large-Scale Data Visualization Using Parallel Data Streaming. *IEEE Computer Graphics and Applications*, 21:34–41, August 2001.

[2] J. Ahrens, N. Desai, P. McCormick, K. Martin, and J. Woodring. A Modular Extensible Visualization System Architecture for Culled Prioritized Data Streaming. In *Society of Photo-Optical Instrumentation Engineers (SPIE) Conference Series*, volume 6495, January 2007.

[3] J. Ahrens, J. Woodring, D. DeMarle, J. Patchett, and M. Maltrud. Interactive Remote Large-Scale Data Visualization Via Prioritized Multi-Resolution Streaming. In *Proceedings of the 2009 Workshop on Ultrascale Visualization (UltraVis '09)*, pages 1–10. ACM, 2009.

[4] J. Biddiscombe, B. Geveci, K. Martin, K. Moreland, and D. Thompson. Time Dependent Processing in a Parallel Pipeline Architecture. *IEEE Transactions on Visualization and Computer Graphics*, 13:1376–1383, November 2007.

[5] Yi-Jen Chiang, C. Silva, and W. Schroeder. Interactive Out-Of-Core Isosurface Extraction. *Proceedings of IEEE Conference on Visualization (Vis'98)*, pages 167–174, 1998.

[6] Yi-Jen Chiang and C.T. Silva. I/O Optimal Isosurface Extraction. *Proceedings of IEEE Conference on Visualization (Vis97)*, pages 293–300, 1997.

[7] Hank Childs, Eric S. Brugger, Kathleen S. Bonnell, Jeremy S. Meredith, Mark Miller, Brad J. Whitlock, and Nelson Max. A Contract Based System For Large Data Visualization. *Proceedings of IEEE Conference on Visualization 2005 (Vis05)*, pages 190–198, 2005.

[8] W. Corrêa, J. Klosowski, and C. Silva. iWalk: Interactive Out-of-Core Rendering of Large Models. Technical report, University of Utah, 2002.

[9] Michael Cox and David Ellsworth. Application-Controlled Demand Paging for Out-of-Core Visualization. In *Proceedings of the IEEE Conference on Visualization (Vis'97)*, pages 235–244, Los Alamitos, CA, USA, 1997. IEEE Computer Society Press.

[10] D. DeMarle, J. Woodring, and J. Ahrens. Multi-Resolution Streaming in VTK and ParaView. *The Kitware Source*, 1(15):13–15, Oct. 2010.

[11] D. Duke, R. Borgo, M. Wallace, and C. Runciman. Huge Data But Small Programs: Visualization Design via Multiple Embedded DSLs. In

Proceedings of the 11th International Symposium on Practical Aspects of Declarative Languages (PADL '09), pages 31–45, 2009.

[12] D. Duke, M. Wallace, R. Borgo, and C. Runciman. Fine-Grained Visualization Pipelines and Lazy Functional Languages. *IEEE Transactions on Visualization and Computer Graphics*, 12:973–980, 2006.

[13] Enrico Gobbetti and Fabio Marton. Far Voxels: a Multiresolution Framework for Interactive Rendering of Huge Complex 3D Models on Commodity Graphics Platforms. In *ACM SIGGRAPH ASIA 2008 Courses*, pages 32:1–32:8, 2008.

[14] Robert B. Haber, David A. McNabb, and Robert A. Ellis. Eliminating Distance in Scientific Computing: An Experiment in Televisualization. *International Journal of Supercomputing Applications and High Performance Engineering*, 4(4):71–89, December 1990.

[15] L. Ibáñez, W. Schroeder, L. Ng, and J. Cates. Streaming Large Data. In *The ITK Software Guide*, pages 743–748. Kitware, 2nd edition, 2005.

[16] M. Isenburg, P. Lindstrom, and H. Childs. Parallel and Streaming Generation of Ghost Data for Structured Grids. *IEEE Computer Graphics and Applications*, 30:32–44, May 2010.

[17] M. Isenburg, P. Lindstrom, and J. Snoeyink. Streaming Compression of Triangle Meshes. In *Proceedings of the EuroGraphics Symposium on Geometry Processing*, pages 111–118, July 2005.

[18] M. Isenburg, Y. Liu, J. Shewchuk, and J. Snoeyink. Streaming Computation of Delaunay Triangulations. In *ACM SIGGRAPH 2006 Papers*, SIGGRAPH '06, pages 1049–1056, 2006.

[19] Martin Isenburg and Peter Lindstrom. Streaming Meshes. *Proceedings of IEEE Conference on Visualization (Vis05)*, pages 231–238, 2005.

[20] J. Kim, S. Lee, and L. Kobbelt. View-Dependent Streaming of Progressive Meshes. In *Proceedings of the Shape Modeling International 2004*, pages 209–220, 2004.

[21] C. Charles Law, W. Schroeder, K. Martin, and J. Temkin. A Multi-Threaded Streaming Pipeline Architecture for Large Structured Data Sets. In *Proceedings of the IEEE Conference on Visualization (Vis'99)*, pages 225–232, 1999.

[22] D. Lipsa, R. Laramee, R. Bergeron, and T. Sparr. Techniques for Large Data Visualization. *International Journal of Research and Reviews in Computer Science*, 2:315–322, 2011.

[23] P. Moran and C. Henze. Large Field Visualization with Demand-Driven Calculation. In *Proceedings of the IEEE Conference on Visualization (Vis'99)*, pages 27–33, 1999.

[24] Patrick J Moran. Field Model: An Object-Oriented Data Model for Fields. Technical report, NASA Ames Research Center, 2001.

[25] Renato Pajarola. Stream Processing Points. *Proceedings of IEEE Conference on Visualization (Vis05)*, pages 239–246, 2005.

[26] V. Pascucci and R. Frank. Hierarchical Indexing for Out-of-Core Access to Multi-Resolution Data. Technical report, LLNL, 2001.

[27] S. Prohaska, A. Hutanu, R. Kahler, and H. Hege. Interactive Exploration of Large Remote Micro-CT Scans. In *Proceedings of the IEEE Conference on Visualization (Vis'04)*, pages 345–352, 2004.

[28] Szymon Rusinkiewicz and Marc Levoy. QSplat: a Multiresolution Point Rendering System for Large Meshes. In *Proceedings of SIGGRAPH*, pages 343–352, 2000.

[29] Szymon Rusinkiewicz and Marc Levoy. Streaming QSplat: a Viewer for Networked Visualization of Large, Dense Models. In *Proceedings of the Symposium on Interactive 3D Graphics (I3D'01)*, pages 63–68, 2001.

[30] W. Schroeder, H. Martin, and B. Lorensen. *The Visualization Toolkit.* Kitware, 1996.

[31] C. Silva, Y. Chiang, J. El-sana, and P. Lindstrom. Out-of-Core Algorithms for Scientific Visualization and Computer Graphics. In *IEEE Visualization Course Notes*, 2002.

[32] Deyang Song and Eric Golin. Fine-Grain Visualization Algorithms in Dataflow Environments. In *Proceedings of the IEEE Conference on Visualization (Vis93)*, pages 126–133, 1993.

[33] Robert Stephens. A Survey of Stream Processing. *Acta Informatica*, 34:491–541, 1997. 10.1007/s002360050095.

[34] A. Varchola, A. Vaško, V. Solčány, L. Dimitrov, and M Šrámek. Processing of Volumetric Data by Slice- and Process-Based Streaming. In *Proceedings of the 5th International Conference on Computer Graphics, Virtual Reality, Visualisation and Interaction in Africa*, AFRIGRAPH '07, pages 101–110, 2007.

[35] Jeffrey Scott Vitter. External Memory Algorithms and Data Structures: Dealing with Massive Data. *ACM Computing Surveys*, 33:209–271, June 2001.

[36] H. Vo, D. Osmari, B. Summa, J. Comba, V. Pascucci, and C. Silva. Streaming-Enabled Parallel Dataflow Architecture for Multicore Systems. *Computer Graphics Forum*, 29(3):1073–1082, 2010.

[37] Michael Wand, Alexander Berner, Martin Bokeloh, Arno Fleck, Mark Hoffmann, Philipp Jenke, Benjamin Maier, Dirk Staneker, and Andreas Schilling. Interactive Editing of Large Point Clouds. In *Proceedings from the Symposium on Point-Based Graphics*, pages 37–46, 2007.

[38] J. Woodring, J. Ahrens, J. Figg, J. Wendelberger, S. Habib, and K. Heitmann. In-situ Sampling of a Large-Scale Particle Simulation for Interactive Visualization and Analysis. *Computer Graphics Forum*, 30(3):1151–1160, 2011.

[39] J. Woodring, J. Ahrens, S. Mniszewski, C. Brislawn, and D. DeMarle. Revisiting Wavelet Compression using JPEG 2000 to Ensure Precision for Large Scale Data. In *Proceedings of the IEEE Symposium on Large Data Analysis and Visualization (LDAV)*, pages 31–38, 2011.

[40] Tian Xia and Eric Shaffer. Streaming Tetrahedral Mesh Optimization. In *Proceedings of the ACM Symposium on Solid and Physical Modeling (SPM08)*, pages 281–286, 2008.

Part III

Advanced Architectural Challenges and Solutions

Chapter 11

GPU-Accelerated Visualization

Marco Ament

VISUS, Universität Stuttgart

Steffen Frey

VISUS, Universität Stuttgart

Christoph Müller

VISUS, Universität Stuttgart

Sebastian Grottel

VISUS, Universität Stuttgart

Thomas Ertl

VISUS, Universität Stuttgart

Daniel Weiskopf

VISUS, Universität Stuttgart

11.1 Introduction 224
11.2 Programmable Graphics Hardware 224
11.2.1 High-Level Shader Languages 226
11.2.2 General Purpose Computing on GPUs 227
11.2.3 GPGPU Programming Languages 228
11.3 GPU-Accelerated Volume Rendering 229
11.3.1 Basic GPU Techniques 229
11.3.1.1 2D Texture-Based Rendering 229
11.3.1.2 3D Texture-Based Rendering 230
11.3.1.3 Ray Casting 230
11.3.2 Advanced GPU Algorithms 231
11.3.3 Scalable Volume Rendering on GPU-Clusters 233
11.3.3.1 Sort-Last Volume Rendering 233
11.3.3.2 Sort-First Volume Rendering 234
11.4 Particle-Based Rendering 235
11.4.1 GPU-Based Glyph Rendering 236
11.4.2 Large Molecular Dynamics Visualization 238
11.4.3 Iterative Surface Ray Casting 239

11.5 GPGPU High Performance Environments 239
11.5.1 New Challenges in GPGPU Environments 240
11.5.2 Distributed GPU Computing 241
11.5.3 Distributed Heterogeneous Computing 242
11.6 Large Display Visualization 242
11.6.1 Flat Panel-Based Systems 243
11.6.2 Projection-Based Systems 244
11.6.3 Rendering for Large Displays 246
References 248

The visualization of large data is a computationally demanding task. The increase in performance and the flexible programmability have made graphics processing units (GPUs) an attractive platform to address large data visualization. The parallel architecture of GPUs and the low costs, coupled with high availability, have paved the way for this significant field of research. This chapter reviews the fundamental principles of modern graphics hardware, then summarizes the latest research in GPU-based visualization techniques for standalone and cluster-based systems.

11.1 Introduction

Over the last decade, graphics hardware has enabled compelling performance gains in visualization. The development of the programmable graphics pipeline [63] can be considered as one of the most important technological advances for the success of graphics hardware, since it allows programmers to execute user-written code on the GPU. Both the high performance and the ease of programming support the rapid design of stunning visual effects; however, they also facilitate interactive visualization of large scientific data sets.

However, the architecture of GPUs is different from typical multi-core platforms. The latter are suited well for coarse and heavy-weight threads that usually interact closely with the operating system. The performance of a single thread is fairly high when executed on the CPU, which is designed to process an individual workload as fast as possible. In contrast, the GPU's strategy follows an orthogonal approach. Modern graphics accelerators are designed as optimized data-parallel [42] streaming processors that enable fine and light-weight threads along with a high number of computational units and very fast memory access. The performance of a single thread is relatively poor, but the high parallelism allows the GPU to concurrently execute a large number of such threads.

11.2 Programmable Graphics Hardware

Graphics processors implement the computation of a rasterized image as a pipeline consisting of a sequence of operations that are applied to the input data. The main stages of the programmable graphics pipeline, based on OpenGL 4.2 [105], are illustrated in Figure 11.1. In general, a shader is a set of software instructions used to implement customized algorithms at different stages within the pipeline. The current shader model supports five different kinds of shaders, i.e., vertex, tessellation control, tessellation evaluation, geometry, and fragment shaders; each handling different tasks in the rendering pipeline. In the following, the functionality of the main stages of the pipeline and the associated shader programs are summarized.

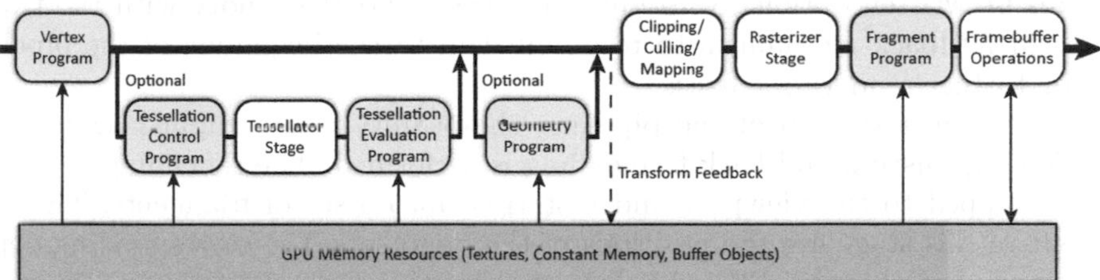

FIGURE 11.1: The simplified graphics pipeline based on OpenGL 4.2 with the programmable vertex, tessellation control, tessellation evaluation, geometry, and fragment programs.

The input of the pipeline is a stream of vertices and associated connectivity information describing the polygonal geometry of a scene. In the first step, these vertices undergo transformations like translation, rotation, and scaling. Model coordinates are transformed into camera space by applying the model-view matrix and finally into screen space by applying the projection matrix. At this point, a custom vertex shader operates on each incoming vertex and its associated data. The most common functions of a vertex shader are to manipulate vertex positions, apply per-vertex lighting, or assign texture coordinates. Afterwards, the transformed and lit vertices are assembled to geometric primitives like points, lines, or triangles.

OpenGL 4 introduced two new optional programmable tessellation stages in the pipeline. The tessellation control shader (also called hull shader) transforms input coordinates into a new surface representation and computes the required tessellation level, depending on the distance to the camera. The output of the control shader determines how the fixed-function tessellator generates new vertices in the following stage. Afterwards, the tessellation evaluation shader, also known as the domain shader. computes displacement positions of the incoming vertices and the new vertex data is handed over to the next step in the pipeline. In general, the tessellation shaders are the most valuable if the number of vertices changes a lot, for example, by applying dynamic level-of-detail. In this way, scene geometry can be tremendously augmented without affecting bandwidth requirements or geometry storage.

The geometry shader is executed optionally on each output primitive from the previous stage. The geometry shader has access to all vertices of the incoming primitive and it can emit new vertices, which are assembled into a new set of output primitives. Alternatively, the geometry shader can also discard vertices or entire primitives and provide some sort of culling. Before OpenGL 4, the geometry shader was also used for mesh refinement, but the tessellation shaders now provide a more suitable and dedicated tool for this purpose. Typical applications of the geometry shader are simpler transformations for which the ratio of input and output vertices is not too large, for example, the construction of a sprite from a vertex. The output of this stage can be written optionally to buffer objects in GPU memory with the transform feedback mechanism to resubmit data to the beginning of the pipeline for further rendering passes.

In the next step of the pipeline, the primitives are clipped against the viewing frustum and back face culling is performed. The remaining geometry is mapped to the viewport and rasterized into a set of fragments. The output of this stage are the positions of the fragments, the corresponding depth values, and the interpolated values of the associated attributes, e.g., normals and texture coordinates. The data is then passed to the fragment shader, which can perform computations based on the fragment position and its interpolated attributes. However, this position is fixed for each fragment, i.e., a fragment shader cannot manipulate attributes of other fragments. The output of the fragment program is the color and the alpha value of each fragment and optionally a modified depth value. Typical applications of the fragment shader include per-pixel lighting, texture mapping, or casting rays through a volumetric data set.

11.2.1 High-Level Shader Languages

In early times of GPU programming, shaders were developed in assembly-like languages [63] like the OpenGL Assembly Language, which is a standard low-level instruction set for programmable graphics processors. However, as GPU programming evolved rapidly, high-level languages became necessary to cope with the growing complexity and to reduce development time in a manner similar to CPU programming. Modern shader languages support common built-in data types and operators as well as functions and control structures. Each shader language provides access to basic library functions such as elementary math functions or GPU-specific features like texture access functions. However, as the scope of this chapter is not an introduction to GPU programming, we refer the reader to additional literature. A good overview of shader programming can be found, in GPU Gems 3 [82], for example, which covers a wide range of practical examples in different shading languages.

The OpenGL Shading Language (GLSL) was designed by the OpenGL

Architecture Review Board (ARB) to offer developers a syntax based on the C programming language. Originally, GLSL had been introduced as an extension to OpenGL 1.4, but has been added formally to the OpenGL 2.0 core. GLSL offers platform-independent development of shader programs on multiple operating systems, including Linux, Windows, and Mac OS X. Furthermore, GLSL can also be used on any hardware vendor's graphics board that supports OpenGL 1.4 or higher. The shader code itself is a set of strings that is compiled and linked by a vendor-specific GLSL compiler, allowing one to create optimized GPU code. Details of the GLSL language can be found in the official specification [50] or in comprehensive literature, such as *The Orange Book* [98] or the *OpenGL 4.0 Shading Language Cookbook* [128].

The High Level Shading Language (HLSL) is a proprietary shading language by Microsoft and part of the Direct3D API since the release of DirectX 9.0. Similar to GLSL, HLSL is also based on the C programming language but works solely on Windows platforms. To overcome this limitation, the Cg (C for graphics) language [71] was developed by NVIDIA in close collaboration with Microsoft [31]. The Cg language has the same syntax as HLSL with some minor modifications. However, while HLSL only compiles into DirectX code, Cg can compile to DirectX and OpenGL shader programs and is platform-independent. (For a comprehensive introduction to the Cg language, see *The Cg Tutorial* [31].)

11.2.2 General Purpose Computing on GPUs

For the past five years, GPUs have been evolving towards flexible and fully programmable computation devices based on massive thread-level parallelism. The addition of programmable stages and high-precision arithmetic to the pipeline encouraged developers to exploit the computational power of graphics hardware for general purpose computing on GPUs (GPGPU). In this context, the GPU is considered as a generic streaming multiprocessor, following the single instruction, multiple data (SIMD) paradigm, which is applicable to a wide range of computational problems. The high availability and the rather low costs of graphics hardware provide attractive options for high performance computing and the recent developments have even led to dedicated GPGPU-cluster systems, which are discussed in 11.5. General purpose computing is also of high interest in the visualization community because many advanced algorithms do not map well to the stages of the shader pipeline.

The concept of the programmable pipeline is further generalized to offer an abstract interface that is independent of a specific task. Figure 11.2 illustrates a simplified model of a GPGPU architecture. Similar to a shader in the graphics pipeline, a compute program (or kernel) executes user-written code on the GPU [53]. However, a compute program is not dedicated to a specific operation and it does not require a graphics API, such as DirectX or OpenGL. Instead, the host code triggers a GPU kernel by setting up a thread configuration and by calling the compute program like a regular function. Once

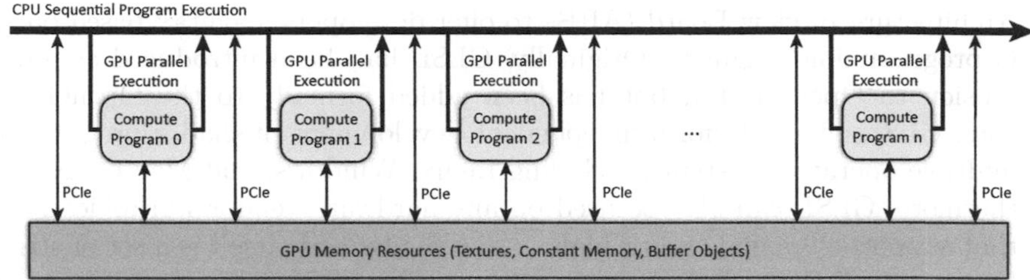

FIGURE 11.2: GPGPU architecture with multiple compute programs called in sequence from the CPU, but running each in parallel on the GPU.

the kernel is deployed, the GPU executes the code in parallel with full read and write access to arbitrary locations in GPU memory. Furthermore, input and output data of the kernels can be transferred over the PCIe bus between system memory and GPU resources.

11.2.3 GPGPU Programming Languages

In the beginning of general purpose computing on graphics hardware (GPGPU), plenty of research focused on the development of data structures and special shader algorithms to solve non-graphics related problems [87]. At that time, expert knowledge in graphics and shader programming was essential for developing hardware-accelerated code. To overcome this limitation, new abstract programming interfaces were developed.

The Compute Unified Device Architecture (CUDA), developed by NVIDIA [84], implements a GPGPU architecture and allows access to the memory and the parallel computational elements of CUDA-capable GPUs with a high-level API. The runtime environment is available for Windows, Linux, and Mac OS X platforms and does not require any graphics library. The programming of compute kernels is achieved with an extension to the C/C++ programming language or alternatively with a Fortran binding. In contrast to shaders, CUDA kernels have access to arbitrary addresses in GPU memory and still benefit from cached texture reading with hardware-accelerated interpolation. Furthermore, fast shared memory is available for thread cooperation or for the implementation of a user-managed cache. The addition of double precision floating-point arithmetic and the development of additional high-level libraries are essential capabilities for general purpose computing. Some of the high-level libraries include: cuFFT for fast Fourier transforms, cuBLAS for basic linear algebra, or cuSPARSE for sparse matrix operations. A comprehensive introduction to CUDA programming is provided in the textbooks by Sanders and Kandrot [101] and by Kirk and Hwu [53]. By now, applications using CUDA cover a wide range including, but not limited to computational fluid dynamics, molecular dynamics, or computational finance. (For a good overview of practical applications, see *The GPU Computing Gems* [46].)

A limitation of CUDA is vendor dependency in the choice of GPUs. Therefore, the Khronos group [51], a non-profit industry consortium, made an effort to develop an independent standard for programming heterogeneous computing devices including various kinds of GPUs and multi-core CPUs. The Open Computing Language (OpenCL) provides parallel computing using task-based and data-based parallelism. The OpenCL syntax is also based on the C programming language and the architecture is very similar to CUDA, allowing an easy transfer of knowledge. As OpenCL is an open specification, every processor vendor can decide to implement the standard for their products. Details on OpenCL programming and thorough code examples can be found, in Munshi et al.'s *OpenCL Programming Guide* [80].

11.3 GPU-Accelerated Volume Rendering

After having reviewed the fundamental architecture of current GPUs, this section discusses GPU-based algorithms for direct volume rendering (DVR). First, the section will summarize the basic techniques to introduce the principles of hardware-accelerated rendering. Subsequently, the section examines more advanced algorithms from the latest research before moving on to distributed volume rendering, which focuses on the scalability on GPU-cluster systems.

11.3.1 Basic GPU Techniques

Chapter 4 provided an introduction to the basic techniques of direct volume rendering. The first class of GPU-based algorithms that is discussed is an object-order approach based on 2D or 3D texturing. The basic principle of these algorithms is to render a stack of polygon slices and to utilize texture mapping for assigning color and opacity values from the transfer function. For the final image, the rendered slices are composited using alpha blending.

11.3.1.1 2D Texture-Based Rendering

Early GPU-based approaches relied on 2D textures and bilinear interpolation. In this case, the volumetric data set is stored in three object-aligned stacks of 2D texture slices, one for each major axis. This is necessary to allow interactive rotation of the data set. Depending on the viewing direction, one of these stacks is chosen for rendering so that the angle between the slice normal and the viewing ray is minimized. Once the correct stack is determined, the proxy geometry is rendered back-to-front and texture sampling can be implemented in a fragment shader. The main advantages of this algorithm are a simple implementation and high performance because graphics hardware is highly optimized for 2D texturing and bilinear interpolation. However, there are severe drawbacks concerning image quality and data scalability. Although sampling artifacts can be reduced by using multi-texturing and trilinear interpolation [95], flickering still remains when switching between the stacks. Furthermore, valuable graphics memory is wasted because of needing to store three copies of the same volume data, a problem overcome with 3D texture-based methods.

11.3.1.2 3D Texture-Based Rendering

The introduction of hardware support for 3D textures [14, 21] enabled a new class of slicing algorithms. The volume data is stored in a 3D texture representing a uniform grid. View-aligned polygons are rendered for generating fragments and 3D texture coordinates are used for trilinear resampling of data values. Gradients can be either precomputed for each grid point or calculated on-the-fly to provide local illumination [117, 29]. In a similar way, Westermann and Ertl [126] presented diffuse shading of texture-based isosurfaces.

Unlike 3D textures, real-world volume data is not always arranged on a uniform grid. The projected tetrahedra algorithm was adapted to 3D texturing by Röttger et al. [100] for rendering unstructured data, occurring in computational fluid dynamics, for example. Independent from the grid structure, high frequency components in the transfer function can cause severe artifacts because of an improper sampling rate. To avoid expensive oversampling, preintegration was introduced [58, 30] to handle high frequencies in the transfer function [8], occurring when isosurfaces or similar localized features were classified with sharp peaks in the transfer function. Assuming a piecewise linear reconstruction, the contribution of a ray segment can be precomputed accurately and the result is stored in a lookup table, which is later used at runtime.

The rendering of large data sets was addressed by Weiler et al. [123], who developed a multiresolution approach based on 3D textures to achieve an adaptive level of detail. In numerical simulations, even larger amounts of 4D volume data is generated. Lum et al. [64] employed compression and hardware texturing to render time-varying data sets efficiently.

Simple lighting models often do not exhibit sufficient details when exploring volume data. However, solving full radiative transfer is usually too expensive in visualization. Therefore, Kniss et al. [56] developed an approximate solution, based on half-angle slicing, which incorporates effects such as volumetric shadows and forward-directed scattering. Furthermore, the exploration of volume data is also inhibited by occlusion. Weiskopf et al. [125] presented a method for volume clipping using arbitrary geometries to unveil hidden details in the data set. Depending on the transfer function, significant areas of the volume can be transparent or opaque, leading to many samples that do not contribute to the final image. Li et al. [62] show how empty space skipping and occlusion clipping can be adapted to texture-based volume rendering.

11.3.1.3 Ray Casting

The second fundamental DVR algorithm is GPU-based ray casting, an image-order technique that directly solves the discrete volume rendering equation by tracing eye rays through the volume. The resampled data values are mapped to optical properties based on the transfer function. With the development of advanced shader capabilities, image-order techniques soon dominated over the slice-based methods. The most prominent form of ray casting [61] was introduced to GPU-based volume rendering by Röttger et al. [99], followed by an alternative implementation by Krüger and Westermann [59]. At a similar time, Weiler et al. [122] adapted Garrity's [33] ray propagation algorithm for GPU-based traversal of tetrahedral meshes. Up until then, all implementations relied on multiple rendering passes; but, as soon as shaders allowed branching operations, Stegmaier et al. [108] developed a single-pass algorithm in a fragment shader.

Ray casting can be advantageous compared to slice-based rendering because acceleration techniques like empty space skipping, early ray termination, or adaptive sampling are easily facilitated. Similar methods can also be used for rendering isosurfaces on the GPU. Hadwiger et al. [40] demonstrated how ray casting can be used to refine intersections with discrete isosurfaces. The introduction of GPGPU programming languages further simplified GPU-based ray casting. Maršálek et al. [72] presented an optimized CUDA implementation. Compared to shader implementations, the generation of eye rays does not require the rendering of proxy geometry. Instead, ray origin and direction are calculated directly from camera parameters for each pixel in the compute kernel.

For further reading about DVR and GPU-accelerated visualization methods, see the books by Engel et al. [29] and Weiskopf [124].

11.3.2 Advanced GPU Algorithms

The majority of state-of-the-art DVR algorithms are based on ray casting, but despite the growing computational power of GPUs, sampling remains a major bottleneck. High-quality images require thorough data reconstruction to avoid visible artifacts, leading to a bandwidth-limited problem.

Knoll et al. [57] introduced a peak finding algorithm to render isosurfaces, arising from sharp spikes in the transfer function. Usually, post-classified rendering requires very high sampling rates to achieve sufficient quality and while preintegration solves many deficiencies with fewer samples, opacity is scaled improperly, depending on data topology. Therefore, Knoll et al. [57] developed an algorithm that analyzes the transfer function, searching for peaks in a preprocessing step. Similar to preintegration, a 2D lookup table is constructed, storing possible isovalues for a set of data segments in a 2D texture. During ray casting, the entry and exit values of each ray segment are used to lookup if it contains an isosurface, avoiding expensive over-sampling. In addition, color and opacity of the isosurface are not scaled because classification only depends on the isovalue from the lookup table.

Preintegration assumes a linear function between two samples, but uniform sampling with trilinear interpolation usually leads to nonlinear behavior. Under these circumstances, voxel-based sampling leads to piecewise cubic polynomials along a ray [67]. Ament et al. [4] exploited this observation and developed a CUDA-based algorithm to reconstruct Newton polynomials efficiently with four trilinear samples in each cell, allowing a piecewise analytic representation of the scalar field. The linearization of the cubic polynomials with respect to their local extrema guarantees crack-free rendering in conjunction with preintegration. The problem of scaled opacity is addressed by a modified visualization model, allowing the user to classify volume data independent of its topology and its dimensions in the spatial domain.

The previous approach achieved bandwidth reduction under the assumption of trilinear reconstruction. However, higher-order techniques further push bandwidth requirements. Lee et al. [60] introduced a GPU-accelerated algorithm to approximate the scalar field between uniform samples with third-order Catmull-Rom splines [15] to provide virtual samples by evaluating the polynomial functions arithmetically. The control points of the splines are sampled with tricubic B-spline texture filtering, by using the technique from Sigg and Hadwiger [106] with eight trilinear texture fetches. Compared to full tricubic 4× oversampling, the computational evaluation leads to an improved performance of about 2.5×–3.3× at comparable rendering quality. In addition to intensity reconstruction, the approach can also be used to calculate virtual samples for gradient estimation.

The quality of volume shading strongly depends on gradient estimation, especially in the surrounding area of specular lobes. Also, preintegration with normals requires at least four additional dimensions in the lookup table, which may not be feasible. Guetat et al. [38] introduced nonlinear gradient interpolation for preintegration without suffering from excessive memory consumption. Their first contribution is the development of an energy conserving interpolation scheme for gradients between two ray samples, guaranteeing normalized gradients and hence conservation of specular components along the entire segment. In addition, the authors were able to reduce dependency of the normals from preintegration. With their approach, it only takes four two-dimensional lookup tables to implement high-quality shading with preintegration.

The previously discussed GPU methods focuses on the improvement of performance and rendering quality. However, accelerated DVR is a valuable tool for interactive exploration of volume data, for example, a user should be able to examine and classify relevant features of a data set. Traditionally, the transfer function serves this purpose, but it can be a cumbersome task to find proper parameters. Moreover, standard DVR may not always be the best choice to visualize volume data. Bruckner and Gröller [13] presented maximum intensity difference accumulation (MIDA), a hybrid visualization model combining characteristics from DVR and maximum intensity projection (MIP). Their approach modifies standard DVR compositing by introducing a weighting coefficient that accounts for new maxima along a view ray. Specifically, when the maximum changes from a low to a high value, the corresponding sample contributes a higher intensity compared to other samples. This behavior shows the same essential features as MIP, but still allows opacity accumulation controlled by a transfer function. As MIDA is based on ray casting, the implementation is well-suited for GPU acceleration and interactive exploration.

The idea of classifying data with a modified DVR model was further examined by Marchesin et al. [68], who introduced view-dependent per-pixel opacity modulation. In this approach, opacity is not controlled by a transfer function, but by a relevance function, weighting a sample's contribution based on its importance relative to other contributions. The relevance function can depend on arbitrary parameters, like scalar value, gradient magnitude, or other feature-based rules. At rendering time, each ray sample is classified with the same extinction coefficient, optimized for the highest visibility of all features according to the relevance function. As this extinction coefficient is calculated for each individual pixel, opacity adapts itself to the view direction and offers automatic classification together with optimized visibility. The algorithm is implemented on the GPU with a two-pass ray casting. In the first step, the relevance function is accumulated, which is necessary to compute the per-pixel extinction coefficient. The second pass implements final rendering by mapping color from the transfer function to the samples.

11.3.3 Scalable Volume Rendering on GPU-Clusters

Thus far, the chapter has discussed single-GPU DVR algorithms that benefit from hardware-acceleration in terms of sampling, rendering quality, and interactive exploration. With growing data size, more memory and performance resources are required than a single GPU can provide. Using more GPUs to increase rendering speed and memory is an attractive solution due to its price-to-performance ratio. Distributed rendering algorithms on CPU-based clusters were studied thoroughly in the past; however, the introduction of GPUs to rendering nodes exhibits new challenges in data and performance scalability because graphics accelerators have less memory, but more computational power than CPUs. The subsections following will discuss recently developed techniques for distributed GPU-clusters focusing on the sort-last and the sort-first paradigms [73].

11.3.3.1 Sort-Last Volume Rendering

In sort-last rendering, object-space is decomposed into spatially disjoint subvolumes that are distributed and rendered separately on the cluster nodes. The resulting intermediate images typically overlap each other in image-space, requiring a final compositing step (see Chap. 5).

Although sort-last configurations inherently provide good data scalability, the partitioned data set can still be larger than one single GPU can handle. Strengert et al. [109] developed a hierarchical algorithm based on wavelet compression to decrease memory requirements and thereby increase the effective data size that can be rendered. The decomposition level is chosen adaptively, depending on the view direction and the data set, so the system can meet memory constraints and provide a rendering speed.

A static distribution of subvolumes can lead to significant performance imbalance, as typical classification parameters lead to heterogeneous visibility. By using empty space skipping, sparse bricks are rendered faster than dense bricks, potentially wasting valuable computational resources due to load imbalance. Dynamic load balancing aims to redistribute workload evenly on the nodes to achieve optimal performance. Müller et al. [79] and Marchesin et al. [69] utilize a kd-tree decomposition of the volume to detect imbalances among the render nodes. Based on the rendering times of the previous frame, imbalances can be identified either by hierarchical aggregation [69] or by direct communication between the nodes [79]. Using a kd-tree guarantees that all subvolumes remain a convex shape when transferring slices of voxels to adjacent leafs across the subdivision plane, guaranteeing correct order when blending the images.

In a distributed environment, communication over an interconnection network is often a performance bottleneck, especially when rendering high-resolution images. The network workload and latency can be reduced to some amount by using a lower number of nodes but with multiple GPUs in each node [70]. In such a configuration, the overall communication expense is distributed over the system memory bus and the network layer. It was experimentally shown that multi-GPU nodes can be advantageous compared to a higher number of single-GPU nodes, given the same total amount of accelerators [70].

11.3.3.2 Sort-First Volume Rendering

The second major class of scalable DVR algorithms relies on sort-first rendering. In this approach, image-space is decomposed into a set of disjoint tiles, each rendered by a different node. In contrast to sort-last, no additional compositing step is required for the final display. However, it is a challenging task to achieve data scalability with a sort-first approach because of view-dependent data requirements according to the frustum of each node. Typically, sort-first methods for parallel GPU-clusters either replicate the data set on each GPU [1] or transfer data over the network [10].

Sort-first partitioning is well-suited for high-resolution displays in conjunction with ray-coherent algorithms; for example, early ray termination where all information along a ray is required on a single node. Moloney et al. [75] developed a dynamic-load balancing approach that achieves improved data scalability compared to previous methods. In their approach, each GPU node decides, dynamically, which part of the data set is required for the next frame,

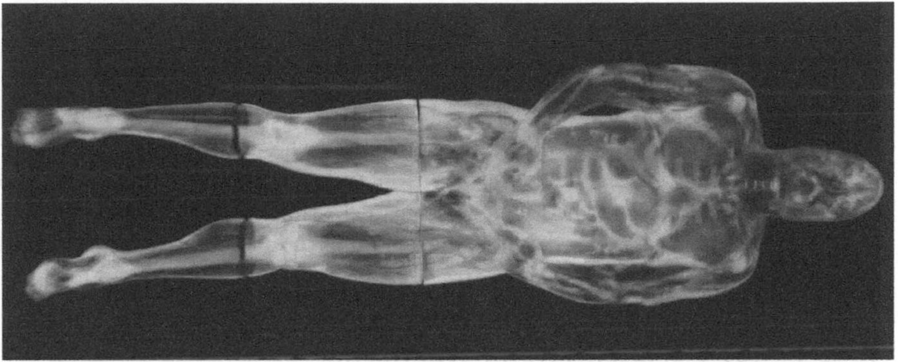

FIGURE 11.3: Interactive sort-first volume rendering of the Visible Human ($2048 \times 1024 \times 1878$). Image source: Moloney et al., 2011 [74], data courtesy of the National Library of Medicine.

allowing them to visualize data sets that are larger than the amount of video memory of one GPU. The load-balancing algorithm is based on a kd-tree decomposition of the image-space and a per-pixel cost estimation for each ray, allowing an accurate prediction of the computational expense. A summed-area table is then used to calculate the expected overall rendering costs and to rebalance the kd-tree. Later on, this approach was extended [74] with a volumetric shadow algorithm, using a sort-first decomposition of the light's image-space for coherent shadow rays. This leads to a hybrid partitioning scheme because such a decomposition is essentially a sort-last distribution from the camera's point of view. It was demonstrated that data scalability of a sort-first can be improved significantly compared to full data replication on each GPU.

11.4 Particle-Based Rendering

Particle-based simulations are a widely used tool in many research fields, ranging from molecular dynamics in biology and computational physics to material sciences. The resolution of these models increases as the computational platforms drop in price and increase in performance. The data set sizes, tens of billions of particles per timestep, are not uncommon, which poses a significant challenge for real-time rendering [36].

In most cases, the visualization of simulation output by simply rendering points at the positions of the particles is inadequate. Additional information about each particle can be conveyed by means of more complex glyphs with higher semantic density. Such glyphs could be ellipsoids in the simplest case or glyphs composed of several primitives like dipole glyphs or ball-and-stick representations, which are widely used for illustrating molecules.

Particle data without topology information can also be obtained from real-world measurements; ellipsoidal glyphs are used for visualizing the outcome of diffusion tensor imaging (DTI), which is a magnetic resonance imaging (MRI) technique for investigating the structure of the white matter of the brain.

All of the aforementioned applications share the commonality that in the first instance the particle itself (its position and properties, without any topological information) is in the focus of interest. In that, particle-based rendering is set apart from point-based rendering—namely point set surfaces [2]—which uses points to represent and also directly render surface shapes.

11.4.1 GPU-Based Glyph Rendering

Specifically for molecular dynamics data, a variety of visualization tools exist, of which Chimera [90], PyMOL [104], and VMD [43] are most widely known. More generic visualization tools like AVS [115, 5] or Amira [119] usually also provide modules for performing molecular visualization. While these applications offer many analysis features, their rendering performance does not allow for interactive rendering of very large data sets comprised of hundreds of thousands of atoms or more. This limitation is due to the use of polygon-based representations, without level-of-detail techniques, which result in large data transfer volumes from the main memory to the GPU, as well as high geometry transformation load on the graphics card itself. Using textured sprites results in significantly faster rendering at the cost of decreasing the visual quality. The approach is thereby limited to simple glyphs like spheres that may be represented by flat sprites.

However, the advent of programmable shader units set the stage for glyph rendering directly on the GPU, solving some of the aforementioned performance problems. Loops and branches in vertex and fragment shaders allow for the ray casting of a variety of shapes from their implicit surface description, thus drastically reducing the amount of data per glyph required on the graphics card while maintaining superior visual quality when zooming into the data set at the same time.

Ellipsoids are one of the simplest types of 3D glyphs. The first method for ray casting ellipsoids correctly uses perspective projection on the GPU, as presented by Gumhold [39]. His approach uses splats made of four vertices for each glyph. In contrast, Klein and Ertl [55] show how to unfold the bounding box of the projected silhouette from one point primitive. Thus, it is possible to limit the amount of data transferred to the GPU, to the necessary minimum: the center point $\vec{c}$ of the ellipsoid encoded as a 3D vertex position, its extents $\vec{h}$, and its orientation described by a quaternion $\vec{q}$. The latter can be attached as vertex attributes like the normal vector and the vertex color.

Then, the workflow for splatting the ellipsoid consists of two major steps: one in the vertex shader and one in the fragment shader. First, a bounding rectangle of the screen space silhouette is determined to adjust the size of the point primitive. This operation is performed in the vertex shader by projecting the vertices of the bounding box of the ellipsoid into clip space. The longer edge of the screen space bounding rectangle of this projection is then used as the point size. A large number of superfluous fragments may result, since points in OpenGL are rendered as squares, especially for anisotropic glyphs. This problem can, however, be addressed by computing better silhouettes made of n-sided polygons in the geometry shader [54]. The geometry shader, however, has a significant impact on the overall rendering performance, wherefore it is highly data-dependent on whether the reduced number of fragments outweighs the performance penalty of this additional pipeline stage.

Second, all fragments generated by the point must be tested whether they lie within the silhouette of the ellipsoid and then they must be shaded. Following, a ray from the eye position through the position of the fragment in question has to be intersected with the implicit representation of the ellipsoid. This test is best performed in the local coordinate system of the ellipsoid (object space), in which its surface points are represented by $\{\vec{x_o} | \|\vec{x_o}\|^2 = 1\}$. The eye position $\vec{e_o} = M^{-1}(\vec{e} - \vec{c})$ is constant for all fragments of one ellipsoid. Using $M = QH$ defined by the rotation matrix $Q = R(\vec{q})$ and scaling matrix $H = \frac{1}{2}\text{diag}(\vec{h})$, the eye position can be computed in the vertex shader and passed through to all pixels. However, the ray direction $\vec{d_o}$ has to be computed individually from the screen space position of each pixel. Inserting the ray equation $\vec{x_o} = \vec{e_o} + \lambda \vec{d_o}$ into the surface equation then yields the condition $\vec{d_o}^T \vec{d_o} \lambda^2 + 2\vec{e_o}^T \vec{d_o} \lambda + \vec{e_o}^T \vec{e_o} - 1 = 0$. If its discriminant is less than zero, the ray misses the ellipsoid and the fragment is discarded. Otherwise, the equation is solved for the intersection parameter λ_s, which can be used to compute the actual intersection point. The position vector of the intersection point in object space is identical to the surface normal and can therefore be used to perform per-pixel Phong shading after the transformation into world space. To ensure correct depth sorting, the intersection point must be transformed back to world space and pass the transformation pipeline again, which eventually yields the correct depth value of the fragment.

By combining different types of ray casted primitives, this approach can be extended to render more complex glyphs like dipoles [93] or a ball-and-stick representation of molecules [107].

11.4.2 Large Molecular Dynamics Visualization

While leveraging the increasing computational power and flexibility of GPUs, ray casting hundreds of thousands of glyphs at interactive frame rates becomes possible. However, visualizing an even larger number of particles requires a careful optimization of data transfer and rendering. As CPUs and GPUs become faster, the data transfer from main memory to graphics memory becomes a bottleneck, which poses a challenge for visualizing large time-dependent data sets, requiring continuous data uploads to the GPU.

Choosing the right way of uploading the data to the graphics card is, therefore, an important factor, especially since OpenGL provides a vast choice of different methods ranging from the original immediate mode over vertex arrays to vertex buffer objects, which in turn support a plethora of performance hints. In an extensive empirical study, Grottel et al. [37] concluded that for most of the scenarios they tested, vertex arrays provide the best performance for transferring the particles.

Optimizing the vertex upload and per-glyph rendering performance increases the number of particles that can be visualized interactively. Data sets consisting of tens of millions of atoms, like the laser ablation simulation in Figure 11.4, need a more sophisticated handling on current hardware, though. The visualization of dense particle clouds, as they can be found in simulations in material sciences, especially can benefit from a two-level occlusion culling approach presented by Grottel et al. [36]. For discarding large chunks

FIGURE 11.4: Visualization of a laser ablation simulation with 48 million atoms rendered interactively on a standard workstation computer. Image source: Grottel et al., 2010 [36].

of data, they suggest subdividing the data set using a regular grid and they use hardware occlusion queries to test those against the depth field based on the data of the previous frame. The subdivision grid is additionally employed to reduce the bus bandwidth in two other ways: first for caching the vertex buffer objects containing the data of a cell and second for reducing the storage requirements through cell-relative quantized coordinates. For all vertices of a cell, the overhead introduced by the massive overdraw in dense data sets is further diminished by implementing a hierarchical depth buffer using a MIP Map [127] of the z-values from the previous rendering pass. The need for that arises from the early z-test being disabled since the fragment shader emits a depth value for each pixel.

11.4.3 Iterative Surface Ray Casting

At a certain point, the mathematical description of a glyph might become too complex to compute an analytical solution in the fragment shader. High Angular Resolution Diffusion Imaging (HARDI) glyphs, which can be defined by means of a function residing on a sphere, are an example of such glyphs. They are used to visualize the data from HARDI, which is an advancement of DTI in that it can resolve more than the primary direction of the diffusion pattern. Peeters et al. [89] presented a method for rendering these glyphs solely on the GPU. They use spherical harmonics and perform an iterative approximation of the intersection point between the function and the view ray. Computing the normals at the intersection points on the surface is also not as simple as for ellipsoids. Expressing the spherical function as a volume density in spherical coordinates is the suggested solution here: the normals can then be obtained from the gradient of the respective isosurface in the volume.

Similar iterative approaches in the fragment shader can be employed for rendering metaballs, which are useful to express the surface of clusters of particles directly from their center points. Müller et al. [78] explored the possibility of a direct, particle-centric implementation as well as another one working in parallel on all pixels in a depth peeling-like manner. By combining a per-ray list of active metaballs—Bézier clipping and depth peeling for searching the intersection between viewing rays and the surface points—the frame rates for rendering hundreds of thousands of metaballs come close to being interactive [48]. By using a fitted bounding volume hierarchy in a hybrid CPU–GPU implementation, Gourmel et al. [35] were able to accelerate the rendering further, provided that the data set is not too dense. For very dense scenes, however, the construction of their hierarchy negatively impacts the interactivity of their algorithm.

11.5 GPGPU High Performance Environments

In recent years, not only small and medium size cluster environments, but also, large scale high performance computation systems increasingly feature GPUs for general purpose computations. In June 2011, three out of the five fastest systems in the TOP500 [114] feature GPUs. While for small and medium scale systems the floating-point operations per second (FLOPS) per dollar ratio is usually the biggest advantage, the large systems also particularly benefit from the superior energy efficiency (FLOPS per watt) and the consequential lower heat emission.

11.5.1 New Challenges in GPGPU Environments

The downside of this trend is the increasing complexity to develop software efficiently for these systems. In simplified terms, "classic" CPU-only systems feature groups of cores that communicate over shared memory, whereas intergroup communication is done via network interconnects. From a programmer's perspective, these systems can be programmed quite easily using the Message Passing Interface (MPI) [76] or a combination of MPI and the Open Multi-Processing (OpenMP) [86] API to explicitly model the layer hierarchy for maximum efficiency. Adding GPUs introduces a new device type at an additional hierarchy level and thus introduces a number of new challenges.

1. GPUs contain their own memory to which data from the main CPU memory needs to be transferred for computation. The deeper memory hierarchy introduces additional data transfers and the size of the available GPU memory is approximately one order of magnitude smaller than the available CPU memory. Thus, new approaches for data handling are required, especially because GPUs also bring other caches in addition to GPU (main) memory that need to be considered for achieving maximum performance. Traditionally, data transfers between GPUs are routed through CPU memory, but newly introduced functionality (i.e., NVIDIA GPUDirect) also includes transfers directly between GPUs.

2. GPGPU systems are typically heterogeneous systems in that they feature both capable CPUs and GPUs. This raises the question on which device type different types of tasks should be processed. In the cases in which tasks can be executed alternatively by CPUs and GPUs, elaborate scheduling mechanisms need to be used that consider many different factors (most importantly estimated execution and data transfer time).

3. GPUs need to be programmed differently, although efforts with OpenCL and other projects [88] exist to close this gap to a great extent. Additionally, GPU kernel calls typically require a number of parameters to be tuned manually in order to achieve maximum performance (see 14.2).

All these challenges make GPGPU environments harder to use efficiently, in comparison to the traditional CPU-only architectures and demand a lot of effort and knowledge from the application developer. To decrease complexity for the programmer, the hardware shift towards GPUs also requires new software approaches and infrastructures. The requirements regarding these software infrastructures or frameworks heavily depend on the usage scenario.

11.5.2 Distributed GPU Computing

Going from one GPU to multiple GPUs can be seen as simple extension of the GPU layer hierarchy. One layer is added respectively for groups of host memory connected GPUs and network connected GPU groups.

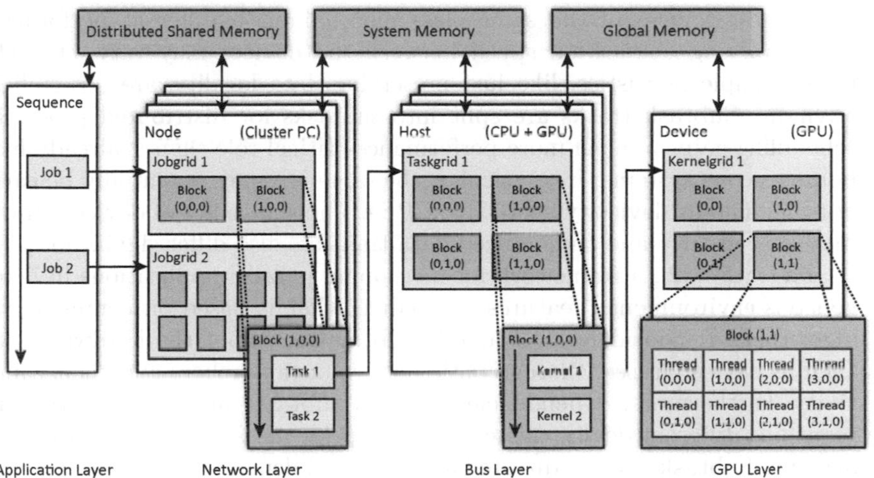

FIGURE 11.5: Schematic overview of the abstraction layers of the CUDASA programming environment. The topmost level is placed left, with a decreasing level of abstraction from left to right. Image source: Müller et al., 2009 [77].

The Compute Unified Device and Systems Architecture (CUDASA), by Müller et al. [77], is a development environment for distributed GPU computing that models and handles these layers explicitly (Fig. 11.5) and thus enables workload distribution capability for multiple GPUs. The lowest level utilizes the parallel architecture of a single GPU, while the next higher level represents the parallelism of multiple GPUs within a shared-memory system. The third layer handles distributed cluster environments, by scaling with the number of participating cluster nodes. The topmost layer represents the sequential portion of an application that issues function calls. To account for the fact that data transfer is expensive with respect to time and energy, a scheduling mechanism is employed that considers data locality. CUDASA introduces a minimal set of extensions for the additional abstraction layers that generalizes the present programming paradigm.

Other approaches dealing with multiple GPUs include a technique that runs an unmodified OpenCL source code for a single GPU system on multiple GPUs [52]. A multi-GPU scheduling technique has also been proposed that is based on work stealing with the goal to support scalable rendering [129]. Good timing estimates, when moving single-GPU applications to multiple GPUs, can be provided using a recently presented modeling framework [103]. Stuart and Owens [110] leverage a multi-GPU cluster to distribute and process large-scale data in parallel by implementing the Map-Reduce paradigm.

11.5.3 Distributed Heterogeneous Computing

In heterogeneous hardware environments, there may be different classes of devices, devices of the same class may exhibit a different performance, and a varying connection speed between the devices may vary, etc. Thus, rather simple heuristics, like just preserving data locality, are not sufficient anymore. Subtask (tasks are split into subtasks for distributed processing) scheduling gains a much more performance-critical role than it already has in homogeneous systems. In particular, the same job may potentially be carried out on different device types (GPU/CPU). The suitability of devices for a job may vary greatly and the induced data transfers also differ significantly.

PaTraCo [32], a framework for developing parallel applications in heterogeneous environments, features a scheduler that is based on common graph algorithms. It considers all available information about the hardware setup, such as device classes, characteristic performance numbers and their connectivity. Furthermore, dependencies between subtasks and timing estimations regarding different device classes based on prior measurements are taken into account. Subtasks form a directed and weighted dependency graph. According to the basic critical path method [49], PaTraCo's scheduling heuristic iteratively assigns the fastest possible device combination to the longest path in the graph until all subtasks have been assigned to a device.

One fundamental idea of PaTraCo is that the distribution of parallel applications is based on the graph structure describing the data dependencies. Basic research in this area has been done by Diekmann [24] amongst others. Sunderam [111] proposed a programming environment for parallel applications consisting of many interacting components which is intended to operate on a collection of heterogeneous computing elements. Wang et al. [121] present a simple task scheduling algorithm for single machine CPU–GPU environments that does not just use a first idle compute device, but the algorithm chooses the fastest device from all idling devices. Teresco et al. [113] worked on a distributed system in which every CPU requests tasks from the scheduler that are then sized according to the device's measured performance score. Resource-aware distributed scheduling strategies for large-scale grid/cluster systems were proposed by Viswanathan et al. [120].

11.6 Large Display Visualization

There is an ever-increasing size of data sets, which results in visually more complex representations of the data sets. Therefore, the interest in larger displays with increased resolution arises, which usually refers to an increase of the total number of pixels in context of tiled displays rather than an increase of pixel density.

Although large, high-resolution displays are still expensive to build and rather difficult to operate, real-world application areas have existed for quite some time. One of the most prominent applications is vehicle design, where the goal is to have a model at an original scale for designers and engineers to work with. Control rooms that are used to manage telecommunications or electricity networks are another example. Physically large displays are also the key to immersive applications, which are often used for training and real-world simulation purposes. Finally, many high-resolution installations can be found in academia—not the least for researching the construction and operation of those systems themselves, though. The benefits of a large display area and a large number of pixels have also been subject of numerous studies, namely for navigational tasks, multitasking, memorizing, and general usability [22, 112, 6]. Ni et al.'s [83] survey provides a comprehensive summary of usability aspects as well as hardware configurations and applications of tiled displays.

11.6.1 Flat Panel-Based Systems

With consumer graphics cards having two or more video connectors, the easiest way to increase display space is by attaching multiple monitors to a GPU [3]. With a graphics cluster, this setup can be extended to large, wall-sized, tiled displays. The use of commodity hardware makes the construction of systems reaching more than 300 megapixels affordable [81].

Besides a lower original cost, compared to the formerly projection-based tiled display installations, liquid crystal display (LCD) arrays have the advantage of lower maintenance costs, mostly thanks to expensive lamp bulb replacements being unnecessary [83, 23, 81]. Furthermore, the requirements regarding the infrastructure around the system itself are much lower: LCDs do not have a throwing distance and hence require less space, and their heat dissipation is not as high as of a large number of projectors. Finally, color correction and geometric alignment are much easier to achieve, not in the least because of the bezels between the screens making pixel-exact registration unnecessary.

While those bezels have even been found beneficial for organizing different tasks on a desktop, they nevertheless hamper immersion and introduce visual distortion or missing information when images cross them [97]. Using a hardware setup resembling Baudisch et al.'s focus-and-context screen [7], Ebert et al. tried to overcome this problem by projecting low-resolution imagery on the bezels [27].

11.6.2 Projection-Based Systems

Tiled displays that do not suffer from discontinuities caused by screen bezels have traditionally been built using an array of video projectors. The challenges of building such a system have barely changed over the last decade [41]: projectors and a matching screen material must be chosen, the devices must be mounted stably, and the whole system must be calibrated to form a seamless, continuous display.

Over the years, a variety of projection technologies have been developed. Early installations, like the first CAVE [20], were built using CRT projectors, which offer freely-scalable image geometry, but they suffer from low brightness. Later, LCD projectors became the prevalent video projection technology used in nearly every conference room, making them cheap commodity hardware. Closely related to LCD is Liquid Crystal on Silicon (LCoS), which uses a reflective silicon chip with the liquid crystals on it. LCoS enables commercially-available devices with resolutions up to 4096×2100 pixels (4K), and research prototypes with 33 megapixels exist [47]. A six-sided CAVE installation with 24 LCoS projectors and a total of 100 megapixels has the highest resolution amongst rear-projection systems nowadays [85]. Contrariwise, Digital Light Processing (DLP) uses a myriad of micromirrors for reflecting the light of each pixel separately. These mirrors can individually be moved very quickly, creating the impression of gray scales by reflecting the light either out of the projector or not. Color is added either through time multiplexing via a color wheel, or through a separate chip for each color channel. The mirrors can actually be toggled so fast that the frame rates required for active stereo projections are reached, which is a clear advantage for building immersive installations.

Stereoscopic displays show two perspectively different images, one being only visible to the left eye of the user and the other only to the right eye [20]. This channel separation can be achieved by different means, most notably active shutters (time multiplexing), polarized light, or interference filters. All of these technologies require the user to wear matching stereo glasses, which complement the filters built into the projector. While stereo displays can also be built from flat panels, their full resolution cannot be brought to bear to each eye, especially in the case of autostereoscopic displays, which remove the need for wearing stereo glasses completely [102].

Aside from projector technology, the screen material is a crucial component that affects the quality of a display environment. However, choosing the screen material usually means finding a compromise and is also dependent on the projectors and, if applicable, the technology used for stereoscopy. While having a nearly Lambertian surface for the screen makes calibration easier and, therefore, is desirable for a large number of tiles [12], such a screen results in reduced image sharpness, hot spots, and it might be inappropriate to use because it may affect the polarization of light. Likewise, tinted screens have become popular, since they increase the contrast of the image, but obvi-

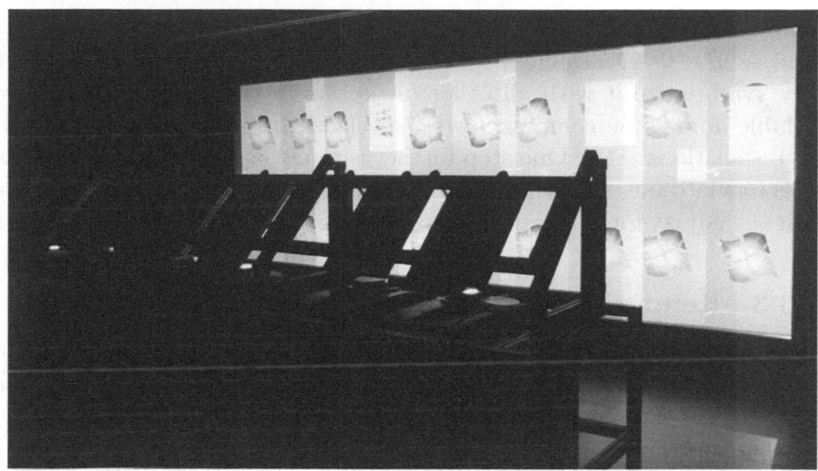

FIGURE 11.6: A rear-projection tiled display without photometric calibration.

ously they require brighter projectors. An acrylic glass, or a float glass, are the two most common support materials as they provide the rigidity needed for building systems with very small pixels. Thanks to increasing demand, those materials are currently also available as seamless pieces in sizes required for large screens.

What makes the installation of a high-resolution rear-projection display difficult is the need for different types of alignment—geometric registration and photometric calibration—that have to be carried out thoroughly to make the display seamless. A satisfactory geometric calibration is difficult to achieve using hardware due to nonlinear lens distortions, which cannot be compensated by using positioning devices with six degrees of freedom. Hence, a lot of effort has been put into distorting the images so that the effects of an incomplete mechanical registration cancel out [19, 17, 92]. The goal of photometric calibration is to remove variations of luminance and chrominance between different projectors and within a projector. Madjumder and Stevens [66] identified the reasons for those variations and also stated that aligning luminance is the most important factor for achieving the impression of a seamless display—a problem that is specifically relevant because overlapping projections cause areas of exceptionally high brightness (Fig. 11.6). Matching the color gamuts of the projectors after that is a computationally expensive operation [9], but less important because chrominance does not vary significantly between devices of the same type [65]. Therefore, a per-pixel correction of luminance can suffice to create the impression of a seamless screen. For geometric and photometric calibration, cameras or spectroradiometers are typically used to build a feedback loop and automate the process. Brown et al. [12] give a good overview of both fields, also addressing the fact that physically large displays might require the camera to be moved or might require simply using more cameras [92, 19, 17]. Devices that can measure their light flux and automatically synchronize it with all other projectors in the array are commercially available from some companies who specialize in building immersive projection installations [118]. One step further is the idea of combining cameras and projectors into one smart projector that can sense its environment and automatically adjust for arbitrarily shaped and colored projection surfaces [11].

11.6.3 Rendering for Large Displays

Interactive visualization on large tiled displays usually involves the use of a graphics cluster, simply because each of the display devices requires an input, and the number of outputs a single machine can provide is limited. This hardware setup predetermines a natural image-space subdivision, which leads to a bias towards the sort-first class of Molnar et al.'s taxonomy [73]. While their taxonomy applies to parallel rendering of 3D scenes in general, Chen et al. [16] introduce a classification targeted at rendering for high-resolution display walls. They identify three classes of data transfers: control data, which enable multiple instances of the same application to run in a synchronized way, primitives, which are rasterized separately on each machine, or simply pixels of the final image. From that they derived two models of program execution: master–slave, which basically implements the primitive's distribution pattern, and synchronized execution [18].

The synchronized execution model usually minimizes the amount of data to be transferred and can be implemented on an application or system level. The former is often used in tools solving a specific application problem or even targeting a specific hardware constellation. At system level the goal is to synchronize applications transparently by coordinating buffer swaps, timers and I/O operations.

WireGL [44] and its successor Chromium [45] are typical representatives of the master–slave pattern. Their goal is total application transparency to enable unmodified OpenGL applications running on clusters by intercepting and distributing all OpenGL API calls and executing those on the remote machines. An equivalent for 2D graphics in X-Window systems is Distributed Multihead X (DMX) [25]. It implements an X server that packs the API commands and sends them to back end machines. Again, the remote machines perform the actual rendering, which results in a large X11 desktop spanned over multiple machines.

As fully transparent solutions introduce performance penalties, due to their generality and implementing custom applications from scratch is cumbersome. A lot of research has been done in developing generic, efficient, and easy-to-use middleware layers. Raffin and Soares [91] and Ni et al. [83] give a good overview of those. Often, these frameworks do not strictly fall in one of Chen et al.'s classes nor do they implement both, like Aura [116]. Aura is a retained mode graphics API used to build visualization applications for tiled displays on a graphics cluster that offers two different communication modes: one mode is that multiple copies implement the synchronized execution pattern, and the second mode is that broadcast replicates the scene to all rendering machines thus implementing master–slave rendering. On top of it, the VIRPI toolkit provides controls and event handling for building user interfaces in virtual reality environments [34].

The Cross Platform Cluster Graphics Library (CGLX) by Doerr and Kuester [26] exposes a callback-based interface like the widely used OpenGL Utility Toolkit (GLUT) to facilitate porting existing applications to high-resolution tiled display environments. CGLX synchronizes events raised by user interaction from the master instance to its slave nodes and provides a means for passing user-defined messages between the nodes. The distributed OpenGL contexts are managed by intercepting API calls, which is, in contrast to Chromium, not fully transparent, but still must be adapted manually by replacing the OpenGL function calls with pass-through functions of the framework.

Only a few of the frameworks have found wider usage, such as commercially supported products like the CAVElib or Equalizer [28]. The latter is an object-oriented rendering framework that supports building image-space and object-space task subdivisions including distributed compositing algorithms. Equalizer allows for specifying task decomposition strategies by means of trees that are automatically run in a synchronized execution manner by the framework.

The broad availability of high-speed network interconnects like 10 Gigabit Ethernet or InfiniBand makes the implementation of frameworks distributing pixels feasible. The OptIPuter project aims to leverage high-speed optical networks for tightly coupling remote storage, computing and visualization resources [23]. Therefore, the Scalable Adaptive Graphics Environment (SAGE) [94] has been developed, which is the implementation of a distributed high-resolution desktop for displaying image streams from various sources. It comes with a variety of streaming source providers including ones for videos, 2D imagery, 3D renderings, and remote desktops via Virtual Network Computing (VNC) [96]. Like windows on a desktop, all streams can be freely moved and resized in real-time.

References

[1] Frederico Abraham, Waldemar Celes, Renato Cerqueira, and Joao Luiz Campos. A Load-balancing Strategy for Sort-first Distributed Rendering. In *Proceedings of Brazilian Symposium on Computer Graphics and Image Processing*, pages 292–299. IEEE Computer Society, 2004.

[2] Marc Alexa, Johannes Behr, Daniel Cohen-Or, Shachar Fleishman, David Levin, and Claudio T. Silva. Point Set Surfaces. In *Proceedings of IEEE Visualization*, pages 21–28. IEEE Computer Society, 2001.

[3] AMD. Eyefinity Technology, http://www.amd.com/eyefinity, 2011.

[4] Marco Ament, Daniel Weiskopf, and Hamish Carr. Direct Interval Volume Visualization. *IEEE Transactions on Visualization and Computer Graphics*, 16(6):1505–1514, 2010.

[5] AVS. Advanced Visual Systems Inc., 2011.

[6] Robert Ball and Chris North. An Analysis of User Behavior on High-Resolution Tiled Displays. In *Proceedings of Interact*, pages 350–363. Springer-Verlag, 2005.

[7] Patrick Baudisch, Nathaniel Good, and Paul Stewart. Focus Plus Context Screens: Combining Display Technology with Visualization Techniques. In *Proceedings of ACM Symposium on User Interface Software and Technology*, pages 31–40. ACM Press, 2001.

[8] Steven Bergner, Torsten Möller, Daniel Weiskopf, and David J. Muraki. A Spectral Analysis of Function Composition and its Implications for Sampling in Direct Volume Visualization. *IEEE Transactions on Visualization and Computer Graphics*, 12(5):1353–1360, 2006.

[9] Marshall Bern and David Eppstein. Optimized Color Gamuts for Tiled Displays. In *Proceedings of the Symposium on Computational Geometry*, pages 274–281. ACM Press, 2003.

[10] E. Wes Bethel, Greg Humphreys, Brian Paul, and J. Dean Brederson. Sort-first, Distributed Memory Parallel Visualization and Rendering. In *Proceedings of IEEE Symposium on Parallel and Large Data Visualization and Graphics*, pages 41–50. IEEE Computer Society, 2003.

[11] Oliver Bimber, Andreas Emmerling, and Thomas Klemmer. Embedded Entertainment with Smart Projectors. *Computer*, 38(1):48–55, 2005.

[12] Michael Brown, Aditi Majumder, and Ruigang Yang. Camera-based Calibration Techniques for Seamless Multiprojector Displays. *IEEE*

Transactions on Visualization and Computer Graphics, 11(2):193–206, 2005.

[13] Stefan Bruckner and M. Eduard Gröller. Instant Volume Visualization using Maximum Intensity Difference Accumulation. *Computer Graphics Forum*, 28(3):775–782, 2009.

[14] Brian Cabral, Nancy Cam, and Jim Foran. Accelerated Volume Rendering and Tomographic Reconstruction using Texture Mapping Hardware. In *Proceedings of Symposium on Volume Visualization*, pages 91–98. ACM Press, 1994.

[15] Edwin Catmull and Raphael Rom. A Class of Local Interpolating Splines. In *Computer Aided Geometric Design*, pages 317–326. Academic Press, 1974.

[16] Han Chen, Yuqun Chen, Adam Finkelstein, Thomas Funkhouser, Kai Li, Zhiyan Liu, Rudrajit Samanta, and Grant Wallace. Data Distribution Strategies for High-resolution Displays. *Computers and Graphics*, 25(5):811–818, 2001.

[17] Han Chen, Rahul Sukthankar, Grant Wallace, and Kai Li. Scalable Alignment of Large-format Multi-projector Displays using Camera Homography Trees. In *Proceedings of IEEE Visualization*, pages 339–346. IEEE Computer Society, 2002.

[18] Yuqun Chen, Han Chen, Douglas W. Clark, Zhiyan Liu, Grant Wallace, and Kai Li. Software Environments for Cluster-based Display Systems. In *Proceedings of IEEE/ACM International Symposium on Cluster Computing and the Grid*, pages 202–210. IEEE Computer Society, 2001.

[19] Yuqun Chen, Douglas W. Clark, Adam Finkelstein, Timothy C. Housel, and Kai Li. Automatic Alignment of High-Resolution Multi-Projector Displays Using An Un-Calibrated Camera. In *Proceedings of IEEE Visualization 2000*, pages 125–130. IEEE Computer Society, 2000.

[20] Carolina Cruz-Neira, Daniel J. Sandin, and Thomas A. DeFanti. Surround-screen Projection-based Virtual Reality: The Design and Implementation of the CAVE. In *Proceedings of ACM SIGGRAPH*, pages 135–142, 1993.

[21] Timothy J. Cullip and Ulrich Neumann. Accelerating Volume Reconstruction with 3D Texture Hardware. Technical report, University of North Carolina at Chapel Hill, 1994.

[22] Mary Czerwinski, Greg Smith, Tim Regan, Brian Meyers, and Gary Starkweather. Toward Characterizing the Productivity Benefits of Very Large Displays. In *Proceedings of Interact*, pages 9–16. IOS Press, 2003.

[23] Thomas A. DeFanti, Jason Leigh, Luc Renambot, Byungil Jeong, Alan Verlo, Lance Long, Maxine Brown, Daniel J. Sandin, Venkatram Vishwanath, Qian Liu, Mason J. Katz, Philip Papadopoulos, Joseph P. Keefe, Gregory R. Hidley, Gregory L. Dawe, Ian Kaufman, Bryan Glogowski, Kai-Uwe Doerr, Rajvikram Singh, Javier Girado, Jürgen P. Schulze, Falko Kuester, and Larry Smarr. The OptIPortal, a Scalable Visualization, Storage, and Computing Interface Device for the OptiPuter. *Future Generation Computer Systems*, 25(2):114–123, 2009.

[24] Ralf Diekmann. *Load Balancing Strategies for Data Parallel Applications.* PhD thesis, Universität Paderborn, Germany, 1998.

[25] DMX. Distributed Multihead X Project. `http://dmx.sourceforge.net`, Last accessed October 17, 2011.

[26] Kai-Uwe Doerr and Falko Kuester. CGLX: A Scalable, High-Performance Visualization Framework for Networked Display Environments. *IEEE Transactions on Visualization and Computer Graphics*, 17(3):320–332, 2011.

[27] Achim Ebert, Sebastian Thelen, Peter-Scott Olech, Joerg Meyer, and Hans Hagen. Tiled++: An Enhanced Tiled Hi-Res Display Wall. *IEEE Transactions on Visualization and Computer Graphics*, 16(1):120–132, 2010.

[28] Stefan Eilemann, Maxim Makhinya, and Renato Pajarola. Equalizer: A Scalable Parallel Rendering Framework. *IEEE Transactions on Visualization and Computer Graphics*, 15(3):436–452, 2009.

[29] Klaus Engel, Markus Hadwiger, Joe Kniss, Christof Rezk-Salama, and Daniel Weiskopf. *Real-Time Volume Graphics.* A. K. Peters, 2006.

[30] Klaus Engel, Martin Kraus, and Thomas Ertl. High-quality Pre-integrated Volume Rendering using Hardware-accelerated Pixel Shading. In *Proceedings of the ACM SIGGRAPH/EUROGRAPHICS Workshop on Graphics Hardware*, pages 9–16. ACM Press, 2001.

[31] Randima Fernando and Mark Kilgard. *The CG Tutorial: The Definitive Guide to Programmable Real-Time Graphics.* Addison-Wesley Professional, 2003.

[32] Steffen Frey and Thomas Ertl. PaTraCo: A Framework Enabling the Transparent and Efficient Programming of Heterogeneous Compute Networks. In *Proceedings of Eurographics Symposium on Parallel Graphics and Visualization*, pages 131–140. Eurographics Association, 2010.

[33] Michael P. Garrity. Raytracing Irregular Volume Data. *ACM SIGGRAPH Computer Graphics*, 24(5):35–40, 1990.

[34] Desmond Germans, Hans J. W. Spoelder, Luc Renambot, and Henri E. Bal. VIRPI: A High-level Toolkit for Interactive Scientific Visualization in Virtual Reality. In *Proceedings of Immersive Projection Technology/Eurographics Virtual Environments Workshop*, pages 109–120, 2001.

[35] Olivier Gourmel, Anthony Pajot, Mathias Paulin, Loïc Barthe, and Pierre Poulin. Fitted BVH for Fast Raytracing of Metaballs. *Computer Graphics Forum*, 29(2):281–288, 2010.

[36] Sebastian Grottel, Guido Reina, Carsten Dachsbacher, and Thomas Ertl. Coherent Culling and Shading for Large Molecular Dynamics Visualization. *Computer Graphics Forum*, 29(3):953–962, 2010.

[37] Sebastian Grottel, Guido Reina, and Thomas Ertl. Optimized Data Transfer for Time-dependent, GPU-based Glyphs. In *Proceedings of IEEE Pacific Visualization Symposium*, pages 65–72. IEEE Computer Society, 2009.

[38] Amel Guetat, Alexandre Ancel, Stéphane Marchesin, and Jean-Michel Dischler. Pre-integrated Volume Rendering with Non-linear Gradient Interpolation. *IEEE Transactions on Visualization and Computer Graphics*, 16(6):1487–1494, 2010.

[39] Stefan Gumhold. Splatting Illuminated Ellipsoids with Depth Correction. In *Proceedings of Workshop on Vision, Modelling, and Visualization*, pages 245–252. AKA, 2003.

[40] Markus Hadwiger, Christian Sigg, Henning Scharsach, Katja Bühler, and Markus Gross. Real-time Ray-casting and Advanced Shading of Discrete Isosurfaces. *Computer Graphics Forum*, 24(3):303–312, 2005.

[41] Mark Hereld, Ivan R. Judson, Joseph Paris, and Rick L. Stevens. Developing Tiled Projection Display Systems. In *Proceedings of the International Immersive Projection Technology Workshop*, 2000.

[42] W. Daniel Hillis and Guy L. Steele. Data Parallel Algorithms. *Communications of the ACM*, 29(12):1170–1183, 1986.

[43] William Humphrey, Andrew Dalke, and Klaus Schulten. VMD: Visual Molecular Dynamics. *Journal of Molecular Graphics*, 14(1):33–38, 1996.

[44] Greg Humphreys, Matthew Eldridge, Ian Buck, Gordon Stoll, Matthew Everett, and Pat Hanrahan. WireGL: A Scalable Graphics System for Clusters. In *Proceedings of ACM SIGGRAPH*, pages 129–140. ACM Press, 2001.

[45] Greg Humphreys, Mike Houston, Ren Ng, Randall Frank, Sean Ahern, Peter D. Kirchner, and James T. Klosowski. Chromium: A Stream-processing Framework for Interactive Rendering on Clusters. *ACM Transactions on Graphics*, 21(3):693–702, 2002.

[46] Wen-Mei W. Hwu. *GPU Computing Gems (Applications of Gpu Computing)*. Morgan Kaufmann Publishers, 2011.

[47] JVC. JVC Develops 1.75-inch 8K4K D-ILA Device, 2008. `http://pro.jvc.com/pro/pr/2008/releases/JVC_Develops_1.75-inch_8K4K_D-ILA_Device_f.pdf`, Last accessed October 17, 2011.

[48] Yoshihiro Kanamori, Zoltan Szego, and Tomoyuki Nishita. GPU-based Fast Ray Casting for a Large Number of Metaballs. *Computer Graphics Forum*, 27(2):351–360, 2008.

[49] James E. Kelley and Morgan R. Walker. Critical-path Planning and Scheduling. In *Proceedings of the Eastern Joint Computer Conference*, pages 160–173. ACM Press, 1959.

[50] John Kessenich, Dave Baldwin, and Randi J. Rost. The OpenGL Shading Language, 2011.

[51] Khronos. The Khronos Group, 2011. `http://www.khronos.org/`, Last accessed October 25, 2011.

[52] Jungwon Kim, Honggyu Kim, Joo Hwan Lee, and Jaejin Lee. Achieving a Single Compute Device Image in OpenCL for Multiple GPUs. In *Proceedings of ACM Symposium on Principles and Practice of Parallel Programming*, pages 277–288. ACM Press, 2011.

[53] David B. Kirk and Wen-mei W. Hwu. *Programming Massively Parallel Processors: A Hands-on Approach*. Morgan Kaufmann, 2010.

[54] Thomas Klein. *Exploiting Programmable Graphics Hardware for Interactive Visualization of 3D Data Fields*. PhD thesis, Universität Stuttgart, Germany, 2008.

[55] Thomas Klein and Thomas Ertl. Illustrating Magnetic Field Lines using a Discrete Particle Model Simulation of Particles in a Magnetic Field. In *Proceedings of Workshop on Vision, Modelling, and Visualization*, pages 387–394. IOS Press, 2004.

[56] Joe Kniss, Simon Premoze, Charles Hansen, Peter Shirley, and Allen McPherson. A Model for Volume Lighting and Modeling. *IEEE Transactions on Visualization and Computer Graphics*, 9(2):150–162, 2003.

[57] Aaron Knoll, Younis Hijazi, Rolf Westerteiger, Mathias Schott, Charles Hansen, and Hans Hagen. Volume Ray Casting with Peak Finding and Differential Sampling. *IEEE Transactions on Visualization and Computer Graphics*, 15(6):1571–1578, 2009.

[58] Martin Kraus, Wei Qiao, and David S. Ebert. Projecting Tetrahedra without Rendering Artifacts. In *Proceedings of IEEE Visualization*, pages 27–34. IEEE Computer Society, 2004.

[59] Jens Krüger and Rüdiger Westermann. Acceleration Techniques for GPU-based Volume Rendering. In *Proceedings of IEEE Visualization*, pages 287–292. IEEE Computer Society, 2003.

[60] Byeonghun Lee, Jihye Yun, Jinwook Seo, Byonghyo Shim, Yeong-Gil Shin, and Bohyoung Kim. Fast High-quality Volume Ray-casting with Virtual Samplings. *IEEE Transactions on Visualization and Computer Graphics*, 16(6):1525–1532, 2010.

[61] Marc Levoy. Display of Surfaces from Volume Data. *IEEE Computer Graphics and Applications*, 8(3):29–37, 1988.

[62] Wei Li, Klaus Mueller, and Arie Kaufman. Empty Space Skipping and Occlusion Clipping for Texture-based Volume Rendering. In *Proceedings of IEEE Visualization*, pages 317–324. IEEE Computer Society, 2003.

[63] Erik Lindholm, Mark J. Kligard, and Henry Moreton. A User-programmable Vertex Engine. In *Proceedings of ACM SIGGRAPH*, pages 149–158. ACM Press, 2001.

[64] Eric Lum, Kwan-Liu Ma, and John Clyne. Texture Hardware Assisted Rendering of Time-varying Volume Data. In *Proceedings of IEEE Visualization*, pages 263–270. IEEE Computer Society, 2001.

[65] Aditi Majumder, Zhu He, Herman Towles, and Greg Welch. Achieving Color Uniformity across Multi-projector Displays. In *Proceedings of IEEE Visualization*, pages 117–124. IEEE Computer Society, 2000.

[66] Aditi Majumder and Rick Stevens. Color Nonuniformity in Projection-based Displays: Analysis and Solutions. *IEEE Transactions on Visualization and Computer Graphics*, 10(2):177–88, 2003.

[67] Stéphane Marchesin and Guillaume Colin de Verdière. High-Quality, Semi-Analytical Volume Rendering for AMR Data. *IEEE Transactions on Visualization and Computer Graphics*, 15(6):1611–1618, 2009.

[68] Stéphane Marchesin, Jean-Michel Dischler, and Catherine Mongenet. Per-pixel Opacity Modulation for Feature Enhancement in Volume Rendering. *IEEE Transactions on Visualization and Computer Graphics*, 16(4):560–570, 2010.

[69] Stéphane Marchesin, Catherine Mongenet, and Jean-Michel Dischler. Dynamic Load Balancing for Parallel Volume Rendering. In *Proceedings of Eurographics Symposium on Parallel Graphics and Visualization*, pages 51–58. Eurographics Association, 2006.

[70] Stéphane Marchesin, Catherine Mongenet, and Jean-Michel Dischler. Multi-GPU Sort-last Volume Visualization. In *Proceedings of Eurographics Symposium on Parallel Graphics and Visualization*, pages 1–8. Eurographics Association, 2008.

[71] William R. Mark, R. Steven Glanville, Kurt Akeley, and Mark J. Kilgard. Cg: A System for Programming Graphics Hardware in a C-like Language. *ACM Transactions on Graphics*, 22(3):896–907, 2003.

[72] Lukas Marsalek, Armin Hauber, and Philipp Slusallek. High-speed Volume Ray Casting with CUDA. In *IEEE Symposium on Interactive Ray Tracing*, pages 185–185. IEEE Computer Society, 2008.

[73] Steve Molnar, Michael Cox, David Ellsworth, and Henry Fuchs. A Sorting Classification of Parallel Rendering. *IEEE Computer Graphics and Applications*, 14(4):23–32, 1994.

[74] Brendan Moloney, Marco Ament, Daniel Weiskopf, and Torston Möller. Sort First Parallel Volume Rendering. *IEEE Transactions on Visualization and Computer Graphics*, 17(8):1164–1177, 2011.

[75] Brendan Moloney, Daniel Weiskopf, Torsten Möller, and Magnus Strengert. Scalable Sort-First Parallel Direct Volume Rendering with Dynamic Load Balancing. In *Proceedings of Eurographics Symposium on Parallel Graphics and Visualization*, pages 45–52. Eurographics Association, 2007.

[76] MPI. The Message Passing Interface, 2011. `http://www.mcs.anl.gov/mpi`, Last accessed November 2, 2011.

[77] Christoph Müller, Steffen Frey, Magnus Strengert, Carsten Dachsbacher, and Thomas Ertl. A Compute Unified System Architecture for Graphics Clusters Incorporating Data Locality. *IEEE Transactions on Visualization and Computer Graphics*, 15(4):605–617, 2009.

[78] Christoph Müller, Sebastian Grottel, and Thomas Ertl. Image-Space GPU Metaballs for Time-Dependent Particle Data Sets. In *Proceedings of Workshop on Vision, Modelling, and Visualization*, pages 31–40. Max-Planck-Institut für Informatik, 2007.

[79] Christoph Müller, Magnus Strengert, and Thomas Ertl. Optimized Volume Raycasting for Graphics Hardware-based Cluster Systems. In *Proceedings of Eurographics Symposium on Parallel Graphics and Visualization*, pages 59–66. Eurographics Association, 2006.

[80] Aaftab Munshi, Dan Ginsburg, Timothy G. Mattson, and Benedict Gaster. *OpenCL Programming Guide*. Addison-Wesley Professional, 2011.

[81] Paul A. Navrátil, Brandt Westing, Gregory P. Johnson, Ashwini Athalye, Jose Carreno, and Freddy Rojas. A Practical Guide to Large Tiled Displays. In *Proceedings of Advances in Visual Computing 2009*, pages 970–981. Springer-Verlag, 2009.

[82] Hubert Nguyen. *GPU Gems 3*. Addison-Wesley Professional, 2007.

[83] Tao Ni, Greg S. Schmidt, Oliver G. Staadt, Mark A. Livingston, Robert Ball, and Richard May. A Survey of Large High-Resolution Display Technologies, Techniques, and Applications. In *Proceedings of IEEE Virtual Reality Conference*, pages 223–236. IEEE Computer Society, 2006.

[84] NVIDIA. Compute Unified Device Architecture (CUDA), 2011. `http://developer.nvidia.com/category/zone/cuda-zone`, Last accessed October 25, 2011.

[85] James Oliver, Chiu-Shui Chan, Eve Wurtele, Mark Bryden, and Mike Krapfl. The Most Realistic Virtual Reality Room in the World, 2006. `http://www.public.iastate.edu/~nscentral/news/2006/may/c6update.shtml`, Last accessed October 17, 2011.

[86] OpenMP. The Open Multi-Processing API, 2011. `http://www.openmp.org`, Last accessed November 2, 2011.

[87] John D. Owens, David Luebke, Naga Govindaraju, Mark Harris, Jens Krüger, Aaron E. Lefohn, and Timothy J. Purcell. A Survey of General-Purpose Computation on Graphics Hardware. *Computer Graphics Forum*, 26(1):80–113, 2007.

[88] Alexandros Panagiotidis, Daniel Kauker, Steffen Frey, and Thomas Ertl. DIANA: A Device Abstraction Framework for Parallel Computations. In *Proceedings of the International Conference on Parallel, Distributed, Grid and Cloud Computing for Engineering*. Paper 20. Civil-Comp, 2011.

[89] Tim H.J.M. Peeters, Vesna Prckovska, Markus van Almsick, Anna Vilanova, and Bart M. ter Haar Romeny. Fast and Sleek Glyph Rendering for Interactive HARDI Data Exploration. In *Proceedings of IEEE Pacific Visualization Symposium*, pages 153–160. IEEE Computer Society, 2009.

[90] Eric F. Pettersen, Thomas D. Goddard, Conrad C. Huang, Gregory S. Couch, Daniel M. Greenblatt, Elaine C. Meng, and Thomas E. Ferrin. UCSF Chimera – A Visualization System for Exploratory Research and Analysis. *Journal of Computational Chemistry*, 25:1605–1612, 2004.

[91] B. Raffin, L. Soares, R. Ball, G.S. Schmidt, M. A. Livingston, O. G. Staadt, and R. May. PC Clusters for Virtual Reality. In *Proceedings of IEEE Virtual Reality Conference*, pages 215–222. IEEE Computer Society, 2006.

[92] Ramesh Raskar, Jeroen van Baar, and Jin Xiang Chai. A Low-cost Projector Mosaic with Fast Registration. In *Proceedings of Asian Conference on Computer Vision*, pages 161–168, 2002.

[93] Guido Reina and Thomas Ertl. Hardware-accelerated Glyphs for Mono- and Dipoles in Molecular Dynamics Visualization. In *Proceedings of Joint Eurographics/IEEE VGTC Symposium on Visualization*, pages 177–182. Eurographics Association, 2005.

[94] Luc Renambot, Arun Rao, Rajvikram Singh, Jeon Byungil, Naveen Krishnaprasad, Venkatram Vishwanath, Vaidya Chandrasekhar, Nicolas Schwarz, Allan Spale, Charles Zhang, Gideon Goldman, Jason Leigh, and Andrew Johnson. Sage: The Scalable Adaptive Graphics Environment. In *Proceedings of WACE*, 2004.

[95] Christof Rezk-Salama, Klaus Engel, Michael Anthony Vincent Bauer, Gunther Greiner, and Thomas Ertl. Interactive Volume Rendering on Standard PC Graphics Hardware using Multi-textures and Multi-stage Rasterization. In *Proceedings of ACM SIGGRAPH/Eurographics Workshop on Graphics Hardware*, pages 109–118. ACM Press, 2000.

[96] Tristan Richardson, Quentin Stafford-Fraser, Kenneth R. Wood, and Andy Hopper. Virtual Network Computing. *IEEE Internet Computing*, 2(1):33–38, 1998.

[97] George Robertson, Mary Czerwinski, Patrick Baudisch, Brian Meyers, Daniel Robbins, Greg Smith, and Desney Tan. The Large-display User Experience. *IEEE Computer Graphics and Applications*, 25(4):44–51, 2005.

[98] Randi J. Rost. *OpenGL Shading Language (3rd ed.)*. Addison-Wesley Professional, 2009.

[99] Stefan Röttger, Stefan Guthe, Daniel Weiskopf, Thomas Ertl, and Wolfgang Strasser. Smart Hardware-accelerated Volume Rendering. In *Proceedings of Symposium on Data Visualisation*, pages 231–238. Eurographics Association, 2003.

[100] Stefan Röttger, Martin Kraus, and Thomas Ertl. Hardware-accelerated Volume and Isosurface Rendering Based on Cell-projection. In *Proceedings of IEEE Visualization*, pages 109–116. IEEE Computer Society, 2000.

[101] Jason Sanders and Edward Kandrot. *CUDA by Example: An Introduction to General-Purpose GPU Programming*. Addison-Wesley Professional, 2010.

[102] Daniel J. Sandin, Todd Margolis, Jinghua Ge, Javier Girado, Tom Peterka, and Thomas A. DeFanti. The Varrier Autostereoscopic Virtual Reality Display. *ACM Transactions on Graphics*, 24(3):894–903, 2005.

[103] Dana Schaa and David Kaeli. Exploring the multiple-gpu design space.

[104] Schrödinger. The PyMOL Molecular Graphics System, 2011. `http://www.pymol.org/`, Last accessed October 29, 2011.

[105] Mark Segal and Kurt Akeley. The OpenGL Graphics System: A Specification (Version 4.2 (Core Profile)), 2011.

[106] Christian Sigg and Markus Hadwiger. Fast Third-Order Texture Filtering. In *GPU Gems 2: Programming Techniques for High Performance Graphics and General Purpose Computation*, pages 313–329. Addison-Wesley Professional, 2005.

[107] Christian Sigg, Tim Weyrich, Mario Botsch, and Markus Gross. GPU-based Ray-casting of Quadratic Surfaces. In *Proceedings of Eurographics Symposium on Point-Based Graphics*, pages 59–65. Eurographics Association, 2006.

[108] Simon Stegmaier, Magnus Strengert, Thomas Klein, and Thomas Ertl. A Simple and Flexible Volume Rendering Framework for Graphics Hardware-based Raycasting. In *Proceedings of the International Workshop on Volume Graphics*, pages 187–195. Stony Brook, NY, 2005.

[109] Magnus Strengert, Marcelo Magallón, Daniel Weiskopf, Stefan Guthe, and Thomas Ertl. Hierarchical Visualization and Compression of Large Volume Datasets using GPU Clusters. In *Proceedings of Eurographics Symposium on Parallel Graphics and Visualization*, pages 41–48. Eurographics Association, 2004.

[110] Jeff A. Stuart and John D. Owens. Multi-GPU MapReduce on GPU Clusters. In *Proceedings of IEEE International Symposium on Parallel and Distributed Processing*, pages 1068–1079. IEEE Computer Society, 2011.

[111] Vaidy S. Sunderam. VM: A Framework for Parallel Distributed Computing. *Concurrency: Practice and Experience*, 2(4):315–337, 1990.

[112] Desney S. Tan, Darren Gergle, Peter Scupelli, and Randy Pausch. With Similar Visual Angles, Larger Displays Improve Spatial Performance. In *Proceedings of the SIGCHI Conference on Human Factors in Computing Systems*, pages 217–224. ACM Press, 2003.

[113] James D. Teresco, Jamal Faik, and Joseph E. Flaherty. Resource-aware Scientific Computation on a Heterogeneous Cluster. *Computing in Science and Engineering*, 7(2):40–50, 2005.

[114] TOP500. Ranking of Supercomputers According to the LINPACK Benchmark, 2011. `http://www.top500.org/list/2011/06/100`, Last accessed October 27, 2011.

[115] Craig Upson, Thomas Faulhaber Jr., David Kamins, David H. Laidlaw, David Schlegel, Jeffrey Vroom, Robert Gurwitz, and Andries van Dam. The Application Visualization System: A Computational Environment for Scientific Visualization. *Computer Graphics and Applications*, 9(4):30–42, July 1989.

[116] Tom Van der Schaaf, Luc Renambot, Desmond Germans, Hans Spoelder, and Henri Bal. Retained Mode Parallel Rendering for Scalable Tiled Displays. In *Proceedings of Immersive Projection Technologies Symposium*, 2002.

[117] Allen Van Gelder and Kwansik Kim. Direct Volume Rendering with Shading via Three-dimensional Textures. In *Proceedings of Symposium on Volume Visualization*, pages 23–30. IEEE Press, 1996.

[118] Els Van Loocke. Barco's Stereoscopic Dome Projection: A World's First at the IMAX Tycho Brahe Planetarium, Copenhagen, 2006. `http://www.barco.com/en/pressrelease/1678/en`, Last accessed October 17, 2011.

[119] Visage Imaging GmbH. Amira, http://www.amira.com, 2011.

[120] Sivakumar Viswanathan, Bharadwaj Veeravalli, and Thomas G. Robertazzi. Resource-aware Distributed Scheduling Strategies for Large-Scale Computational Cluster/Grid Systems. *IEEE Transactions on Parallel and Distributed Systems*, 18(10):1450–1461, 2007.

[121] Lei Wang, Yong-Zhong Huang, Xin Chen, and Chun-Yan Zhang. Task Scheduling of Parallel Processing in CPU-GPU Collaborative Environment. In *Proceedings of the International Conference on Computer Science and Information Technology*, pages 228–232. IEEE Computer Society, 2008.

[122] Manfred Weiler, Martin Kraus, Markus Merz, and Thomas Ertl. Hardware-based Ray Casting for Tetrahedral Meshes. In *Proceedings of IEEE Visualization*, pages 333–340. IEEE Computer Society, 2003.

[123] Manfred Weiler, Rüdiger Westermann, Chuck Hansen, Kurt Zimmermann, and Thomas Ertl. Level-of-detail Volume Rendering via 3D Textures. In *Proceedings of Symposium on Volume Visualization*, pages 7–13. ACM Press, 2000.

[124] Daniel Weiskopf. *GPU-based Interactive Visualization Techniques.* Springer, 2006.

[125] Daniel Weiskopf, Klaus Engel, and Thomas Ertl. Interactive Clipping Techniques for Texture-based Volume Visualization and Volume Shading. *IEEE Transactions on Visualization and Computer Graphics*, 9(3):298–312, 2003.

[126] Rüdiger Westermann and Thomas Ertl. Efficiently using Graphics Hardware in Volume Rendering Applications. In *Proceedings of ACM SIGGRAPH*, pages 169–177. ACM Press, 1998.

[127] Lance Williams. Pyramidal Parametrics. *ACM SIGGRAPH Computer Graphics*, 17(3):1–11, 1983.

[128] David Wolff. *OpenGL 4.0 Shading Language Cookbook*. Packt Publishing, 2011.

[129] Kun Zhou, Qiming Hou, Zhong Ren, Minmin Gong, Xin Sun, and Baining Guo. RenderAnts: Interactive Reyes Rendering on GPUs. *ACM Transactions on Graphics*, 28(5):155:1–155:11, 2009.

Chapter 12

Hybrid Parallelism

E. Wes Bethel
Lawrence Berkeley National Laboratory

David Camp
Lawrence Berkeley National Laboratory

Hank Childs
Lawrence Berkeley National Laboratory

Christoph Garth
University of Kaiserslautern

Mark Howison
Brown University

Kenneth I. Joy
University of California, Davis

David Pugmire
Oak Ridge National Laboratory

12.1 Introduction 262
12.2 Hybrid Parallelism and Volume Rendering 264
12.2.1 Background and Previous Work 264
12.2.2 Implementation 264
12.2.2.1 Shared-Memory Parallel Ray Casting 266
12.2.2.2 Parallel Compositing 266
12.2.3 Experiment Methodology 267
12.2.4 Results 268
12.2.4.1 Initialization 268
12.2.4.2 Ghost Data/Halo Exchange 269
12.2.4.3 Ray Casting 269
12.2.4.4 Compositing 272
12.2.4.5 Overall Performance 272
12.3 Hybrid Parallelism and Integral Curve Calculation 275
12.3.1 Background and Context 275

12.3.2 Design and Implementation 276
12.3.2.1 Parallelize Over Seeds 276
12.3.2.2 Parallelize Over Blocks 277
12.3.3 Experiment Methodology 278
12.3.3.1 Factors Influencing Parallelization Strategy 278
12.3.3.2 Test Cases 279
12.3.3.3 Runtime Environment 279
12.3.3.4 Measurements 280
12.3.4 Results 280
12.3.4.1 Parallelization Over Seeds 280
12.3.4.2 Parallelization Over Blocks 282
12.4 Conclusion and Future Work 283
References 287

Hybrid parallelism refers to a blend of distributed- and shared-memory parallel programming techniques within a single application. This chapter presents results from two studies. They aimed to explore the thesis that hybrid parallelism offers performance advantages for visualization codes on multi-core platforms. The findings show that, compared to a traditional distributed-memory implementation, the hybrid parallel approach uses a smaller memory footprint, performs less interprocess communication, has faster execution speed, and, for some configurations, performs significantly less data I/O.

12.1 Introduction

A *distributed-memory parallel* computer is made up of multiple nodes, with each node containing one or more cores. Each instance of a parallel program is called a task (or sometimes a *Processing Element* or PE). A pure distributed-memory program has one task for each core on each node of the computer. This is not necessary, however. Hybrid parallel programs have fewer tasks per node than cores. They make use of the remaining cores by using *threads*, which are lightweight programs controlled by the task. Threads can share memory amongst themselves and between the main thread associated with the task; allowing for optimizations that are not possible with distributed-memory programming. For example, consider a distributed-memory parallel computer with eight quad-core nodes. A pure distributed-memory program would have thirty-two tasks running and none of these tasks would make use of shared-memory techniques (although some cores would reside on the same node). A hybrid configuration could have eight tasks, each running with four threads, sixteen tasks, each running with two threads, or even configurations where the number of tasks and threads per node varies.

This chapter defines and uses the following terminology and notation. *Traditional parallelism*, or P^T, refers to a design and implementation that uses

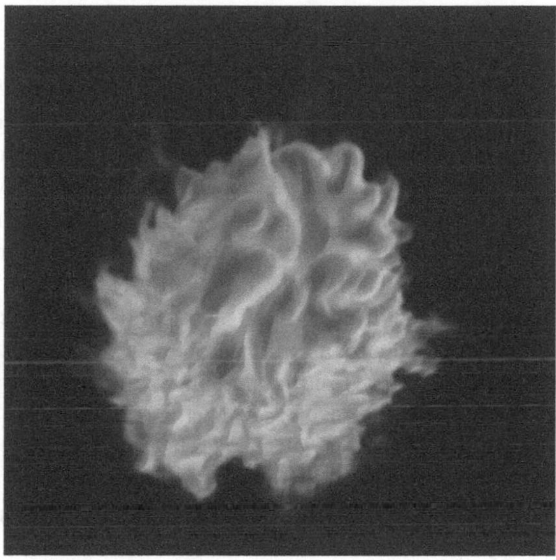

FIGURE 12.1: 4608^2 image of a combustion simulation result, rendered by hybrid parallel MPI+pthreads implementation running on 216,000 cores of the JaguarPF supercomputer. Image source: Howison et al., 2011 [14]. Combustion simulation data courtesy of J. Bell and M. Day (LBNL).

only MPI for parallelism, regardless of whether the parallel application is run on a distributed- or shared-memory system. *Hybrid parallelism*, or P^H, refers to a design and implementation that uses both MPI and some other form of shared-memory parallelism like POSIX threads [4], OpenMP [3], OpenCL [12], CUDA [9], and so forth.

The main focus of this chapter is to present results from two different experiments within the field of high performance visualization that aim to study the extent to which visualization algorithms can benefit from hybrid parallelism when applied to today's largest data sets and on today's largest computational platforms. The studies presented in this chapter use a P^H design, whereby each MPI task will in turn invoke some form of shared-memory parallelism on multi-core CPUs and many-core GPUs.

One experiment studies a hybrid parallel implementation of ray casting volume rendering at extreme-scale concurrency. The other studies a hybrid parallel implementation of integral curve computation using two different approaches to parallelization. The material in this chapter consolidates information from earlier publications on hybrid parallelism for volume rendering [13, 14] and streamline/integral curve computations [6, 21]. Both of these studies show that a P^H implementation runs faster, uses less memory, and performs less communication and data movement than its P^T counterpart. In some cases, the difference is quite profound, and reveals many insurmountable obstacles if P^T paths continue to be used on exascale-class computational platforms.

There are several factors that influence scalability at extreme levels of concurrency. The cost of initialization itself may become a limiting factor. One of the studies in this chapter reveals that there is a cost, in terms of memory footprint, associated with each MPI task that grows at a nonlinear rate with the concurrency level. Another factor is the overhead associated with synchronizing processes. Traditional message-based approaches that have worked well in distributed-memory parallel environments may prove to be too costly when there are hundreds to thousands of cores per chip. More generally, communication patterns and associated overhead may likely suffer from similar scalability limits. Load balancing, which refers to the process of having each of the individual processes perform about the same amount of work, has been the subject of much research over the years in distributed-memory parallel environments. Adding in the complexity of scores, hundreds to thousands of cores per chip, adds to that complexity, which, in turn, can also limit scalability.

12.2 Hybrid Parallelism and Volume Rendering

12.2.1 Background and Previous Work

Volume rendering (see 4.4) is a common technique used for displaying 2D projections of 3D sampled data [15, 10] and it is computationally, memory, and data I/O intensive. The study focuses on an P^H implementation at extreme concurrencies in order to take advantage of multi- and many-core processor architectures.

This study's P^H implementation makes use of a design pattern common in many parallel volume rendering applications that use a mixture of both object- and pixel-level parallelism [17, 24, 16, 2]. The design employs an object-order partitioning to distribute source data blocks to processors where they are rendered using ray casting [15, 22, 10, 26]. Then, within a processor, an image-space decomposition, similar to Nieh and Levoy [18], is used to allow multiple rendering threads to cooperatively generate partial images that are later combined, via compositing, into a final image [10, 15, 26]. This design approach, which uses a blend of object- and pixel-level parallelism, has proven successful in achieving scalability and tackling large data sizes.

The most substantial difference between the work in this study and previous work in parallel volume rendering is the exploitation of P^H parallelism at an order of magnitude of greater concurrency than any previous studies. This work also performs an in-depth study to better understand scalability characteristics as well as potential performance gains of the P^H approach.

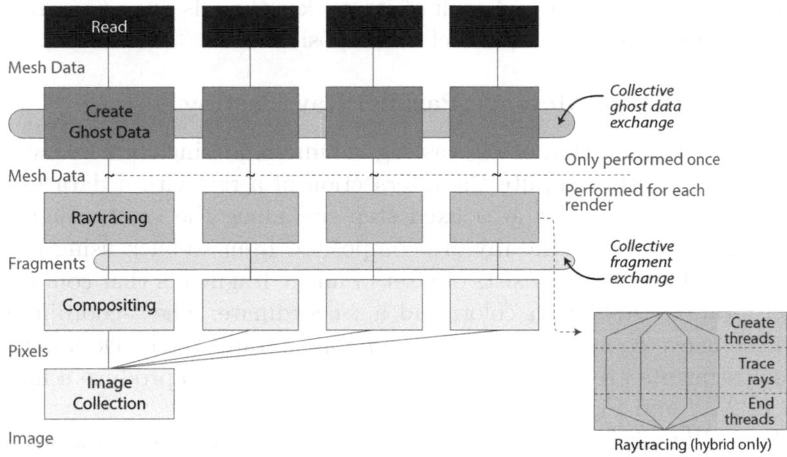

FIGURE 12.2: P^Hvolume rendering system architecture. Image courtesy of Mark Howison, E. Wes Bethel, and Hank Childs (LBNL).

12.2.2 Implementation

The study's parallel volume rendering implementation uses a design pattern similar to that in previous works (e.g., [17, 24, 16, 2]). Given a source data volume S and n parallel tasks, each task reads in $1/n$ of S, performs ray casting volume rendering on this data subdomain to produce a set of image fragments, and, then participates in a compositing stage in which fragments are exchanged and combined into a final image. The completed image is gathered to the root task for display or I/O for storage. Figure 12.2 provides a block-level view of this organization.

The P^Timplementation is written in C/C++, using the MPI [23] library. The portions of the implementation that are shared-memory parallel are written using a combination of C/C++ and either POSIX threads [4], OpenMP [7], or CUDA (version 3.0) [19] so that the study is comparing three P^Himplementations referred to as hybrid/pthreads, hybrid/OpenMP, and hybrid/CUDA.

The $P^T$and P^Himplementations differ in several key respects. First, the ray casting volume rendering algorithm is a serial process in the P^Timplementation; this serial process is replicated across N processors. In contrast, in the P^Himplementation, the ray casting algorithm runs in a shared-memory parallel fashion rather than as a serial process—an idea presented in more detail in 12.2.2.1. Second, the communication topology in the compositing stage differs slightly between the $P^T$and P^Himplementations—the details of which are the subject of 12.2.2.2. The third difference is how data is partitioned across the tasks. In the P^Timplementation, each task loads and operates on a disjoint block of data. In the P^Himplementation, each task loads a disjoint block of data and each of its worker threads operate in parallel on that data using an image-parallel decomposition [18].

12.2.2.1 Shared-Memory Parallel Ray Casting

The implementation of ray casting volume rendering code follows Levoy's method [15]: first, compute the intersection of a ray with a data block, and then, compute the color at a fixed step size along the ray through the volume. All colors along the ray are composited front-to-back using the "over" operator [20]. Output consists of a set of image fragments that contain an x, y pixel location; R, G, B, α color; and a z-coordinate. The z-coordinate is the location in eye coordinates, where the ray penetrates the block of data. Later, these fragments are composited in the correct order to produce a final image (cf. 12.2.2.2 and Fig. 4.2).

Each P^Ttask invokes a serial ray caster that operates on its own disjoint block of data. Since this code is processing structured rectilinear grids, all data subdomains are spatially disjoint, and the z-coordinate for sorting during the subsequent composition step is the ray's entry point into the data block.

In contrast, the P^Htasks invoke a ray caster with shared-memory parallelism that consists of T threads, executing concurrently to perform the ray casting on a shared block of data. As in Nieh and Levoy [18], the authors use an image-space partitioning: each thread is responsible for ray casting a portion of the image. In the pthreads and OpenMP ray casting implementations, the image-space partitioning is interleaved, with the image divided into many tiles that are distributed amongst the threads. The CUDA ray casting implementation is slightly different because of the data-parallel nature of the language. The image is treated as a 2D CUDA grid, which is divided into CUDA thread blocks. Each thread block corresponds to an image tile, and the individual CUDA threads, within each block, are mapped to individual pixels in the image. There are a number of CUDA-centric issues and considerations with this implementation. For more information, see Howison et al. 2011 [14].

12.2.2.2 Parallel Compositing

While Chapter 5 discusses parallel image compositing methods, this particular implementation uses an approach that is somewhat different. Compositing begins by partitioning the pixels of the final image across the tasks. Next, an all-to-all communication step exchanges each fragment from the task where it was generated in the ray casting phase to the task that owns the region of the image in which the fragment lies. This exchange is done using an `MPI_Alltoallv` call. After the exchange, each task then performs the final compositing for each pixel, in its region of the image, using the "over" operator [20]. The final image is then gathered on the root task. (See the original studies for additional details [13, 14].)

The $P^H$and P^Tcompositing differ in one key way that results in fewer, but larger messages in the P^Himplementation, which in turn, results in P^Hhaving a better performance. In both the $P^H$and P^Timplementations, each of the MPI tasks will participate in the fragment exchange communication process. Because there are far fewer MPI tasks in the P^Himplementation, those MPI tasks will exchange fewer and larger messages. In the P^Himplementation, one thread gathers fragments from all other threads in the shared-memory parallel CPU or GPU and performs the communication rather than having all threads participate in the communication. The overall effect of this design choice is an improvement in communication characteristics, as presented later in 12.2.4.4.

12.2.3 Experiment Methodology

The study's methodology is designed to test the hypothesis that a P^H implementation exhibits better performance and resource utilization than the P^T implementation.

The study uses two systems, a large Cray XF5 system, JaguarPF, located at Oak Ridge National Laboratory, and a large GPU cluster, Longhorn, located at the Texas Advanced Computing Center. In 2009, JaguarPF was ranked number one on the list of TOP500 fastest supercomputers with a peak theoretical performance of 2.3 petaflops [25]. Each of its 18,688 nodes has two sockets, and each socket has a six-core 2.6GHz AMD Opteron processor, for a total of 224,256 compute cores. With 16GB per node (8GB per socket), the system has 292TB of aggregate memory and roughly 1.3GB per core.

Longhorn has 256 host nodes with dual socket, quad-core Intel Nehalem CPUs and 24GB of memory. They share 128 NVIDIA OptiPlex 2200 external quad-GPU enclosures for a total of 512 FX5800 GPUs. Each GPU has a clock speed of 1.3GHz, 4GB of device memory, and can execute 30 CUDA thread blocks concurrently. The study treats the FX5800 as a generic "many-core" processor with a data-parallel programming model (CUDA) that serves as a surrogate for anticipating what future many-core clusters may look like. In terms of performance, the study positions the FX5800 relative to the Opteron in terms of its actual observed runtime for this particular application rather than relying on an *a priori* architectural comparison.

The study includes three types of scalability experiments:

- **strong scaling**, in which the image size is fixed at 4608^2 and the data set size at 4608^3 (97.8 billion cells) for all concurrency levels;
- **weak-data set scaling**, with the same fixed 4608^2 image, but also, a data set size increasing with concurrency up to 23040^3 (12.2 trillion cells) at 216,000-way parallel; and
- **weak scaling**, in which both the image and the data set increase in size up to 23040^2 and 23040^3, respectively, at 216,000-way parallel.

Note that at the lowest concurrency level, all three cases coincide with a 4608^2 image and 4608^3 data set size. The source data for this experiment consisted of an output from a combustion simulation code that aims to perform a laboratory-scale modeling of a flame. Figure 12.1 shows a sample image produced by these runs. (See the original studies for additional details of the experimental methodology configuration, including details of the source data used in the experiment, how data is partitioned across processors on each platform, and data/image sizes for each of the three scaling experiments [13, 14].)

The size and shape of the image tiles for image-space partitioning in shared-memory ray casting is a tunable algorithmic parameter. The process and methodology for finding the best-performing set of tunable algorithmic parameters is described in Chapter 14.

12.2.4 Results

The study compares the cost of MPI runtime overhead and corresponding memory footprint in 12.2.4.1; the absolute amount of memory required for data blocks and ghost (halo) exchange in 12.2.4.2; the scalability of the ray casting and compositing algorithms in 12.2.4.3 and 12.2.4.4; and the communication resources required during the compositing phase in 12.2.4.4. 12.2.4.5 concludes with a comparison of results from the six-core CPU system and the many-core GPU system to understand how the balance of P^H vs. P^T parallelism affects the overall performance.

12.2.4.1 Initialization

Because there are fewer P^H tasks, they incur a smaller aggregate memory footprint for the MPI runtime environment and program-specific data structures that are allocated per task. Table 12.1 shows the memory footprint of the program as measured directly after calling `MPI_Init` and reading in command-line parameters.[1] Memory usage was sampled only from tasks 0 through 6, but those values agreed within 2% of each other. Therefore, the per-task values reported in Table 12.1 are from task 0 and the per-node and aggregate values are calculated from the per-task value.[2]

The P^T implementation uses twelve MPI tasks per node while the P^H one uses only two. The P^H implementation's two MPI tasks per node each go six-way parallel, so that all cores on both node's multi-core CPUs are fully occupied in both the P^H and P^T implementations. At a 216,000-way concurrency, the runtime overhead per P^T task is more than 2× the overhead per P^H task. The per-node and aggregate memory usage is 6× larger for the P^T implementation, because it uses 6× as many tasks. Thus, the P^T implementation uses nearly 12× as much memory per-node and in-

[1]The authors collected the `VmRSS`, or "resident set size," value from the `/proc/self/status` interface.

[2]In Table 12.1, the value in the Per Task column multiplied by twelve, the number of cores per node, which consists of two, six-core CPUs, does not always equal the value in the Per Node column due to the round-off error that results from presenting values as integral numbers of MB or GB.

TABLE 12.1: Comparison of memory usage at MPI Initialization for P^H and P^T volume rendering implementations. At 216,000-way concurrency, the P^T implementation uses twelve times the memory of the P^H version.

CPU cores	Implementation	Tasks	Per Task (MB)	Per Node (MB)	Agg. (GB)
1728	P^H	288	67	133	19
1728	P^T	1728	67	807	113
13824	P^H	2304	67	134	151
13824	P^T	13824	71	857	965
46656	P^H	7776	68	136	518
46656	P^T	46656	88	1055	4007
110592	P^H	18432	73	146	1318
110592	P^T	110592	121	1453	13078
216000	P^H	36000	82	165	2892
216000	P^T	216000	176	2106	37023

aggregate than the P^H one for initializing the MPI runtime at a 216,000-way concurrency.

12.2.4.2 Ghost Data/Halo Exchange

Two layers of ghost data are required in the ray casting phase: the first layer is for trilinear interpolation of sampled values, and the second layer is for computing the gradient field using central differences (gradients are not precomputed for this data set). In the P^H configuration, since the source data is partitioned into fewer, larger blocks, the P^H version requires less exchange and storage of ghost data by roughly 40% across all concurrency levels and for both strong and weak scaling compared to the P^T version—Figure 12.3 illustrates these results.

12.2.4.3 Ray Casting

All of the scaling studies demonstrate good scaling for the ray casting phase, since no message passing is involved (Fig. 12.4). The authors used trilinear interpolation for data sampling along the ray as well as a Phong-style shader for these runs and timings. The final ray casting time is essentially the runtime of the thread that takes the most integration steps. This behavior is entirely dependent on the view. The study's approach, which is aimed at understanding "average" behavior, uses ten different views and reports an average runtime.

In the strong scaling study, performance for the ray casting phase proves to be linear up to a 216,000-way concurrency with P^T (see Fig. 12.4). The P^H implementation exhibited different scaling behavior because of its different

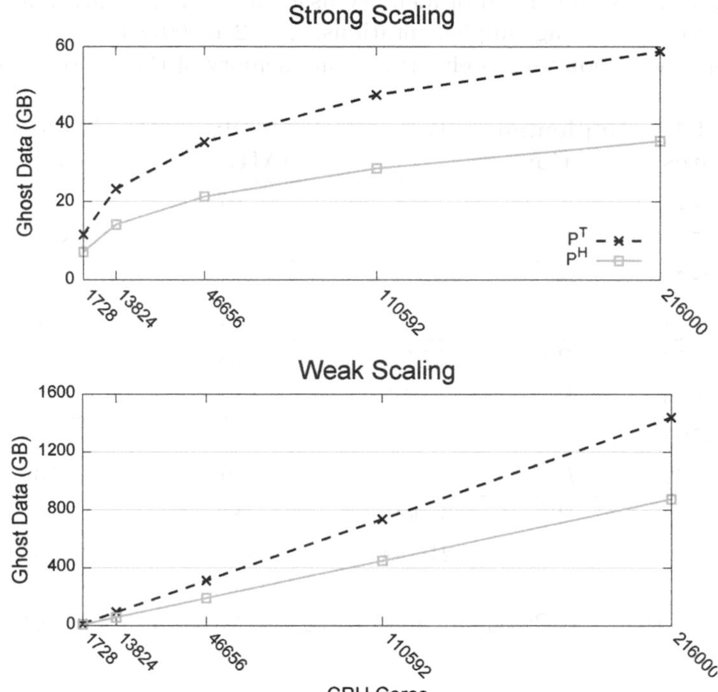

FIGURE 12.3: Because it uses fewer and larger data blocks, the P^H volume renderer requires 40% less memory for ghost data than the P^T implementation. As a result, the P^H code uses less interprocessor communication for ghost data exchange than the P^T version. Image source: Howison et al., 2011 [14].

decomposition geometry: the P^T data blocks had a perfectly cubic decomposition, but the P^H version uses $1 \times 2 \times 3$ cubic blocks, resulting in a larger rectangular block than in the P^T version (see Howison et al. 2011 [14], Table 4, for additional details). The smaller size of the GPU cluster limited the feasible concurrencies for the hybrid/CUDA implementation, leading to similarly irregular blocks in that case.

The interaction of the decomposition geometry and the camera direction determines the maximum number of ray integration steps, which is the limiting factor for the ray casting time. At lower concurrencies, this interaction benefited the P^H implementation by as much as 11% (see Howison et al. 2011 [14], Table 4). At higher concurrencies, the trend flips and the P^T implementation outperforms the P^H one by 10% in the ray casting phase. The authors expected that if they were able to run the P^H implementation with cubic-shaped data blocks (such as $2 \times 2 \times 2$ on an eight-core system), the ray casting phase of both implementations would scale identically. They also note that at 216,000 cores, ray casting is less than 20% of the total runtime (see Fig. 12.6), and

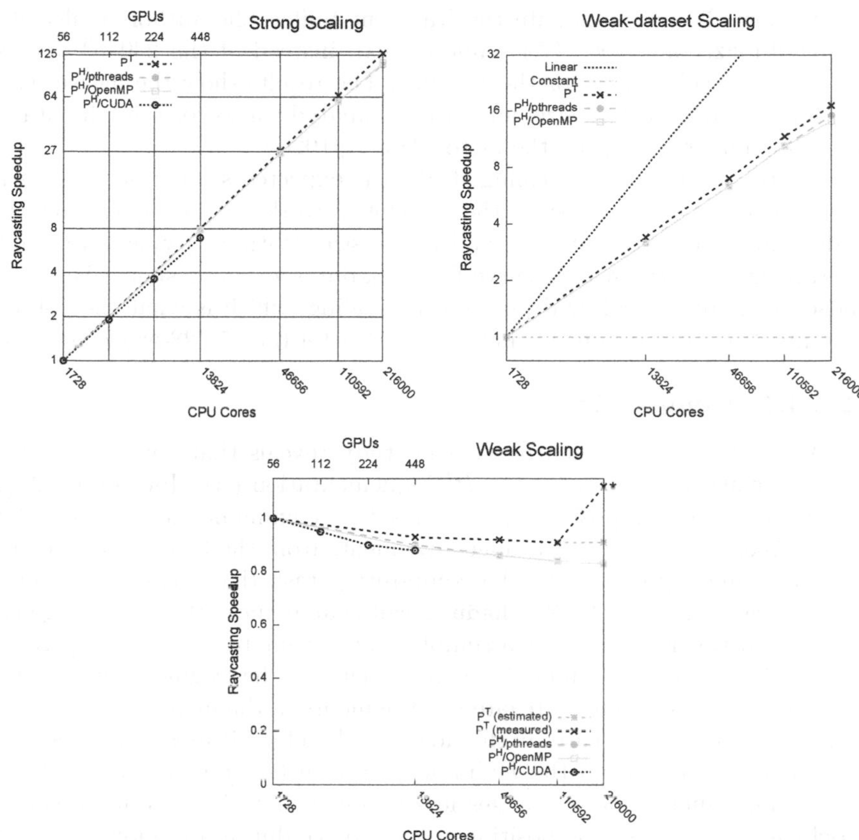

FIGURE 12.4: The speedups (referenced to 1,728 cores) for both the ray casting phase and the total render time (ray casting and compositing). The ray casting speedup is linear for the P^T version, but is sublinear for the P^H version: this effect is caused by the difference in decomposition geometries (cubic vs. rectangular). Image source: Howison et al., 2011 [14].

the P^H implementation is over 50% faster because of gains in the compositing phase, which is the subject of the next subsection.

For weak scaling, the P^H implementation maintained 80% scalability out to 216,000 cores. Overall ray casting performance is only as fast as the slowest thread. And because of perspective projection, the number of samples each thread must calculate varies. This variation becomes larger at a higher concurrency, since each core is operating on a smaller portion of the overall view frustum, which accounts for the 20% degradation.

The P^T result at a 216,000-way concurrency appears (misleadingly) to be superlinear, but that is because the authors could not maintain the data size per core. Although they could maintain it for the weak-data set scaling, increasing the image size to 23040^3 in the weak scaling study caused the temporary buffers for the image fragments (the output of the ray casting phase) to overflow. To accommodate the fragment buffer, the authors scaled down to a data size of 19200^3 (7.1 trillion cells), instead of the 23040^3 data size (12.2 trillion cells) used for the P^H runs. The results show the measured values for this smaller data size, and also, estimated values for the full data size, assuming linear scaling by the factor $23040^3/19200^3$.

For the weak-data set scaling study, the expected scaling behavior is neither linear nor constant, since the amount of work for the ray casting phase is dependent on both data size and image size. With a varying data size but fixed image size, the scaling curve for weak-dataset scaling should lie between those of the pure weak and pure strong scaling, which is what they observe. Overall, 216,000-way concurrency was 10× faster than 1,728-way concurrency.

12.2.4.4 Compositing

Above 1,728-way concurrency, the study reveals that compositing times are systematically better for the P^H implementation (see Howison et al. [14], Figure 7). The compositing phase has two communication costs: (1) the `MPI_Alltoallv` call that exchanges fragments from the task where they originated during ray casting to the compositing task that owns their region of image space; and (2) the `MPI_Reduce` call that reduces the final image components to the root task for assembly and output to a file or display. (See Fig. 12.6 for a breakdown of these costs.) During the fragment exchange, the P^H implementation can aggregate the fragments in the memory, shared by all worker threads, in this case six threads, and therefore, it uses on average about 6× fewer messages than the P^T implementation (see Fig. 12.5). In addition, the P^H implementation exchanges less fragment data because its larger data blocks allow for more compositing to take place during ray integration. Similarly, one-sixth as many P^H tasks participate in the image reduction, which leads to better performance.

12.2.4.5 Overall Performance

In the strong scaling study at a 216,000-way concurrency, the best compositing time with P^H (0.35 seconds, 4500 compositors) was 3× faster than with P^T (1.06 seconds, 6750 compositors). Furthermore, at this scale, compositing time dominated rendering time, which was roughly 0.2 seconds for both implementations. Thus, the total render time was 2.2× faster with P^H (0.56 seconds vs. 1.25 seconds). Overall, the strong scaling study shows that the advantage of P^H over P^T becomes greater as the number of cores increases (Fig. 12.6), primarily due to the P^H's faster compositing time.

For weak-data set and weak scaling, P^H still shows gains over the P^T, but they are less pronounced because the ray casting phase dominates. Although the 216,000-way breakdown for weak scaling looks like it favors P^T, this is actually an artifact of the reduced data size (19200^3); a reduced size data set was required for the P^T implementation to avoid out-of-memory errors.

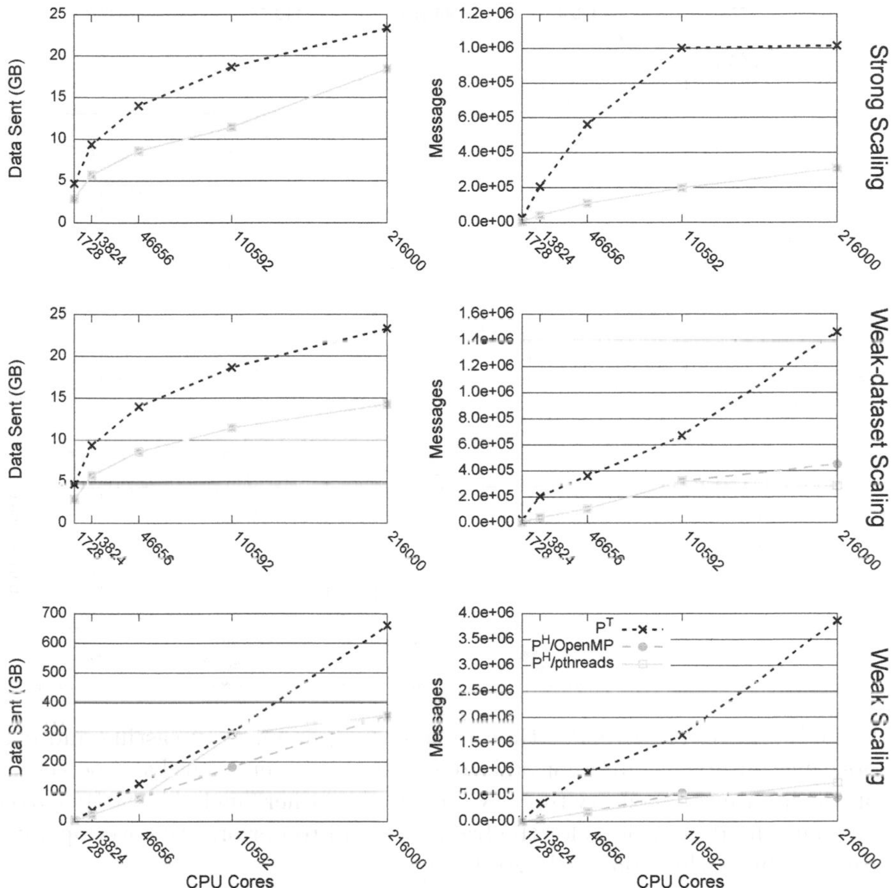

FIGURE 12.5: Comparison of the number of messages and total data sent during the fragment exchange in the compositing phase for P^H and P^T runs. The P^H version uses fewer messages and requires less total communication during compositing than the P^T implementation. Image source: Howison et al., 2011 [14].

Comparing an estimated value for a 23040^3 P^T data size suggests that the P^H implementation would be slightly faster.

On Longhorn, using 448 GPUs and processing the 4608^3 data set, the hybrid/CUDA implementation averaged 1.18 seconds for ray casting and exhibited a minimum compositing time of 0.15 seconds (with a median of 0.19 seconds). The ray casting performance positions this run close to the 46,656-way run on JaguarPF (see Howison et al. 2011 [14], Fig. 10). However, the compositing time for the hybrid/CUDA run on Longhorn (0.15 seconds) is

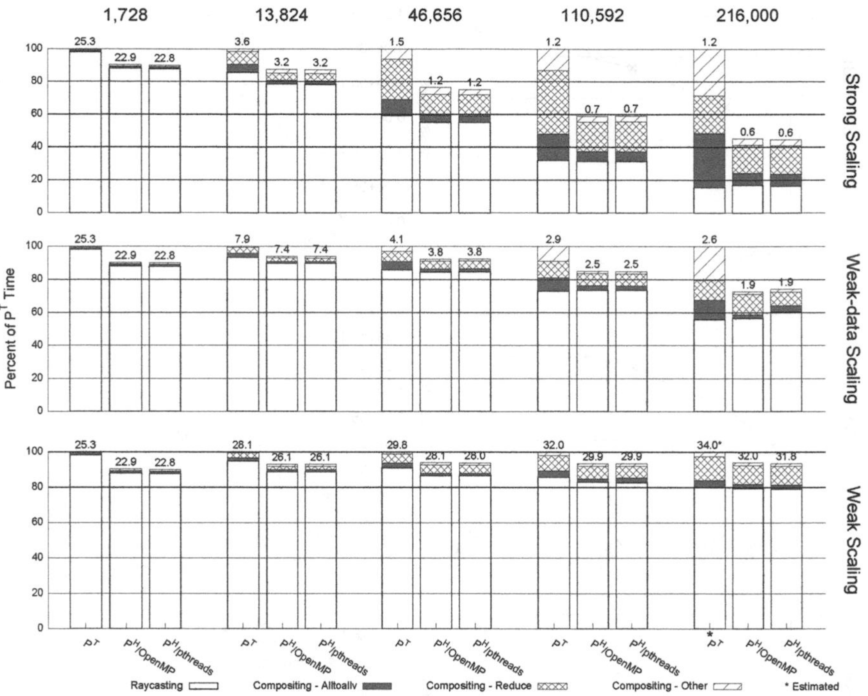

FIGURE 12.6: Total render time in seconds split into ray casting and compositing components and normalized at each concurrency level as a percentage of the distributed-memory time. "Compositing-Other" includes the time to coordinate the destinations for the fragments and to perform the over operator. Image source: Howison et al., 2011 [14].

half that of hybrid-memory runs at the 46,656-way concurrency on JaguarPF (0.31–0.33 seconds).

This result from Longhorn points more generally to the increasing potential of P^H for ray casting volume rendering at higher core-to-node ratios. The shared-memory parallelism used by the dynamically-scheduled, image-based decomposition in the ray casting phase, scales well to high concurrencies per node—up to thousands of threads in the case of CUDA. In turn, decreasing the absolute number of nodes improves the communication performance during the compositing phase.

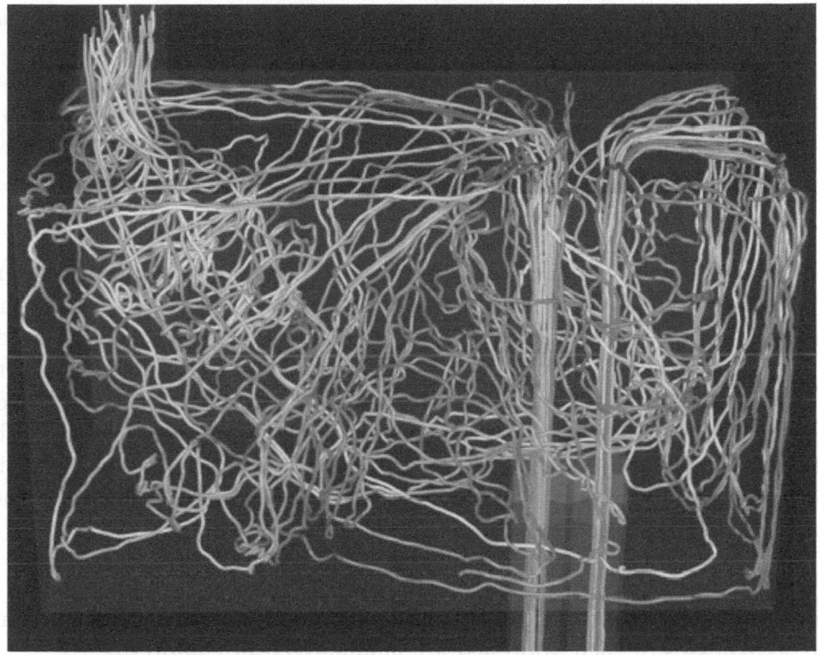

FIGURE 12.7: Integral curve computation lies at the heart of visualization techniques like streamlines, which are very useful for seeing and understanding complex flow-based phenomena. This image shows an example from computational thermal hydraulics. Algorithmic performance is a function of many factors, including the characteristics of the flow field in the underlying data set. This data set has different characteristics than those shown in Figure 6.3. Image courtesy of Hank Childs and David Camp (LBNL).

12.3 Hybrid Parallelism and Integral Curve Calculation

12.3.1 Background and Context

Chapter 6 provides much of the background and context for the hybrid parallel *IC* work in this chapter. There is one key concept from Chapter 6 worth repeating here to set the context for the *IC* hybrid parallel work, namely parallelization strategies.

There are two primary ways of performing a parallel *IC* computation: parallelize over data blocks (POB), and parallelize over seeds (POS). In POB, data is divided and distributed as evenly as possible across all processors. Each processor computes *IC*s that fall within the data block it owns. As an *IC* leaves one data block, a processor must communicate the *IC* to another processor, where the *IC* computation will continue. In POS, the *IC*'s themselves are distributed as evenly as possible among processors, and each

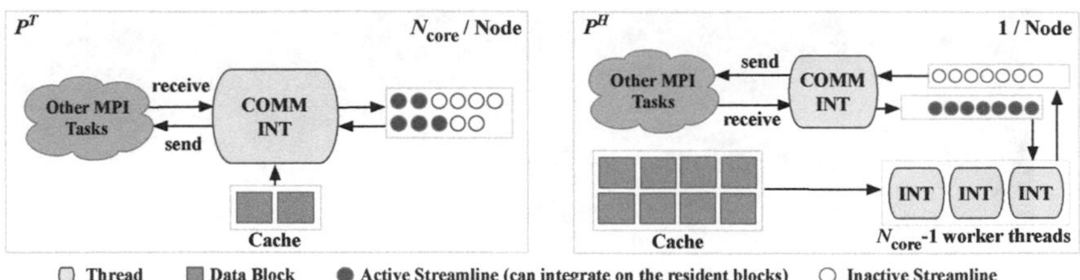

FIGURE 12.8: Comparison of P^T(left) and P^H(right) implementations of the *parallelize-over-seeds* algorithm. In the P^Tversion, each task performs the IC computation, an integration (INT), and manages its own cache by loading blocks from a disk (I/O), whereas N_{core} worker threads share a cache in the P^Himplementation. In the P^Hversion, multiple I/O threads manage the cache and observe which ICs can (*active*) and cannot (*inactive*) continue with the resident data blocks. Future blocks to load are determined from the list of inactive ICs. MPI communication, not shown here, is limited to gathering results. Image source: Camp et al., 2011 [6].

processor computes an entire IC, loading data blocks as needed. The strengths and weaknesses of each approach are summarized in Chapter 6.

In contrast to Chapter 6, this chapter focuses on how each of these two classes of algorithms differ between P^T and P^H implementation and presents the results from a study that explores the characteristics of each [6]. Considering limits to scalability, this study evaluates performance in terms of absolute runtime, the amount of data moved at runtime between processors, the amount of I/O required by each configuration, and load balance characteristics.

12.3.2 Design and Implementation

12.3.2.1 Parallelize Over Seeds

The parallelize-over-seeds (POS) algorithm, which is the subject of 6.3.2, assigns $1/n$ of the IC seeds to each parallel task. Each task will then perform a numerical integration over its seed set, loading data blocks as needed. In general, the POS algorithm does not require any inter-task communication at runtime, although it may incur redundant data block loads if the ICs in different tasks enter the same data block. Figure 12.8 provides an overview of both $P^T$and P^Himplementations of this algorithm.

In the P^Tversion, a single thread, corresponding to the task's process, performs all operations: data block I/O and IC integration. There is a single queue containing the ICs that are owned by a task. A task will begin work on the IC at the head of the queue, load any data blocks needed for the IC's current location, and continue integration until the IC exits all data blocks currently in the task's cache. At that point, that IC goes to the tail of the queue, and processing resumes on the *IC* now at the head of the queue. This process repeats until all *IC*s have been integrated.

In contrast, the P^H implementation maintains *IC*s in two sets, or queues. *IC*s in the *active* set can be integrated using the blocks currently residing in the cache, while *inactive* *IC*s require access to data blocks not resident in the cache for computation to proceed. The P^H implementation uses two pools of threads for execution. In the first thread pool, I/O threads identify which data blocks need to be loaded to satisfy the needs of the *IC* calculation in the inactive set, and then, initiate I/O if there is room in the cache. After a block is loaded, the *IC*s waiting on it are migrated to the active set. In the second thread pool, a worker thread fetches *IC*s from the active set, performs integration on each one using the cached data blocks, and then retires them to the inactive set when the *IC* exits all currently loaded data block domains. *IC*s for which integration has been completed, are sent to a separate list and the algorithm terminates once both active and inactive sets are empty. Access to active and inactive sets, as well as the block cache, is synchronized through standard mutex and condition variable primitives.

The performance of the POS scheme depends primarily on the data loads and cache size N_{blocks}; if the cache is too small, blocks must be repeatedly brought in from external storage. Recent work has studied the effects of alternate strategies for caching data blocks in systems with an extended memory hierarchy [5]. The idea is to expand the data block cache into a two-level hierarchy that would include both primary memory, as discussed here, as well as node-local storage devices, such as a local hard drive or a local solid-state drive.

The P^H implementation has three main advantages over the P^T version: (1) the P^H version will have a larger shared cache and will perform less redundant I/O; (2) when a significant number of *IC*s are inside the same data block, each P^T task must load that block separately, whereas the P^H version will perform only one read and immediately share it among its threads; and, (3) since I/O and *IC* integration are performed by separate thread pools, those two operations can execute asynchronously and simultaneously—algorithm performance will likely improve as long as I/O and *IC* integration can proceed concurrently.

12.3.2.2 Parallelize Over Blocks

The parallelize-over-blocks (POB) algorithm, the subject of 6.3.3, assigns $1/n$ of the data blocks to each parallel task. Seed points for each *IC* are distributed to the tasks that own the data block containing the location of the seed. Tasks then perform numerical integration of each *IC*. When an *IC* exits a data block owned by a task, the *IC* is sent to the task that owns the data block where the *IC* will go next. (The processing details are explained in the original study [6].) Figure 12.9 shows an overview of both the P^T and P^H implementations for this algorithm.

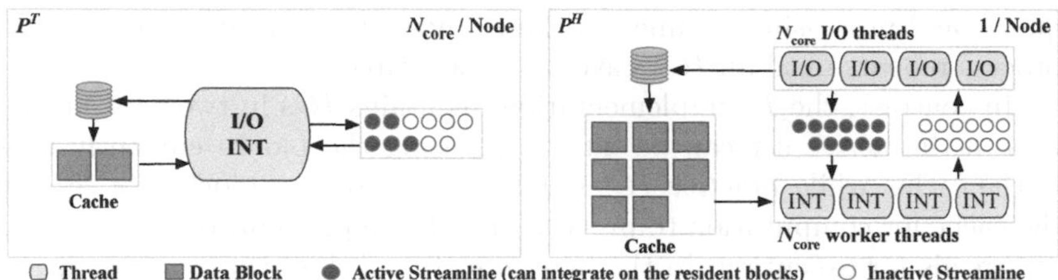

FIGURE 12.9: Comparison of P^T (left) and P^H (right) implementations of the *parallelize over blocks* algorithm. In the P^T version, each task performs the IC computation, an integration (INT), and communicates with other tasks to send and receive streamlines as they leave or enter different data partitions (COMM). In the P^H version, $N_{\text{core}}-1$ worker threads perform IC computation over a set of data blocks. There is one additional thread that performs both IC computation, as well as communication operations. I/O is limited to an initial load of data blocks assigned to each task. Image source: Camp et al., 2011 [6].

The POB algorithm performs minimal I/O: before integration commences, every task loads all blocks assigned to it, leveraging maximal parallel I/O bandwidth. The trade-off is a recurring inter-task communication phase, where ICs are exchanged between tasks as they migrate through the computational domain during integration.

As in POS algorithm, the P^T version consists of a single thread that maintains a set of ICs to integrate. Each thread performs integration of ICs that lie within the spatial region of data blocks in the cache.

The P^H implementation maintains ICs in two queues. Newly received ICs are stored in the *active* queue, where workers fetch from it, integrate, and put ICs in the *inactive* queue when they exit the data blocks loaded by this task. From there, the ICs are sent off to other tasks by the supervisor thread, or retired to a separate list when complete. All tasks' supervisor threads also maintain a global count of active and complete ICs to determine when to terminate. Synchronization between threads is performed as in the POS implementation.

12.3.3 Experiment Methodology

12.3.3.1 Factors Influencing Parallelization Strategy

The parallel IC problem is complex and challenging. To design an experimental methodology that provides robust coverage of different aspects of algorithmic behavior (some of which is data set dependent), one must take into account several factors that influence parallelization strategy in designing an effective and robust performance test design methodology. These factors, which are a way to classify particle-based problems, are discussed in detail earlier in 6.2.1.

12.3.3.2 Test Cases

To cover a wide range of potential problem characteristics, this study uses four tests that address all combinations of seed set size (small or large) and seed point distribution (sparse or dense) for each of the three data sets. Two of the data sets in the study are the same as those described earlier in Chapter 6 and 6.3.1. The third data set for this study comes from a computational thermal hydraulics application.

In the thermal hydraulics simulation, twin inlets pump water into a box, with a temperature difference between the water inserted by each inlet. Eventually, the water exits through an outlet. The mixing behavior and the temperature of the water at the outlet are of interest. Suboptimal mixing can be caused by long-lived recirculation zones that effectively isolate certain regions of the domain from heat exchange. The simulation was performed using the NEK5000 code [11] on an unstructured grid comprised of twenty-three million hexahedral elements. Streamlines are seeded according to two application scenarios. First, sparse seeds are distributed uniformly through the volume to show areas of high velocity, areas of stagnation, and areas of recirculation. Second, seeds are placed densely around one of the inlets to examine the behavior of particles entering through it. The resulting streamlines illustrate the turbulence in the immediate vicinity of the inlet. Small seed sets contain 1,500 seed points with an integration time of 12 units and the large sets consist of 6,000 seed points propagated for 3 time units. Figure 12.7 shows the streamlines computed for this data set.

12.3.3.3 Runtime Environment

The study conducted all tests on the NERSC Cray XT4 system *Franklin*. The 38,288 processor cores available for scientific applications are provided by 9,572 nodes, equipped with one quad-core AMD Opteron processor and 8GB of memory (2GB per core). Compute nodes are connected through HyperTransport for high performance, low-latency communication for MPI and SHMEM jobs. This platform supports parallel I/O, via a Lustre file system, which provides access to approximately 436TB of user disk space.

The authors implemented the two P^H algorithms into the VisIt [1, 8] visualization system, which is available on Franklin and routinely used by application scientists (see Chap. 16). P^T variants of both POB and POS algorithms were already implemented in recent VisIt releases and were also instrumented to provide the measurements discussed below.

The authors ran each benchmark run using 128 cores (32 nodes). For the P^T algorithm tests, they used 128 MPI tasks, one per core. The P^H test configuration was such that the total number of worker threads over all nodes was 128. One configuration, for example, used 32 MPI tasks, one per node, with each MPI task, in turn, spawning four worker threads. Note, the additional I/O or communication thread running per-MPI task in the P^H approach is not counted in this scheme; here, the study focuses on employing a constant number of worker threads for performing actual integration work to determine the impact of P^H.

12.3.3.4 Measurements

While absolute runtime is important, the authors added an instrumentation code so as to obtain more fine-grained information about different algorithmic steps. The steps include, for instance, the start and end of integration for a particular IC for worker threads, the begin and end of an I/O operation for corresponding threads, and time spent performing communication using MPI calls. Timestamps are taken as elapsed wall-clock time since the start of the MPI task. These event logs are carefully implemented to a have negligible impact on the overall runtime behavior.

The pure integration time, T_{int}, represents the actual computational workload and it is the sum of all times taken by the worker threads to perform the IC computation. This time should be almost independent across all runs for a specific test case, since the integration workload in terms of the number of integration steps taken over all ICs is identical in each case. Similarly, $T_{\text{I/O}}$ and T_{comm} contain time spent doing I/O and communication, respectively. To gain a deeper insight into the role of the block cache, the authors count the number of blocks loaded, N_{load}, and purged, N_{purged}. The amount of MPI communication between tasks in bytes is shown as N_{comm}. Finally, as a derived measure of efficiency, they examine the integration ratio, R_{int}, as the fraction of the total algorithm runtime that was used to compute ICs.

In total, twelve tests of the form X(YZ) were run per algorithm in hybrid and non-hybrid variants, where X indicates the data set (**A**stro, **T**hermal Hydraulics, **F**usion), Y denotes the seed set size (**L**arge or **S**mall), and Z denotes the seed set density (**S**parse or **D**ense).

12.3.4 Results

Overall, the results show that for each of the two parallelization approaches, P^H has better performance characteristics than its P^T counterpart. However, the reasons for the performance differences vary between the different parallelization approach.

12.3.4.1 Parallelization Over Seeds

A P^H POS algorithm has three fundamental advantages:

Larger data cache. Both the P^H and P^T algorithms use a memory-based cache to store data blocks/domains for later reuse to avoid performing redundant I/O. For the P^H case, the cache is shared amongst worker threads, hence, the worker threads can be larger than the P^T cache by an amount proportional

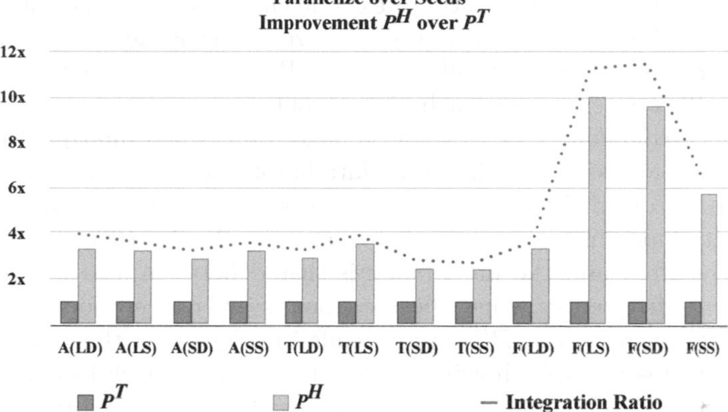

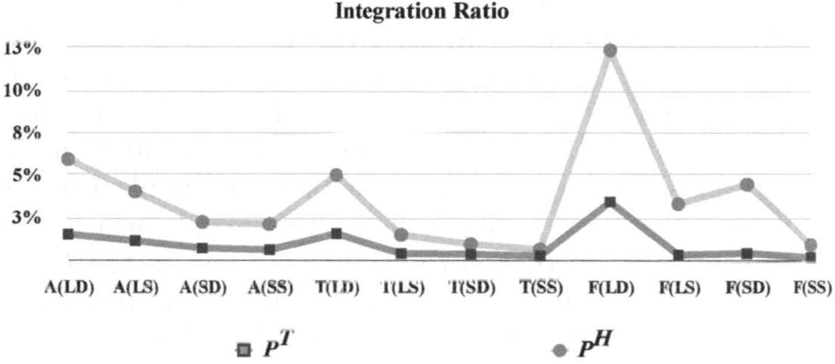

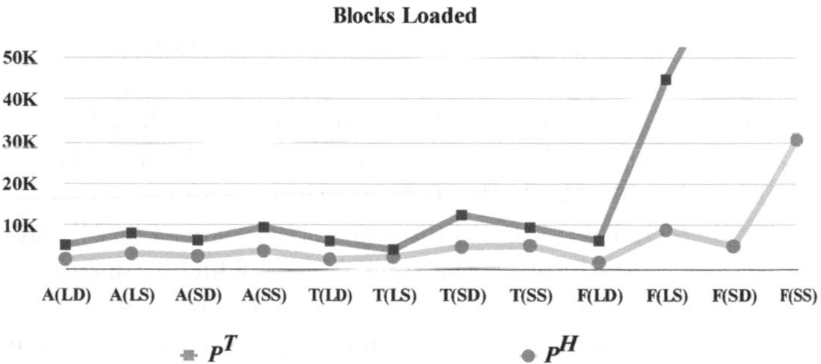

FIGURE 12.10: Comparison of $P^H$and P^Tperformance for the parallelize-over-seeds IC computation approach. the P^Hversion is faster in all cases because fewer blocks are loaded, allowing for an increased integration ratio, along with better load balancing due to there being more processors to work on a given number of ICs. (See 12.3.2.1 for a detailed discussion.) Image source: Camp et al., 2011 [6].

to the threading factor. For these experiments, the P^Hcache was four times larger than the P^Tcache, though cache size is a runtime, tunable parameter. Figure 12.10 shows that in all cases, the P^Hversion loads fewer data blocks than its P^Tcounterpart for each given problem.

Avoids redundant I/O. For the P^Talgorithm, when multiple tasks on the same node need to access the same data block, they must each read the same data block from disk. In effect, they are performing redundant loads of the same data block into the same node's memory. In the P^Hcase, only a single read is required and the data block is shared between all threads. This use case occurs frequently when the seed points are densely located in a small region. On average, the P^Hversion loads 64% less data than the P^Tversion. Figure 12.10 is a graph showing the number of data block loads.

Better load balancing. The time to calculate each IC varies from curve to curve because of the data dependent nature of the advection step. The study refers to ICs that take longer to compute as "slow ICs." Since ICs are permanently assigned to tasks, the tasks that process slow ICs will take longer to execute. Towards the end of the calculation, the tasks with slow ICs will still be executing, while the tasks with fast ICs will be finished, meaning there is a relatively poor load balance. Figure 12.11 shows that the P^Himplementation has less load imbalance between tasks. With the P^Himplementation, a larger number of ICs are shared between the worker threads, which creates a more even distribution of slow ICs. For example, if a task in the P^Tcase received many slow ICs, there is only one thread to handle them. In the P^Hcase, there will be more worker threads to advance these slow ICs.

12.3.4.2 Parallelization Over Blocks

A P^H POB algorithm has two fundamental advantages:

Better load balancing. Because data is partitioned over tasks, the only task that can advance a given IC is the task that owns the data block in which the IC resides. When many ICs traverse the same block, the corresponding task becomes a bottleneck. With the P^Halgorithm, more tasks can be used to relieve the bottleneck. Figure 12.13 illustrates this point. The P^Timplementation has only two tasks working on the longest ICs, which translates to two cores. The P^Himplementation also has two tasks working on these ICs, but since each task has four worker threads, that equates to a total of eight worker threads vs. two worker threads for the P^Timplementation. The additional workers allow the P^Hcase to finish more quickly. This factor can be quantified by measuring the percentage of time each core spends doing integration. The integration ratio R_{int} is very low for both implementations, but it is higher for the P^Himplementation.

Less communication. The communication cost of moving ICs between MPI processes is much lower in most of the P^Hcases of the POB method, which can be seen in the communication time and data transmitted between tasks (see Fig. 12.12). The P^HPOB implementation has four times fewer tasks than

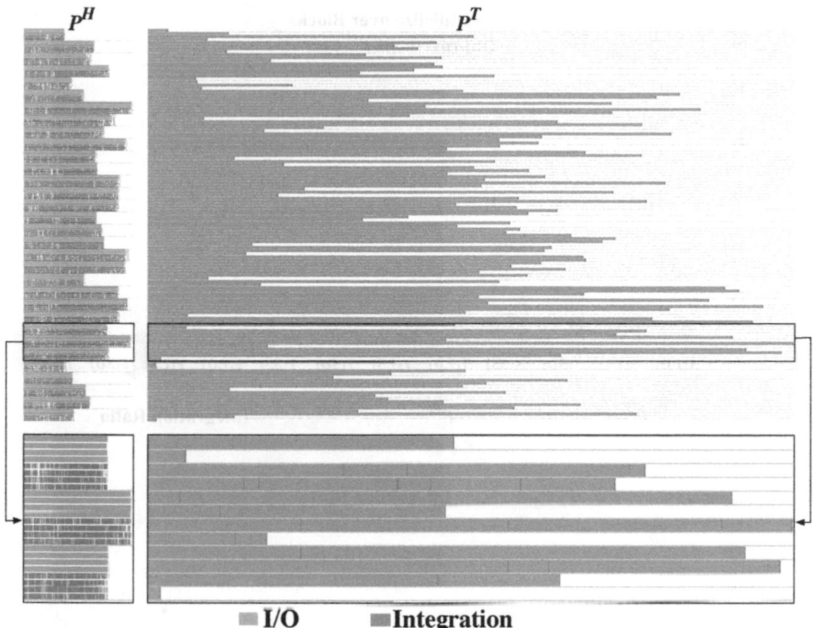

FIGURE 12.11: This Gantt chart shows a comparison of integration and I/O performance/activity of the *parallelize-over-seeds* P^T and P^H versions for one of the benchmark runs. Each line represents one thread (left column) or task (right column). The P^H approach outperforms the P^T one by about 10×, since the four I/O threads in the P^H can supply new data blocks to the four integration threads at an optimal rate. However, work distribution between nodes is not optimally balanced. In the P^T implementation, the I/O wait time dominates the computation by a large margin, due to redundant data block reads, and work being distributed less evenly. This can be easily seen in the enlarged section of the Gantt chart. (See 12.3.2.1 for more details.) Image source: Camp et al., 2011 [6].

P^T. Since each P^H task holds four times more data, it is less likely to send the IC to another task, which results in less overall inter-task communication.

12.4 Conclusion and Future Work

The two studies presented in this chapter, both of which compare the performance of traditional MPI-only with MPI-hybrid implementations of two staple visualization algorithms, reveal a consistent theme: on current multi- and many-core platforms, the MPI-hybrid design and implementation enjoys clear performance advantages.

First, both MPI-hybrid implementations require a significantly smaller

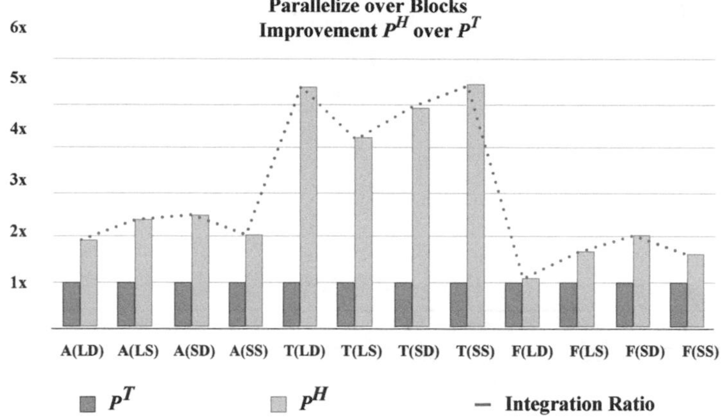

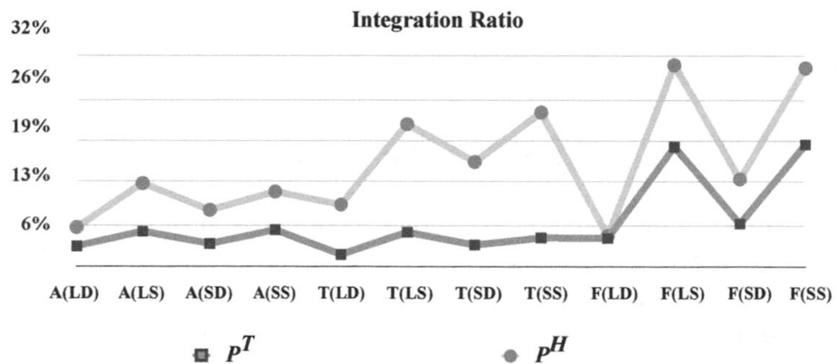

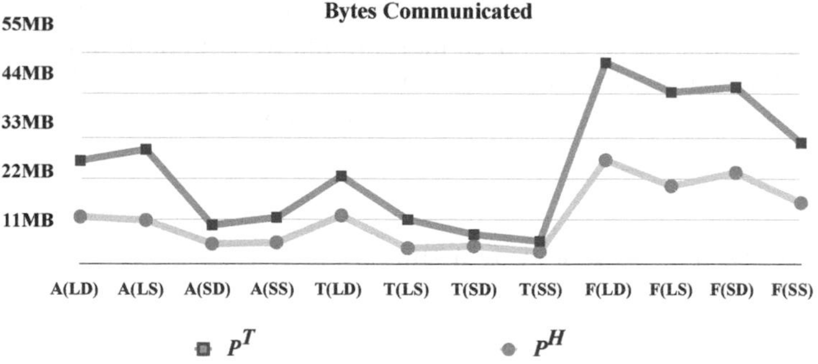

FIGURE 12.12: Performance comparison of the P^H and P^T variants of the *parallelize-over-blocks* algorithm. The P^H version has a much lower runtime for each test due to better load balancing, and because less communication enables a higher integration ratio. (See 12.3.2.2 for a detailed discussion.) Image source: Camp et al., 2011 [6].

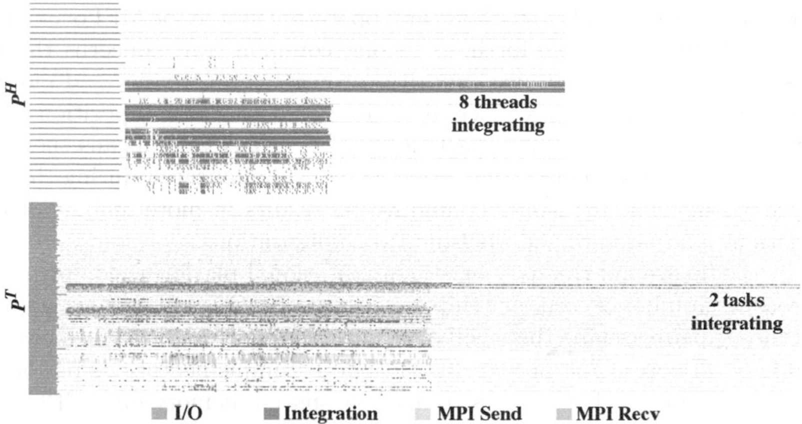

FIGURE 12.13: This Gantt chart shows a comparison of integration, I/O, `MPI_Send`, and `MPI_Recv` performance/activity of the *parallelize-over-blocks* P^T and P^H versions for one of the benchmark runs. Each line represents one thread (top) or task (bottom). The comparison reveals that the initial I/O phase, using only one thread, takes about 4× longer. The successive integration is faster, since multiple threads can work on the same set of blocks, leading to less communication. Towards the end, the eight threads are performing *IC* integration in the P^H approach, as opposed to only two tasks in the P^T model. (See 12.3.2.2 for more details.) Image source: Camp et al., 2011 [6]

memory footprint at initialization. Per-chip MPI overhead, in the form of internal buffers and so forth, expands as a function of the number of per-core MPI tasks, and, as the volume rendering study shows, a difficult-to-predict scaling factor that is likely related to how vendors implement MPI. Extrapolating from the volume rendering study results shown here, it seems clear that continuing along the path of an MPI-only implementation will eventually exhaust all available memory at extreme concurrency simply for the application initialization. The implication is clear: future high performance visual data analysis and exploration applications must, by necessity, use some form of hybrid-parallel design to make effective use of computational platforms, comprised of multiple or many cores per chip.

Second, the MPI-hybrid implementations perform significantly less inter-chip communication. The reasons for the lighter communication load vary, depending on the algorithm, and in some cases the particular problem or data set. Nonetheless, one common trait that results in less communication is the fact that the hybrid-parallel implementations use multiple processing threads, typically one per core, that all have access to a common shared memory. Communication between processing threads on a CPU that has global shared memory visible to all its cores takes place through that memory. And, depending on the underlying shared-memory programming model, that communication can take place without any operating system or MPI overhead like message buffering and so forth. A second common trait concerns the use of ghost zones. With a data decomposition consisting of smaller and more data partitions, there is more surface area compared to a decomposition that results in fewer and larger data partitions. Less surface area means there is less information—ghost zones—that needs to be communicated during the course of processing. The MPI-only configuration results in more and smaller data partitions compared to the MPI-hybrid configurations.

In the future, all trends suggest computational platforms comprised of an increasing number of cores per chip (see Chap. 15). One unknown is whether or not those future architectures will continue to support a shared memory that is visible to all cores. The present hybrid-parallel implementations perform well because all threads have access to a single shared memory on a CPU chip. If future architectures eliminate this shared memory, future research will need to explore alternative algorithmic formulations and implementations that both exploit available architectural traits as well as achieve low memory footprint utilization and reduced communication when compared to traditional, MPI-only designs and implementations.

References

[1] VisIt – Software that delivers Parallel, Interactive Visualization. http://visit.llnl.gov/.

[2] C. Bajaj, I. Ihm, G. Joo, and S. Park. Parallel Ray Casting of Visibly Human on Distributed Memory Architectures. In *VisSym '99 Joint EUROGRAPHICS-IEEE TVCG Symposium on Visualization*, pages 269–276, 1999.

[3] OpenMP Architecture Review Board. OpenMP Application Program Interface Version 3.1, July 2011. http://www.openmp.org/wp/openmp-specifications.

[4] David R. Butenhof. *Programming with POSIX threads*. Addison-Wesley Longman Publishing Co., Inc., Boston, MA, USA, 1997.

[5] David Camp, Hank Childs, Amit Chourasia, Christoph Garth, and Kenneth I. Joy. Evaluating the Benefits of an Extended Memory Hierarchy for Parallel Streamline Algorithms. In *Proceedings of the IEEE Symposium on Large-Scale Data Analysis and Visualization (LDAV)*, Providence, RI, USA, October 2011.

[6] David Camp, Christoph Garth, Hank Childs, Dave Pugmire, and Ken Joy. Streamline Integration Using MPI-Hybrid Parallelism on a Large Multi-Core Architecture. *IEEE Transactions on Visualization and Computer Graphics*, 17(11):1702–1713, November 2011.

[7] Robit Chandra, Leonardo Dagum, Dave Kohr, Dror Maydan, Jeff McDonald, and Ramesh Menon. *Parallel Programming in OpenMP*. Morgan Kaufmann Publishers Inc., San Francisco, CA, USA, 2001.

[8] Hank Childs, Eric S. Brugger, Kathleen S. Bonnell, Jeremy S. Meredith, Mark Miller, Brad J. Whitlock, and Nelson Max. A Contract-Based System for Large Data Visualization. In *Proceedings of IEEE Visualization*, pages 190–198, 2005.

[9] NVIDIA Corporation. What is CUDA? http://www.nvidia.com/object/what_is_cuda_new.html, 2011.

[10] Robert A. Drebin, Loren Carpenter, and Pat Hanrahan. Volume rendering. *SIGGRAPH Computer Graphics*, 22(4):65–74, 1988.

[11] P. Fischer, J. Lottes, D. Pointer, and A. Siegel. Petascale Algorithms for Reactor Hydrodynamics. *Journal of Physics: Conference Series*, 125:1–5, 2008.

[12] Khronos Group. OpenCL – The Open Standard for Parallel Programming of Heterogeneous Systems. `http://www.khronos.org/opencl/`, 2011.

[13] Mark Howison, E. Wes Bethel, and Hank Childs. MPI-hybrid Parallelism for Volume Rendering on Large, Multi-core Systems. In *Eurographics Symposium on Parallel Graphics and Visualization (EGPGV)*, Norrköping, Sweden, May 2010. LBNL-3297E.

[14] Mark Howison, E. Wes Bethel, and Hank Childs. Hybrid Parallelism for Volume Rendering on Large, Multi- and Many-core Systems. *IEEE Transactions on Visualization and Computer Graphics*, 99(PrePrints), 2011.

[15] Marc Levoy. Display of Surfaces from Volume Data. *IEEE Computer Graphics and Applications*, 8(3):29–37, May 1988.

[16] Kwan-Liu Ma. Parallel Volume Ray-Casting for Unstructured-Grid Data on Distributed-Memory Architectures. In *PRS '95: Proceedings of the IEEE Symposium on Parallel Rendering*, pages 23–30, New York, NY, USA, 1995. ACM.

[17] Kwan-Liu Ma, James S. Painter, Charles D. Hansen, and Michael F. Krogh. A Data Distributed, Parallel Algorithm for Ray-Traced Volume Rendering. In *Proceedings of the 1993 Parallel Rendering Symposium*, pages 15–22. ACM Press, October 1993.

[18] Jason Nieh and Marc Levoy. Volume Rendering on Scalable Shared-Memory MIMD Architectures. In *Proceedings of the 1992 Workshop on Volume Visualization*, pages 17–24. ACM SIGGRAPH, October 1992.

[19] NVIDIA Corporation. *NVIDIA CUDA^{TM} Programming Guide Version 3.0*, 2010. `http://developer.nvidia.com/object/cuda_3_0_downloads.html`.

[20] Thomas Porter and Tom Duff. Compositing Digital Images. *Computer Graphics*, 18(3):253–259, 1984. Proceedings of ACM/Siggraph.

[21] Dave Pugmire, Hank Childs, Christoph Garth, Sean Ahern, and Gunther H. Weber. Scalable Computation of Streamlines on Very Large Datasets. In *Proceedings of Supercomputing (SC09)*, Portland, OR, USA, November 2009.

[22] Paolo Sabella. A Rendering Algorithm for Visualizing 3D Scalar Fields. *SIGGRAPH Computer Graphics*, 22(4):51–58, 1988.

[23] Marc Snir, Steve Otto, Steven Huss-Lederman, David Walker, and Jack Dongarra. *MPI – The Complete Reference: The MPI Core, 2nd ed.* MIT Press, Cambridge, MA, USA, 1998.

[24] R. Tiwari and T. L. Huntsberger. A Distributed Memory Algorithm for Volume Rendering. In *Scalable High Performance Computing Conference*, Knoxville, TN, USA, May 1994.

[25] The Top 500 Supercomputers, 2011. `http://www.top500.org`.

[26] Craig Upson and Michael Keeler. V-buffer: Visible Volume Rendering. In *SIGGRAPH '88: Proceedings of the 15th Annual Conference on Computer Graphics and Interactive Techniques*, pages 59–64, New York, NY, USA, 1988. ACM.

Chapter 13

Visualization at Extreme Scale Concurrency

Hank Childs
Lawrence Berkeley National Laboratory

David Pugmire
Oak Ridge National Laboratory

Sean Ahern
Oak Ridge National Laboratory

Brad Whitlock
Lawrence Livermore National Laboratory

Mark Howison
Brown University & Lawrence Berkeley National Laboratory

Prabhat
Lawrence Berkeley National Laboratory

Gunther Weber
Lawrence Berkeley National Laboratory

E. Wes Bethel
Lawrence Berkeley National Laboratory

13.1 Overview—Pure Parallelism 292
13.2 Massive Data Experiments 293
 13.2.1 Varying over Supercomputing Environment 296
 13.2.2 Varying over I/O Pattern 297
 13.2.3 Varying over Data Generation 298
13.3 Scaling Experiments 299
 13.3.1 Study Overview 299
 13.3.2 Results 300
13.4 Pitfalls at Scale 300
 13.4.1 Volume Rendering 301

13.4.2 All-to-One Communication 303
13.4.3 Shared Libraries and Start-up Time 304
13.5 Conclusion 305
References 306

The chapters from Part II describe techniques for processing massive data sets while minimizing computation, memory footprint, and/or I/O. But these techniques' benefits come at the cost of increased complexity, especially when compared with the "pure parallelism" technique described in Chapter 2. This chapter contributes to the motivation for these more complex techniques, by asking several related questions: Will it be possible to use the simpler pure parallelism technique to process tomorrow's data? Can pure parallelism scale sufficiently to process massive data sets? And, restated, are the techniques described in Part II needed at all?

To answer these questions, the researchers performed a series of experiments, originally published in *IEEE Computer Graphics and Applications* [1] and forming the basis of this chapter, that studied the scalability of pure parallelism in visualization software on massive data sets. These experiments utilized multiple visualization algorithms and were run on multiple architectures. There were two types of experiments performed. The first experiment examined performance at a massive scale: 16,000 or more cores and one trillion or more cells. The second experiment studied whether the approach can maintain a fixed amount of time to complete an operation when the data size is doubled and the amount of resources is doubled, also known as weak scalability. At the time of their original publication, these experiments represented the largest data set sizes ever published in visualization literature. Further, their findings still continue to contribute to the understanding of today's dominant processing paradigm (pure parallelism) on tomorrow's data, in the form of scaling characteristics and bottlenecks at high levels of concurrency and with very large data sets.

13.1 Overview—Pure Parallelism

Pure parallelism, introduced in Chapter 2, is the brute force approach: data parallelism with no optimizations to reduce the amount of data read. In this paradigm, the simulation writes data to a disk and the visualization software reads this data at full-resolution, storing it in primary memory. To deal with large data, parallel processing is used. The visualization software partitions data over its tasks, with each task working on a piece of the larger problem. Through parallelization, the visualization software has access to more I/O bandwidth (to load data faster), more memory (to store more data), and more compute power (to execute its algorithms more quickly).

The majority of visualization software for large data, including much of the production visualization software that serves large user communities, utilizes the pure parallelism paradigm. Some examples of tools that rely heavily on this processing technique include, VisIt (discussed in Chap. 16), ParaView (Chap. 18), and EnSight (Chap. 21).

The study this chapter describes sought to better understand how pure parallelism will perform on more and more cores, with larger and larger data sets. How does this technique scale? What bottlenecks are encountered? What pitfalls are encountered with running production software at a massive scale? In short, will pure parallelism be effective for the next generation of data sets? As you know from reading the chapters in Part II, these questions are especially important because pure parallelism is not the only data processing paradigm. Where pure parallelism is heavily dependent on I/O bandwidth and large memory footprints, alternatives de-emphasize these traits.

The principal finding of this study was that pure parallelism at extreme scale works, that algorithms such as contouring and rendering performed well, and that I/O times for massive data dominated execution time. These I/O bottlenecks persisted over many supercomputers and also over I/O pattern (collective and noncollective I/O). These findings are discussed in 13.2. Another important finding was a validation of the weak scaling of pure parallelism when processing data sizes within the supercomputer's I/O bandwidth capabilities, and is described in 13.3. Finally, the study itself encountered common pitfalls at high concurrency that are the subject of 13.4.

13.2 Massive Data Experiments

The basic experiment for massive data at high levels of concurrency had a parallel program read in a data set with trillions of cells, apply a contouring algorithm ("Marching Cubes" [2]), and render the resulting surface as a 1024×1024 pixel image (see Fig. 13.1).

The study originally set out to perform volume rendering as well, but encountered difficulties (see 13.4 on Pitfalls). An unfortunate reality in experiments of this nature is that running large jobs on the largest supercomputers in the world is a difficult and opportunistic undertaking. Where the initial set of experiments demonstrated the problem, it was not possible for the authors to re-run data on these machines, after improvements were made to the volume rendering code. Further, these runs were undoubtedly affected by real-world issues, like I/O and network contention. That said, the study still had great value since isocontouring is representative of the typical visualization operations: loading data, applying an algorithm, and rendering.

The variations of this experiment fell into three categories:

- *Diverse supercomputing environments*, to test the viability of these techniques with different operating systems, I/O behavior, compute power (e.g., FLOPs), and network characteristics. These tests were performed

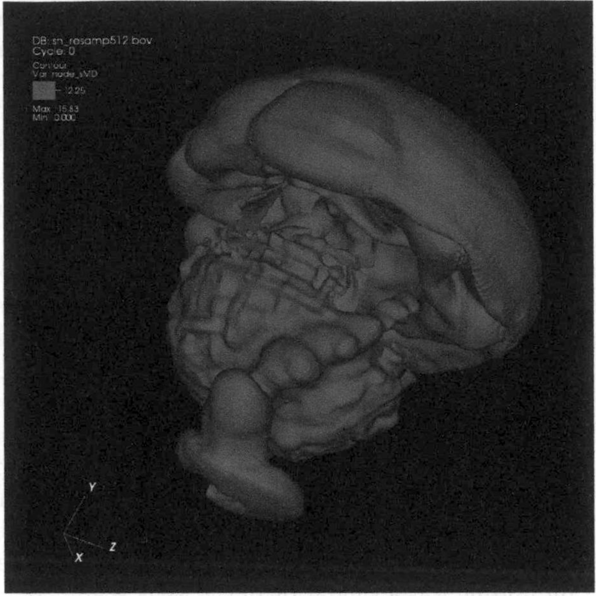

FIGURE 13.1: Contouring of two trillion cells, visualized with VisIt on Franklin using 32,000 cores. Image source: Childs et al., 2010 [1].

on two Cray XT machines (Oak Ridge National Laboratory's JaguarPF and Lawrence Berkeley National Laboratory's Franklin), a Sun Linux machine (the Texas Advanced Computing Center's Ranger), a CHAOS Linux machine (Lawrence Livermore National Laboratory's Juno), an AIX machine (LLNL's Purple), and a BlueGene/P machine (LLNL's Dawn). For five of the six machines, the experiment consisted of 16,000 cores and visualizing one trillion cells. Runs on the Purple machine were limited to 8,000 cores and one half trillion cells, because the full machine has only 12,208 cores, and only 8,000 were easily obtainable for large jobs. For the machines with more than 16,000 cores available, like JaguarPF and Franklin, additional tests were added to perform a weak scaling study, maintaining a ratio of one trillion cells for every 16,000 cores. More information about the machines can be found in Table 13.1.

- *I/O pattern*, to understand the impact of collective and noncollective communication patterns at scale. Collective communication refers to an activity where there is coordination between the tasks; noncollective communication requires no coordination. For the noncollective tests, the data was stored as compressed binary data (gzipped). The study used ten files for every task and every file contained 6.25 million data points, for a total of 62.5 million data points per task. The study aimed to approximate real-world conditions, as simulation codes often write out

Machine Name	Machine Type/OS	System Type	Top 500 Rank (as of 11/2009)
JaguarPF	Cray	XT5	#1
Ranger	Sun Linux	Opteron Quad	#9
Dawn	BG/P	PowerPC	#11
Franklin	Cray	XT4	#15
Juno	Commodity (Linux)	Opteron Quad	#27
Purple	AIX	POWER5	#66

Machine Name	Total # of Cores	Memory Per Core	Clock Speed	Peak FLOPS
JaguarPF	224,162	2GB	2.6GHz	2.33PFLOPs
Ranger	62,976	2GB	2.0GHz	503.8 TFLOPs
Dawn	147,456	1GB	850MHz	415.7 TFLOPs
Franklin	38,128	1GB	2.6GHz	352 TFLOPs
Juno	18402	2GB	2.2GHz	131.6 TFLOPs
Purple	12,208	3.5GB	1.9GHz	92.8 TFLOPs

TABLE 13.1: Characteristics of supercomputers used in a trillion cell performance study.

one file per task and visualization codes receive, at most, one-tenth of the tasks of the simulation code. Of course, reading many small files is not optimal for I/O access, so the study also considered a separate test where all tasks use collective access on a single, large file via MPI-IO.

- *Data generation.* No simulations produced meshes with trillions of cells at the time of the study, so the experimenters created synthetic data. The primary mechanism for generating this data was to upsample data by interpolating a scalar field from a smaller mesh onto a high resolution rectilinear mesh. However, to offset concerns that upsampled data may be unrepresentatively smooth, the study included a second experiment where the large data set was a many times over replication of a small data set. The data set came from a core-collapse supernova simulation, using the CHIMERA code on a curvilinear mesh of more than three and one half million cells.[1] The use of synthetic data, while not ideal, was not a large concern for the experiment, since it was sufficient for meeting the study's primary objective: to better understand the performance and functional limits of parallel visualization software.

[1]Sample data came courtesy of Tony Mezzacappa and Bronson Messer (ORNL), Steve Bruenn (Florida Atlantic University) and Reuben Budjiara (University of Tennessee).

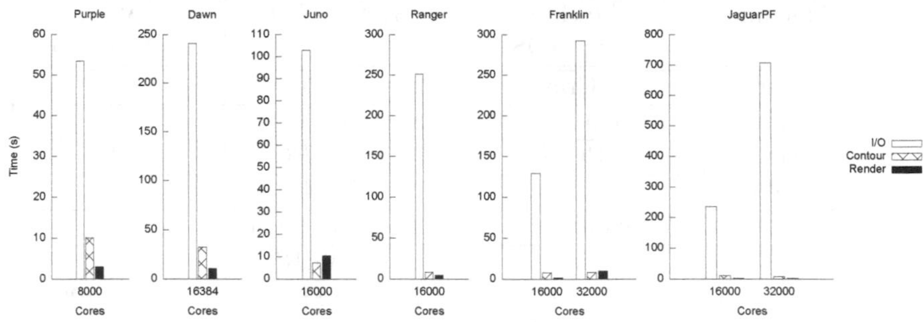

FIGURE 13.2: Plots of execution time for the I/O, contouring, and rendering phases of the trillion cell visualizations over six supercomputing environments. I/O was by far the slowest portion. Image source: Childs et al., 2010 [1].

13.2.1 Varying over Supercomputing Environment

The first variant of the experiment was designed to understand differences from supercomputing environment. The experiment consisted of running an identical problem on multiple platforms, keeping the I/O pattern and data generation fixed, and using noncollective I/O and upsampled data generation. Results can be found in Figure 13.2 and Table 13.2.

Machine	Cores	Data set size	I/O	Contour	TPE	Render
Purple	8000	0.5 TCells	53.4s	10.0s	63.7s	2.9s
Dawn	16384	1 TCells	240.9s	32.4s	277.6s	10.6s
Juno	16000	1 TCells	102.9s	7.2s	110.4s	10.4s
Ranger	16000	1 TCells	251.2s	8.3s	259.7s	4.4s
Franklin	16000	1 TCells	129.3s	7.9s	137.3s	1.6s
JaguarPF	16000	1 TCells	236.1s	10.4s	246.7s	1.5s
Franklin	32000	2 TCells	292.4s	8.0s	300.6s	9.7s
JaguarPF	32000	2 TCells	707.2s	7.7s	715.2s	1.5s

TABLE 13.2: Performance across diverse architectures. "TPE" is short for total pipeline execution (the amount of time to generate the surface). Dawn's number of cores is different from the rest since that machine requires all jobs to have core counts that are a power of two.

There were several noteworthy observations:

- I/O striping refers to transparently distributing data over multiple disks to make them appear as a single fast, large disk; careful consideration of the striping parameters was necessary for optimal I/O performance on Lustre filesystems (Franklin, JaguarPF, Ranger, Juno, & Dawn). Even though JaguarPF had more I/O resources than Franklin, its I/O performance did not perform as well in these experiments, because its default stripe count was four. In contrast, Franklin's default stripe count of two was better suited for the I/O pattern which read ten separate compressed files per task. Smaller stripe counts often benefit file-per-task I/O because the files were usually small enough (tens of MB) that they would not contain many stripes, and spreading them thinly over many I/O servers increases contention.

- Because the data was stored on disk in a compressed format, there was an unequal I/O load across the tasks. The reported I/O times measure the elapsed time between a file open and a barrier, after all the tasks were finished reading. Because of this load imbalance, I/O time did not scale linearly from 16,000 to 32,000 cores on Franklin and JaguarPF.
- The Dawn machine has the slowest clock speed (850MHz), which was reflected in its contouring and rendering times.
- Some variation in the observations could not be explained by slow clock speeds, interconnects, or I/O servers:
  - For Franklin's increase in rendering time from 16,000 to 32,000 cores, seven to ten network links failed that day and had to be statically re-routed, resulting in suboptimal network performance. Rendering algorithms are "all reduce" type operations that are very sensitive to bisection bandwidth, which was affected by this issue.
  - The experimenters concluded Juno's slow rendering time was similarly due to a network problem.

13.2.2 Varying over I/O Pattern

This variant was designed to understand the effects of different I/O patterns. It compared collective and noncollective I/O patterns on Franklin for a one trillion cell upsampled data set. In the noncollective test, each task performed ten pairs of `fopen` and `fread` calls on independent gzipped files without any coordination among tasks. In the collective test, all tasks synchronously called `MPI_File_open` once, then called `MPI_File_read_at_all` ten times on a shared file (each read call corresponded to a different piece of the data set). An underlying collective buffering, or "two phase" algorithm, in Cray's MPI-IO implementation aggregated read requests onto a subset of 48 nodes (matching the 48 stripe count of the file) that coordinated the low-level I/O workload, dividing it into 4MB stripe-aligned `fread` calls. As the 48 aggregator nodes filled their read buffers, they shipped the data using message passing to their final destination among the 16,016 tasks. A different number of tasks was used for each scheme (16,000 versus 16,016), because the collective communication scheme could not use an arbitrary number of tasks; the closest value to 16,000 possible was picked. Performance results are listed in Table 13.3.

I/O pattern	Cores	Total I/O time	Data read	Read bandwidth
Collective	16016	478.3s	3725.3GB	7.8GB/s
Noncollective	16000	129.3s	954.2GB	7.4GB/s

TABLE 13.3: Performance with different I/O patterns. The bandwidth for the two approaches are very similar. The data set size for collective I/O corresponds to four bytes for each of the one trillion cells. The data read is less than 4000GB because, 1GB is 1,073,741,824 bytes. The data set size for noncollective I/O is much smaller because it was compressed.

Both patterns led to similar read bandwidths, 7.4 and 7.8GB/s, which are about 60% of the maximum available bandwidth of 12GB/s on Franklin. In the noncollective case, load imbalances, caused by different compression factors, may account for this discrepancy. For the collective I/O, coordination overhead between the MPI tasks may be limiting efficiency. Of course, the processing would still be I/O dominated, even if perfect efficiency was achieved.

13.2.3 Varying over Data Generation

This variant was designed to understand the effects of source data. It compared upsampled and replicated data sets, with each test processing one trillion cells on 16,016 cores of Franklin using collective I/O. Performance results are listed in Table 13.4.

Data generation	Total I/O time	Contour time	TPE	Rendering time
Upsampling	478.3s	7.6s	486.0s	2.8s
Replicated	493.0s	7.6s	500.7s	4.9s

TABLE 13.4: Performance across different data generation methods. "TPE" is short for total pipeline execution (the amount of time to generate the surface).

The contouring times were nearly identical, likely since this operation is dominated by the movement of data through the memory hierarchy (L2 cache to L1 cache to registers), rather than the relatively rare case where a cell contains a contribution to the isosurface. The rendering time, which is proportional to the number of triangles in the isosurface, nearly doubled, because the isocontouring algorithm run on the replicated data set produced twice as many triangles.

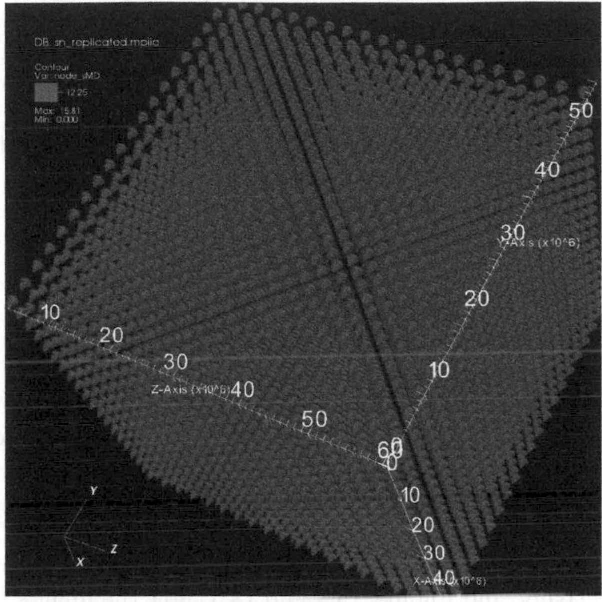

FIGURE 13.3: Contouring of replicated data (one trillion cells total), visualized with VisIt on Franklin using 16,016 cores. Image source: Childs et al., 2010 [1].

13.3 Scaling Experiments

Where the first part of the experiment [1] informed performance bottlenecks of pure parallelism at an extreme scale, the second part sought to assess its weak scaling properties for both isosurface generation and volume rendering. Once again, these algorithms exercise a large portion of the underlying pure parallelism infrastructure and indicates a strong likelihood of weak scaling for other algorithms in this setting. Further, demonstrating weak scaling properties on high performance computing systems met the accepted standards of "Joule certification," which is a program within the U.S. Office of Management and Budget to confirm that supercomputers are being used efficiently.

13.3.1 Study Overview

The weak scaling studies were performed on an output from Denovo, which is a 3D radiation transport code from ORNL that models radiation dose levels in a nuclear reactor core and its surrounding areas. The Denovo simulation code does not directly output a scalar field representing effective dose. Instead, this dose is calculated at runtime through a linear combination of 27 scalar fluxes. For both the isosurface and volume rendering tests, VisIt read in 27 scalar fluxes and combined them to form a single scalar field representing radiation dose levels. The isosurface extraction test consisted of extracting six evenly spaced isocontour values of the radiation dose levels and rendering an 1024×1024 pixel image. The volume rendering test consisted of ray casting with 1000, 2000 and 4000 samples per ray of the radiation dose level on a 1024×1024 pixel image.

These visualization algorithms were run on a baseline Denovo simulation consisting of 103,716,288 cells on 4,096 spatial domains, with a total size on disk of 83.5GB. The second test was run on a Denovo simulation nearly three times the size of the baseline run, with 321,117,360 cells on 12,720 spatial domains and a total size on disk of 258.4GB. These core counts are large relative to the problem size and were chosen because they represent the number of cores used by Denovo. This matching core count was important for the Joule study and is also indicative of performance for an *in situ* approach.

13.3.2 Results

Tables 13.5 and 13.6 show the performance for contouring and volume rendering respectively, and Figures 13.4 and 13.5 show the images they produced. The time to perform each phase was nearly identical over the two concurrency levels, which suggests the code has favorable weak scaling characteristics. Note that I/O was not included in these tests.

Algorithm	Cores	Minimum Time	Maximum Time	Average Time
Calculate radiation	4,096	0.18s	0.25s	0.21s
Calculate radiation	12,270	0.19s	0.25s	0.22s
Isosurface	4,096	0.014s	0.027s	0.018s
Isosurface	12,270	0.014s	0.027s	0.017s
Render (on task)	4,096	0.020s	0.065s	0.0225s
Render (on task)	12,270	0.021s	0.069s	0.023s
Render (across tasks)	4,096	0.048s	0.087s	0.052s
Render (across tasks)	12,270	0.050s	0.091s	0.053s

TABLE 13.5: Weak scaling study of isosurfacing. **Isosurface** refers to the execution time of the isosurface algorithm, **Render (on task)** indicates the time to render that task's surface, while **Render (across tasks)** indicates the time to combine that image with the images of other tasks. **Calculate radiation** refers to the time to calculate the linear combination of the 27 scalar fluxes.

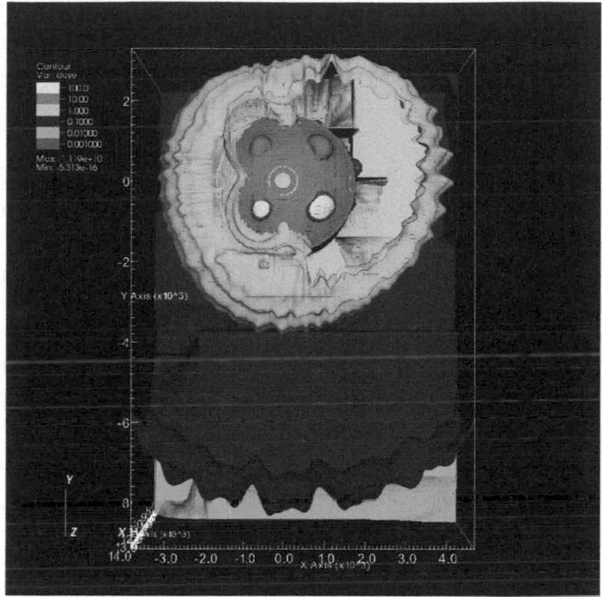

FIGURE 13.4: Rendering of an isosurface from a 321 million cell Denovo simulation, produced by VisIt using 12,270 cores of JaguarPF. Image source: Childs et al., 2010 [1].

13.4 Pitfalls at Scale

Algorithms that work well on the order of hundreds of tasks, can become extremely slow with tens of thousands of tasks. The common theme of this section is how implementations that were appropriate at modest scales became unusable at extreme scales. The problematic code existed at various levels of the software, from core algorithms (volume rendering), to code that supported the algorithms (status updates), to foundational code (plug-in loading).

13.4.1 Volume Rendering

VisIt's volume rendering code uses an all-to-all communication phase to redistribute samples along rays according to a partition with dynamic assignments. An "optimization" for this phase was to minimize the number of samples that needed to be communicated by favoring assignments that kept samples on their originating task. This "optimization" required an $O(n^2)$ buffer that contained mostly zeroes. Although this "optimization" was, indeed, effective for small task counts, the coordination overhead caused VisIt to run out of memory at this scale. Removing the optimization—by simply assigning pixels to tasks without concern of where individual samples lay—significantly improved performance. The authors concluded that, as the number of samples

Cores	Samples Per Ray: 1000	2000	4000
4,096	7.21s	4.56s	7.54s
12,270	6.53s	6.60s	6.85s

TABLE 13.6: Weak scaling study of volume rendering. 1000, 2000, and 4000 represent the number of samples per ray. The algorithm demonstrates super-linear performance, because the number of samples per task (which directly affects work performed) is smaller at 12,270 task, while the number of cells per task is constant. The anomaly where performance increases at 2000 samples per ray requires further study. The times for each operation are similar at the two concurrency levels, showing favorable weak scaling characteristics.

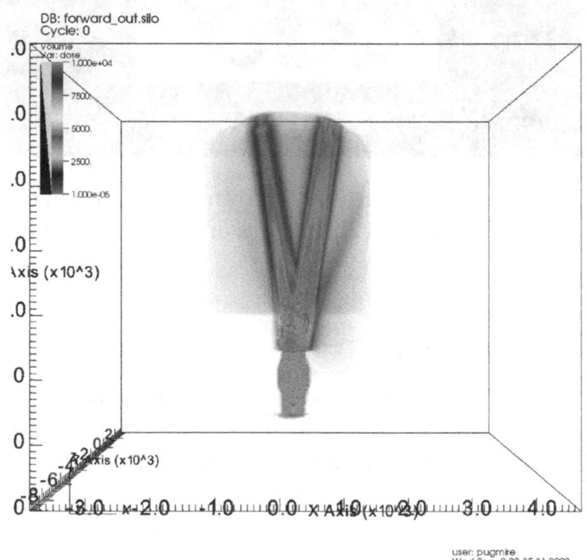

FIGURE 13.5: Volume rendering of data from a 321 million cell Denovo simulation, produced by VisIt using 12,270 cores on JaguarPF. Image source: Childs et al., 2010 [1].

gets smaller with larger task counts, coordination costs outweigh the benefits that might come from keeping samples on the task where it was calculated.

The authors did not produce a comprehensive performance study for the one trillion cell data sets. However, they observed that after removing the coordination costs, ray casting performance was approximately five seconds per frame for a 1024×1024 image. The resulting image is shown in Figure 13.6.

After implementing this improvement, the authors were able to re-run the

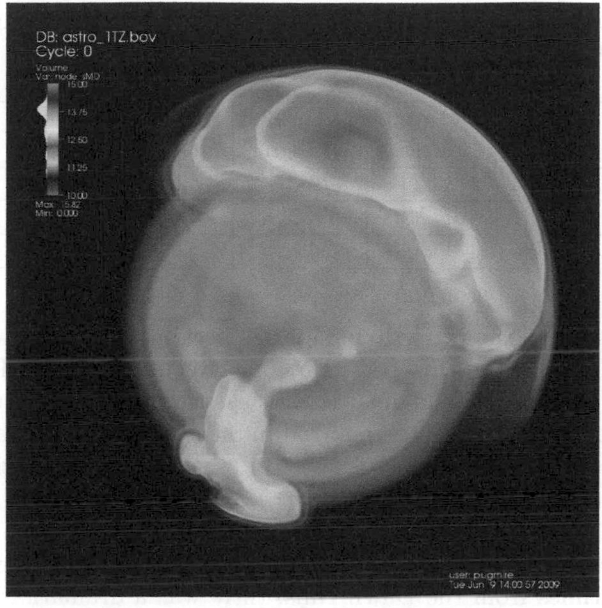

FIGURE 13.6: Volume rendering of one trillion cells, visualized by VisIt on JaguarPF using 16,000 cores. Image source: Childs et al., 2010 [1].

experiment on the Denovo data, however. They saw an approximately 5× speedup running with 4,096 cores (see Table 13.7).

Cores	Date run	Samples Per Ray: 1000	2000	4000
4,096	Spring 2009	34.7s	29.0s	31.5s
4,096	Summer 2009	7.21s	4.56s	7.54s

TABLE 13.7: Volume rendering of Denovo data at 4,096 cores before and after speedup.

13.4.2 All-to-One Communication

Upon completing a pipeline execution, each task reports its status (success or failure), as well as some metadata (extents, etc). These statuses and extents were being communicated from each task to task #0, through point-to-point communication. However, having every task send a message to task #0 led to a significant delay, as shown in Table 13.8. This problem was corrected subsequently with a tree communication scheme. Table 13.8 shows the extent of the delay using experiments run on LLNL's Dawn machine in June 2009 (using the old all-to-one status scheme) and August 2009 (using the new tree communication scheme).

All-to-one status	Cores	Data set size	Total I/O time	Contour time	TPE	TPE minus contour & I/O
yes	16384	1 TCells	88.0s	32.2s	368.7s	248.5s
yes	65536	4 TCells	95.3s	38.6s	425.9s	294.0s
no	16384	1 TCells	240.9s	32.4s	277.6s	4.3s

TABLE 13.8: Performance with old status checking code vs. new status checking code. "TPE" stands for Total Pipeline Execution. The "TPE minus contour & I/O" time approximates the time spent waiting for status and extents updates. Note, the other runs reported in this experiment had a status checking code disabled and the last Dawn run was the only reported run with a new status code.

Another "pitfall" was the difficulty in obtaining consistent results. Looking at the I/O times from the Dawn runs, there was a dramatic slowdown from June to August. This is because, in July, the I/O servers backing the file system became unbalanced in their disk usage. This caused the algorithm, that assigns files to servers, to switch from a "round robin" scheme to a statistical scheme, meaning files were no longer assigned uniformly across I/O servers. This scheme makes sense from an operating system perspective by leveling out storage imbalance, but it hampers access times for end users.

13.4.3 Shared Libraries and Start-up Time

The experimenters observed that VisIt's start-up time was longer than expected on Dawn, beginning at 4,096 tasks and worsening with higher task counts. They concluded it was because each task was reading plug-in information from the filesystem, creating contention for I/O resources. They partially addressed the problem during their experiments by modifying VisIt's plug-in infrastructure to load plug-in information on task #0 and to broadcast the information to other tasks. This change made plug-in loading nine times faster. However, start-up time was still quite slow, taking as long as five minutes.

VisIt's design made use of shared libraries because it encourages the development of new plug-ins that augment existing capabilities. Using shared libraries allows new plug-ins to access symbols that are not used by any current VisIt routines; linking statically removes these symbols. After this study, the developers decided that the performance penalties were too extreme at high task counts and so they added an option for static linking. Dynamic linking is still common, especially at low task counts—but static linking is now available for high task counts.

13.5 Conclusion

This chapter began by asking whether or not pure parallelism, the dominant paradigm for production visualization tools, could be successful for tomorrow's data sets or whether the techniques discussed in Part II would be required. To answer this question, the chapter presented the results of a study designed to answer two questions about pure parallelism: (1) Can pure parallelism be successful on extreme data sets at extreme concurrency on a variety of architectures? And (2) does pure parallelism exhibit weak scaling properties? Although the results provided evidence that pure parallelism scales well, they also showed that the technique is only as as good as its supporting I/O infrastructure and that I/O limitations are already prevalent on many supercomputers.

Insufficient I/O bandwidth will only increase in the future, as supercomputing budgets are inordinately devoted to FLOPs at the cost of I/O performance. Improvements can come from either software or hardware solutions, or some combination of the two. On the software side, query-driven visualization (Chap. 7), multiresolution processing (Chap. 8), and *in situ* processing processing (Chap. 9) are approaches that all significantly reduce I/O. On the hardware side, emerging I/O technologies, such as solid-state drives (SSDs), offer significantly faster read times than the spinning disks. If simulations could stage their data on these SSDs for visualization programs, bypassing the spinning disk, it could increase access times by an order of magnitude and possibly make pure parallelism viable.

References

[1] Hank Childs, David Pugmire, Sean Ahern, Brad Whitlock, Mark Howison, Prabhat, Gunther Weber, and E. Wes Bethel. Extreme Scaling of Production Visualization Software on Diverse Architectures. *IEEE Computer Graphics and Applications*, 30(3):22–31, May/June 2010.

[2] William E. Lorensen and Harvey E. Cline. Marching Cubes: A High Resolution 3D Surface Construction Algorithm. *SIGGRAPH Computer Graphics*, 21:163–169, August 1987.

Chapter 14

Performance Optimization and Auto-Tuning

E. Wes Bethel

Lawrence Berkeley National Laboratory

Mark Howison

Lawrence Berkeley National Laboratory

14.1 Introduction 308
14.2 Optimizing Performance of a 3D Stencil Operator on the GPU 310
14.2.1 Introduction and Related Work 310
14.2.2 Design and Methodology 312
14.2.3 Results 313
14.2.3.1 Algorithmic Design Option: Width-, Height-, and Depth-Row Kernels 313
14.2.3.2 Device-Specific Feature: Constant Versus Global Memory for Filter Weights 314
14.2.3.3 Tunable Algorithmic Parameter: Thread Block Size 314
14.2.4 Lessons Learned 317
14.3 Optimizing Ray Casting Volume Rendering on Multi-Core GPUs and Many-Core GPUs 317
14.3.1 Introduction and Related Work 317
14.3.2 Design and Methodology 318
14.3.3 Results 320
14.3.3.1 Tunable Parameter: Image Tile Size/CUDA Block Size 320
14.3.3.2 Algorithmic Optimization: Early Ray Termination 322
14.3.3.3 Algorithmic Optimization: Z-Ordered Memory 324
14.3.4 Lessons Learned 325
14.4 Conclusion 325
References 327

In the broader computational research community, one subject of recent research is the problem of adapting algorithms to make effective use of multi- and many-core processors. Effective use of these architectures, which have complex memory hierarchies with many layers of cache, typically involves a careful examination of how an algorithm moves data through the memory hierarchy. Unfortunately, there is often a nonobvious relationship between algorithmic parameters like blocking strategies, and their impact on memory utilization, and, in turn, the relationship with runtime performance. Auto-tuning is an empirical method used to discover optimal values for tunable algorithmic parameters under such circumstances. The challenge is compounded by the fact that the settings that produce the best performance for a given problem and a given platform may not be the best for a different problem on the same platform, or the same problem on a different platform.

The high performance visualization research community has begun to explore and adapt the principles of auto-tuning for the purpose of optimizing codes on modern multi- and many-core processors. This chapter focuses on how performance optimization studies reveal a dramatic variation in performance for two fundamental visualization algorithms: one based on a stencil operation having structured, uniform memory access, and the other is ray casting volume rendering, which uses unstructured memory access patterns. The two case studies highlighted in this chapter show the extra effort required to optimize such codes by adjusting the tunable algorithmic parameters can return substantial gains in performance. Additionally, these case studies also explore the potential impact of and the interaction between algorithmic optimizations and tunable algorithmic parameters, along with the potential performance gains resulting from leveraging architecture-specific features.

14.1 Introduction

Most HPC research in the last decade or so, including high performance visualization, has focused on the distributed-memory parallel platform where each node consists of a single-core processor. Until recently, such platforms became increasingly more powerful due to increases in clock speeds at a rate that, more or less, follow Moore's Law. Now, however, heat and power constraints limit gains in processor speeds; therefore, multi-core CPUs and many-core GPUs have emerged as the industry's solution to provide more computational capacity per processor without increasing the clock rate.

Achieving optimal algorithm performance on these modern platforms can be challenging due to characteristics that are both platform specific and difficult, if not impossible, to estimate *a prioi* with a predictive performance model. For example, on many-core GPUs, performance can be highly sensitive to the degree by which threads in a thread block all execute the same code: performance degrades when there is conditional branching and threads execute different code paths. Even on multi-core CPUs, recent work has shown there is not necessarily a clear correlation between the amount of memory traffic and absolute runtime [9].

When considering the overall issue of performance optimization, there are three significant issues to consider. First, what combination(s) of tunable algorithmic parameters produce the best performance for a given code on a given platform? For data-dependent problems, like volume rendering (see 14.3), overall algorithmic performance may also be influenced by the data values themselves. Second, are there any algorithmic optimizations that might improve performance through the reuse of memory or the continuation of computations by more efficiently using the memory hierarchy, or by eliminating unnecessary computations? Finally, are there any device specific capabilities one might leverage to achieve a faster runtime? For example, faster, programmer-accessible cache memory might offer the opportunity for the application to make explicit use of the memory hierarchy in advantageous ways.

To find the best performing combinations of tunable algorithmic parameters, the computational science community developed a method known as auto-tuning. Its objective is to find values, or settings, for tunable algorithmic parameters that result in the best performance for an algorithm on a particular platform, for a particular problem size. It has been used with success to optimize the performance of stencil-based, numerical solver codes on multi-core CPUs and many-core GPUs [31, 13, 17, 9, 7, 8]. Auto-tuning is based on the idea that one can enumerate all possible tunable configurations of an algorithm. As the number of potential configuration permutations can be quite large, search strategies have emerged as a specific research area to avoid performing an exhaustive search for all possible configurations.

The high performance visualization research community recently began to adapt these principles and practices to visualization algorithms. This chapter provides an overview of the performance optimization issues from a visualization perspective and highlights results from two studies where substantial performance gains result from a combination of using auto-tuning principles, algorithmic optimizations, and leveraging device-specific capabilities.

The first study examines the performance optimization of a GPU implementation using a structured memory access, stencil-based algorithm that is common in many visualization, analysis, and image processing/computer vision algorithms. The study reports that there can be as much as a 7.5× performance gain, depending on the setting of one tunable algorithmic parameter for a given problem on a given platform, and an additional 4× performance gain through a combination of algorithmic optimization and use of device-specific capabilities, for a maximum performance gain of about 30× [2].

The second study examines the performance of multi-core CPU and many-core GPU implementations of an unstructured memory access code, ray casting volume rendering, which is a staple visualization algorithm. That study shows there can be as much as a 2.5× performance gain just by tuning one algorithmic parameter. An additional performance gain of 1.5× is possible through an algorithmic optimization. Also, a varying amount of gain is possible through a data-dependent algorithmic optimization [3]. Together, these two studies suggest that taking the time to find the appropriate algorithmic parameters that produce the best performance is a challenge, but also worth the effort in terms of potential performance gains: they reflect the performance gain one might expect between a naïve and a well-tuned implementation.

14.2 Optimizing Performance of a 3D Stencil Operator on the GPU

14.2.1 Introduction and Related Work

The fundamental idea behind a stencil operation is that, in order to compute a value at grid point G_i, the computation of G_i must include the neighboring grid points. For example, $G_i = f(G_{i-1}, G_i, G_{i+1})$, where f combines the three different input values by multiplying each by a weight. f may become quite complex, as is the case when stencil-based operations are used as the basis for partial differential equation solvers. In this way, G_i is replaced with a weighted average of itself and its nearest neighbors. This idea is often generalized into two, three, or more dimensions. It may take into account n nearby neighbors, not just the next-door neighbors. The code that implements this type of operation is known as a *stencil kernel* or *convolution kernel.* While the stencil kernel is fundamental in many different classes of numerical solvers, it is also fundamental in many visualization, analysis, and computer vision/image processing algorithms.

In visualization, for example, isocontouring algorithms like Marching Cubes [20] produce triangles as outputs. Each triangle vertex contains information about its x, y, z position as well as an estimate of a surface normal vector, $\vec{N}$. $\vec{N}$ plays an important role in rendering (Chap. 4) and is often computed using a finite-difference approach that is a six-point stencil operation in three dimensions. The finite-difference approach is not limited to isocontouring algorithms. The approach is also prevalent in many different visualization and rendering algorithms for generating surface normals.

Components in the visualization processing pipeline are not strictly limited to generating or rendering geometry; they also include operators that perform computations on data. Some common examples of operators are thresholding, contrast/gain adjustment, and scaling operators. These types of algorithms are not stencil-based algorithms. They operate on one data point at a time, without any influence of neighboring points. Some types of operations include filtering, which may be needed to reduce noise in experimental data or to reduce "noise" in simulation data containing a substantial degree of high-frequency characteristics. Such noise reductions aid in subsequent visual interpretations and they also aid in facilitating better class operations in analysis codes, which may otherwise be sensitive to the "noise" in data.

Noise reduction, or smoothing, is a fundamental operation in many types of visualization and image processing algorithms. One of the simplest approaches to smoothing is to take the average of nearby points to compute an estimate of the denoised signal. A "box filter" computes an estimate by using equal weights for all the nearby sample points. The best estimate for the average would be to afford greater weights to the closest nearby points and smaller weights to the farthest points. The Gaussian low-pass filter computes such an average by using a set of weights defined over a normal distribution so that points near the target sample point have a greater contribution to the average than points far away from the sample point. This type of smoothing is isotropic in the sense that the filter application performs independently of the underlying signal. The result is that the Gaussian low-pass filter smoothes equally in all directions, which gives rise to an unfortunate side effect called blurring edges.

Bilateral filtering, as defined by Tomasi [30], aims to perform anisotropic smoothing by using a low-cost, non-iterative formulation. In bilateral filtering, the output at each image pixel, $d(i)$, is the weighted average of the influence of nearby image pixels, $\bar{i}$, from the source image, s, at a location, i. The "influence" is computed as the product of a geometric spatial component, $g(i, \bar{i})$, and a signal difference, $c(i, \bar{i})$.

$$d(i) = \frac{1}{k(i)} \sum g(i, i)c(i, \bar{i}), \tag{14.1}$$

where $k(i)$ is the normalization factor that is the sum of all weights $g(i, \bar{i})$ and $c(i, \bar{i})$. $k(i)$ is computed as:

$$k(i) = \frac{1}{\sum g(i, \bar{i})c(i, \bar{i})}. \tag{14.2}$$

It is possible to precompute the portions of $k(i)$ that are contributed by $g(i, \bar{i})$, since $g(i, \bar{i})$ depends only on the 3D Gaussian probability distribution function (PDF). However, the set of contributions from $c(i, \bar{i})$ are not known *a priori* since they depend on the actual set of photometric differences observed across the neighborhood of $c(i, \bar{i})$. $c(i, \bar{i})$ will vary depending on the source image contents and target location i.

Tomasi defines g and c to be Gaussian functions that attenuate the influence of nearby points so that those nearby points in geometric or signal space have a greater influence, while those farther away, in geometric or signal space, have a lesser influence according to a Gaussian distribution [30]. Therefore, the Gaussian function g, yields,

$$g(i, \bar{i}) = e^{-\frac{1}{2}\left(\frac{d(i, \bar{i})}{\sigma_d}\right)^2}. \tag{14.3}$$

Here, $d(i, \bar{i})$ is the distance between pixels i and $\bar{i}$. The photometric similarity influence weight, $c(i, \bar{i})$, uses a similar formulation but $d(i, \bar{i})$ is the absolute difference $\|s(i) - s(\bar{i})\|$ between the source pixel, $s(i)$, and the nearby pixel, $s(\bar{i})$.

The intent is to perform more smoothing in regions where there is a "homogeneous" signal, and to perform less smoothing in regions where there is

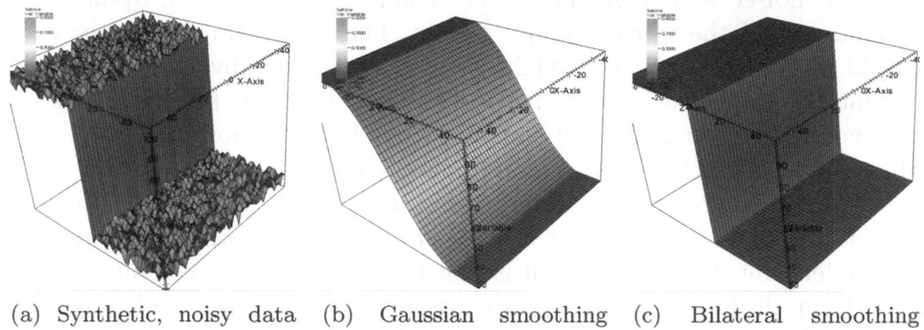

(a) Synthetic, noisy data set.

(b) Gaussian smoothing ($r = 16$).

(c) Bilateral smoothing ($r = 16, \sigma = 15\%$).

FIGURE 14.1: These images compare the results of smoothing to a synthetic data set (a), which contains added noise. While Gaussian smoothing blurs the sharp edge in this data set (b), bilateral filtering removes noise (c), while preserving the sharp edge. Image source: Bethel, 2009 [1].

an abrupt change in the signal. Figure 14.1 shows how bilateral filtering is a preferable operation for anisotropic smoothing, in comparison to the traditional Gaussian filtering of a 3D, synthetic data set with added noise.

The bilateral filter kernel is a good candidate to use in a performance optimization study on multi- and many-core platforms because it is quite computationally expensive compared to other common stencil kernels, e.g., Laplacian, divergence, or gradient operators [16]. Also, the size of the stencil, or bilateral filter window, varies according to its user preference and application, whereas most other stencil-based solvers tend to consider only the nearest-neighbor points.

The subsections that follow present a summary of a study by Bethel [2]. The study reports there can be as much as a 7.5× performance gain possible, depending on the setting of one tunable algorithmic parameter for a given problem on a given platform, and an additional 4× gain possible, through a combination of algorithmic optimizations and the use of device-specific capabilities, for a total of about a 30× the overall possible performance gain.

14.2.2 Design and Methodology

The main focus of this study is to examine how an algorithmic design option, a device-specific feature, and one tunable algorithmic parameter can combine to produce the best performance. The algorithmic design choice tells how to structure the CUDA implementation inner loop of the 3D bilateral filter so as to iterate over the voxels in memory (see 14.2.3.1). The device-specific feature makes use of local, high-speed memory for storing filter weights so as to avoid recomputing them over and over again. These filter weights remain constant for the course of each problem run (see 14.2.3.2). Finally,

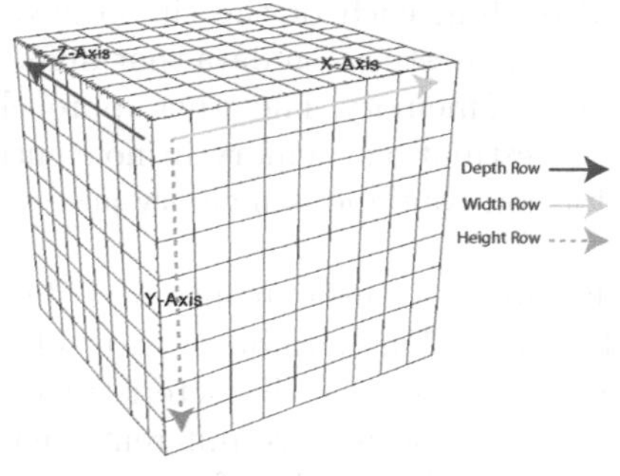

(a) Three potential thread memory access patterns: depth-, width-, and height-row.

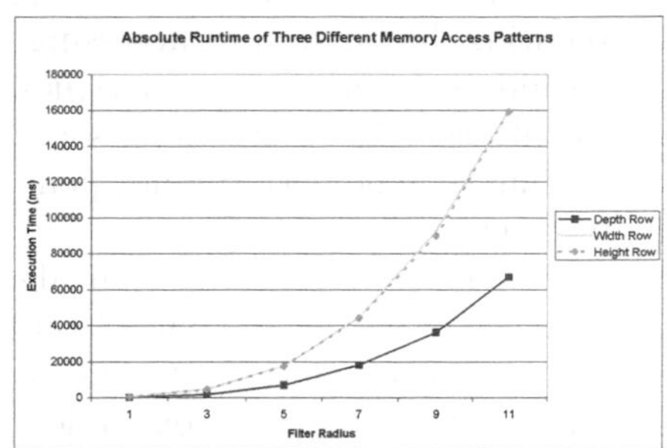

(b) Runtime performance for each of the depth-, width-, and height-row traversals.

FIGURE 14.2: Three different 3D memory access patterns, (a) have markedly different performance characteristics on a many-core GPU platform (b), where the depth-row traversal is the clear winner by a factor of about 2×. Image source: Bethel, 2009 [1].

the tunable algorithmic parameter is the size and shape of the CUDA thread block (see 14.2.3.3).

14.2.3 Results

14.2.3.1 Algorithmic Design Option: Width-, Height-, and Depth-Row Kernels

One of the first design choices in any parallel processing algorithm is to decide on a work partitioning strategy. For this particular problem, there are many potential partitionings of workload: they all come down to deciding how to partition the 3D volume into smaller pieces that are assigned to different processors.

In this particular study, the authors aim was to determine the memory access pattern that yields the best performance. To that end, they implemented three different memory access strategies. The first, called *depth-row* processing, has each thread process all voxels along all depths for a given `(i,j,*)` location. This approach appears as the dark, solid line and arrow on the 3D grid shown in Figure 14.2a. A *width-row* order has each thread process all voxels along a width at a given `(*,j,k)` location, and appears as a light gray, solid line and arrow in Figure 14.2a. Similarly, a *height-row* traversal has each thread process all voxels along a height at a given `(i,*,k)` location, and this is shown as a medium-gray, dashed line and arrow in Figure 14.2a.

The results of the memory access experiment, shown in Figure 14.2b, indicate the depth-row traversal offers the best performance by a significant amount, about a 2× performance gain. The best performance stems from the depth-row algorithm's ability to achieve so-called "coalesced memory" accesses. In the case of the depth-row algorithm, each of the simultaneously executing threads will be accessing source and destination data that differs by one address location, e.g., contiguous blocks of memory. The other approaches, in contrast, will be accessing source and destination memory in noncontiguous chunks. The performance penalty for non-contiguous memory accesses is apparent in Figure 14.2b.

By simply altering how an algorithm iterates through memory, this result shows a significant impact on performance. Furthermore, this result will likely vary from platform to platform, depending on the characteristics of the memory subsystem. For example, different platforms have different memory prefetch strategies, different sizes of memory caches, and so forth.

14.2.3.2 Device-Specific Feature: Constant Versus Global Memory for Filter Weights

Another design consideration is how to take advantage of and leverage device-specific features such as device memory types and memory speeds. The NVIDIA GPU presents multiple types of memory, some slower and uncached, others faster and cached. According to the NVIDIA CUDA Programming Guide [23], the amount of global (slower, uncached) and texture (slower, cached) memory varies as a function of specific customer options on the graphics card. Under CUDA v1.0 and higher, there is 64KB of *constant* memory and 16KB of *shared* memory that resides on-chip and visible to all threads. Generally speaking, device memory (global and texture memory) has a higher latency and lower bandwidth than an on-chip memory (constant and shared memories).

The authors wondered if storing, rather than recomputing, portions of the problem in on-chip, high speed memory (a device-specific feature) would offer any performance advantage. The portions of the problem they stored, rather than recomputed, were the filter weights, which are essentially a discretization of a 3D Gaussian and a 1D Gaussian function.

The performance question is then: "How is performance impacted by having all of the filter weights resident in on-chip rather than in device memory?"

The results, shown in Figure 14.3, indicate that the runtime is about 2× faster when the filter weights are resident in a high-speed, on-chip constant memory, as opposed to when the weights reside in the device's (slower) global shared memory. This result is not surprising given the different latency and bandwidth characteristics of the two different memory subsystems.

14.2.3.3 Tunable Algorithmic Parameter: Thread Block Size

In this particular problem, the tunable algorithmic parameter is the size and shape of the CUDA thread block. The study's objective is to find the combination of block size parameters—a tunable algorithmic parameter—that produce optimal performance.

A relatively simple example, shown in Figure 14.4, reveals that there is

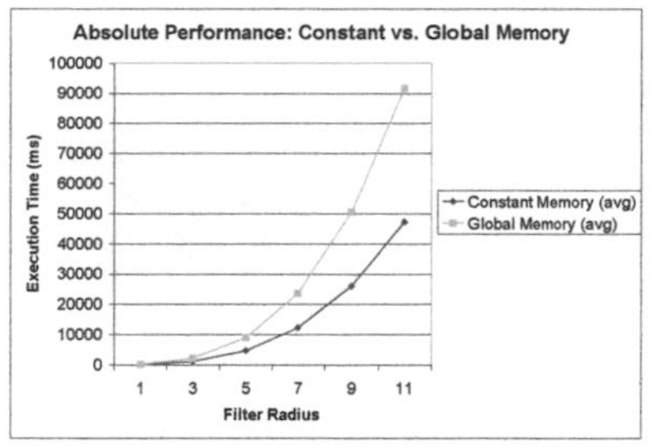

(a) Absolute performance (smaller is better).

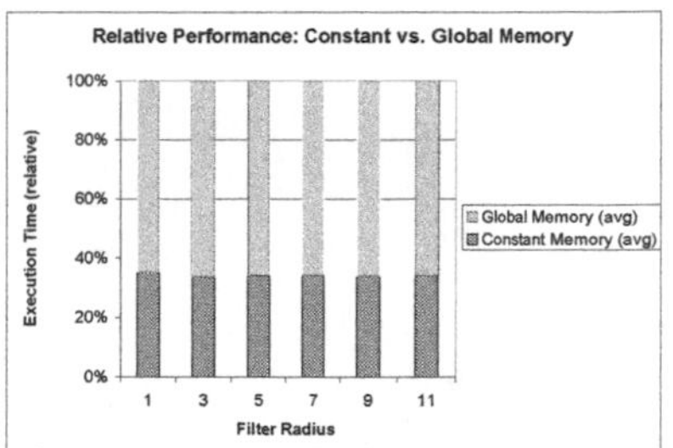

(b) Relative performance (smaller is better).

FIGURE 14.3: These results compare the absolute and relative performance difference when precomputed filter weights are resident in device memory vs. constant memory (2× faster). In all ranges of filter sizes, the constant memory implementation performs about twice as fast as the device memory implementation. Image source: Bethel, 2009 [1].

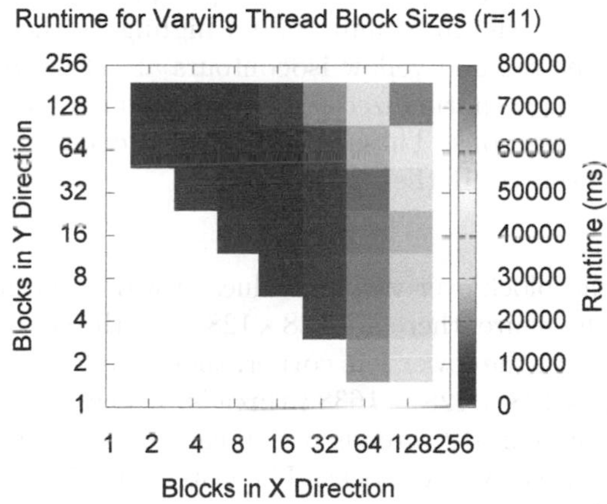

FIGURE 14.4: Runtime (ms) at radius $r = 11$ for different thread-block configurations on the GPU exhibit a 7.5× variation in performance: the coordinate, (8, 128), has a runtime of 9,550 ms while (128, 128) is much slower at 71,537 ms. Image courtesy of Mark Howison (LBNL and Brown University).

up to a 7.5× performance difference between the best- and worst-performing configurations for a filter radius width of $r = 11$, which corresponds to a very large 23^3-point stencil configuration.[1] The experiment varies the size of

[1] A filter radius r corresponds to a filter "box size" of $2r + 1$. So, a filter radius of 1 corresponds to a 3×3 stencil in 2D, or a $3 \times 3 \times 3$ stencil in 3D.

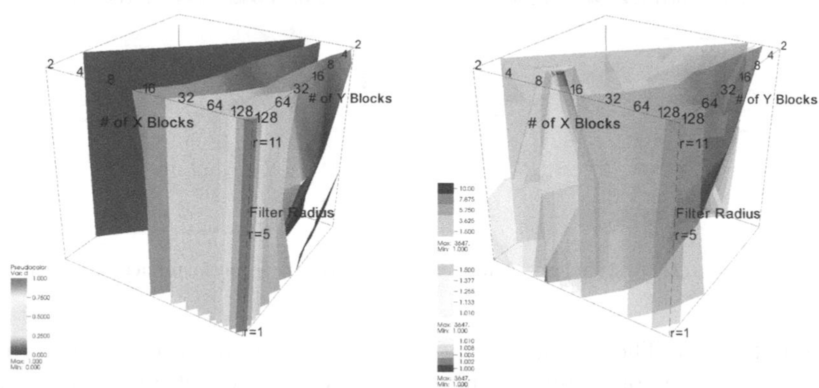

(a) Runtimes normalized by maximum highlight the poorest performing configurations.

(b) Runtime normalized by minimum highlight the best performing configurations.

FIGURE 14.5: Visualization of performance data collected by varying the number and size of the GPU thread blocks for the 3D bilateral filter are shown at three different filter sizes, $r = \{1, 5, 11\}$. In (a), the performance data (normalized to the maximum value) highlights the poorest performing configurations; the red and yellow isocontours are close to the viewer. In (b), the performance data (normalized to the minimum value) highlights the best performing configurations. These appear as the cone-shaped red/yellow isocontours. Image source: Bethel, 2009 [1].

the CUDA thread block over varying values of width and height. In the upper right corner of the figure, there are 128×128 total thread blocks, each of which has four threads. In the lower left corner, there are 2×2 total thread blocks, each of which has $128 \times 128 = 16384$ threads, which is not valid since CUDA permits a maximum of 512 threads per thread block; all invalid configurations appear in the figure as white blocks. The fastest runtime in this $r = 11$ search was about 9.5s for the $(8, 128)$ configuration, while the slowest was about 71.5s for the $(128, 128)$ configuration.

One question that arises is whether or not the configuration that works best at $r = 11$ also works best at other values of r. The authors ran more tests at $r = \{1, 5, 11\}$ to produce the 3D set of results shown in Figure 14.5. In that figure, the invalid configurations—those with more than 512 threads per thread block—are located "behind the blue curtain" in the back corner of Figure 14.5a. In Figure 14.5, performance data is normalized by the maximum so that values range from $(0.0, 1.0]$. That approach then highlights poorly performing configurations. As in the $r = 11$ case, the configurations that are closest to $(128, 128)$, which have relatively few threads per thread block, perform poorly. These poorly performing regions are shown as yellow/red isocontours in Figure 14.5a. The relative performance of thread-block configurations across all three filter radii appears more or less consistent.

Figure 14.5b shows a slightly different view of the same data. Here, performance data is normalized by the minimum so values range from $[1.0, \infty)$: minimum execution time normalizes to 1.0 and all other values are larger. This image shows the "sweet spot" across all configurations as the region enclosed by the red surface, which contains values close to or equal to 1.0. The xy-plane closest to the viewer corresponds to the largest filter size, while the one furthest from the viewer corresponds to the smallest filter size. The "sweet spot" appears to be much smaller for the largest filter size (close to the viewer) than for the smallest filter size.

14.2.4 Lessons Learned

This optimization study looks at three different dimensions of the performance optimization problem. What values for tunable algorithmic parameters produce the best performance? Are there algorithmic optimizations that might make a difference in performance? And, are there any device or platform specific capabilities that might be used to an advantage?

The study shows that there is a 2× performance gain to be had simply by altering how the individual CUDA threads iterate through memory. One configuration proves to be a lot better than the other two. Another 2× performance gain results from using high speed, on chip cache memory that is directly accessible to the application through the CUDA programming language. Finally, a 7.5× performance gain results from the careful selection of tunable algorithmic parameters. These combine to produce about a 30× speedup over an untuned implementation.

14.3 Optimizing Ray Casting Volume Rendering on Multi-Core GPUs and Many-Core GPUs

14.3.1 Introduction and Related Work

Ray casting volume rendering, (see Chap. 4), is a staple visualization algorithm used for creating images of 3D scalar data sets. Parallelization of this algorithm typically follows one of two approaches. In *image order* parallelism, each processor is responsible for rendering a subset of image pixels, and all processors have access to the same data set. This approach is more commonly used in shared-memory parallel environments. *Object order* parallelism refers to having each processor render a subset of the overall data set, with the resulting images combining together in a final image via compositing (see Chapter 5). This approach is more commonly used in distributed-memory parallel environments.

Over the years, there have been a number of studies that examine how various algorithmic optimizations, parallelization strategies, and tunable algorithmic parameters can improve performance. Nieh and Levoy [22] study alternative work assignment strategies for threads on a shared-memory parallel machine. Palmer et al. [25] studied two parallel partitioning and dynamic load balancing algorithms to explore the trade-offs between their memory hierarchy performance on shared- and distributed-memory parallel architectures. Grimm et al. [12] explored the resulting performance impact when varying the size of data blocks allocated to processors in object-order parallel volume rendering on a shared-memory, multi-core CPU platform. In their study of the radix-k algorithm, Peterka et al. [27] measure substantial variations in performance for a given problem on a given platform simply by varying the value of one tunable algorithmic parameter, the $\vec{k}$ vector. Peterka et al. [28] observe a 2$\times$ performance difference in I/O rates in a parallel volume rendering application by tuning the read buffer size to the size of the data set's netCDF record size.

Marsalek et al. [21] implement a ray casting volume rendering kernel in CUDA that uses an image-parallel partitioning, where each CUDA performs ray integration through a data volume for a single pixel. Their results suggest that CUDA kernels perform at a rate commensurate with earlier fragment-based approaches [18, 29, 11, 10]. They achieve a performance boost through a process of manual experimentation with thread block size—the number of threads per thread block—to find a good balance between register and shared memory use.

An interesting algorithmic optimization to consider is in-memory data layout strategies. The layout strategy is applicable in many types of data-intensive computing algorithms, in general, and, in particular, visualization algorithms. With the aim of exploring a memory layout that is more "cache friendly," Pascucci and Frank [26] use an indexing scheme based upon the Lebesgue, or Z-order, space filling curve to improve and optimize progressive visualization algorithms. This data layout has desirable spatial locality properties: at any point in the mesh, nearby points are nearby in memory or storage.

A recent study by Bethel and Howison [3] explores the effect of varying the size of the image-tile partitioning, and two algorithmic optimizations in a shared-memory volume renderer on modern multi-core CPU and many-core GPU platforms. Two such algorithmic optimizations are early ray termination, and an alternative memory layout based on the Z-order indexing. The study's objective is to present the large performance variation, or gain, by carefully selecting tunable algorithmic parameters. The study also focuses on the interplay between two different algorithmic optimizations on multiple, modern multi- and many-core platforms.

14.3.2 Design and Methodology

The study's implementation is a ray casting volume renderer that follows Levoy's formulation [19], which is parallelized by using a shared-memory programming model: POSIX threads [4] and OpenMP [6] on multi-core CPUs, and CUDA [24] on many-core GPUs. In this code, each thread is responsible for casting rays into the volume, performing color and opacity integration along its ray, and writing the result into a final image buffer. For this shared-memory implementation, there is one single copy of the volume data, and work is divided using an image-space decomposition: each thread is responsible for operating on a different part of the overall image.

The study looks at how the performance is affected by the choice of one tunable algorithmic parameter and two algorithmic optimizations. The tunable algorithmic parameter chosen here is the amount of work assigned to each processing thread. One algorithmic optimization is whether or not early ray termination (ERT) is enabled. The objective for ERT is to discontinue the inner ray integration loop when, during a front-to-back sampling of the source data, the opacity reaches full saturation. When that condition occurs, no further processing is necessary since further calculations will not contribute anything further to the final pixel color. The other algorithmic optimization is a choice of memory layouts. The idea is to compare the traditional, array-order layout with the Z-order layout, which is known to be more "cache friendly" in determining the two different memory strategy performance impacts on different platforms and different problem configurations. The implementation here uses the indexing scheme described by Pascucci and Frank [26].

The intent of this study is to better understand the relationships between algorithmic design choices and tunable parameters, and their impact on performance. The study focuses on two measures of performance. The first measure is absolute runtime. Absolute runtime is the time elapsed to create an image from an in-memory volume. The study does not take into account the I/O time that may be needed to load a source data set. These assumptions are appropriate for the use scenario where an application creates several images from an already loaded data set, as would be the case in interactive visualization. The second measure is memory efficiency, or memory utilization. This metric can be obtained by measuring L2 cache misses. The idea is that the more L2 cache misses, the less efficient or effective the algorithm utilizes the memory hierarchy.

Unlike the stencil-based code, where the same operation is performed on all data points, the volume rendering algorithm's performance depends, to some degree, on the data values themselves, as well as problem-specific parameters like viewpoint. For example, the degree to which ERT reduces runtime will vary from data set to data set. The runtime performance will also vary as a function of the viewpoint: far away views will result in the data covering fewer pixels than close-up views. In the latter case, the algorithm must perform a significant amount of work, so absolute runtime will be higher. To account for

Platform	Concurrency	Array-Order	Z-Order	Array + ERT	Z + ERT
Intel/Nehalem	1	10.2%	7.7%	2.8%	2.6%
	2	7.4%	10.1%	9.7%	9.1%
	4	32.8%	28.7%	30.3%	27.7%
	8	55.1%	52.8%	50.9%	46.5%
AMD/MagnyCours	6	102.7%	100.7%	93.6%	92.4%
	24	247.2%	254.1%	241.5%	242.5%
NVIDIA/Fermi	-	74.8%	255.0%	78.9%	265.1%

TABLE 14.1: Percent variation in runtime (averaged over 10 views) across all block sizes, broken down by levels of concurrency, memory layout, and ERT. Block size has a dramatic impact on performance, particularly as concurrency increases to all available CPU cores. On the GPU, variation was greatest for Z-ordered memory, because the performance of the fastest block configurations increased considerably with Z-ordered access.

these variances, the study's test battery uses several different viewpoints to create diversity in memory access and stride patterns. The test battery also uses two different transfer functions, which to create opportunities for more ERT-based performance gains.

14.3.3 Results

14.3.3.1 Tunable Parameter: Image Tile Size/CUDA Block Size

In the CPU tests, the authors explore 64 different image tile sizes where width and height both vary over $\{1, 2, 4, 8, 16, 32, 64, 128\}$. On the AMD/MagnyCours, a complete test battery consists of approximately 5,120 tests: ten views, sixty-four block sizes, with and without early ray termination (ERT), two memory layouts, two concurrency levels, and only the dynamic work assignment. On the Intel/Nehalem, the test battery was similar to the study, with three work assignment methods but only 8-way concurrency, for a total of 7,680 tests. In addition, the authors reran another 7,680 tests on the Intel/Nehalem at 1-way, 2-way, and 4-way concurrency, but with only the dynamic work assignment.

On the GPU, the study consisted of 880 test runs, over twenty-two different thread block sizes, with a width and height varying over $\{1, 2, 4, 8, 16\}$ (excluding the 1×1, 1×2, and 2×1 blocks), across two memory layouts, and with and without ERT. The authors excluded blocks having fewer than four threads because the NVIDIA/Fermi requires at least four threads per block to saturate the computational throughput of a warp of execution.

The study determines if the block size impacts performance, and if so, by how much, as measured overall by percent variation. The summary results, shown in Table 14.1, indicate a substantial variation in runtime performance results vs. variations in block size: as much as 254.1% on multi-core CPUs and

		Array Order					Z Order				
		1	2	4	8	16	1	2	4	8	16
NoERT	1			1.00	0.77	0.74			0.90	0.58	0.44
	2		0.98	0.71	0.63	0.68		0.86	0.51	0.35	0.30
	4	1.04	0.70	0.59	0.61	0.74	0.89	0.51	0.32	0.26	0.33
	8	0.81	0.63	0.61	0.68	0.73	0.56	0.34	0.25	0.27	0.32
	16	0.83	0.71	0.72	0.68	0.71	0.42	0.30	0.27	0.27	0.32
ERT	1			0.92	0.70	0.66			0.83	0.53	0.40
	2		0.90	0.64	0.57	0.60		0.79	0.47	0.32	0.27
	4	0.95	0.64	0.53	0.54	0.65	0.82	0.46	0.29	0.23	0.29
	8	0.73	0.56	0.54	0.60	0.63	0.51	0.31	0.23	0.24	0.28
	16	0.74	0.62	0.62	0.59	0.61	0.38	0.27	0.24	0.24	0.28

(a) Runtime (s)

		Array Order					Z Order				
		1	2	4	8	16	1	2	4	8	16
NoERT	1			50	115	481			25	28	42
	2		36	57	238	632		17	17	22	67
	4	52	59	179	517	907	14	14	16	43	240
	8	139	243	527	839	873	14	17	43	182	202
	16	636	724	909	827	834	29	63	176	166	166
ERT	1			43	95	401			22	24	35
	2		29	48	193	530		15	15	19	56
	4	38	47	144	429	773	12	12	14	37	198
	8	106	192	436	711	741	12	14	37	153	163
	16	506	596	765	696	701	24	53	148	137	130

(b) L2 cache misses (millions)

FIGURE 14.6: Parallel ray casting volume rendering performance measures on the NVIDIA/Fermi GPU include absolute runtime (a), and L2 cache miss rates, (b) averaged over ten views for different thread block sizes. Gray boxes indicated thread blocks with too few threads to fill a warp of execution. Surprisingly, the best performing configurations do not correspond to the best use of the memory hierarchy on that platform. Image source: Bethel and Howison, 2012 [3].

265.1 % on multi-core GPUs. The significance of the variation is that these differences may be representative of the difference in performance between an untuned implementation and one that is well tuned.

The study's detailed results show that the runtime and L2 cache miss rates for each block size and algorithmic optimization. Although it is not shown here, the multi-core CPU platforms, reveal several interesting conclusions. First, the amount of variation increases with increasing concurrency on the CPUs. This result is not surprising since there is an increased opportunity for reusing cache-resident data that is shared among multiple threads. Block sizes that better utilize the cache hierarchy can perform much better, increasing the variation in performance when compared to less optimal block sizes. Second, for array- and Z-ordered memory layouts, block size has a nearly equal effect on variation for both approaches. Although Z-ordering improves the spatial locality of data within the same 3D neighborhood, it is still sensitive to changes in spatial locality caused by different block sizes. Third, on the GPU, the variation is greater for Z-ordering by more than a factor of three when compared to array-ordering (see 14.3.3.3 for more details).

The NVIDIA/Fermi (GPU) test results show a wide variation in runtime across different block sizes (Fig. 14.6). Both Z-ordering and ERT show benefits for the more optimal block sizes. The best configurations on the NVIDIA/Fermi are the 4×8 and 8×4 thread blocks with Z-ordering and ERT. Performance appears to vary with the total number of threads in a block, leading to the diagonal striping in Figure 14.6.

Even though CUDA thread blocks require a minimum of thirty-two threads to saturate computational throughput, many of the block sizes with sixteen or fewer threads still perform well because of the branching nature of the algorithm, which causes divergence among CUDA threads. A thread block is executed in a single-instruction-multiple-thread (SIMT) fashion in which warps of thirty-two threads are executed across four clock cycles in subsets of eight threads that share a common instruction. If those eight threads do not share a common instruction, such as when conditionals cause branching code paths, the threads diverge and are executed serially.

This situation occurs frequently in this algorithm. For example, suppose a thread block owns a region of the image that only partially covers the data volume. Some of the threads in that block exit immediately due to the empty-space skipping optimization in the algorithm, while the other threads proceed to cast rays through the volume. However, the threads that proceed together with ray casting may also have rays of different lengths, which will cause divergence and load imbalance.

Since a warp must be scheduled across at least four clock cycles, using fewer than four threads per thread block will guarantee under-utilization. The configurations that met these requirements were excluded from the parameter sweep. The study shows empirically that the sweet spot for a thread block size is 16 or 32 threads, depending on the memory ordering and whether ERT is enabled. Many block sizes with 16 threads perform well, even though this number is less than the warp size of 32 threads, indicating the complex interaction of the CUDA runtime and warp scheduler in handling the branching for this particular algorithm and problem. It is also likely that larger thread blocks exhibit greater load imbalance because, the variation in ray lengths tends to increase with block size.

Surprisingly, the small thread blocks that display the worst performance also exhibit the fewest L2 cache misses (see Figure 14.6). Note, the converse of this is not true: the most optimal block sizes do not show the most L2 cache misses. Instead, L2 cache misses appear to rise uniformly with the total number of threads in a block, leading to the same diagonal striping as seen in the runtime plot. The study suggests that achieving the best performance on the GPU is a trade-off between using enough threads to saturate a warp and using fewer than enough threads to maintain a good cache utilization. The study also finds that, as in the CPU tests, L2 cache misses are systematically less when using the Z-ordered memory layout on the GPU because of the improved spatial locality.

Interestingly, the *NVIDIA CUDA Programming Guide* [24] says: "The effect of execution configuration on performance for a given kernel call generally depends on the kernel code. Experimentation is therefore recommended." These experiments show a wide variation in performance depending upon thread block size. While such variation isn't all that surprising, the amount of variation—as much as 265%—is somewhat unexpected, as is the fact that the optimal block size for one problem is not the same as for another problem when run on the same platform.

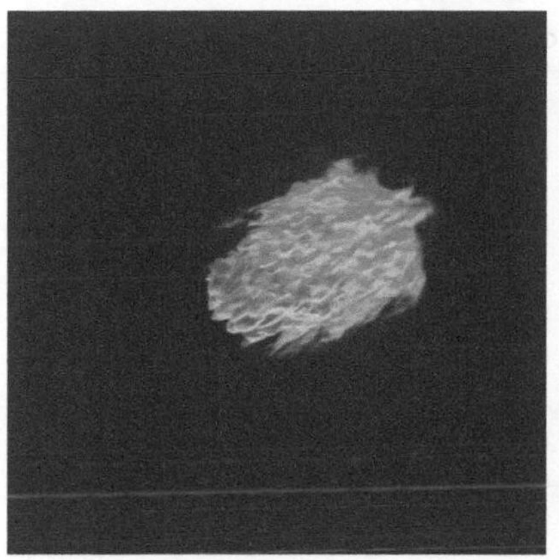

(a) Transfer function "A"

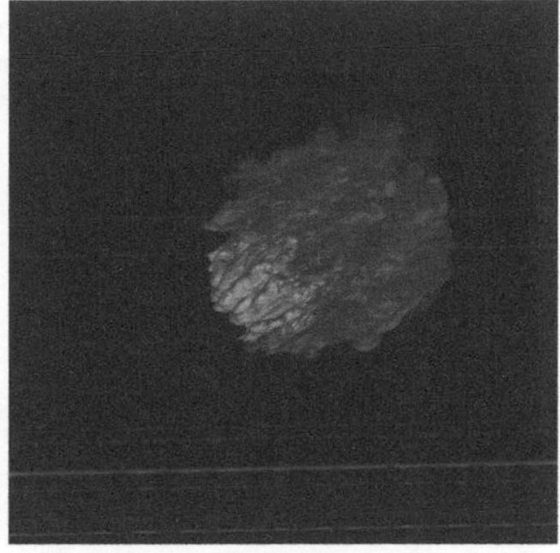

(b) Transfer function "B"

FIGURE 14.7: Two different transfer functions have different benefits from early ray termination. They also yield images that accentuate different features of the underlying data set. The performance tests in this study use transfer function "A." Image source: Bethel and Howison, 2012 [3].

14.3.3.2 Algorithmic Optimization: Early Ray Termination

The study includes evaluating the performance impact of an algorithmic optimization called early ray termination (ERT). In ERT, there is a conditional within the inner-most loop that iterates along an individual ray tests. It tests whether or not the ray has reached full opacity, meaning that any further calculations would not contribute anything further to the final pixel color. The effect of this optimization depends on both the input data and on the transfer function that determines how data values map to color and opacity. For the study's specific data set and transfer function, the authors saw approximately 10% fewer integration steps when ERT was enabled, which is directly reflected in the reported runtimes. In the cases where ERT was enabled, there was a visible 10–15% improvement in absolute runtime.

To demonstrate the relationship between the transfer function and the benefits of using ERT, the study included an additional test with a "shallower" transfer function (see Fig. 14.7b) that did not penetrate as deep into the volume. With ERT enabled on the shallower transfer function, the algorithm performed 19.7% fewer integration steps and ran 19.1% faster. The reduced runtime reflects the reduced amount of computational work.

Even though the use of ERT made the problem run faster, there may be a qualitative cost associated with the reduced runtime. The image rendered with transfer function "B" shows significant qualitative differences in many areas. The "interior" features of the data set that are visualized may or may not be appropriate according to the application.

Since the benefits of ERT depend highly on the scene characteristics and

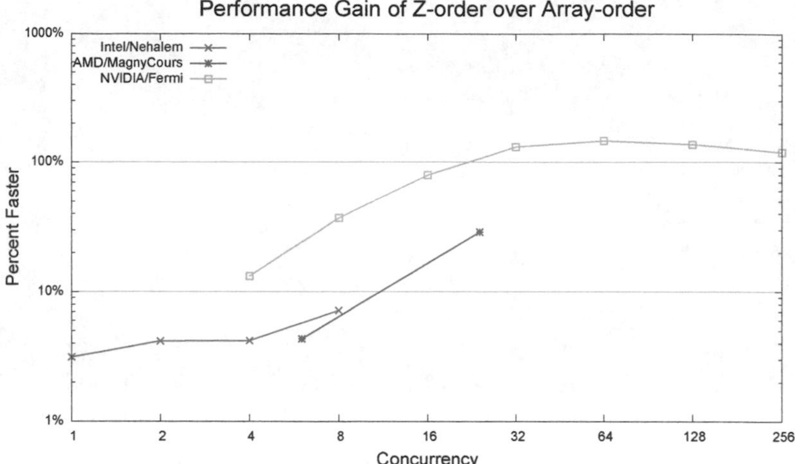

FIGURE 14.8: The performance gains (averaged across thread block size, with ERT) from using Z-ordered memory instead of array-ordered increase with concurrency. The gains are most notable on the NVIDIA/Fermi, where the number of threads per block corresponds to the level of "per-processor" concurrency. Image source: Bethel and Howison, 2012 [3].

transfer function, it is impossible to say with certainty how much ERT will help algorithm performance. Earlier work investigating the use of ERT in ray casting volume rendering [18] reports a a 300% improvement in runtime for an opaque volume, but incurs a 30% performance penalty when run on a semi-transparent volume. This effect attributes to the extra overhead when evaluating the conditional in the inner integration loop. The conditional can have an adverse impact on performance, particularly for GPU implementations where divergent code paths are costly. The effects—benefits or drawbacks—of ERT are highly variable. There are combinations of data sets and transfer functions where ERT has no effect, and others where it leads to a dramatic reduction in the number of integration steps.

14.3.3.3 Algorithmic Optimization: Z-Ordered Memory

The study shows that configurations using the Z-order memory layout outperform the array-order layout ones on all platforms and on all concurrency levels, and also, the performance gains increase with the increase of concurrency (see Fig. 14.8). At higher concurrency, the benefits of Z-ordering are most likely due to the larger penalties for non-contiguous access and cache misses in shared-memory systems that service many cores, such as the AMD/-MagnyCours and NVIDIA/Fermi. While the Intel/Nehalem and AMD/MagnyCours show modest gains from Z-ordered memory, ranging from 3% to 29%, the NVIDIA/Fermi exhibits a gain of 146% at a 64-way concurrency, that is, 64 threads per thread block.

The improved locality of Z-ordered memory access and the corresponding, lower L2 miss rates becomes increasingly beneficial as more cores access memory through a shared-memory controller and subsystem. The NVIDIA/Fermi, whose memory subsystem must service 14 multi-processors (each with many thread blocks in flight), realizes the biggest performance gains in Z-ordering. While the slower block configurations improve somewhat with Z-ordering, the best configurations improve greatly, leading to a larger variation for Z-ordering than array-ordering.

In practice, since most data sets are not stored in a Z-order layout, there would be a cost associated with reordering data. That cost is not included in the study. However, in the case where multiple visualization images would be generated from the same data set, that one-time cost is amortized across many renderings.

14.3.4 Lessons Learned

This optimization study looks at two different dimensions of the performance optimization problem on three different platforms. What values for tunable algorithmic parameters produce the best performance? Are there algorithmic optimizations that might make a difference in performance? And how do these performance characteristics vary, and why, on different multi-core CPU and many-core GPU platforms?

The study shows there is about a 2.5× performance gain to be had through careful selection of a tunable algorithmic parameter: the thread block/image tile size. This variation reflects differences in load balance and relatively better, or worse, use of the memory hierarchy. This magnitude of performance gain appears to be consistent across all three platforms.

The memory layout algorithmic optimization has a positive benefit on performance, but the amount of performance gain varies depending on platform and on the settings of the tunable algorithmic parameter. The study shows performance gains ranging from 3–29% on multi-core CPU platforms, and up to 146% on a many-core GPU platform. The large difference in performance between these two platforms reflects the difference in performance characteristics in the memory subsystems.

The benefit of the other algorithmic optimization—early ray termination—depends completely on the data or problem. In some problem configurations, the extra cost of the conditional to implement the ERT optimization incurs a performance penalty. In other configurations, ERT can result in a substantial performance gain. These penalties or gains seem to be largely independent of memory layout or algorithmic tuning.

14.4 Conclusion

The main point of these studies was to explore the relationship between tunable algorithmic parameters and known algorithmic optimizations, and also, to explore the resulting impact on performance for two different visualization algorithms on modern multi-core and many-core platforms. The studies show that the algorithmic parameters that produce the best performance vary from problem to problem and platform to platform, often in a non-obvious way. The GPU volume rendering study results emphasize this conclusion: optimal performance results in what appears to be a crossover point between memory cache utilization and thread warp size. By empirically measuring the performance of different algorithmic parameter settings, the study's authors were able to find the best performing configurations. Their study's approach is both successfully and widely used in the computational science community [17, 9, 7].

Some algorithms, like ray casting volume rendering with perspective projection, use an unstructured, or irregular memory access pattern. Each ray will, in effect, execute a different traversal path through memory. The set of paths and the number of computational steps required for each ray is a function of runtime parameters, such as the viewpoint, data set, and color transfer function. Other visualization algorithms exhibit similar memory access characteristics, like parallel streamline computation [5], and can benefit from this performance optimization methodology. It may be difficult, if not impossible, to develop a performance model for such algorithms—the ones with unpredictable memory access patterns. Therefore, auto-tuning methodology is a promising approach for finding the optimal settings of tunable algorithmic parameters.

The performance improvements to be realized can have tangible impacts: the results of the volume rendering study provided the basis for adjusting the tunable algorithmic parameters for a set of extreme-concurrency runs [14, 15] that require millions of CPU hours. By finding and using optimal settings for tunable algorithmic parameters, the tunable algorithmic parameters, in effect, saved millions of additional CPU hours that would have been spent executing an application in a non-optimal configuration.

References

[1] E. Wes Bethel. High Performance, Three-Dimensional Bilateral Filtering. Technical Report LBNL-1601E, Lawrence Berkeley National Laboratory, 2009.

[2] E. Wes Bethel. Exploration of Optimization Options for Increasing Performance of a GPU Implementation of a Three-Dimensional Bilateral Filter. Technical Report (In patent review), Lawrence Berkeley National Laboratory, Berkeley, CA, USA, 94720, 2012.

[3] E. Wes Bethel and Mark Howison. Multi-core and Many-core Shared-memory Parallel Raycasting Volume Rendering Optimization and Tuning. *International Journal of High Performance Computing Applications*, (In press), 2012.

[4] David R. Butenhof. *Programming with POSIX threads*. Addison-Wesley Longman Publishing Co., Inc., Boston, MA, USA, 1997.

[5] David Camp, Christoph Garth, Hank Childs, Dave Pugmire, and Kenneth I. Joy. Streamline Integration Using MPI-Hybrid Parallelism on a Large Multicore Architecture. *IEEE Transactions on Visualization and Computer Graphics*, 17:1702–1713, 2011.

[6] Robit Chandra, Leonardo Dagum, Dave Kohr, Dror Maydan, Jeff McDonald, and Ramesh Menon. *Parallel Programming in OpenMP*. Morgan Kaufmann Publishers Inc., San Francisco, CA, USA, 2001.

[7] Kaushik Datta, Shoaib Kamil, Samuel Williams, Leonid Oliker, John Shalf, and Katherine Yelick. Optimization and Performance Modeling of Stencil Computations on Modern Microprocessors. *SIAM Review*, 51(1):129–159, 2009.

[8] Kaushik Datta, Mark Murphy, Vasily Volkov, Samuel Williams, Jonathan Carter, Leonid Oliker, David Patterson, John Shalf, and Katherine Yelick. Stencil Computation Optimization and Auto-tuning on State-of-the-art Multicore Architectures. In *SC '08: Proceedings of the 2008 ACM/IEEE Conference on Supercomputing*, pages 1–12, Piscataway, NJ, USA, 2008. IEEE Press.

[9] Kaushik Datta, Sam Williams, Vasily Volkov, Jonathan Carter, Leonid Oliker, John Shalf, and Katherine Yelick. Auto-tuning the 27-point Stencil for Multicore. In *4th International Workshop on Automatic Performance Tuning (iWAPT)*, 2009.

[10] T. Fogal and J. Krüger. Tuvok, an Architecture for Large Scale Volume Rendering. In *Proceedings of the 15th International Workshop on Vision, Modeling, and Visualization*, pages 139–146, November 2010.

[11] Enrico Gobbetti, Fabio Marton, and José Antonio Iglesias Guitián. A Single-pass GPU Ray Casting Framework for Interactive Out-of-core Rendering of Massive Volumetric Datasets. *The Visual Computer*, 24(7):797–806, 2008.

[12] Sören Grim, Stefan Bruckner, Armin Kanistar, and Eduard Gröller. A Refined Data Addressing and Processing Scheme to Accelerlate Volume Raycasting. *Computers and Graphics*, 5(28):719–729, 2004.

[13] J. Hollingsworth and A. Tiwari. End-to-end Auto-tuning with Active Harmony. In David H. Bailey, Robert F. Lucas, and Samuel W. Williams, editors, *Performance Tuning of Scientific Applications*. CRC Press, Boca Raton, FL, USA, 2010.

[14] Mark Howison, E. Wes Bethel, and Hank Childs. MPI-hybrid Parallelism for Volume Rendering on Large, Multi-core Systems. In *Eurographics Symposium on Parallel Graphics and Visualization (EGPVG)*, Norrköping, Sweden, May 2010.

[15] Mark Howison, E. Wes Bethel, and Hank Childs. Hybrid Parallelism for Volume Rendering on Large, Multi- and Many-core Systems. *IEEE Transactions on Visualization and Computer Graphics*, 99(PrePrints), 2011.

[16] S. Kamil, C. Chan, S. Williams, L. Oliker, J. Shalf, M. Howison, E. W. Bethel, and Prabhat. A Generalized Framework for Auto-tuning Stencil Computations. In *Proceedings of Cray User Group Conference*, Atlanta GA, USA, May 2009. LBNL-2078E.

[17] Shoaib Kamil, Cy Chan, Leonid Oliker, John Shalf, and Sam Williams. An Auto-tuning framework for Parallel Multicore Stencil Computations. In *International Parallel & Distributed Processing Symposium (IPDPS)*, 2010.

[18] Jens Krüger and Rüdiger Westermann. Acceleration Techniques for GPU-based Volume Rendering. In *Proceedings IEEE Visualization 2003*, 2003.

[19] Marc Levoy. Display of Surfaces from Volume Data. *IEEE Computer Graphics and Applications*, 8(3):29–37, May 1988.

[20] William E. Lorensen and Harvey E. Cline. Marching Cubes: A High Resolution 3D Surface Construction Algorithm. *SIGGRAPH Computer Graphics*, 21(4):163–169, August 1987.

[21] Lukas Marsalek, Armin Hauber, and Philipp Slusallek. High-speed Volume Ray Casting with CUDA. In *IEEE Symposium on Interactive Ray Tracing*, 2008. Poster.

[22] Jason Nieh and Marc Levoy. Volume Rendering on Scalable Shared-Memory MIMD Architectures. In *Proceedings of the 1992 Workshop on Volume Visualization*, pages 17–24. ACM Siggraph, October 1992.

[23] NVIDIA Corporation. *NVIDIA CUDA^{TM} Version 2.1 Programming Guide*, 2008.

[24] NVIDIA Corporation. *NVIDIA CUDA^{TM} Programming Guide Version 3.2 RC*, 2010. `http://developer.nvidia.com/object/cuda_3_2_toolkit_rc.html`.

[25] Michael E. Palmer, Brian Totty, and Stephen Taylor. Ray Casting on Shared-Memory Architectures: Memory-Hierarchy Considerations in Volume Rendering. *IEEE Concurrency*, 6(1):20–35, 1998.

[26] Valerio Pascucci and Randall J. Frank. Global Static Indexing for Real-time Exploration of Very Large Regular Grids. In *Proceedings of the 2001 ACM/IEEE Conference on Supercomputing (CDROM)*, Supercomputing '01, New York, NY, USA, 2001. ACM.

[27] Tom Peterka, David Goodell, Robert Ross, Han-Wei Shen, and Rajeev Thakur. A Configurable Algorithm for Parallel Image-Compositing Applications. In *Proceedings of Supercomputing 2009*, Portland OR, November 2009.

[28] Tom Peterka, Hongfeng Yu, Robert Ross, Kwan-Liu Ma, and Rob Latham. End-to-End Study of Parallel Volume Rendering on the IBM Blue Gene/P. In *Proceedings of ICPP 09*, Vienna, Austria, 2009.

[29] S. Stegmaier, M. Strengert, T. Klein, and T. Ertl. A Simple and Flexible Volume Rendering Framework for Graphics-Hardware–based Raycasting. In *Proceedings of the International Workshop on Volume Graphics '05*, pages 187–195, 2005.

[30] C. Tomasi and R. Manduchi. Bilateral Filtering for Gray and Color Images. In *ICCV '98: Proceedings of the Sixth International Conference on Computer Vision*, page 839, Washington, DC, USA, 1998. IEEE Computer Society.

[31] S. Williams, K. Datta, L. Oliker, J. Carter, J. Shalf, and K. Yelick. Auto-tuning Memory-Intensive Kernels for Multicore. In David H. Bailey, Robert F. Lucas, and Samuel W. Williams, editors, *Performance Tuning of Scientific Applications*. CRC Press, Boca Raton, FL, USA, 2010.

Chapter 15

The Path to Exascale

Sean Ahern

Oak Ridge National Laboratory & University of Tennessee, Knoxville

15.1 Introduction 331
15.2 Future System Architectures 332
15.3 Science Understanding Needs at the Exascale 335
15.4 Research Directions 338
15.4.1 Data Processing Modes 338
15.4.1.1 *In Situ* Processing 338
15.4.1.2 Post-Processing Data Analysis 339
15.4.2 Visualization and Analysis Methods 341
15.4.2.1 Support for Data Processing Modes 341
15.4.2.2 Topological Methods 342
15.4.2.3 Statistical Methods 343
15.4.2.4 Adapting to Increased Data Complexity .. 343
15.4.3 I/O and Storage Systems 344
15.4.3.1 Storage Technologies for the Exascale 345
15.4.3.2 I/O Middleware Platforms 346
15.5 Conclusion and the Path Forward 346
References 349

The hardware and system architectural changes that will occur over the next decade, as high performance computing (HPC) enters the exascale era, will be dramatic and disruptive. Not only are scientific simulations forecasted to grow by many orders of magnitude, but also the current methods by which HPC systems are programmed and data are stored are not expected to survive into the exascale. Most of the algorithms outlined in this book have been designed for the petascale—not the exascale—and simply increasing concurrency is insufficient to meet the challenges posed by exascale computing. Changing the fundamental methods by which scientific understanding is obtained from HPC simulations is daunting. This chapter explores some research directions for addressing these formidable challenges.

15.1 Introduction

In February 2011, the Department of Energy Office of Advanced Scientific Computing Research convened a workshop to explore the problem of scientific understanding of data from HPC at the exascale. The goal of the workshop was to identify the research directions that the data management, analysis, and visualization community must take to enable scientific discovery for HPC at this extreme scale (1 exaflop = 1 quintillion floating point calculations per second). Projections from the international TOP500 list [9] place the availability of the first exascale computers at around 2018–2019.

Extracting scientific insight from large HPC facilities is of crucial importance for the United States and the world. The scientific simulations that run on supercomputers are only half of the "science"; scientific advances are made only once the data produced by the simulations is processed into output that is understandable by a scientist. As mathematician Richard Hamming said, "The purpose of computing is insight, not numbers" [17]. It is precisely the visualization and analysis community that provides the algorithms, research, and tools to enable that critical insight.

The hardware and software changes that will occur as HPC enters the exascale era will be dramatic and disruptive. Scientific simulations are forecasted to grow by many orders of magnitude, and also the methods by which current HPC systems are programmed and data is extracted are not expected to survive into the exascale. Changing the fundamental methods by which scientific understanding is obtained from HPC simulations is a daunting task. Dramatic changes to concurrency will reformulate existing algorithms and workflows and cause a reconsideration of how to provide the best scalable techniques for scientific understanding. Specifically, the changes are expected to affect: concurrency, memory hierarchies, GPU and other accelerator processing, communication bandwidth, and, finally, I/O.

This chapter provides an overview of the February 2011 workshop [1], which examines potential research directions for the community as computing leaves the petascale era and enters the uncharted exascale era.

15.2 Future System Architectures

For most of the history of scientific computation, Moore's Law [28] predicts the doubling of transistors per unit of area and cost every 18 months, which has been reflected in increased scalar floating point performance and increased processor core count while a fairly standard balance is maintained among memory, I/O bandwidth, and CPU performance. However, the exascale will usher in an age of significant imbalances between system components—sometimes by several orders of magnitude. These imbalances will necessitate a significant transformation in how all scientists use HPC resources.

Although extrapolation of current hardware architectures allows for a fairly strong prediction that one exaflop will be reached by some worldwide computational resource by the year 2020, the particulars of the system architecture are not yet assured. Currently, there are two schools of thought: multi-core and many-core. The multi-core design primarily uses traditional multi-core processors (albeit increased beyond the petascale) with a large number of system nodes. The many-core design is more disruptive in nature, using a large number of processors akin to today's GPUs with very wide vector execution. Although some blend of the two designs could emerge, the trends with many-core and multi-core may be usefully extrapolated. Table 15.1 (from *The Scientific Grand Challenges Workshop: Architectures and Technology for Extreme Scale Computing* [34]) shows potential multi-core and many-core exascale system architectures.

System Parameter	2011	"2018" Multi-core	"2018" Many-core	Factor Change
System Peak FLOPS	2 PF/s	1 EF/s		500
System Power	6 MW	$\leq$20 MW		3
Total System Memory	0.3 PB	32–64 PB		100–200
Total I/O Capacity	15 PB	300–1,000 PB		20–67
Total I/O Bandwidth	0.2 TB/s	20–60 TB/s		10–30
Total Concurrency	225K	1B$\times$10	1B$\times$100	40,000–400,000
Node Performance	125 GF	1 TF	10 TF	8–80
Node Memory Bandwidth	25 GB/s	400 GB/s	4 TB/s	16–160
Node Concurrency	12	1,000	10,000	83–830
Network Bandwidth	1.5 GB/s	100 GB/s	1,000 GB/s	66–660
System Size (nodes)	18,700	1,000,000	100,000	50–500

TABLE 15.1: Expected exascale architecture parameters and comparison with current hardware. From *The Scientific Grand Challenges Workshop: Architectures and Technology for Extreme Scale Computing.*

An examination of the figures in Table 15.1 leaves one strong conclusion: a machine capable of one exaflop of raw compute capability will not simply be a petascale machine scaled in each parameter by a factor of 1,000. The primary reason the scaling will not be consistent across all parameters is because of the need to control the power consumption of such a machine. An increase of a factor of 1,000 in power consumption is simply not sustainable due to the commensurate increase in cost. In considering how to exploit an exascale machine, certain implications become clear.

One implication is that the total concurrency of scientific simulations will increase by a factor of 40,000–400,000, but available system memory will increase by only a factor of 100. Consequently, the memory available per thread of execution will fall from around 1GB to 0.1–1MB—a factor of 1,000 less than in petascale systems. This dichotomy is likely to change how the systems are programmed, with memory management becoming much more dominant in the system architecture. Current application codes will no longer be able to exploit simple weak scaling as machines grow in size. Instead, codes, including algorithms for scientific data understanding, will have to greatly increase their exploitation of on-node parallelism using very scarce memory resources.

The second implication is that, primarily for reasons of energy efficiency, the locality of data will become much more important. On an exascale-class architecture, the most expensive operation from a power perspective will be moving data. The further the data is moved, the more expensive the process will be. Therefore, approaches that maximize locality as much as possible and pay close attention to data movement are likely to be the most successful. Although this is also the case at the petascale, it will become a much more dominant factor at the exascale. As well as locality between nodes, it will also be essential to pay attention to on-node locality, as the memory hierarchy is likely to be deeper. The importance of locality also implies that global synchronization will be very expensive, and the level of effort required to manage varying levels of synchronization will be high.

Finally, the last implication is that the growth in the external secondary storage system on an exascale machine, both in capacity and bandwidth, will be dramatically less than the growth in floating point operations per second (FLOPS) and concurrency. The relative decrease in I/O will have dramatic impacts upon the way data is moved off the HPC system. The relative performance of an I/O system often can be judged by measuring how long it will take to "checkpoint" the entire machine, that is, write the entire contents of the memory to the secondary storage system (spinning disk). Over the past 15 years of HPC, that time has grown steadily, from 5 minutes in 1997 to over 26 minutes in 2008 [7]. Extrapolations to the exascale vary between 40 and 100 minutes. Clearly, it will no longer be practical to quickly "dump" the current state of a simulation to disk for later analysis by a visualization tool. Both storing and analyzing simulation results are thus likely to require entirely new approaches. One immediate conclusion is that much analysis of simulation data is likely to be performed *in situ* with the simulation to minimize communication and I/O bandwidth to secondary storage. (See 15.4.1.1 for a more in-depth discussion.)

One bright note rings out that possibly mitigates some of these issues. New non-volatile memory technologies are emerging that may facilitate some of the dramatic changes needed for I/O optimization. Probably the most well known of these is NAND flash, because of its ubiquity in consumer electronics devices such as music players, phones, and cameras. The last few years have seen the first exploration of its use in HPC systems [6]. Because it enables significant improvements in read and write latency, non-volatile memory holds the promise of improving relative I/O bandwidth for small I/O operations. Adding a non-volatile memory device to the nodes of an HPC system provides the underlying hardware resource necessary to both improve checkpoint performance and provide a fast "swap" capability for these memory-constrained nodes.

Current visual data analysis and exploration platforms, such as VisIt [20] and ParaView [32], are well positioned to exploit current memory-parallel platforms and can scale adequately to today's leadership computing platforms [7]. But, given the new prominence of hybrid architectures in HPC, as well as the push toward very large core counts in the next several years, it is critical that the analysis community redesign existing visualization and analysis algorithms, and the production tools where they are delivered, for much better utilization of GPUs and multi-core systems. Some early work toward this goal has begun; see Chapter 12 and [26, 22, 29].

Heterogeneous architectures such, as GPUs, FPGAs, and other accelerators (e.g., Intel's Many Integrated Core, or MIC, architecture), are beginning to show promise for significant time savings over traditional CPUs for certain analysis algorithms. Future analysis approaches must take these emerging computational architectures into account even though there are substantial programmatic challenges in exploiting them. Scalable algorithms must consider hybrid cores, nonuniform memory, deeper memory hierarchies, and billion-way parallelism. MPI is likely to be inadequate at billion-way concurrency and does not provide portability to these new classes of computing architectures (e.g., GPUs, accelerators).

Transient hardware failures are expected to become much more common as core count goes up and system complexity increases. Consequently, like every other area of computer science, visualization and analysis may need to adopt some measure of fault-tolerant computing. Increasing core count is also likely to require changes in programming models.

15.3 Science Understanding Needs at the Exascale

Research leading toward data understanding and visualization at the exascale cannot be done in a vacuum; it must have a firm grounding in the expressed needs of the computational science communities. Eight reports from the Scientific Grand Challenges Workshop Series [40, 3, 43, 35, 31, 13, 33, 27] present a cross section of the grand challenges of their science domains. They contain the recommendations of domain-specific workshops for addressing the most significant technical barriers to scientific discovery in each field. Each report found that scientific breakthroughs are expected as a result of the dramatic expansion of computational power; but each also identified specific challenges to visualization, analysis, and data management. Some domains have certain unique needs, but a number of cross-cutting and common themes span the science areas:

- The widening gap between system I/O and computational capacity dramatically constrains science applications, requiring new methods that enable analysis and visualization to be performed while data is still resident in memory.
- Every scientific domain will produce more data than ever before, whether

for higher spatio-temporal resolution, expanded use of multivariate data to represent greater scientific complexity, or an increased use of ensemble simulation for the purpose of uncertainty quantification.

- Additional multiple orders of magnitude of computing capability will allow some science applications to generate data types not seen before in visualization and analysis.
- Many science areas will require dedicated computational infrastructure to facilitate experimental analysis.
- Science areas that employ ensembles, explore "what if?" scenarios, and perform parameter studies will require advances in scientific workflow management systems.

The need for *in situ* processing capabilities for huge data sets was specifically noted in at least two reports. In high energy physics research, the entire software repertoire must be updated significantly to produce and analyze 100PB data sets. Scientists in that domain recognize the necessity of a change from FLOPS-heavy petascale systems to data-intensive engines with a strong focus on end-to-end analysis and *in situ* visualization. There is a particular need for on-the-fly analysis and strategies to overcome I/O bottlenecks. Fusion energy research is another field in which *in situ* processing has the potential for a dramatic positive impact. The fusion community also has specific I/O challenges due to a few successful codes that are dramatically increasing their rate of data production. The high output is already placing severe stress on current technologies for data processing. The Basic Energy Sciences community cites a need for real-time data processing as well.

Several fields express a need for higher resolution modeling of complex systems. In climate science, for example, higher spatial resolution—required to correctly model important climate systems like cloud formation and atmospheric energy transfer—will increase mesh resolutions by at least four orders of magnitude. Temporal resolution is expected to increase by one to two orders of magnitude as climate scientists explore decadal predictive capabilities and paleoclimate studies. Accurate climate simulation requires multiscale modeling, from processes taking minutes—birth of cloud drops, ice crystals, and aerosols—to circulation systems taking significantly longer times. The nuclear energy research community, also, is considering a wide range of modeling efforts—including safety simulation, materials behavior, reactor core modeling, and full systems integration—likely to generate data sets that will challenge processing technologies. The simulation time scales range from femtoseconds (atomistic formation of defect clusters) to years (full engineering simulations). In computational biology, researchers face challenges ranging from modeling the connections between biomolecules and the work of entire organisms, to the simulation and understanding of the human brain.

In a similar way, biological research, because of the emergence of high-throughput DNA sequencing, has long focused on data-intensive analysis.

Large data integration and validation, with data volumes from sequencing equipment as well as HPC data sets, are a core need for extreme-scale computational biology.

Climate simulation was one of the first fields to significantly embrace uncertainty quantification through the collection of simulation runs called ensembles, which increase the resultant data set size and analysis load. For nuclear energy research, the use of HPC to create extremely large ensemble collections, some with as many as one million members, goes far beyond current capabilities to perform uncertainty quantification. In addition, the use of large energy groups to accurately model neutron energy transport in fission-based power systems is not well supported by current analysis tools, and the significant increase in energy group discretization will challenge existing techniques in multivariate analysis.

Scientists in many fields need to harness HPC to facilitate the analysis of experimental data and to develop simulations validated by experiment results. The fusion community seeks to develop a predictive simulation capability for magnetically confined fusion plasmas that are validated against physical experiments. The challenges are time-sensitive, as the success of efforts such as the international ITER project hinge upon achieving the desired plasma performance. In addition to the analysis difficulties common to the other science areas, fusion simulation brings new requirements for experiment-to-simulation data comparison and real-time computational monitoring. These scientists have also expressed a need for simulation-to-experiment data validation.

Many in the scientific community express confidence that exascale computing has the potential to revolutionize the process of discovering new materials. However, its members anticipate that new methods will be necessary for image processing, advanced statistical analysis, automated data mining, and multi-dimensional histogramming. Similar concerns are voiced in the nuclear physics report, which warns that exascale modeling will uniquely stretch the field of visualization and analysis in the representation and processing of high-dimensional data spaces.

The development of *in situ* processing techniques to mitigate I/O shortfalls will not obviate post-processing capabilities. Computational science communities must continue to store and manage data for further analysis and/or for collaborative research. Computational climate research, particularly, is subject to intense scrutiny and collaboration, and the climate data sets produced by HPC simulation are ultimately used by a large international community. This community predicts its data sets will collectively range into the hundreds of exabytes by 2020. Because much of the analysis of this data will occur long after simulation ensembles have been calculated, *in situ* processing does not solve the data explosion problem. In addition, the challenges of analyzing and visualizing climate simulation results are not limited to the HPC centers where the results are generated. As climate researchers, analysts, and policy makers are distributed worldwide, so too must the many exabytes of simulation results. Distributing such amounts of data in a federated storage and analysis infrastructure will tax future networks. And managing the dramatic increase in data complexity will require significant advances in data provenance tracking, metadata management, and data extraction techniques.

15.4 Research Directions

The following sections outline the results of the breakout groups at the exascale workshop, which explored the driving forces behind exascale analysis needs and possible solutions. Many of the preceding chapters of this book flesh out some of these ideas further. In the next subsections: 15.4.1 summarizes the challenges that the exascale poses for current visualization and analysis interaction modes and gives a synopsis of the push toward concurrent or *in situ* processing; 15.4.2 explores the gaps in understanding due to changes in user needs and the new types of visualization and analysis methods required to understand new forms of data generated by scientific simulation codes; and 15.4.3 considers dramatic changes to the storage hierarchy of HPC systems, including techniques to include storage considerations in any data analysis framework.

15.4.1 Data Processing Modes

The architectural changes inherent in the drive to exascale, as described in 15.2, are likely to have the most significant impact on the methods by which scientific discovery is performed, that is, *how* data from HPC systems is processed. In *post-processing*, the traditional method of discovery, simulation data files are exported from the simulation software to secondary storage; applications that organize, analyze, and visualize the data sets are run after the simulation has finished. Complete data sets from simulations can be stored indefinitely for further analysis and visualization. *In situ processing* integrates the simulation and the data analysis and visualization operations, running them concurrently on the same computer. Data files may not be saved to storage, or may be stored at a much coarser resolution than the simulation resolution. One of the strongest messages from the February 2011 workshop was that hardened methods for processing data concurrently with simulations will become essential in the next 5–8 years. However, concurrent processing will not make post-processing analysis obsolete. Indeed, scientific applications like climate simulation are fundamentally dependent on post-processing for scientific discovery long after the simulation has run. The following two subsections discuss the two methods.

15.4.1.1 *In Situ* Processing

Post-processing is by far the dominant method for scientific discovery on today's supercomputers. However, the dramatic drop-off of relative I/O bandwidth, as detailed in 15.2, will severely constrain this paradigm, because it will force simulations to reduce the amount of data written out to disk. In addition to the relative reduction in I/O, exascale power limitations will discourage moving data between nodes. These constraints point to moving analysis methods as "close" to the data as possible: the point when it is being generated on the source computational system.

Indeed, because data reduction must occur in the scientific analysis pipeline in extreme-scale computing, only during the simulation will all relevant data about the simulated field, computational domain, and related fields be available at the highest fidelity. This scenario is key to *in situ* processing—that data must be intelligently reduced, analyzed, transformed, and indexed while still resident in memory, before leaving the machine.

Though *in situ* processing has achieved notable successes over the past 20 years [16, 18, 24, 44, 41], the method is not a panacea. There are significant barriers to the successful adoption and the fundamental limitation of this method. Moving to a tightly-coupled software model is a significant departure from current methods, one where the simulation code and the analysis code share the same memory space and threads of execution. It involves substantial software development costs, including instrumenting the simulation and developing *in-situ*-appropriate analysis routines. In addition, the *in situ* analysis code will necessarily consume memory, computation, and network bandwidth that currently are consumed solely by the simulation. And the current state of memory-, computation-, and network-limited analysis routines is not mature. As is always the case with tight code coupling, resilience is an issue: analysis routines may create additional failure modes. More fundamentally, *in situ* analysis, by its nature, is limited to operations that we know to perform *a priori*. It is not open to serendipitous discovery or analysis after the simulation has terminated, because the data generated will no longer be available.

However, at a larger level, *in situ* analysis can be considered part of a pipeline of processing that starts at the simulation code and continues through the network and on to secondary storage for subsequent analysis. Considered in this light, *in situ* analysis can be an ideal deployment paradigm for general and domain-specific data reduction that is later written to disk for further analysis. A workflow that successfully blends *in situ* analysis with data post-processing is shown in Figure 15.1.

15.4.1.2 Post-Processing Data Analysis

Among the many advantages of post-processing, or "offline" processing, data analysis are ease of implementation compared with *in situ*, support for serendipitous discovery, and accommodation of any type of data analysis technique. Post-processing is not going away; however, the dramatic disparity between expected FLOPS versus memory capacity and I/O bandwidth expected from exascale architectures requires that this successful model evolve. Future

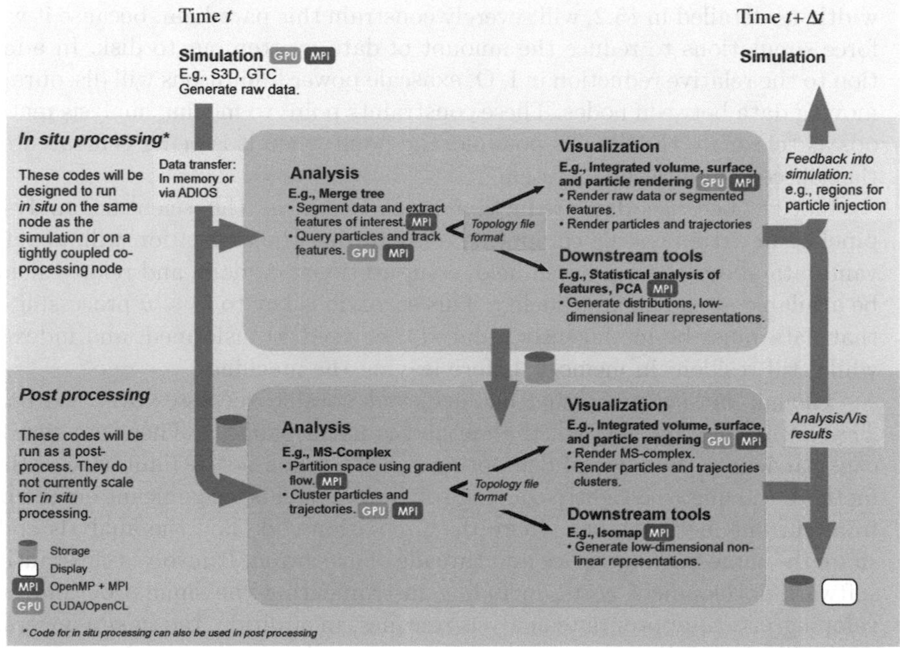

FIGURE 15.1: A workflow that blends *in situ* and post-processing elements for processing combustion and fusion simulations. This diagram was adapted from slides presented at the exascale workshop by Dr. Jacqueline Chen, Sandia National Laboratories.

analytic approaches will need to consider power efficiency and scalability at previously unexplored levels.

Note that the per-node architectural changes that will enable exascale computing (e.g., high levels of concurrency, low memory per thread, low I/O per core) will also be found on smaller resources such as desktops and laptops. Thus the changes to post-processing necessary on the largest HPC resources will also be required for processing even smaller data sets on local computers.

There are a few research techniques that hold the promise of alleviating some of the burdens of future architectures.

One technique uses *out-of-core methods*, which operate in constrained memory spaces by directly managing secondary storage as part of the analysis process, saving intermediate values of a computation to disk and reading them back as needed and thereby conserving relatively scarce memory resources. As these methods are generally bound to a system's I/O performance, they may not adapt to future architectures in which relative I/O will dramatically decrease. However, the advent of solid-state drives and other local secondary storage mechanisms have the potential for low-power, low-latency, and high-capacity I/O within a single node. As these types of intermediate storage devices appear in HPC systems, out-of-core analytics may see a resurgence at the node level.

The use of *index-based computation* has been demonstrated to reduce the amount of computation necessary for data selection (e.g., identifying all field values that fall within a specified range) and processing by several orders of magnitude [22] (and see 7.4.2). Simple statistical calculations such as mean, median, maximum, and minimum can be easily incorporated into these methods. And, pre-generating 1D histograms using indexing has enabled scientific discovery with a significant drop in I/O cost [42].

The huge increase in FLOPS at the exascale will spur an increase in ensemble calculations to explore the stability and uncertainty of simulation results. Measuring the extent of physical response to a range of input conditions lands firmly in the field of *uncertainty quantification*. As this analysis is necessarily performed after the simulation set has finished, global uncertainty quantification can only be performed in a post-processing modality. The best example of the need for uncertainty quantification at the petascale is climate simulation (see 15.3). Nuclear energy simulation is also expected to employ this technique of data exploration, and its ensembles are expected to have as many as one million members. Though uncertainty visualization has been a topic of research for over a decade, visualizing uncertainty and rigorous uncertainty quantification is still in its infancy and will be stressed even further by the exascale data explosion.

15.4.2 Visualization and Analysis Methods

The increase in computational capacity expected over the next decade is expected to allow exploration of scientific disciplines and methods previously not approachable through computational science. For example, in computational climate simulation, global climate models will include more than just the carbon cycle, adding methane and nitrogen cycles for a more complete modeling of the world's atmosphere. In addition, sea ice, an important factor for understanding sea level rise and ocean circulation modeling, will be a noticeable addition to global computational climate models. New areas of scientific exploration always require new methods for analysis. This subsection discusses some new analysis methods expected to apply to both the new areas of scientific discovery and the coming hardware architectures.

15.4.2.1 Support for Data Processing Modes

Algorithm development must consider the changes in modes of data processing during the transition to the exascale. *In situ* processing is a necessary technology, especially given the scientific need for real-time monitoring and analysis. Co-scheduling is a similar deployment mechanism that retains many direct coupling characteristics, including full access to all simulation data. In this mode, analysis and visualization algorithms run at the same time as the

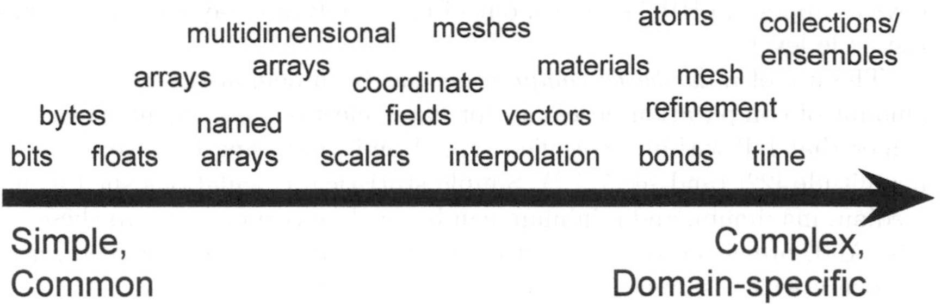

bits floats arrays scalars interpolation bonds time
bytes named fields vectors refinement
arrays arrays coordinate materials mesh ensembles
multidimensional meshes atoms collections/
Simple, Common
Complex, Domain-specific

FIGURE 15.2: A spectrum of scientific data model elements. A shared data model can contain a representation of a range of scientific concepts, depending on how domain-specific the data model must be. Elements more commonly shared among disparate data models are on the left, and less common and more domain-specific elements progress to the right without end.

simulation code but on separate nodes of the HPC platform, with data transferring across the high-speed interconnect between them. Co-scheduling has been identified as a key research area and appears to be fertile ground for I/O middleware (see 15.4.3.2). Finally, post-processing will remain a fundamental exploratory mode (see 15.4.1.2). When data is generated for post-processing, I/O is critical. As simulations become increasingly constrained by the time to write, the written formats may not be optimal for future analysis [30]. And generally, little, if any, metadata is written to assist the analysis. Performing the appropriate data abstraction rapidly requires data formats that are friendly to both simulation I/O and visualization and analysis algorithms. The formats must have an efficient layout for I/O and hierarchical layouts for better scales of navigation [2]. Thus, research into general and domain-specific scientific data formats is critical for the exascale.

Scientific data has less value unless a common data model is shared between the code writing the data to the secondary storage and the analysis algorithms. This common data model, sometimes called a schema, spans a spectrum of concepts, from the convention that bytes are made up of eight binary bits on one end of the spectrum, all the way to very domain-specific concepts on the other (see Fig. 15.2). At the exascale, as the scientific concepts simulated become more complex and the data becomes more distributed, these data formats and models will become even more critical. They are the fundamental underpinnings of the middleware, I/O subsystems, and analysis algorithms that enable scientific discovery.

15.4.2.2 Topological Methods

In recent years, new data analysis techniques based on the mathematics of topology have risen to prominence for scientific understanding as a result of their ability to perform feature extraction in a robust fashion [19, 15, 8,

4, 5, 25]. The methods are generally insensitive to numerical approximation and noise and achieve tremendous data reduction while still representing full families of scientific feature spaces. Though the prospects for their application to exascale data are great, much fundamental research is needed to realize their potential. For example, only recently has parallelization of these methods been attempted, as many topological constructs require global propagation of information; global propagation is increasingly expensive at greater levels of concurrency. The extension of the scalar Morse theory [10, 11] to more general vector fields could be valuable in many areas of scientific exploration as a means for analysis. Advances in topological methods will require the development of scientific data formats, providing access to compact feature space representations that can be computed *in situ* for later exploration.

15.4.2.3 Statistical Methods

Although the focus of this chapter is visualization, adapting statistical analysis routines to the exascale can contribute greatly to scientific understanding. Statistical sampling theory is a mature field that may allow for rigorous uncertainty quantification while reducing data loads, but it needs further research in order to apply it to future computing architectures. Often, statistical analysis algorithms cover both the spatial and the temporal domains, making the application to *in situ* processing more difficult. This is true of both Bayesian and frequentist methods. These algorithms will require careful re-engineering to run in a single pass through data while preserving mathematical utility and rigor.

Scientific visualization and statistical analysis are very complementary in the analysis of large data sets. Ideally, these two fields will increasingly see close coupling to provide an even richer arsenal of techniques for data understanding. Given this coupling, as well as the expected I/O limitations of the exascale, it is likely that there will be convergences, like the visualization of statistics.

15.4.2.4 Adapting to Increased Data Complexity

The size and the complex structures of data sets produced by increasingly high-resolution simulations pose challenges that require new abstraction tools. Many of the science application areas explored in 15.3, for example, express an increased need for multiphysics and multiscale modeling. Petascale computing has already resulted in an increase in code coupling, running multiple simulation codes or algorithmic cores that operate at different physical or temporal scales or computing different physics. These coupled codes, which provide input to each other, and existing analysis infrastructures often do not have the capability to explore the coupled results; most parallel visualization tools are designed for deep analysis of a single simulation model, not many at one time. As is always the case, scientific understanding is limited by the ability of the analysis and visualization algorithms to expose the complexity and richness of the underlying data model. Research into novel techniques such as machine learning, statistical analysis, feature identification and tracking, and dimensionality reduction would greatly facilitate data understanding.

Current methods for exploring data abstractions have several limitations. Fusion science, for example, requires better techniques for filtering, searching, and analyzing particles from particle-in-cell simulation codes. Nuclear energy communities simulating multigroup radiation fields—fields that define a function at every grid location—do not have sufficient multivariate exploration tools. Members of the Basic Energy Science community have expressed a need for tools to perform automated data mining, multi-dimensional histogramming, data inversion, and image reconstruction. Many scientific communities are requiring much deeper adaptive mesh refinement (AMR) structures, even to the point of refining individual mesh elements rather than employing block-structured AMR. High-order fields have also become more common as physics algorithms become more complex. Researchers are beginning to generate non-mesh data structures such as graphs and hierarchical trees and apply them to scientific simulations. Scientists are increasingly requiring analysis algorithms that span time periods, exposing time-varying phenomena (e.g., in climate simulations). To provide scientific understanding across such complex data structures, new techniques are needed, including tensors, topology, information theory, and data mining.

The exascale is not unique in ushering in new data models. To be sure, the history of computing has been one of developing new ways of mathematically modeling natural and man-made phenomena. But the dramatic growth of generated data, combined with relative I/O constraints, makes the need for multiphysics data understanding more keenly felt.

15.4.3 I/O and Storage Systems

As system FLOPS scale up by hundreds of times, I/O capacity increases by only factors of a few tens at best. Thus the ability of I/O to make data available is decreasing dramatically, relative to the size of data sets. Even at the petascale, I/O is already a significant challenge for scientific applications and is starting to severely limit the amount of "science" that can be performed on the largest machines. In a real sense, scientific understanding can only be achieved if the output of the calculations, the generated simulation data, exits the HPC resource and enters the mind of a curious scientist. Limitations on the data output of a supercomputer directly impact the information rate that the scientist can employ. Reducing the effects of I/O limitations on calculations is a significant challenge for researchers: reducing the I/O load allows greater computation, but it means less data is available for subsequent visualization and analysis. Balancing the two is a major challenge that will increase immensely into the exascale, and the ratio between FLOPS (and thus, computational capability) and I/O bandwidth changes dramatically for the worse (see 15.2).

The field of I/O research has been profitable at the terascale and petascale. It is possible to create self-describing file formats that have straightforward APIs for simulation codes (e.g., Silo [21], HDF5 [37], Parallel NetCDF [38], and ADIOS [23]), and many of the underlying communication methods use I/O middleware such as MPI-IO [36] to provide high performance. Important considerations for the future are ease of user adoption, percentage of peak I/O bandwidth for both read and write operations, resiliency to failure, and overall energy use.

Current and future storage systems, I/O middleware that bridges between simulations code and the underlying storage systems, and ways of coupling these with scientific data models will all play a part in managing the impacts of diminishing I/O to FLOPS ratios.

15.4.3.1 Storage Technologies for the Exascale

At the exascale, the expected data set sizes will require greater attention to the large-scale organization and management of the data produced by scientific simulations. The exascale infrastructure will require active management at three distinct levels:

1. navigating the various deep memory hierarchies at the *node level*, including persistent NVRAM;
2. managing data movement between nodes along the *interconnect*, considering application performance and power consumption; and,
3. considering I/O load balance across the *entire exascale machine*, with an eye toward managing data hot spots and reducing contention.

The environment of an exascale node is expected to contain multiple memory sockets for main memory: multiple memory coherence domains (i.e., "clusters" on a chip); multiple types of memory, including NVRAM; and disjoint memory accessible across in-node communication buses. The memory and power trade-offs between data movement on-node vs. movement off-node for subsequent analysis are areas relevant for future research. *In situ* analysis allows for power savings through reduced data movement but may degrade performance of simulation codes. Other solutions to consider are moving select data to other on-node memory (e.g., persistent, discrete) before analysis is performed on-node. In most situations, there is an opportunity to perform data reduction operations on data in flight as it is being moved [23, 39]. The fundamental questions then remain: *which* operations to apply *where* on-node and *when*, that is, whether in concert with the simulation or asynchronously during data movement or on other resources.

Analysis algorithms that move data across the large-scale interconnect incur significant power and performance costs. To reduce those costs, it may be advantageous to move data to a smaller number of nodes prior to analysis. This method, *in situ* data staging or co-scheduling, represents a middle ground between *in situ* analysis and data post-processing. It offers an opportunity to provide different hardware for analysis, such as larger memory and substantial NVRAM. Such specialized data staging nodes, although they encompass a small portion of an HPC system, may allow: a better use of the interconnect, more complex analysis than is feasible on-node (because memory is larger), and the potential ability to operate on multiple time steps at once.

The secondary storage system for HPC systems is frequently a large parallel file system attached to the periphery of the large compute resources or residing on separate network-connected machines. The exascale is not likely to bring a major architectural change in this component. However, the ratio of aggregate FLOPS and bandwidth to the secondary storage system is expected to drop precipitously (see Table 15.1). Thus, the need to move data analysis "closer" to the source of the data is of prime importance over the next several years.

15.4.3.2 I/O Middleware Platforms

Visualization researchers have deployed solutions at the desktop for many decades and are beginning to deliver solutions *in situ* within the simulation codes themselves. But another platform for algorithm deployment is beginning to emerge. Libraries like ADIOS [23] and GLEAN [39] provide simple APIs that bridge the gap between simulation codes and the underlying storage subsystem. These middleware libraries, employing parallel communication and data conversion, can be leveraged for analysis, visualization, and data reduction while data is "in flight." (See the discussion of Adaptors in 9.3.)

As HPC resources grow in size, the task of globally managing I/O across the entire system becomes more complex. I/O contention becomes more common, decreasing the performance of the simulation codes. Future hardware architectures that provide multiple tiers of storage and remote memory access will complicate the task. The responsibility for I/O management falls most readily to these I/O middleware libraries and subsystems. The inclusion of MPI I/O [36] in standard MPI distributions, including MPICH2 [14] and OpenMPI [12], has allowed middleware developers to deliver portable methods of managing both I/O and parallel communication. But there continue to be significant challenges, including I/O staging (see 15.4.3.1), managing system jitter through quality-of-service protocols, and navigating the diverse network topologies expected on future interconnects. Even single-node architectures are expected to have multiple pathways for communication, increasing the need for I/O middleware to appropriately manage the resources. Research into managing I/O must also rest upon efficient data storage methods, allowing efficient layout of data on disk, collectives for large data reduction, and on-the-fly transformation of data into formats suitable for data post-processing.

15.5 Conclusion and the Path Forward

Techniques for scientific discovery, visualization, analysis, and data movement will undergo a disruptive change resulting from the hardware and architectural changes accompanying the move to exascale computing. Only by directly tackling and addressing the limitations of exascale platforms can the scientific research community continue to employ HPC platforms for significant scientific advancement and a competitive advantage.

The high concurrency expected at the exascale will strain the ability of both scientific simulations and analysis and visualization codes to scale to these levels. Therefore, research efforts for scientific discovery should closely track the development of new programming models to evaluate alternatives to the traditional MPI and thread models of parallel programming. As the choice of successful model will not be clear for many years, this will be ongoing work that impacts both post-processing and *in situ* approaches.

The dramatic drop in I/O relative to FLOPS will make it impractical for simulations to frequently write data at full spatial (and temporal) resolution. The conventional post-processing mode of data analysis, allowing the significant decoupling of the simulation and analysis tools, will necessarily become less common for scientific discovery (though still crucial for several scientific domains). Increasingly, analysis and visualization will be necessary while the simulation data resides in memory on the large HPC system. As a result, significant advances in *in situ* visualization and analysis are likely required to facilitate discovery at the exascale. Although there have been notable successes with specific applications, general infrastructures for *in situ* and concurrent analysis must be developed in a broadly applicable manner. These infrastructures will require the ability to constrain themselves to low-memory environments and tight coupling with simulation codes. This requirement also has implications for resiliency and graceful failure.

Although *in situ* analysis likely will increase in prevalence over the next several years, its restrictions limit data exploration, serendipitous discovery, and uncertainty quantification. Traditional post-processing techniques must also evolve to embrace data generated indirectly by *in situ* methods. This co-evolution should see a blending of the two techniques.

Just as the drop in relative I/O will see an increase in the use of *in situ* techniques, it will also cause dramatic changes in how data is moved by simulation codes, especially to secondary storage. Data management research, including frameworks to effectively manage the memory hierarchies of future computing architectures, must focus on minimizing data movement among these hierarchies, especially to external storage. And just as importantly, visualization and analysis practitioners must co-develop these frameworks to provide fertile ground for the deployment of analysis, visualization, and data reduction algorithms.

The reduction in relative I/O constrains the ability to write out "raw" data sets to secondary storage. To counter this limitation, improved algorithms are needed that can massively reduce the data sets generated by simulation codes. Abstract representations of features, statistical and topological representations, and general compression methods are needed to facilitate this massive reduction while retaining the semantic context needed for scientific understanding.

The trends from the Scientific Grand Challenge Workshop Series reports indicate that simulations continue to approach more complete and complex physical models of nature that involve multiple simultaneous temporal and spatial scales as well as multiple coupled physics models. Such advances, although welcome, strain current techniques for visualization and analysis and necessitate continued research into new techniques for visualization and analysis of physical properties. Often, such research may be domain-specific; close collaboration between visualization scientists and domain scientists will increasingly be necessary.

Finally, the history of HPC includes a high degree of "decoupling" of the various elements. The most successful hardware platforms are general in nature, designed independently from the scientific simulation codes. The simulation codes usually store results independently from the analysis codes, leaving analysis options unconstrained. This decoupling allows each field to concentrate on its own discipline, but the predictions of exascale computing threaten all these areas of independence. Science might better be served by focusing more on discovery and less on computer technology: by considering what discoveries need to be made instead of what computations may be performed on a given platform. Therefore, the focus on the end result would then drive all subsequent activities, including: what physics should be included in a given simulation run, how it should be run, what data should be generated, what analyses should be performed on that data, and how the results should be presented to the scientist. Attention to the end scientific result would then drive hardware design, software infrastructures, analysis research, and machine procurement. It has the prospect of changing the very nature of how scientific discovery is achieved on the largest computational architectures.

References

[1] Sean Ahern, Arie Shoshani, Kwan-Liu Ma, Alok Choudhary, Scott Klasky, Valerio Pascucci, Jim Ahrens, E. Wes Bethel, Hank Childs, Jian Huang, Ken Joy, Quincey Koziol, Gerald Lofstead, Jeremy Meredith, Kenneth Moreland, George Ostrouchov, Michael Papka, Venkatram Vishwanath, Matthew Wolf, Nicholas Wright, and Kesheng Wu. Scientific Discovery at the Exascale: Report from the DOE ASCR 2011 Workshop on Exascale Data Management, Analysis, and Visualization, Houston, TX, USA, February 2011.

[2] James P. Ahrens, Jonathan Woodring, David E. DeMarle, John Patchett, and Mathew Maltrud. Interactive Remote Large-Scale Data Visualization via Prioritized Multi-Resolution Streaming. In *Proceedings of the 2009 Workshop on Ultrascale Visualization*, November 2009.

[3] Roger Blandford, Young-Kee Kim, Norman Christ, et al. Challenges for the Understanding the Quantum Universe and the Role of Computing at the Extreme Scale. Technical report, ASCR Scientific Grand Challenges Workshop Series, December 2008.

[4] P.-T. Bremer, G. Weber, V. Pascucci, M. Day, and J. Bell. Analyzing and Tracking Burning Structures in Lean Premixed Hydrogen Flames. *IEEE Transactions on Visualization and Computer Graphics*, 16(2):248–260, 2010.

[5] P.-T. Bremer, G. Weber, J. Tierny, V. Pascucci, M. Day, and J. B. Bell. Interactive Exploration and Analysis of Large Scale Simulations Using Topology-based Data Segmentation. *IEEE Transactions on Visualization and Computer Graphics*, 17(9):1307–1324, September 2011.

[6] David Camp, Christoph Garth, Hank Childs, Dave Pugmire, and Kenneth I. Joy. Streamline Integration Using MPI-Hybrid Parallelism on a Large Multicore Architecture. *IEEE Transactions on Visualization and Computer Graphics*, 17(11):1702–1713, November 2011.

[7] Hank Childs, David Pugmire, Sean Ahern, Brad Whitlock, Mark Howison, Prabhat, Gunther Weber, and E. Wes Bethel. Extreme Scaling of Production Visualization Software on Diverse Architectures. *IEEE Computer Graphics and Applications*, 30(3):22–31, May/June 2010. LBNL-3403E.

[8] M. Day, J. Bell, P.-T. Bremer, V. Pascucci, V. Beckner, and M. Lijewski. Turbulence Effects on Cellular Burning Structures in Lean Premixed Hydrogen Flames. *Combustion and Flame*, 156:1035–1045, 2009.

[9] J. J. Dongarra, H. W. Meuer, H. D. Simon, and E. Strohmaier. TOP500 Supercomputer Sites. `www.top500.org`, (updated every 6 months).

[10] H. Edelsbrunner, D. Letscher, and A. Zomorodian. Topological Persistence and Simplification. *Discrete Computational Geometry*, 28:511–533, 2002.

[11] R. Forman. Combinatorial Vector Fields and Dynamical Systems. *Mathematische Zeitschrift*, 228(4):629–681, 1998.

[12] Edgar Gabriel, Graham Fagg, George Bosilca, Thara Angskun, Jack Dongarra, Jeffrey Squyres, Vishal Sahay, Prabhanjan Kambadur, Brian Barrett, Andrew Lumsdaine, Ralph Castain, David Daniel, Richard Graham, and Timothy Woodall. Open MPI: Goals, Concept, and Design of a Next Generation MPI Implementation. In *Recent Advances in Parallel Virtual Machine and Message Passing Interface*, volume 3241 of *Lecture Notes in Computer Science*, pages 353–377. Springer Berlin/Heidelberg, 2004.

[13] Guilia Galli, Thom Dunning, et al. Discovery in Basic Energy Sciences: The Role of Computing at the Extreme Scale. Technical report, ASCR Scientific Grand Challenges Workshop Series, August 2009.

[14] William Gropp. MPICH2: A New Start for MPI Implementations. In *Recent Advances in Parallel Virtual Machine and Message Passing Interface*, volume 2474 of *Lecture Notes in Computer Science*, pages 37–42. Springer Berlin/Heidelberg, 2002.

[15] Attila Gyulassy, Vijay Natarajan, Mark Duchaineau, Valerio Pascucci, Eduardo M. Bringa, Andrew Higginbotham, and Bernd Hamann. Topologically Clean Distance Fields. *IEEE Transactions on Computer Graphics, Proceedings of Visualization 2007*, 13(6), November/December 2007.

[16] Robert Haimes. pV3: A Distributed System for Large-Scale Unsteady CFD Visualization. In *32nd AIAA Aerospace Sciences Meeting and Exhibit*, Reno, NV, USA, January 1994. paper 94–0321.

[17] Richard Hamming. *Numerical Methods for Scientists and Engineers.* 1962.

[18] C. R. Johnson, S. Parker, and D. Weinstein. Large-Scale Computational Science Applications Using the SCIRun Problem Solving Environment. In *Proceedings of the 2000 ACM/IEEE Conference on Supercomputing*, 2000.

[19] D. Laney, P.-T. Bremer, A. Mascarenhas, P. Miller, and V. Pascucci. Understanding the Structure of the Turbulent Mixing Layer in Hydrodynamic Instabilities. *IEEE Trans. Visualization and Computer Graphics (TVCG) / Proceedings of IEEE Visualization*, 12(5):1052–1060, 2006.

[20] Lawrence Livermore National Laboratory. *VisIt User's Manual*, October 2005. Technical Report UCRL-SM-220449.

[21] Lawrence Livermore National Laboratory. *Silo User's Guide*, 2010. Document Number LLNL-SM-453191.

[22] Li-ta Lo, Christopher Sewell, and James Ahrens. PISTON: A Portable Cross-Platform Framework for Data-Parallel Visualization Operators. In *Eurographics Symposium on Parallel Graphics and Visualization*, 2012.

[23] J. Lofstead, Zheng Fang, S. Klasky, and K. Schwan. Adaptable, Metadata Rich I/O Methods for Portable High Performance I/O. In *IEEE International Symposium on Parallel & Distributed Processing (IPDPS)*, pages 1–10, 2009.

[24] K.-L. Ma, C. Wang, H. Yu, and A. Tikhonova. In Situ Processing and Visualization for Ultrascale Simulations. *Journal of Physics (also Proceedings of SciDAC 2007)*, 78, June 2007.

[25] A. Mascarenhas, R. W. Grout, P.-T. Bremer, E. R. Hawkes, V. Pascucci, and J. H. Chen. *Topological Feature Extraction for Comparison of Terascale Combustion Simulation Data*. Mathematics and Visualization. Springer, 2010.

[26] Jeremy S. Meredith, Robert Sisneros, David Pugmire, and Sean Ahern. A Distributed Data-Parallel Framework for Analysis and Visualization Algorithm Development. In *Proceedings of the 5th Annual Workshop on General Purpose Processing with Graphics Processing Units*, GPGPU-5, pages 11–19, New York, NY, USA, 2012. ACM.

[27] Paul Messina, David Brown, et al. Scientific Grand Challenges in National Security: The Role of Computing at the Extreme Scale. Technical report, ASCR Scientific Grand Challenges Workshop Series, October 2009.

[28] Gordon E. Moore. Cramming More Components onto Integrated Circuits. *Electronics*, 38(8), April 1965.

[29] K. Moreland, U. Ayachit, B. Geveci, and Kwan-Liu Ma. Dax Toolkit: A Proposed Framework for Data Analysis and Visualization at Extreme Scale. In *IEEE Symposium on Large Data Analysis and Visualization (LDAV)*, pages 97 -104, Oct. 2011.

[30] Tom Peterka, Hongfeng Yu, Robert Ross, Kwan-Liu Ma, and Rob Latham. End-to-End Study of Parallel Volume Rendering on the IBM Blue Gene/P. In *Proceedings of ICPP 09*, Vienna, Austria, 2009.

[31] Robert Rosner, Ernie Moniz, et al. Science Based Nuclear Energy Systems Enabled by Advanced Modeling and Simulation at the Extreme Scale. Technical report, ASCR Scientific Grand Challenges Workshop Series, May 2009.

[32] Amy Henderson Squillacote. *The ParaView Guide: A Parallel Visualization Application.* Kitware Inc., 2007.

[33] Rick Stevens, Mark Ellisman, et al. Opportunities in Biology at the Extreme Scale of Computing. Technical report, ASCR Scientific Grand Challenges Workshop Series, August 2009.

[34] Rick Stevens, Andrew White, et al. Architectures and Technology for Extreme Scale Computing. Technical report, ASCR Scientific Grand Challenges Workshop Series, December 2009.

[35] Bill Tang, David Keyes, et al. Scientific Grand Challenges in Fusion Energy Sciences and the Role of Computing at the Extreme Scale. Technical report, ASCR Scientific Grand Challenges Workshop Series, March 2009.

[36] Rajeev Thakur, William Gropp, and Ewing Lusk. On Implementing MPI-IO Portably and with High Performance. In *Proceedings of the 6th Workshop on I/O in Parallel and Distributed Systems*, pages 23–32. ACM Press, 1999.

[37] The HDF Group. HDF5 user guide. `http://www.hdfgroup.org/HDF5/doc/UG/index.html`, 2011.

[38] Unidata. The NetCDF users' guide. `http://www.unidata.ucar.edu/software/netcdf/docs/netcdf/`, 2011.

[39] Venkatram Vishwanath, Mark Hereld, and Michael E. Papka. Simulation-Time Data Analysis and I/O Acceleration on Leadership-Class Systems Using GLEAN. In *Proceedings of the IEEE Symposium on Large Data Analysis and Visualization*, October 2011.

[40] Warren Washington et al. Challenges in Climate Change Science and the Role of Computing at the Extreme Scale. Technical report, ASCR Scientific Grand Challenges Workshop Series, November 2008.

[41] B. Whitlock, J. M. Favre, and J. S. Meredith. Parallel In Situ Coupling of Simulation with a Fully Featured Visualization System. In *Proceedings of Eurographics Parallel Graphics and Visualization Symposium*, April 2011.

[42] K. Wu, S. Ahern, E. W. Bethel, J. Chen, H. Childs, E. Cormier-Michel, C. Geddes, J. Gu, H. Hagen, B. Hamann, W. Koegler, J. Laurent, J. Meredith, P. Messmer, E. Otoo, V. Perevoztchikov, A. Poskanzer, O. Ruebel, A. Shoshani, A. Sim, K. Stockinger, G. Weber, and W.-M. Zhang. FastBit: Interactively searching massive data. *Journal of Physics: Conference Series*, 180(1), 2009.

[43] Glenn Young, David Dean, Martin Savage, et al. Forefront Questions in Nuclear Science and the Role of High Performance Computing. Technical report, ASCR Scientific Grand Challenges Workshop Series, January 2009.

[44] H. Yu, C. Wang, R. W. Grout, J. H. Chen, and K.-L. Ma. In-Situ Visualization for Large-Scale Combustion Simulations. *IEEE Computer Graphics and Applications*, 30(3):45–57, May-June 2010.

Part IV

High Performance Visualization Implementations

Chapter 16

VisIt: An End-User Tool for Visualizing and Analyzing Very Large Data

Hank Childs

Lawrence Berkeley National Laboratory

Eric Brugger, Brad Whitlock

Lawrence Livermore National Laboratory

Jeremy Meredith, Sean Ahern, David Pugmire

Oak Ridge National Laboratory

Kathleen Biagas, Mark Miller, Cyrus Harrison

Lawrence Livermore National Laboratory

Gunther H. Weber, Hari Krishnan

Lawrence Berkeley National Laboratory

Thomas Fogal, Allen Sanderson

University of Utah

Christoph Garth

Technische Universität Kaiserslautern

E. Wes Bethel, David Camp, Oliver Rübel

Lawrence Berkeley National Laboratory

Marc Durant, Jean M. Favre, Paul Navrátil

Tech-X Corporation, Swiss National Supercomputing Center, Texas Advanced Computing Center

16.1 Introduction 358
16.2 Focal Points 359
16.2.1 Enable Data Understanding 359
16.2.2 Support for Large Data 360
16.2.3 Provide a Robust and Usable Product for End Users .. 360

16.3 Design 360
16.3.1 Architecture 361
16.3.2 Parallelism 362
16.3.3 User Interface Concepts and Extensibility 363
16.3.4 The Size and Breadth of VisIt 364
16.4 Successes 364
16.4.1 Scalability Successes 364
16.4.2 A Repository for Large Data Algorithms 365
16.4.3 Supercomputing Research Performed with VisIt 366
16.4.4 User Successes 366
16.5 Future Challenges 368
16.6 Conclusion 368
References 369

VisIt is a popular open source tool for visualizing and analyzing data. It owes its success to its foci of increasing data understanding, large data support, and providing a robust and usable product, as well as its underlying design that fits today's supercomputing landscape. This chapter, which draws heavily from a publication at the *SciDAC Conference* in 2011 by Childs et al. [2], describes the VisIt project and its accomplishments.

16.1 Introduction

A dozen years ago, when the VisIt project started, a new high performance computing environment was emerging. Ever increasing numbers of end users were running simulations and generating large data. This rapidly growing number of large data sets prevented visualization experts from being intimately involved in the visualization process; it was necessary to put tools in the end users' hands. Almost all end users were sitting in front of high-end desktop machines with powerful graphics cards. But their simulations were being run on remote, parallel machines and generating data sets too large to be transferred back to these desktops. Worse, these data sets were too large to even process on their (serial) machines anyways. The types of visualization and analysis users wanted to perform varied greatly; users needed many techniques for understanding diverse types of data, with use cases ranging from confirming that a simulation was running smoothly to communicating the results of a simulation to a larger audience, to gaining insight via data exploration.

VisIt was developed in response to these emerging needs. It was (and is) an open source project for visualizing and analyzing extremely large data sets. The project has evolved around three focal points: (1) enabling data understanding, (2) scalable support for extremely large data, and (3) providing a robust and usable product for end users.

In turn, these focal points have made VisIt a very popular tool for visualizing and analyzing the data sets generated on the world's largest supercomputers. VisIt received a 2005 R&D 100 award for the tool's capabilities in understanding large data sets. It has been downloaded hundreds of thousands of times, and it is used all over the world.

16.2 Focal Points

16.2.1 Enable Data Understanding

In many ways, "VisIt" is a misnomer, as the name implies the tool is strictly about visualization and making pretty pictures. The prospect of lost name recognition makes renaming the tool unpalatable, but it is worthwhile to emphasize that VisIt focuses on five primary use cases:

1. *Visual exploration*: users apply a variety of visualization algorithms to "see" what is in their data.

2. *Debugging*: users apply algorithms to find a "needle in a haystack," for example, such as hot spots in a scalar field or cells that have become twisted over time. The user then asks for debugging information in a representation that is recognizable to their simulation (e.g., cell **X** in computation domain **D** has a NaN).

3. *Quantitative analysis*: users apply quantitative capabilities ranging from simple operations, such as integrating densities over a region to find its mass, to highly sophisticated operations, such as adding a synthetic diagnostic to compare to experimental data.

4. *Comparative analysis*: users compare two related simulations, two time slices from a single simulation, simulation and experiment, etc. The taxonomy of comparative analysis has three major branches, each of which is available in VisIt: image-level comparisons place things side-by-side and has the user detect differences visually. Data-level comparisons put multiple fields onto the same mesh, for example, to create a new field for further analysis that contains the difference in temperature between two simulations. Topological-level comparisons detect features in the data sets and then allow those features to be compared.

5. *Communication*: users communicate properties of their data to a large audience. This may be via movies, via images that are inserted into a PowerPoint presentation, or via line plots or histograms that are placed into a journal article.

16.2.2 Support for Large Data

Twelve years ago, "large data" meant several hundred million cells. Today, "large" means several hundred billion cells. In both cases, the definition of "large" was relative to the resources for processing the data. And this is the target for the VisIt project: data whose full resolution cannot fit into primary memory of a desktop machine. Of course, the amount of data to load varies by situation. Can time slices be processed one at a time? How many variables are needed in a given analysis? Is it necessary to load multiple members of an ensemble simultaneously? For VisIt, the goal was to provide an infrastructure that could support any of these use cases, and it primarily uses parallelism to achieve this goal.

16.2.3 Provide a Robust and Usable Product for End Users

Enabling data understanding for large data is a daunting task requiring a substantial investment. To amortize this cost, the project needed to be delivered to many user communities, across both application areas and funding groups.

The "one big tool" strategy provides benefits to both users and developers. Compared to a smaller, tailored effort, users have access to more functionality and better underlying algorithms for processing data. For developers, the core infrastructure undergoes an economy of scale, where many developers can collectively develop a superior core infrastructure than they would be able to do independently.

But the "one big tool" approach has negative aspects as well. Their user interface tends to provide an overly rich interface where users find many features to be meaningless and simply view them as clutter. Further, developers must deal with a less nimble code base where making functionality changes sometimes leads to unexpectedly large coding efforts.

Further, delivering a product to a large end user community incurs significant cost in and of itself: the VisIt project has almost a thousand pages of manuals, several thousand regression tests that run every night, a sophisticated build process, and a variety of courses designed to teach people to how to use the tool. It requires multiinstitutional coordination for release management, for responses to user requests, and for software development. And, of course, the source code itself must be well documented to reduce barriers to entry for new developers.

The developers of the VisIt project decided to "go big": to pay the costs associated with large user and developer bases in the hopes of writing a tool that would be usable by many and developed by many.

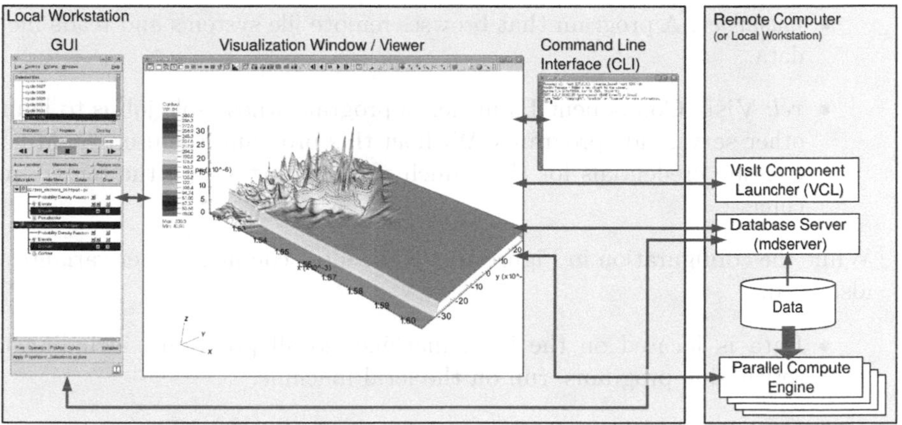

FIGURE 16.1: Diagram of VisIt programs and their communication. Image source: Childs et al., 2011 [2].

16.3 Design

This section describes several facets of VisIt's design, including VisIt's architecture, its parallelism approach, and its user interface concepts.

16.3.1 Architecture

VisIt employs a client-server design, where both client and server are composed of multiple programs (see Fig. 16.1). Client-side programs, typically run on the user's local desktop, are responsible for both user interface and rendering, since interactivity is paramount for these activities. The client-side programs are:

- *gui:* A graphical user interface built using the Qt widget set.
- *cli:* A command line user interface built using the Python language.
- *viewer:* A program responsible for the visual display of data.
- Custom, streamlined user interfaces can also be added to VisIt. The interfaces can either complement the gui and cli or replace them altogether.

Server-side programs, typically run on a remote supercomputer that can access the user's data in a parallel fashion, are responsible for processing data. The server-side programs are:

- *engine:* The program that applies visualization and analysis algorithms to large data sets using parallel processing.

- *mdserver:* A program that browses remote file systems and reads metadata.
- *vcl:* **V**isIt **C**omponent **L**auncher, a program whose sole job is to launch other server-side programs. Without this program, the user would have to issue credentials for the launch of each program on the remote machine.

While the configuration in Figure 16.1 is the most common, other variants are also used:

- Data is located on the local machine, so all programs, including the server-side programs, run on the local machine.
- The client-side programs run on a remote machine. This mode occurs most often in conjunction with graphical desktop sharing, such as VNC.
- Multiple servers are run simultaneously to access data on multiple remote machines.
- VisIt is run entirely in "batch mode." The *gui* program is not used and the *viewer* program runs in a windowless mode.
- VisIt's client-side programs are coupled with a simulation code and data is processed *in situ*. In this case, the simulation embeds a copy of the *engine* program.

16.3.2 Parallelism

VisIt has multiple processing modes—multiresolution processing (Chap. 8), *in situ* processing (Chap. 9), and out-of-core processing (Chap. 10)—but its most frequent mode is pure parallelism (Chap. 2), where the data is partitioned over its MPI tasks and is processed at its native resolution. Most visualization and analysis algorithms are embarrassingly parallel, meaning that portions of the data set can be processed in any order and without coordination and communication. For this case, VisIt's core infrastructure manages the partitioning of the data and all parallelism. For non-embarrassingly parallel cases like streamline calculation or volume rendering, algorithms are able to manage parallelism themselves and can opt to perform collective communication if necessary.

In VisIt's most typical visualization use case, a user loads a large data set, applies operations to reduce the data size, and then transfers the resulting data set to the local client for interactive rendering, using the local graphics card. However, some data sets are so large that their reduced forms are too large for a desktop machine. This case requires a backup plan. VisIt's backup plan is to switch to a parallel rendering mode: data is left on the parallel server, each MPI task renders its own piece, and the resulting subimages are composited together. The final image is then brought back to the *viewer* and placed in the visualization window, as if it was rendered with the graphics card. Although this process sounds cumbersome, the switch to the parallel rendering mode is transparent to end users and frame rates approaching ten frames per second can be achieved.

VisIt was designed for many scales of concurrency. Many users run serial or modestly parallel versions on their desktop machines. When users utilize parallel resources on a supercomputer, they typically run with 32 to 512 tasks. But, for the largest data sets, VisIt servers with thousands or even tens of thousands of tasks are used (see 16.4.1). VisIt demonstrates excellent scalability and performance at each of these scales.

16.3.3 User Interface Concepts and Extensibility

Type	Description	# of instances
Database	How to read from a file	˜115
Operator	How to manipulate data	˜60
Plot	How to render data	˜20
Expression	How to derive new quantities	˜190
Queries	How to extract quantitative and debugging information	˜90

TABLE 16.1: VisIt's five primary user interface concepts.

Table 16.1 shows the five primary user interface concepts in VisIt. A strength of these concepts is their interoperability. Each plot can work on data directly from a file (databases) or from derived data (expressions), and can have an arbitrary number of data transformations or subselections applied (operators). Once the key information is extracted, quantitative or debugging information can be extracted (queries) or the data can be rendered (plots). Consider an example: a user reads from a file (database), calculates the λ-2 metric for finding high vorticity (expressions), isolates out the regions of highest vorticity operators, renders it (plots), then calculates the number of connected components and statistics about them (queries).

VisIt makes it easy to add new types of databases, operators, and plots. The base infrastructure deals with these concepts as abstract types; it only discovers the concrete databases, operators, and plots instances at start-up, by loading them as plug-ins. Thus, adding new functionality to VisIt translates to developing a new plug-in. Further, VisIt facilitates the plug-in development process. It provides an environment for defining a plug-in and also performs code generation. The developer starts by setting up options for the plug-ins, and then VisIt generates attributes for storing the options, user interface components (Python, Qt, and Java), the plug-in bindings, and C++ methods with "dummy" implementations. The developer then replaces the dummy implementations with their intended algorithm, file reading code, etc.

16.3.4 The Size and Breadth of VisIt

Although only briefly discussed in this chapter, VisIt has an extensive list of features. Its ~115 file format readers include support for many HDF5- and NetCDF-based formats, CGNS, and others, including generic readers for some types of binary and ASCII files. Its ~60 operators include transformations (such as projections, scaling, rotation, and translation), data subsetting (such as thresholding and contouring), and spatial thresholding (such as limiting to a box or a plane), among many others. Its ~90 queries allow users to get customizable reports about specific cells or points, integrate quantities, calculate surface areas and volumes, insert synthetic diagnostics/virtual detectors, and much more. Its ~190 expressions go well beyond simple math. For example, the user can create derived quantities like, "if the magnitude of the gradient of density is greater than this, then do this, else do that."

And many features do not fit into the five primary user interface concepts. There is support for positioning light sources, making movies (including MPEG encoding), eliminating data based on known categorizations (e.g., "show me only this refinement level" from an AMR mesh), and rendering effects like shadows and specular highlights, to name a few. In total, VisIt is approximately one and a half million lines of code.

Finally, VisIt makes heavy use of the Visualization ToolKit (VTK) [17]. This library contains an execution model, a data model, and many algorithms for transforming data. VisIt implements its own execution model, but the other two pieces form the foundation of VisIt's data processing. VTK's data model forms the basis of VisIt's data model, although VisIt provides support for mixed material cells, metadata for faster processing, and other concepts not natively supported by VTK. Further, VisIt uses the native VTK algorithm for many embarrassingly parallel visualization algorithms. In short, VTK has provided an important leverage to the VisIt project, allowing VisIt developers to direct their attention to the project's three main focal points.

16.4 Successes

The VisIt project has succeeded in multiple ways: by providing a scalable infrastructure for visualization and analysis, by populating that infrastructure with cutting-edge algorithms, by informing the limits of new hardware architectures, and, most importantly, by enabling successes for the tool's end users. A few noteworthy highlights are summarized in the subsections below.

16.4.1 Scalability Successes

As discussed previously in Chapter 13, a pair of studies were run in 2009 to demonstrate VisIt's capabilities for scalability and large data (see Fig. 16.2). In the first study, VisIt's infrastructure and some of its key visualization algorithms were demonstrated to support weak scaling. This demonstration led to be VisIt being selected as a "Joule code," a formal certification process by the US Office of Management and Budget to ensure that programs running on high-end supercomputers are capable of using the machine efficiently. In the second study, VisIt was scaled up to tens of thousands of cores and used to visualize data sets with trillions of cells per time slice. This study found VisIt itself to perform quite well, although overall performance was limited by the supercomputer's I/O bandwidth. Both studies are further described by Childs et al. [6].

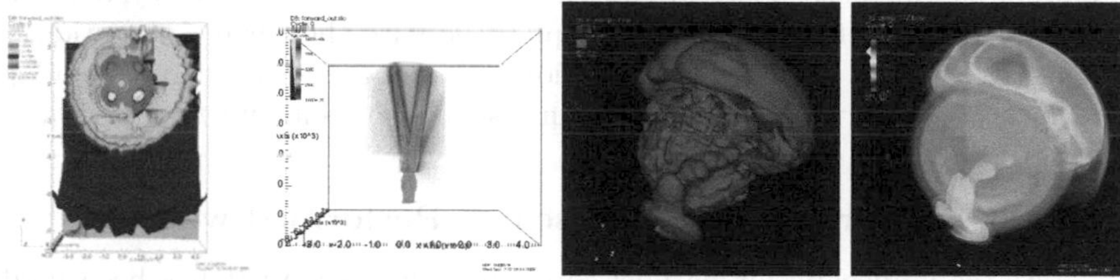

FIGURE 16.2: The two left images show a contouring and a volume rendering from a Denovo radiation transport simulation. They were produced by VisIt using 12,270 cores of JaguarPF as part of the "Joule code" certification, which showed that VisIt is weakly scalable. The two right images show a contouring and a volume rendering of a two trillion cell data set produced by VisIt using 32,000 cores of JaguarPF as part of a study on scalability at high levels of concurrency and on large data sets. The volume rendering was reproduced in 2011 on a one trillion cell version of the data set using only 800 cores of the TACC Longhorn machine. Image source: Childs et al., 2010 [6].

16.4.2 A Repository for Large Data Algorithms

Many advanced algorithms for visualizing and analyzing large data have been implemented inside of VisIt, making them directly available to end users. Notable algorithms include:

- A novel streamline algorithm that melds two different parallelization strategies ("over data" and "over seeds") to retain their positive effects while minimizing their negative ones [13];
- A volume rendering algorithm that handles the compositing complexities inherent to unstructured meshes while still delivering scalable performance [4];

- An algorithm for identifying connected components in unstructured meshes in a distributed-memory parallel setting on very large data sets [9];
- An algorithm for creating crack-free isosurfaces for adaptive mesh refinement data, a common mesh type for very large data [18];
- A well-performing material interface reconstruction algorithm for distributed-memory parallel environments that balances concerns for both visualization and analysis [12]; and
- A method for repeated interpolations of velocity fields in unstructured meshes, to accelerate streamlines [8].

Further, VisIt has been the subject of much systems research, including papers on the base VisIt architecture [5], VisIt's "contract" system which allows it to detect the processing requirements for the current operations and adaptively apply the best optimizations [3], and a description of the adapter layer that allows VisIt to couple with a simulation and run *in situ* [19].

16.4.3 Supercomputing Research Performed with VisIt

As the landscape for parallel computers changes, VisIt has been used to test the benefits of emerging algorithms and hardware features, including:

- Studying modifications to collective communication patterns for ghost data generation, to be suitable for out-of-core processing, thereby improving cache coherency and reducing memory footprint [10];
- Studying the viability of hardware-accelerated volume rendering on distributed-memory parallel visualization clusters powered by GPUs [7];
- Studying the benefits of hybrid parallelism for streamline algorithms [1]; and
- Studying the issues and strategies for porting to new operating systems [14].

16.4.4 User Successes

Of course, the most important measure for the project is helping users better understand their data. Unfortunately, metrics of success in this space are difficult:

- Some national laboratories keep statistics on their user communities: the United States' Lawrence Livermore Lab has approximately 300 regular users, the United Kingdom's Atomic Weapons Establishment (AWE) has approximately 100 regular users, and France's Atomic Energy Commission (CEA) at CESTA has approximately 50 regular users. Other institutions, like Oak Ridge and Lawrence Berkeley, view VisIt as their primary visualization and analysis tool, but do not keep user statistics.

- In terms of monetary support for developing VisIt, the U.S. Department of Energy funds VisIt development through its Office of Science, National Nuclear Security Agency, and Office of Nuclear Energy. Both of the US National Science Foundation (NSF) XD centers on visualization actively deploy and support VisIt as well.
- Another method for measuring usage is studying affiliations of users who ask questions on the mailing list. The majority of these inquiries come from none of the previously mentioned institutions, indicating that usage goes beyond these sites.

FIGURE 16.3: Recent covers of the SciDAC Review Journal created using VisIt.

Tracking individual user successes is difficult, although there is clear evidence with certain types of usage. VisIt is used regularly to make images for journal covers, a high-profile activity (see Fig. 16.3). Further, there have been several notable instances of publications using VisIt to perform novel analysis:

- Analysis of laser wakefield simulations often amounts to finding key particles [15], and query-driven visualization techniques were used to search through terabytes of data to locate these key particles in as little as two seconds.
- Simulations often deal with idealized meshes. VisIt's comparative capabilities were used to quantify the importance of engineering defects when differencing as-built and as-designed models [11].
- VisIt's streamline code was used to find the toroidal magnetic fields found in tokamaks by analyzing the fieldlines through a cross-sectional slice and the topological "islands" they trace out [16].

16.5 Future Challenges

Although VisIt is well suited for today's supercomputing environment, the project will face many challenges in the future. In the short term, I/O limitations will force visualization and analysis activities to de-emphasize I/O. The VisIt development team has invested in pertinent techniques, such as multiresolution processing and *in situ*, but these techniques will need to be further hardened to support production use. In the longer term, power limits will constrain data movement, forcing much processing to occur *in situ* on novel architectures, such as GPU accelerators. Unfortunately, VisIt's existing *in situ* implementation may be mismatched for this many-core future, for two reasons. First, although VisIt can be easily multithreaded, using a pthreads or OpenMP-type approach (see Chap. 12 to further understand the benefits of hybrid parallelism), this approach may not be able to take advantage of these architectures. The many-core future may require CUDA- or OpenCL-type languages; migrating the VisIt code base to this setting would be a substantial undertaking. Second, although VisIt has been demonstrated to work well at high levels of concurrency, some of its algorithms involve large data exchanges. Although these algorithms perform well on current machines, they would violate the data movement constraints on future machines and would need to be redesigned.

16.6 Conclusion

The VisIt project's three focal points—understanding data, large data, and delivering a product—together form a powerful environment for analyzing data from HPC simulations. It is used in a variety of ways: it enables visualization scientists, computational code developers, and the physicists that run these codes to perform a broad range of data understanding activities, including debugging, making movies, and exploring data. The user interface portion of its design provides a powerful paradigm for analyzing data while the data processing portion of its design is well suited for big data. This, in turn, has led to many successes: in scaling up to high levels of concurrency and large data sizes, in providing a "home" for large data algorithms, in understanding how to best use supercomputers, and, most importantly, in helping users understand their data. Further, despite significant upcoming changes in supercomputing architecture, VisIt's future appears bright, as it enjoys vibrant user and developer communities.

References

[1] David Camp, Christoph Garth, Hank Childs, Dave Pugmire, and Kenneth I. Joy. Streamline Integration Using MPI-Hybrid Parallelism on a Large Multicore Architecture. *IEEE Transactions on Visualization and Computer Graphics*, 17:1702–1713, 2011.

[2] Hank Childs, Eric Brugger, Brad Whitlock, Jeremy Meredith, Sean Ahern, Kathleen Bonnell, Mark Miller, Gunther H. Weber, Cyrus Harrison, David Pugmire, Thomas Fogal, Christoph Garth, Allen Sanderson, E. Wes Bethel, Marc Durant, David Camp, Jean M. Favre, Oliver Rübel, Paul Navrátil, Matthew Wheeler, Paul Selby, and Fabien Vivodtzev. VisIt: An End-User Tool For Visualizing and Analyzing Very Large Data. In *Proceedings of SciDAC 2011*, July 2011. `http://press.mcs.anl.gov/scidac2011`.

[3] Hank Childs, Eric S. Brugger, Kathleen S. Bonnell, Jeremy S Meredith, Mark Miller, Brad J Whitlock, and Nelson Max. A Contract-Based System for Large Data Visualization. In *Proceedings of IEEE Visualization*, pages 190–198, 2005.

[4] Hank Childs, Mark Duchaineau, and Kwan-Liu Ma. A Scalable, Hybrid Scheme for Volume Rendering Massive Data Sets. In *Proceedings of Eurographics Symposium on Parallel Graphics and Visualization*, pages 153–162, May 2006.

[5] Hank Childs and Mark Miller. Beyond Meat Grinders: An Analysis Framework Addressing the Scale and Complexity of Large Data Sets. In *SpringSim High Performance Computing Symposium (HPC 2006)*, pages 181–186, 2006.

[6] Hank Childs, David Pugmire, Sean Ahern, Brad Whitlock, Mark Howison, Prabhat, Gunther Weber, and E. Wes Bethel. Extreme Scaling of Production Visualization Software on Diverse Architectures. *IEEE Computer Graphics and Applications*, 30(3):22–31, May/June 2010.

[7] Thomas Fogal, Hank Childs, Siddharth Shankar, J. Krüger, R. D. Bergeron, and P. Hatcher. Large Data Visualization on Distributed Memory Multi-GPU Clusters. In *Proceedings of High Performance Graphics 2010*, pages 57–66, June 2010.

[8] Christoph Garth and Ken Joy. Fast, Memory-Efficient Cell Location in Unstructured Grids for Visualization. *IEEE Transactions on Computer Graphics and Visualization*, 16(6):1541–1550, November 2010.

[9] Cyrus Harrison, Hank Childs, and Kelly P. Gaither. Data-Parallel Mesh Connected Components Labeling and Analysis. In *Proceedings of EuroGraphics Symposium on Parallel Graphics and Visualization*, pages 131–140, April 2011.

[10] Martin Isenburg, Peter Lindstrom, and H. Childs. Parallel and Streaming Generation of Ghost Data for Structured Grids. *IEEE Computer Graphics and Applications*, 30(3):32–44, May/June 2010.

[11] Edwin J. Kokko, Harry E. Martz, Diane J. Chinn, Hank R. Childs, Jessie A. Jackson, David H. Chambers, Daniel J. Schneberk, and Grace A. Clark. As-Built Modeling of Objects for Performance Assessment. *Journal of Computing and Information Science in Engineering*, 6(4):405–417, 12 2006.

[12] Jeremy S. Meredith and Hank Childs. Visualization and Analysis-Oriented Reconstruction of Material Interfaces. *Computer Graphics Forum*, 29(3):1241–1250, 2010.

[13] D. Pugmire, H. Childs, C. Garth, S. Ahern, and G. Weber. Scalable Computation of Streamlines on Very Large Datasets. In *Proceedings of Supercomputing (SC09)*, 2009.

[14] David Pugmire, Hank Childs, and Sean Ahern. Parallel Analysis and Visualization on Cray Compute Node Linux. In *Proceedings of the Cray Users Group Meeting*, 2008.

[15] Oliver Rübel, Prabhat, Kesheng Wu, Hank Childs, Jeremy Meredith, Cameron G. R. Geddes, Estelle Cormier-Michel, Sean Ahern, Gunther H. Weber, Peter Messmer, Hans Hagen, Bernd Hamann, and E. Wes Bethel. High Performance Multivariate Visual Data Exploration for Extemely Large Data. In *Proceedings of Supercomputing (SC08)*, 2008.

[16] A. R. Sanderson, G. Chen, X. Tricoche, D. Pugmire, S. Kruger, and J. Breslau. Analysis of Recurrent Patterns in Toroidal Magnetic Fields. *IEEE Transactions on Visualization and Computer Graphics*, 16(4):1431–1440, 2010.

[17] William J. Schroeder, Kenneth M. Martin, and William E. Lorensen. The Design and Implementation of an Object-Oriented Toolkit for 3D Graphics and Visualization. In *Proceedings of the IEEE Conference on Visualization (Vis96)*, pages 93–100, 1996.

[18] Gunther H. Weber, Vincent E. Beckner, Hank Childs, Terry J. Ligocki, Mark Miller, Brian van Straalen, and E. Wes Bethel. Visualization Tools for Adaptive Mesh Refinement Data. In *Proceedings of the 4th High End Visualization Workshop*, pages 12–25, 2007.

[19] Brad Whitlock, Jean M. Favre, and Jeremy S. Meredith. Parallel In Situ Coupling of Simulation with a Fully Featured Visualization System. In *Proceedings of EuroGraphics Symposium on Parallel Graphics and Visualization*, pages 101–109, April 2011.

Chapter 17

IceT

Kenneth Moreland

Sandia National Laboratories

17.1 Introduction 373
17.2 Motivation 374
17.3 Implementation 374
17.3.1 Theoretical Limitations ... and How to Break Them ... 375
17.3.2 Pixel Reduction Techniques 376
17.3.3 Tricks to Boost the Frame Rate 377
17.4 Application Programming Interface 378
17.4.1 Image Generation 378
17.4.2 Opaque versus Transparent Rendering 379
17.5 Conclusion 379
References 381

The Image Composition Engine for Tiles (IceT) is a high-performance, sort-last parallel rendering library. IceT is designed for use in large-scale, distributed-memory rendering applications. It works efficiently with large numbers of processes, with large amounts of geometry, and with high-resolution images. In addition to providing accelerated rendering for a standard display, IceT also provides the unique ability to generate images for tiled displays. The overall resolution of the display may be several times larger than any viewport rendered by a single machine. IceT is currently being used as the parallel rendering implementation in high performance visualization applications as like VisIt (see Chap. 16) and ParaView (see Chap. 18).

17.1 Introduction

The Image Composition Engine for Tiles (IceT) is an API designed to enable applications to perform sort-last parallel rendering on large displays [8]. The design philosophy behind IceT is to allow very large data sets to be rendered on arbitrarily high resolution displays. Although frame rates can be sacrificed in lieu of scalable polygon/second rendering rates, there are many features in IceT that allow an application to achieve interactive rates. These features include image compression, empty pixel skipping, image reduction, and data replication. Together, these features make IceT a versatile parallel rendering solution that provides optimal parallel rendering under most data size and image size combinations.

IceT is designed to take advantage of spatial decomposition of the geometry being rendered. That is, it works best if all the geometry on each process is located in as small a region of space as possible. When this is true, each process usually projects geometry on only a small section of the screen. This results in less work for the compositing engine.

Overall, IceT demonstrates extraordinary speed and scalability. It is used to render at tremendous rates, such as billions of polygons per second, and on the largest supercomputers in the world [10].

17.2 Motivation

The original motivation for IceT was the need to support high performance rendering for scientific visualization on large format displays [16]. Furthermore, the IceT development group needed to take advantage of the distributed memory rendering clusters that were replacing the more expensive multipipe rendering computers of the day. These requirements still ring true today. Scientific data continues to grow, desktop displays with over 2 megapixels are common, and nearly all high performance scientific visualization is performed on distributed-memory computers.

There are three general classes of parallel rendering algorithms: sort-first, sort-middle, and sort-last [7] (although, it is possible to combine elements of these classes together [14]). When run on a distributed-memory machine, every type of parallel rendering algorithm has some overhead caused by communication. In sort-first and sort-middle algorithms, this overhead is proportional to the amount of geometry being rendered. In the sort-last algorithms, this overhead is proportional the number of pixels on the display.

Although sort-first and sort-middle parallel rendering algorithms efficiently divide screen space and are often used to drive tiled displays, these algorithms simply cannot scale to the size of data that sort-last algorithms are able to support [12, 18]. Because large-scale data is its primary concern, IceT implements sort-last rendering and employs the techniques used to reduce the overhead incurred with high-resolution displays.

17.3 Implementation

The most important aspect of parallel rendering in IceT's implementation is that it performs well with both large amounts of data and high-resolution displays. The sort-last compositing algorithms, described in Chapter 5, ensure that IceT performs well with large amounts of data and large process counts [10]. This chapter primarily describes the techniques IceT uses to effectively render to large-format displays.

17.3.1 Theoretical Limitations ... and How to Break Them

There are a number of theoretical metrics with important practical consequences. These include the number of pixel-blending operations performed by each process (which affects the total time computing), the number of pixels sent or received by each process (which affects how long it takes to transfer data), the number of messages sent at any one time (which can affect network congestion), and the number of sequential messages sent (which can accumulate the effect of the network latency).

With respect to IceT's performance on large images, the most important of these metrics are the number of pixel-blending operations and the number of pixels sent and received. It is easy to show, for example, that the binary-swap algorithm is optimal on both counts. The binary-swap algorithm is described in 5.2.2 as well as previous studies [3, 4].

Binary-swap is a divide-and-conquer algorithm that operates in rounds that pair processes, swap image halves, and recurse in each half. Given p processes compositing an image of n pixels, there must be at least $(p-1) \cdot n$ blending operations (because it takes $p-1$ operations to blend the p versions of each pixel generated by all the processes). A perfectly balanced parallel algorithm will blend $\frac{(p-1) \cdot n}{p}$ pixels in each process.

The binary-swap algorithm has $\log_2 p$ rounds with each round blending $n/2^i$ pixels in each process, where i is the round index starting at 1. The total number of blending operations performed in each process is therefore

$$\sum_{i=1}^{\log_2 p} \frac{n}{2^i} = n - \frac{n}{2^{\log_2 p}} = n - \frac{n}{p} = \frac{(p-1) \cdot n}{p}, \tag{17.1}$$

which is, as previously mentioned, optimal. Likewise, binary-swap has an optimal amount of pixels transferred. Radix-k [13], which is also supported in IceT, is similarly optimal with respect to pixel blending and pixel transfer. In addition, radix-k can also reduce the number of rounds as well as overlap computation and communication [2].

Although this theoretical, optimal solution still grows linearly with respect to the resolution of the image, it is possible, in practice, to perform much better. The previous analysis makes a critical assumption: that every pixel generated by every process contains useful data. Such generation is possible in the worst case, but in practice, many pixels can be ignored.

Consider the example of parallel rendering shown in Figure 17.1. Each process renders a localized cube of data surrounded by an abundance of blank space. Although the example in Figure 17.1 may seem artificial, this case is actually quite common. Sort-last volume rendering requires the geometry to

FIGURE 17.1: An example rendering of a cubic volume by eight processes. The first eight images represent those generated by each process. The image at the right is a fully composited image.

be spatially decomposed in this way [9], and even unstructured data tends to exhibit good spatial decomposition.

If this blank space is identified and grouped, it does not need to be transferred or blended. In this way, the overhead from high-resolution images can be reduced. Furthermore, as more processes are used, the geometry gets divided into even smaller units, thereby further reducing the amount of pixel data. Consequently, the larger overhead from higher-resolution images can be compensated for by adding more processes to the task.

17.3.2 Pixel Reduction Techniques

The first step in reducing the amount of pixel data to be composited is to identify the active pixels, which are those that have been rendered to, and the complementary inactive pixels, which are those that have no render information. IceT first conservatively estimates a region of inactive pixels by projecting the bounding box of the geometry on the screen and declaring everything outside of this inactive. It then checks the remaining pixels for the background depth, or opacity, to find any remaining inactive pixels.

Once inactive pixels are identified, they are marked and their data is removed from the image data storage. IceT uses a form of run-length encoding called *active-pixel encoding* [11]. Active-pixel encoding stores alternating run lengths of inactive and active pixels. Data for inactive pixels are removed, whereas data for active pixels follow their respective run lengths. Active-pixel encoding provides the double benefit of: (1) decreasing the amount of data transferred; and (2) providing a simplified means of skipping over inactive pixels that need not be blended.

Although active-pixel encoding does reduce the overhead of sort-last parallel rendering, the savings are generally not well balanced. When binary-swap or radix-k partitions an image, the resulting subimages are unlikely to contain the same number of active pixels.

A simple but effective method to rebalance the parallel compositing is to interlace the image [7, 17]. In image interlacing, the pixels are shuffled around to distribute local regions throughout the image. When interlacing an image, IceT carefully shuffles regions that match those created by the binary-swap or radix-k algorithm such that each partition binary-swap or radix-k creates contains a block of unshuffled pixels. Thus, the reshuffling back to the original

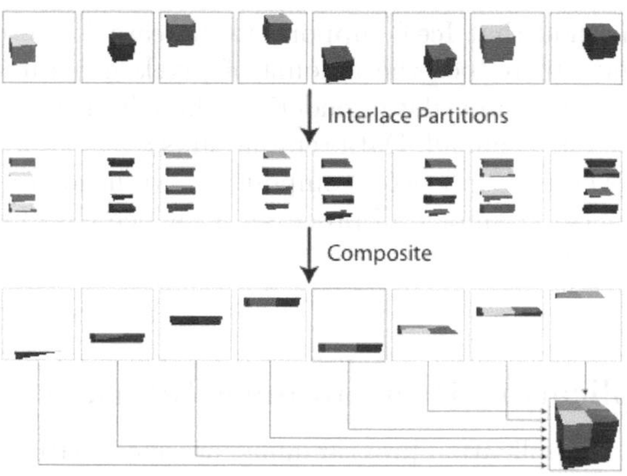

FIGURE 17.2: Image interlacing in IceT.

pixel order is combined with the image partition collection at no extra cost as shown in Figure 17.2

IceT can also take advantage of spatial decomposition in other ways on multitile displays. In such a case, processes tend to render anything only on a small set of tiles. IceT identifies completely blank tiles and removes them from the compositing computation. Special parallel compositing algorithms balance the compositing work for the remaining tiles that contain valid data [11]. The most effective of these algorithms is a reduction algorithm that assigns processes to tiles in proportion to the number of images generated for each tile. All images are sent to a process assigned to the corresponding tile, and subsequently, each process group composites an image for their assigned tile.

17.3.3 Tricks to Boost the Frame Rate

Even with the pixel reduction techniques implemented in IceT, it may be the case that the image composition overhead is still too great to maintain an interactive frame rate. For example, this can occur when there are too few processes driving a large-format display or when the view direction is zoomed on large portions of the geometry.

A straightforward method to reduce the image compositing overhead is to simply reduce the number of pixels in the image. Rather than render a full-resolution image, the user can render a smaller image then resample up to the displayed size. Clearly, this is not a technique that can be used for every render because it loses detail. However, it is often the case in scientific visualization that, when interacting with the data, a reduced level of detail can be used in place of a full-resolution representation [1]. In such a case, you might render a lower-resolution image, then replace it with a full-resolution image after interacting is finished. To better support this technique, IceT can automatically inflate images when drawing to an OpenGL context.

Another method that IceT supports to increase frame rates is the ability to replicate data to reduce the amount of pixels generated. If two or more processes share the same data rendering, then IceT can evenly divide the active pixels among them all. Data replication is a trade-off between rendering data storage overhead and pixel compositing overhead. In the extreme case, the data is replicated among all processes, in which case IceT simply renders images where they are needed.

17.4 Application Programming Interface

IceT has a C application programming interface (API). It is modeled after the OpenGL API [15] because OpenGL remains the most common interface to rendering and graphics hardware. And, although IceT has special methods to simplify binding to OpenGL rendering, its core library is independent of OpenGL and it can use some other rendering system. Like OpenGL, the IceT API is modeled as a state machine where various state variables modify its behavior as it runs [8].

17.4.1 Image Generation

Because sort-last rendering involves two independent and sequential steps—rendering then compositing—it is easy to interface with an otherwise serial rendering system. This ease is reflected in the IceT API. In fact, IceT does not directly render any data, leaving that to whatever rendering subsystem an application provides or uses. IceT only requires the rendering results, that is, images.

For efficiency, IceT also needs to be provided a bounding region around the geometry being rendered. IceT uses this bounding region to quickly estimate where the majority of inactive pixels are located. This bounding region can be specified as either an axis-aligned box or a small set of points that define a convex hull. This region should conservatively contain all geometry that is rendered. A simple but effective axis-aligned bounding box can be determined from coordinate ranges. After rendering, IceT will further search this area for inactive pixels.

Because it can render to multitile displays, IceT needs to change the projection matrix (for rendering the various tiles) as well as invoke multiple renderings with different projection matrices (for rendering geometry that spans multiple tiles). IceT uses a callback mechanism to invoke these renderers. The application using IceT provides IceT with a pointer to a function that is capable of rendering the geometry for a frame. When IceT needs a rendering for a particular region of the screen, it invokes this callback function while providing a modified projection matrix. The callback function returns the rendered image.

17.4.2 Opaque versus Transparent Rendering

IceT supports the rendering of both opaque and transparent objects, typified by surface and volume rendering, respectively. Although the same algorithms are used for compositing both types of objects, slightly different requirements are placed for each.

When rendering opaque geometry, IceT uses a z-buffer to blend pixels together. A z-buffer contains an array of depth values where each value represents the distance between the viewer and the object for each pixel. When blending with a z-buffer, IceT simply picks the color that is closer to the viewer. Z-buffer blending is order independent, so there is no need to have restrictions on the partitioning of data. The renderer is, however, required to provide a z-buffer. Object-order rendering methods, as described in Chapter 4, often build z-buffers anyway, for hidden-surface removal, and image-order rendering methods that can trivially create them.

When rendering transparent geometry, IceT uses a blending operation that mixes colors based on the front image's opacity, represented as an alpha value. When blending colors, order matters. For example, the image in front obscures the image in back by some amount based on its opacity. Thus, to composite transparent images together, the front-to-back ordering must exist and must be known. This so called visibility ordering constrains the partitioning of the geometry [9]. IceT leaves the responsibility for the proper partitioning to the application. The application must give the proper visibility ordering to IceT, which will ensure that the parallel compositing algorithms blend images in the correct order.

IceT's blending options provide more flexibility with regard to the rendering mode. These options satisfy the conditions for surfaces, volume rendering, and the use of multitile displays, as demonstrated in Figures 17.3, 17.4, and 17.5, respectively.

17.5 Conclusion

The IceT library is an encapsulation of efficient parallel rendering algorithms. IceT has proven scalability with respect to the amount of geometry rendered and the number of processors used. It also scales well with the size of displays by reducing pixel data and load balancing the sparse images. IceT also transitions between opaque and translucent rendering. The IceT API is easily integrated into existing applications and is freely available from `http://icet.sandia.gov`.

FIGURE 17.3: An IceT-assisted rendering of an isosurface from a Richtmyer–Meshkov simulation [5]. The detailed surface is represented by 473 million triangles.

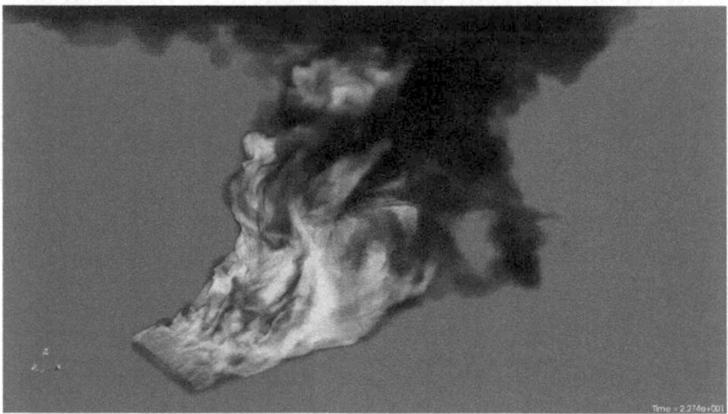

FIGURE 17.4: Results from a coupled SIERRA/Fuego/Syrinx/Calore simulation of objects in a crosswind fire [6]. These 10,000,000 unstructured hexahedra are rendered using IceT's ordered compositing mode.

FIGURE 17.5: An example of using IceT to simultaneously render both surfaces and transparent volumes on a multitile display.

References

[1] Andy Cedilnik, Berk Geveci, Kenneth Moreland, James Ahrens, and Jean Favre. Remote Large Data Visualization in the ParaView Framework. In *Proceedings of the EuroGraphics Symposium on Parallel Graphics and Visualization (EGPGV)*, pages 163–170, May 2006.

[2] Wesley Kendall, Tom Peterka, Jian Huang, Han-Wei Shen, and Robert Ross. Accelerating and Benchmarking Radix-k Image Compositing at Large Scale. In *Proceedings of the EuroGraphics Symposium on Parallel Graphics and Visualization (EGPGV)*, pages 101–110, May 2010.

[3] Kwan-Liu Ma, James S. Painter, Charles D. Hansen, and Michael F. Krogh. A Data Distributed, Parallel Algorithm for Ray-Traced Volume Rendering. In *Proceedings of the 1993 Symposium on Parallel Rendering*, pages 15–22, 1993.

[4] Kwan-Liu Ma, James S. Painter, Charles D. Hansen, and Michael F. Krogh. Parallel Volume Rendering Using Binary-Swap Compositing. *IEEE Computer Graphics and Applications*, 14(4):59–68, July/August 1994.

[5] A. A. Mirin, R. H. Cohen, B. C. Curtis, W. P. Dannevik, A. M. Dimits, M. A. Duchaineau, D. E. Eliason, D. R. Schikore, S. E. Anderson, D. H. Porter, P. R. Woodward, and L. J. Shieh. Very High Resolution Simulation of Compressible Turbulence on the IBM-SP System. In *Proceedings of Supercomputing '99*, 1999.

[6] C. D. Moen, G. H. Evans, S. P. Domino, and S. P. Burns. A Multi-Mechanics Approach to Computational Heat Transfer. In *ASME International Mechanical Engineering Congress and Exposition*, November 2002.

[7] Steven Molnar, Michael Cox, David Ellsworth, and Henry Fuchs. A Sorting Classification of Parallel Rendering. *IEEE Computer Graphics and Applications*, pages 23–32, July 1994.

[8] Kenneth Moreland. IceT Users' Guide and Reference Version 2.1. Technical Report SAND 2011-5011, Sandia National Laboratories, August 2011.

[9] Kenneth Moreland, Lisa Avila, and Lee Ann Fisk. Parallel Unstructured Volume Rendering in ParaView. In *Visualization and Data Analysis 2007, Proceedings of SPIE-IS&T Electronic Imaging*, number 64950F, January 2007.

[10] Kenneth Moreland, Wesley Kendall, Tom Peterka, and Jian Huang. An Image Compositing Solution at Scale. In *Proceedings of the Conference on High Performance Computing, Networking, Storage and Analysis (SC '11)*, November 2011.

[11] Kenneth Moreland, Brian Wylie, and Constantine Pavlakos. Sort-Last Parallel Rendering for Viewing Extremely Large Data Sets on Tile Displays. In *Proceedings of IEEE 2001 Symposium on Parallel and Large-Data Visualization and Graphics*, pages 85–92, October 2001.

[12] Carl Mueller. The Sort-First Rendering Architecture for High-Performance Graphics. In *Proceedings of the 1995 Symposium on Interactive 3D Graphics*, pages 75–84, 1995.

[13] Tom Peterka, David Goodell, Robert Ross, Han-Wei Shen, and Rajeev Thakur. A Configurable Algorithm for Parallel Image-Compositing Applications. In *Proceedings of the Conference on High Performance Computing Networking, Storage, and Analysis (SC '09)*, November 2009.

[14] Rudrajit Samanta, Thomas Funkhouser, Kai Li, and Jaswinder Pal Singh. Hybrid Sort-First and Sort-Last Parallel Rendering with a Cluster of PCs. In *Proceedings of the ACM SIGGRAPH/Eurographics Workshop on Graphics Hardware*, pages 97–108, 2000.

[15] Dave Shreiner, Mason Woo, Jackie Neider, and Tom Davis. *OpenGL Programming Guide.* Addison Wesley, 4th edition, 2004.

[16] Paul H. Smith and John van Rosendale. Data and Visualization Corridors: Report on the 1998 DVC Workshop Series. Technical Report CACR-164, Center for Advanced Computing Research, California Institute of Technology, September 1998.

[17] Akira Takeuchi, Fumihiko Ino, and Kenichi Hagihara. An Improvement on Binary-Swap Compositing for Sort-Last Parallel Rendering. In *Proceedings of the 2003 ACM Symposium on Applied Computing*, pages 996–1002, 2003.

[18] Brian Wylie, Constantine Pavlakos, Vasily Lewis, and Kenneth Moreland. Scalable Rendering on PC Clusters. *IEEE Computer Graphics and Applications*, 21(4):62–70, July/August 2001.

Chapter 18

The ParaView Visualization Application

Utkarsh Ayachit, Berk Geveci

Kitware, Inc.

Kenneth Moreland

Sandia National Laboratories

John Patchett, Jim Ahrens

Los Alamos National Laboratory

18.1 Introduction ... 383
18.2 Understanding the Need ... 384
18.3 The ParaView Framework ... 386
18.3.1 Configurations ... 386
18.4 Parallel Data Processing ... 387
18.5 The ParaView Application ... 390
18.5.1 Graphical User Interface ... 390
18.5.2 Scripting with Python ... 391
18.6 Customizing with Plug-ins and Custom Applications ... 391
18.7 Co-Processing: *In Situ* Visualization and Data Analysis ... 392
18.8 ParaViewWeb: Interactive Visualization for the Web ... 393
18.9 ParaView In Use ... 394
18.9.1 Identifying and Validating Fragmentation in Shock Physics Simulation ... 394
18.9.2 ParaView at the Los Alamos National Laboratory ... 396
18.9.3 Analyzing Simulations of the Earth's Magnetosphere ... 397
18.10 Conclusion ... 397
References ... 399

ParaView is an open-source, multiplatform data analysis and visualization application and framework. ParaView enables users to quickly build visualizations to analyze their data using qualitative and quantitative techniques. The data exploration can be done interactively in 3D or programmatically using ParaView's batch processing capabilities.

18.1 Introduction

This chapter describes the design and features of ParaView [15, 3], a tool that enables visualization and analysis for extremely large data sets. ParaView is a full-featured, general purpose user application with a graphical user interface as well as a scripting interface that can be used to build visualization pipelines for data analysis and rendering. At the same time, ParaView is a framework for developing highly customized, domain-specific applications for the desktop [1] or for the web [8]. Additionally, ParaView can be directly linked into the simulation code for analysis [5] where data processing happens as a step in the simulation code itself.

As a user application, ParaView supports the visualization and rendering of large data sets in parallel, on distributed-memory machines, in diverse configurations, with minimal effort. These configurations includes multitile displays and immersive VR environments such as CAVE [4, 14]. ParaView supports hardware-accelerated parallel rendering using IceT (Chap. 17) and using level-of-detail (LOD) techniques to further improve rendering performance during interaction.

As an application framework, ParaView provides an API for developing highly customized applications that can have domain-specific user interfaces with custom workflows. Developers can also build plug-ins that add complex functionality to the ParaView application, itself. Some examples of the plug-ins include new readers, writers, or even user interface panels and wizards.

As an *in situ* processing framework, ParaView provides a library that is flexible enough to be embedded in various simulation codes with relative ease. The co-processing library can also be easily extended so that users can deploy new analysis and visualization techniques to existing co-processing installations.

This chapter begins with a brief discussion of the requirements and use-cases ParaView was designed to address, while also looking at some of the major features of ParaView that enable high performance visualization. Finally, the chapter explores some of the exciting applications where ParaView serves as a vital visualization and analysis tool.

18.2 Understanding the Need

In the early 2000s, computational sciences started gaining momentum thanks to the advent of advanced HPC resources, which made it possible to solve large problems that were intractable until then. As simulation runs started getting bigger, so did the data results that needed to be analyzed. Therefore, the paradigm of fetching the data locally and then analyzing it started to become impractical. Also, there was a dearth of general purpose visualization tools. Most institutions developed custom solutions for problems at hand, but they were inflexible and hard to maintain or adapt to different

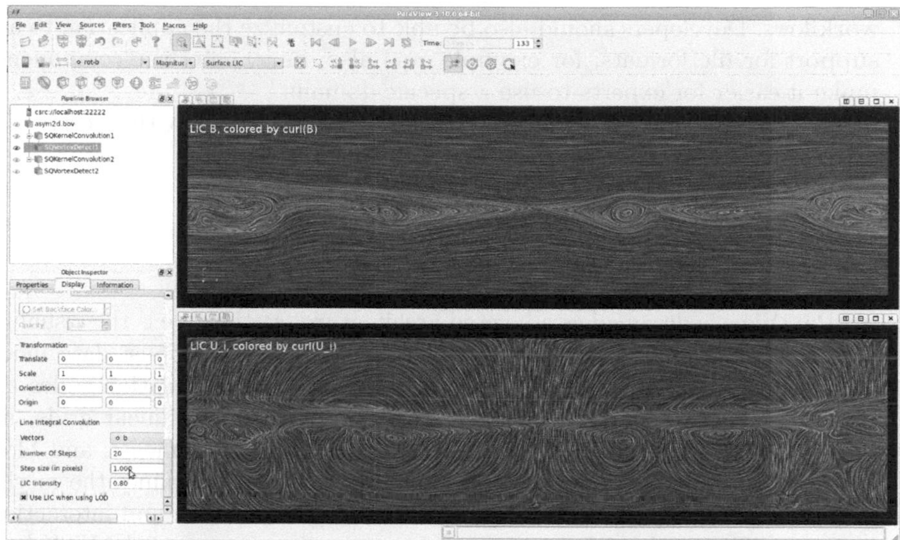

FIGURE 18.1: ParaView being used in the analysis of flow patterns associated with magnetic flux ropes. Image courtesy of Homa Karimabadi (UCSD) and Burlen Loring (LBNL).

problems. There was a need for an easy-to-use, general purpose visualization application that could handle large data sets by leveraging the parallel computing capabilities offered by the HPC resources. Some of the requirements for such an application are described briefly below.

HPC sites often differ in the configurations they use for their setups. Some clusters comprise of identical nodes equipped with graphics cards, while others are set up to do hardware-accelerated rendering only on a smaller subset of nodes. Still others are composed of large $M \times N$ tile displays and other immersive configurations like CAVE. It is necessary that the visualization application supports all these configurations to provide the best possible user experience and utilization of the resources available. Furthermore, different sites have different mechanisms for authenticating and submitting jobs. It is necessary that the application hides these complexities from the user.

The ability to handle large data is also a critical requirement. Large data is defined as data that exceeds the resource limits of a single machine. A visualization application needs to exploit techniques like data streaming and parallelism so that it can visualize such large data. For ensuring that all the processing happens within reasonable performance limits, techniques like multiresolution representations and parallelism can help improve data processing and rendering performance.

Albeit important, visualization is just one of the many tasks a simulation scientist encounters. Thus, it is essential that visualization tools are simple, intuitive, and easy-to-use so that they can be easily integrated into scientific workflows. Developers should also be able to customize the application to add support for file formats, for example, and also, under interface components, make it easier for experts to use a specific domain.

These are some of the major requirements that led to the design and development of ParaView.

18.3 The ParaView Framework

ParaView is designed as a layered architecture. At the core is the visualization toolkit (*VTK*) [13]. VTK provides ParaView with a robust data model that can handle most types of data sets that modern simulation codes can produce including adaptive mesh refinement (AMR), curvilinear grids, unstructured grids with hexes, tets and even higher-order elements, and block hierarchies. VTK also provides the execution model that defines the mechanism for specifying algorithms that process or produce data and interactions between different algorithms connected in a pipeline. On top of the VTK layer sits the ParaView ServerManager, which abstracts the complexity of dealing with remote and distributed environments. The ServerManager provides the client with an unified interface to build visualization pipelines without requiring it to understand the configuration in which ParaView is being operated. The user interface layer is built on top of the ServerManager. This layered approach allows for the rapid development of different applications that leverage the full parallel data processing and visualization capabilities of ParaView. ParaView currently provides three standard interface implementations: the Qt-based GUI that forms the ParaView application, a Python-based scripting interface that supports batch scripting, and a web-based interface that is available as ParaViewWeb.

VTK, and consequently ParaView, is based on what is known as the visualization pipeline [6, 10]. The visualization pipeline comprises filters with inputs and outputs that can be connected together in a data flow pipeline. As the data flows into the input of a filter, it is processed by the filter to produce an output result which is then fed as the input for the next filter in the pipeline and so on. ParaView's ServerManager exposes this underlying pipeline to the application layer. Additionally, ParaView provides higher-level abstractions such as views and representations for rendering the data. Views correspond to a display viewport in which data can be shown. Representations correspond to the data processing and mapping algorithms that convert raw data to forms suitable for presenting in a view. There are several types of views used to generate different renderings of the data including 3D views, line plots, bar plots, parallel coordinate plots, etc.

18.3.1 Configurations

ParaView can be functionally classified into three components: a data-server (where all the data processing and filtering happens), a render-server, and a client (which encapsulates the control and interface components). Different runtime configurations are defined for ParaView, based on which process these three components reside. When the ParaView client starts up, by default all three functions are performed by the same process. This is the simplest case. ParaView can then connect to a remote *pvserver* process that can be running on a distributed machine. In this case, the data-server and render-server roles are taken over by the *pvserver* processes. Alternatively, the remote servers can be launched as a *pvdataserver* acting as the data-server and a *pvrenderserver* acting as the render-server, both of which can be running in parallel. This configuration is suitable in setups where all nodes in the cluster cannot be used for rendering. Finally, ParaView can also be run in batch mode using *pvbatch*. When running in parallel in this mode, the root-node acts as the client, while all the nodes in the group work as the data-server and render-server combined. When using any of the client–server configurations, it's possible to pass command line arguments to the server executables to provide information about tile displays or other immersive setups.

18.4 Parallel Data Processing

ParaView handles large data sets by using data parallelism. In this operational mode, the data is divided amongst participating processes. Each process performs the same operation on its piece of data. Extra communication maybe required to ensure that the results of performing the operation on chunks of data is the same as performing the operation on the whole data set.

For demonstrative purposes, consider the very simplified mesh below.

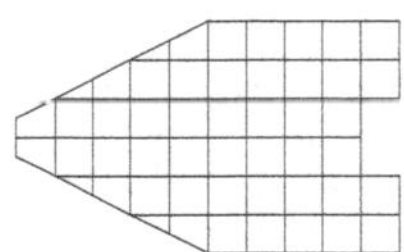

Now suppose the user wants to perform visualizations on this mesh using three processes. The white, gray, and stippled gray regions divide the cells of the mesh as shown below.

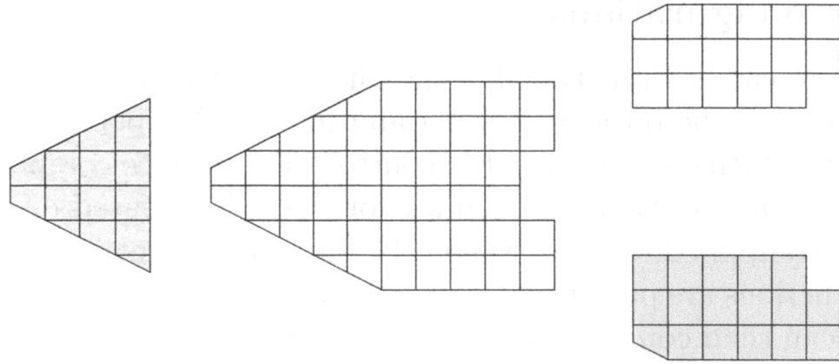

Once partitioned, some visualization algorithms will work by simply allowing each process to independently run the algorithm on its local collection of cells. For example, lets look at a defined clipping plane and give that same plane to each of the processes.

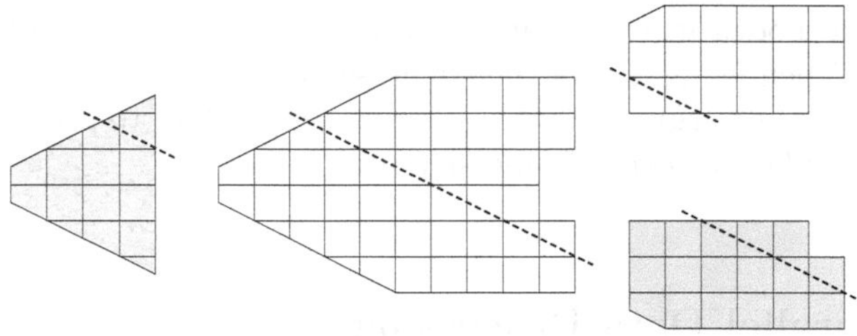

Each process can independently clip its cells with this plane. The end result is the same as if the clipping had been done serially, as shown below. If the cells are brought together, clearly, the clipping operation that took place is correct.

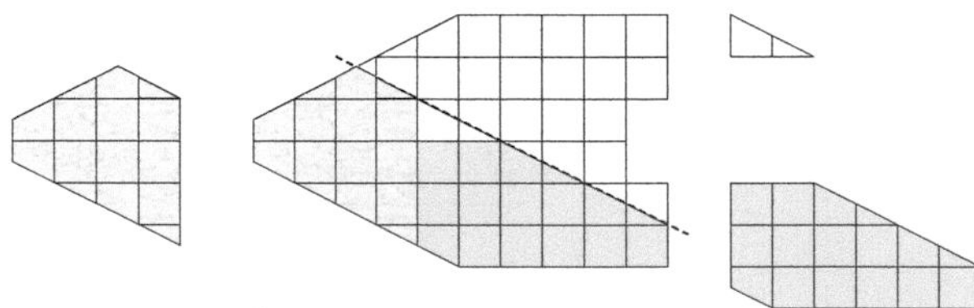

Unfortunately, blindly running visualization algorithms on partitions of cells does not always result in the correct answer. As a simple example, consider the external faces algorithm. The external faces algorithm finds all cell faces that belong to only one cell, thereby identifying the boundaries of the mesh.

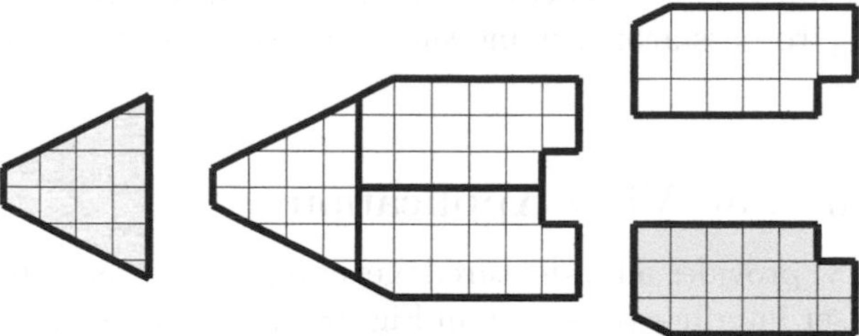

When all the processes ran the external faces algorithm independently, many internal faces were incorrectly identified as being external. This incorrect determination of the external face happens when a cell in one partition has a neighbor in another partition. The processes do not have access to the cells in other partitions, so there is no way of knowing that these neighboring cells exist.

The solution employed by ParaView is to use ghost cells. Ghost cells are cells that belong to one process's partition of the data. However, they are used by other processes' calculations when such calculations cannot be correctly performed without information about the adjacent cells that are not part of that process's partition. To use ghost cells, first, all the neighboring cells in each partition must be identified. These neighboring cells are then copied to the partition and marked as ghost cells, as indicated by the light gray cells in the following example.

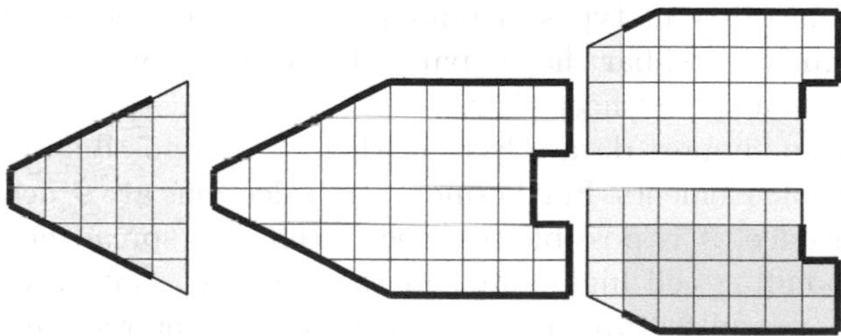

When the external faces algorithm is run with the ghost cells, there are still some internal faces incorrectly identified as external faces. However, all misclassified faces are on ghost cells, and the faces inherit the ghost status of the cell it came from. ParaView then strips the ghost faces and the result is the correct answer.

While ghost cells provide a convenient mechanism for overcoming issues due to data partitioning, they may not be sufficient for all cases. For example, when tracing streamlines, communication between processes is essential to ensure that the streamlines are smooth and connected as the trace travels from one partition to another. ParaView provides communicators that enable communication between processes. These convenient wrappers around MPI make it easy to pass arbitrary messages to processes, including data objects.

18.5 The ParaView Application

ParaView provides a full-featured, general purpose, user application with a graphical user interface (as shown in Fig. 18.1), as well as a scripting front-end that can be used to build visualization pipelines for data analysis and rendering. The Qt-based GUI front-end exposes the underlying capabilities of the ParaView framework in a cross-platform, easy-to-use application. The Python-based scripting API enables users to build visualization pipelines through a scriptable environment, which is suitable for noninteractive modes of operation.

18.5.1 Graphical User Interface

The ParaView application exposes the underlying visualization pipeline to the user. In a typical usage workflow, the user opens one or more data sets for analysis. ParaView supports a large number of file formats and developers can add plug-ins to support custom formats. The user can then transform and process the data by applying any of the filters available. Filters cover a wide array of tasks, from analyzing the input data to computing derived quantities, or statistics to extracting features, contours, streamlines, etc. The results from these filters can then be shown in views. Different types of views are available for producing different types of rendering, including views for rendering 3D geometry, line charts, bar charts, parallel coordinate plots, and even raw data in the form of a spreadsheet view.

Views also support data selection, that is, it is possible to click and drag to select visible elements. Furthermore, the selections are synchronized among all views. Hence, it is possible to select a cell in a spreadsheet view and see the corresponding cell highlighted in the 3D view. Data selection is one of the most powerful features of ParaView. Besides interactive selection, users can select cells or points that satisfy a certain criteria specified in the form of structured queries. Elements can be selected over time to generate plots that show how the variables evolve.

In addition to supporting Python for batch mode, ParaView integrates Python support throughout the application as well. Users can create macros using Python scripts for repeated tasks and can use the integrated Python shell for scripting while using the GUI. Any changes made to the ParaView environment through the Python shell are immediately reflected in the GUI. To understand ParaView's Python API better, users can always generate Python traces to see the Python script corresponding to the actions performed in the GUI. Python can also be used for data processing. Users can employ the Programmable Filter, which runs the user-specified Python code every time the filter executes.

Also, ParaView supports keyframe-based animations by changing filter parameters, or the camera, as the animation progresses. Users can also write custom Python scripts to control the scene in each animation step.

18.5.2 Scripting with Python

```
from paraview.simple import *

# Connect to a remote server.
Connect(''amber11'', 11111)
# Open a data file.
OpenDataFile(''...can.ex2'')
# Create a view to show the data in.
CreateRenderView()
# Show the 3D geometry for the opened dataset
# in the newly created 3D view.
Show()
# Request Render.
Render()
```

FIGURE 18.2: A Python script demonstrating the ParaView scripting interface used to set up a visualization pipeline.

ParaView offers a rich scripting support through Python. Users can write Python scripts to perform almost all the operations that can be done using the graphics interface. Scripting is ideal for batch processing as well as creating macros to automate tasks in the GUI. Besides controlling the visualization pipeline, Python scripts can also be used to produce or process data via programmable sources and filters. ParaView exposes the core components of VTK's data and execution models to Python, so developers can write readers and filters without compiling any C/C++ code.

18.6 Customizing with Plug-ins and Custom Applications

ParaView is designed to be highly customizable. Developers can develop and distribute plug-ins that can add new readers, writers, or algorithms to ParaView. Plug-ins can also add components to the user interface such as dock panels, toolbars, or even new view types and rendering styles. Developers can also create custom applications that are meant for particular domains or provide workflows that are more intuitive for the expected users.

The Computational Model Builder suite (CMB) being developed for the US Army Engineer Research and Development Center is built on ParaView's client-server framework and addresses the pre- and post-processing needs asso-

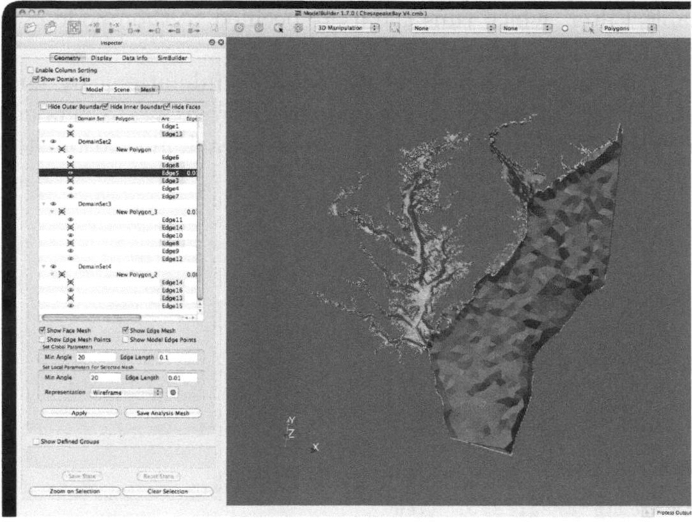

FIGURE 18.3: An application from the Computational Model Builder (CMB) suite based on ParaView's client–server framework being used to develop a suitable mesh of Chesapeake Bay for a surface water simulation.

ciated with the hydrological simulations of ground and surface water problems. Figure 18.3 shows one of the applications from the CMB suite used for mesh editing.

18.7 Co-Processing: *In Situ* Visualization and Data Analysis

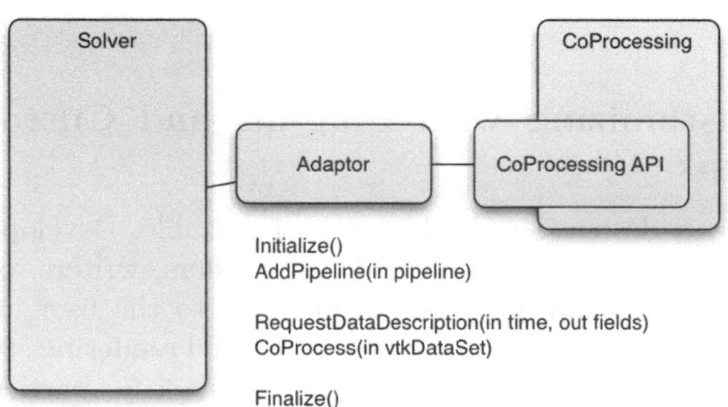

FIGURE 18.4: The ParaView Co-processing Library generalizes to many simulations by using adaptors that map simulation data structures to data structures it can process natively.

Scientific simulation on parallel supercomputers is traditionally performed in three sequential steps: pre-processing (e.g., meshing and partitioning), solution, and post-processing (e.g., data analysis and visualization). Not all of these components are actually run on the supercomputer. In particular, pre-processing and post-processing usually happen on smaller, but more interactive computing resources. However, the previous decade has seen growth in both the need and ability to perform scalable parallel analysis. This need for parallel analysis motivates the coupling of the solver and visualization. The main motivation, however, is that, although, raw compute power for parallel visualization computers has grown, the other aspects of the systems have not—such as networking and file storage. Hence, the time spent writing data to and reading data from disk storage is beginning to dominate the time spent in both the solver and the data analysis [12]. *In situ* analysis can be an effective tool for alleviating the overhead for disk storage [16].

The ParaView Co-processing Library makes it possible to integrate data analysis and visualization with the solver. The library provides a general purpose framework that can be integrated into a variety of solvers. As shown in Figure 18.4, the library relies on adaptors, small pieces of code written for each new linked simulation, to translate data structures between the simulation's code and the Co-processing Library's VTK-based data model. Once an adaptor is developed, the Co-processing Library can be integrated with the simulation code to perform predetermined data extraction and visualization tasks at runtime. The visualization pipeline can also be modified interactively, without interrupting the simulation. Furthermore, the co-processing framework can be used to move simulation data and/or data output generated by *in situ* data extraction algorithms over the network to a parallel ParaView server for runtime co-visualization.

18.8 ParaViewWeb: Interactive Visualization for the Web

The Internet has permeated the modern way of life so much that people are now migrating traditional desktop applications like document editing and finance management to the web. Part of the appeal of the web is its ease of access. One simply types the URL in the web browser to access the application, regardless of location or the type of device. Additionally, the web makes it easier to share and collaborate with other users, regardless of where they are located. ParaViewWeb enables users to access high performance visualization capabilities via a web browser.

ParaViewWeb is a collection of web services that provide interactive visualization to be used in custom web applications. ParaViewWeb provides a Javascript library that can be used within web browsers to access the web services and perform tasks ranging from starting visualization jobs to interacting with the visualization pipeline. It supports interactive rendering using

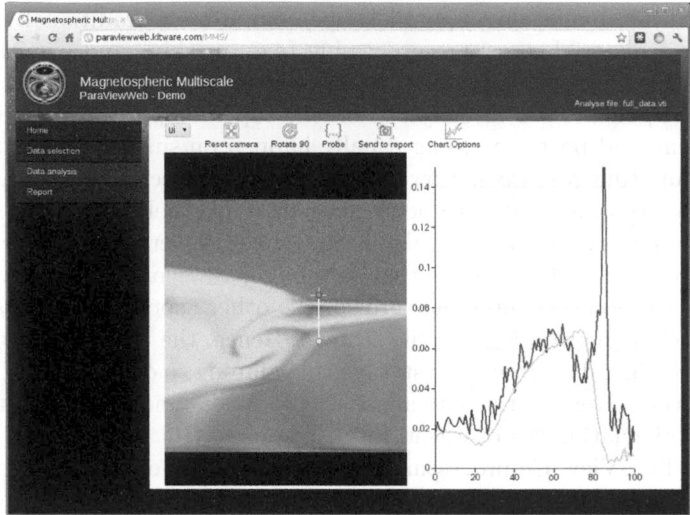

FIGURE 18.5: A prototype web application based on ParaViewWeb for interactive analysis of 3D particle-in-cell (PIC) simulation results for space scientists.

WebGL for in-browser rendering, and using a Java applet, Flash, or Javascript for server-side rendering. ParaViewWeb uses ParaView for all the data processing behind the web server (typically on a HPC resource), while delivering either images or 3D geometry to the web browser for rendering. By providing a set of customizable components, ParaViewWeb makes it possible to build highly focused web applications that can be easily integrated into existing web infrastructure within organizations as shown in Figure 18.5.

18.9 ParaView In Use

ParaView is one of the most widely used applications for high performance visualization. This section discusses some of the scientific domains in which ParaView is successfully used for analysis.

18.9.1 Identifying and Validating Fragmentation in Shock Physics Simulation

The analysis of shock physics, which can involve high energies, high-velocity materials, and highly variable results, is challenging. Very little can be measured, for example, during an experimental explosion; most experimental data is collected in the aftermath. High-fidelity simulations using codes like CTH [7] are possible, but they require a significant amount of post-processing

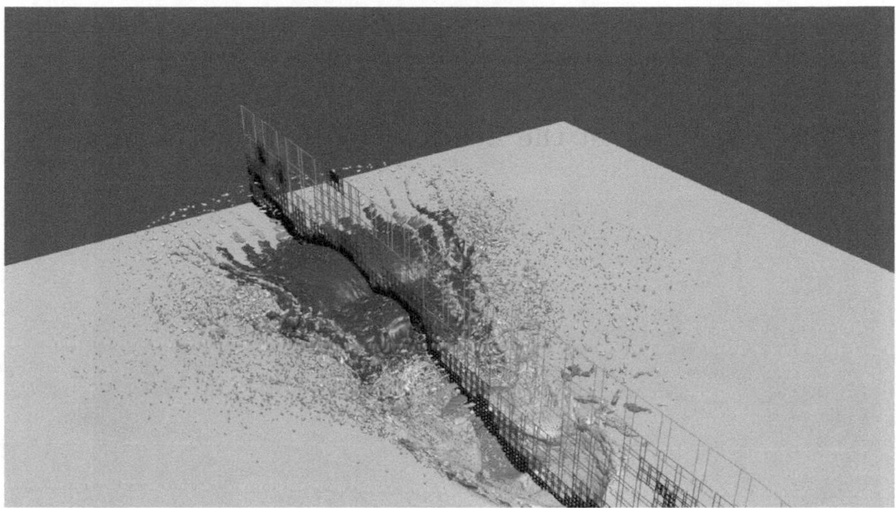

FIGURE 18.6: The fragmentation pattern of a high-speed aluminum ball hitting an aluminum brick.

to properly understand the results. Even the simplest interactions, such as those shown in Figure 18.6, can yield thousands of fragments, many of which are less than one microgram.

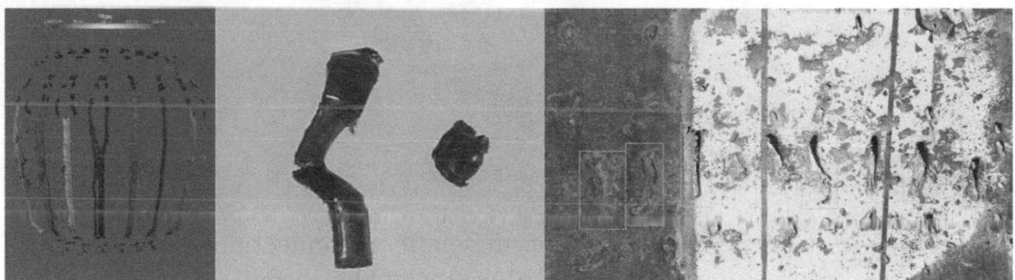

FIGURE 18.7: ParaView's fragment extraction can be used to validate that the fragments in a simulated explosion (left) match collected debris (center) and observed damage (right) from experiments.

Making scientific inquiries from CTH simulation data requires multiple post-processing tasks. Because CTH performs volume fraction computations on an Eulerian grid, object shapes and their metrics are not immediately known. ParaView can identify fragments by isolating connected cells containing a material. Certain quantities are then derived from the shape of the fragments, such as mass and volume [11]. Once these shapes are extracted, they can be visualized and directly compared with those from experiments, as shown in Figure 18.7. Statistics and histograms from the collection of fragments can also be used as valid evidence. The fragmentation can then be used further to hypothesize about other shock physics scenarios.

18.9.2 ParaView at the Los Alamos National Laboratory

FIGURE 18.8: A visualization of magnetic reconnection from the VPIC project at Los Alamos National Laboratory. The visualization is generated from a structured data containing 3.3 billion cells with two vector fields and one scalar field, produced using 256 cores running ParaView in the interactive queue on the Kraken supercomputer at the National Institute for Computational Sciences (`http://www.nics.tennessee.edu/`).

Researchers at the Los Alamos National Laboratory (LANL) use ParaView to visualize and analyze a range of simulation and experimental data on an assortment of supercomputers and desktop computers. In the world of supercomputing, a scientist is often geographically distant from the supercomputer they use for simulation. ParaView's client-server architecture allows for flexibility in connecting a ParaView client to remote sites that are running ParaView servers, which can provide raw data, geometry, or imagery depending on the client's resources and the speed of the network connection. For extremely large data, this feature is convenient, even when the researcher is at the same site as their data.

Nuclear energy, carbon sequestration, cosmology, magnetic reconnection, ocean modeling, forest fire modeling, windblade simulation (wind energy), and visualization science are just some of the areas that have leveraged ParaView to explore, understand, visualize, identify, and quantify features in their data. The data sets range from very small (hundreds of data points) to extremely large data sets, both spatially and temporally. ParaView is also able to support the scientists' models on various levels of sophistication and development. Newly formed or implemented models can be visually explored and even debugged based on visualizations. Features are often first visually and qualitatively located, and, with time, are automatically located and quantified in ParaView. Feature-detecting algorithms developed in ParaView have been incorporated into the scientific code to run at simulation time.

An example of a large visualization produced remotely by a LANL Scientist is in the area of plasma simulation, studied by the VPIC [2] team. Magnetic reconnection is a basic plasma process involving the rapid conversion of magnetic field energy into various forms of plasma kinetic energy, including high-speed flows, thermal heating, and highly energetic particles. Simulation runs were performed on Roadrunner (LANL), Kraken (UTK), and Jaguar(ORNL), computing massive grid sizes of 16 billion cells with multiple variables. The VPIC team saved the data for later post-processing by using the supercomputing platform or an attached visualization cluster, either of which can run ParaView. The ParaView client is regularly used by team members to check the progress of the simulation, even from a home internet connection. They now consider interactive visualization critical to the success of their project (see Fig. 18.8).

18.9.3 Analyzing Simulations of the Earth's Magnetosphere

Researchers at the University of California, San Diego have been using ParaView for analyzing results of 3D global hybrid simulations of the magnetosphere. These simulations that are typically run at full scale on supercomputers (such as the NSF's Kraken) and can generate extremely large data sets (200TB from a single run) that are multivariate and noisy in nature. Using ParaView, they were able to build complex visualization pipelines using physics-based custom filters and algorithms to detect and track events of interest in the data. Figure 18.9 shows visualizations generated using ParaView to find and track *flux ropes* in global hybrid simulations of the Earth's magnetosphere.

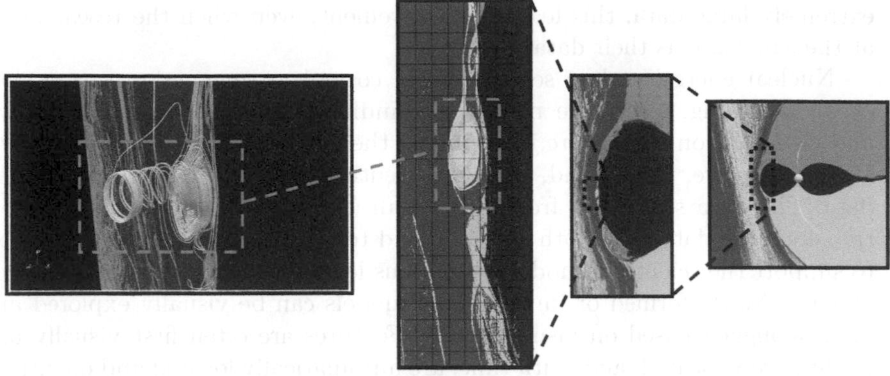

FIGURE 18.9: Finding and tracking of flux ropes in global hybrid simulations of the Earth's magnetosphere using ParaView [9].

18.10 Conclusion

ParaView is one of the widely used open source, end-user applications for high performance visualization. Based on VTK's data processing pipeline, ParaView enables parallel processing and remote visualization under various HPC configurations. ParaView's plug-in-centric design enables developers to add support for custom readers and filters, as well as create highly customized applications for desktops and web browsers. The ParaView *in situ* processing framework can be embedded in various simulation codes to perform visualization and analysis within the simulation run itself.

References

[1] Utkarsh Ayachit and Dave DeMarle. Customizing ParaView. *Proceedings of the REVISE Workshop of the IEEE Visualization Conference*, 2009.

[2] K. J. Bowers, B. J. Albright, L. Yin, W. Daughton, V. Roytershteyn, B. Bergen, and T. J. T. Kwan. Advances in Petascale Kinetic Plasma Simulation with VPIC and Roadrunner. *Journal of Physics: Conference Series*, 180(1):012055, 2009.

[3] Andy Cedilnik, Berk Geveci, Kenneth Moreland, James Ahrens, and Jean Favre. Remote Large Data Visualization in the ParaView Framework. In *Proceedings of the Eurographics Symposium on Parallel Graphics and Visualization (EGPGV)*, pages 163–170, May 2006.

[4] Carolina Cruz-Neira, Daniel J. Sandin, and Thomas A. DeFanti. Surround-Screen Projection-Based Virtual Reality: the Design and Implementation of the CAVE. In *Proceedings of the 20th Annual Conference on Computer Graphics and Interactive Techniques*, SIGGRAPH '93, pages 135–142, 1993.

[5] Nathan Fabian, Kenneth Moreland, David Thompson, Andrew C. Bauer, Pat Marion, Berk Geveci, Michel Rasquin, and Kenneth E. Jansen. The ParaView Coprocessing Library: A Scalable, General Purpose In Situ Visualization Library. In *IEEE Symposium on Large Data Analysis and Visualization (LDAV)*, pages 89–96, 2011.

[6] Paul E. Haeberli. ConMan: A Visual Programming Language for Interactive Graphics. *ACM SIGGRAPH Computer Graphics*, 22(4):103–111, August 1988.

[7] E. S. Hertel, R. L. Bell, M. G. Elrick, A. V. Farnsworth, G. I. Kerley, J. M. Mcglaun, S. V. Petney, S. A. Silling, P. A. Taylor, and L. Yarrington. CTH: A Software Family for Multi-Dimensional Shock Physics Analysis. In *Proceedings of the 19th International Symposium on Shock Waves*, pages 377–382, 1993.

[8] Sebastien Jourdain, Utkarsh Ayachit, and Berk Geveci. ParaViewWeb, A Web Framework for 3D Visualization and Data Processing. *IADIS International Conference on Web Virtual Reality and Three-Dimensional Worlds*, July 2010.

[9] H. Karimabadi, B. Loring, H. X. Vu, Y. Omelchenko, M. Tatineni, A. Majumdar, U. Ayachit, and B. Geveci. Petascale Global Kinetic Simulations Of The Magnetosphere and Visualization Strategies For Analysis Of Very Large Multi-Variate Data Sets. *Numerical Modeling of Space Plasma Flows, ASTRONUM-2010 (ASP Conference Series)*, 444, 2010.

[10] Bruce Lucas, Gregory D. Abram, Nancy S. Collins, David A. Epstein, Donna L. Gresh, and Kevin P. McAuliffe. An Architecture for a Scientific Visualization System. In *Proceedings of the IEEE Conference on Visualization*, pages 107–114, 1992.

[11] Kenneth Moreland, C. Charles Law, Lisa Ice, and David Karelitz. Analysis of Fragmentation in Shock Physics Simulation. In *Proceedings of the 2008 Workshop on Ultrascale Visualization*, pages 40–46, November 2008.

[12] Robert Ross, Tom Peterka, Han W. Shen, Y. Hong, Kwan L. Ma, Hongfeng Yu, and Kenneth Moreland. Visualization and Parallel I/O at Extreme Scale. *Journal of Physics: Conference Series*, July 2008.

[13] Will Schroeder, Ken Martin, and Bill Lorensen. *The Visualization Toolkit: An Object Oriented Approach to 3D Graphics.* Kitware Inc., fourth edition, 2004.

[14] Nikhil Shetty, Aashish Chaudhary, Daniel S. Coming, William R. Sherman, Patrick O'Leary, Eric T. Whiting, and Simon Su. Immersive ParaView: A Community-Based, Immersive, Universal Scientific Visualization Application. In *Proceedings of the IEEE Virtual Reality Conference, VR 2011*, pages 239–240, 2011.

[15] Amy Henderson Squillacote. *The ParaView Guide: A Parallel Visualization Application.* Kitware Inc., 2007.

[16] David Thompson, Nathan Fabian, Kenneth Moreland, and Lisa Ice. Design Issues for Performing In Situ Analysis of Simulation Data. Technical Report SAND2009-2014, Sandia National Laboratories, 2009.

Chapter 19

The ViSUS Visualization Framework

Valerio Pascucci
University of Utah

Giorgio Scorzelli
University of Utah

Brian Summa
University of Utah

Peer-Timo Bremer
University of Utah

Attila Gyulassy
University of Utah

Cameron Christensen
University of Utah

Sujin Philip
University of Utah

Sidharth Kumar
University of Utah

19.1 Introduction 402
19.2 ViSUS Software Architecture 402
19.3 Applications 408
References 412

The ViSUS software framework[1] has been designed as an environment that allows the interactive exploration of massive scientific models on a variety of hardware, possibly over geographically distributed platforms. This chapter is devoted to the description of the scalability principles that are at the basis of

[1]For more information and software downloads see `http://visus.co` and `http://visus.us`.

the ViSUS design and how they can be used in practical applications, both in scientific visualization and other domains such as digital photography or the exploration of geospatial models.

19.1 Introduction

The ViSUS software framework was designed with the primary philosophy that the visualization of massive data need not be tied to specialized hardware or infrastructure. In other words, a visualization environment for large data can be designed to be lightweight, highly scalable, and run on a variety of platforms or hardware. Moreover, if designed generally, such an infrastructure can have a wide variety of applications, all from the same code base. Figure 19.1 details example applications and the major components of the ViSUS infrastructure. The components can be grouped into three major categories: first, a lightweight and fast out-of-core data management framework using multi-resolution space-filling curves. This allows the organization of information in an order that exploits the cache hierarchies of any modern data storage architectures. Second, a data flow framework that allows data to be processed during movement. Processing massive data sets in their entirety would be a long and expensive operation, which hinders interactive exploration. By designing new algorithms to fit within this framework, data can be processed as it moves. The third category is a portable visualization layer, which was designed to scale from mobile devices to Powerwall displays with same the code base. This chapter describes the ViSUS infrastructure, and also explores practical examples in real-world applications.

19.2 ViSUS Software Architecture

Figure 19.1 provides a diagram of the ViSUS software architecture. This section details ViSUS's three major components and how they are used to achieve a fast, scalable, and highly portable data processing and visualization environment.

Data Access Layer. The ViSUS data access layer is a key component allowing an immediate and efficient data pipeline processing that otherwise would be stalled by traditional system I/O cost. In particular, the ViSUS I/O component, and its generalized database component, are focused on enabling the effective deployment of out-of-core and data streaming algorithms. Out-of-core computing [11] specifically addresses the issues of algorithm redesign and data layout restructuring. These are necessary to enable data access patterns having minimal performance degradation with external memory storage. Algorithmic approaches in this area also yield valuable techniques for parallel and distributed computing. In this environment, one typically has to deal with

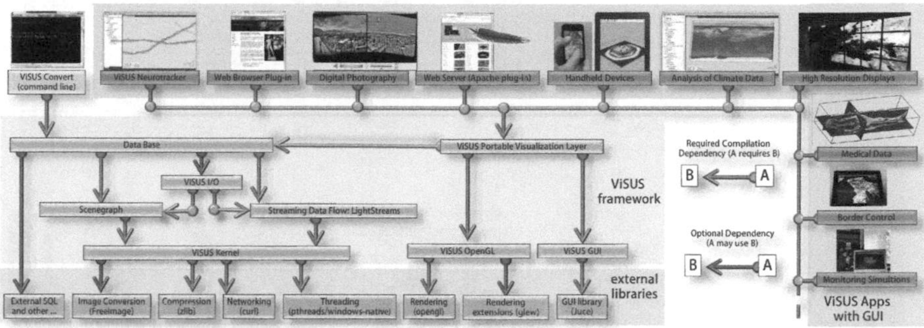

FIGURE 19.1: The architecture of the ViSUS software framework. Arrows denote external and internal dependences of the main software components. Additionally, the architecture shows the relationship with several example applications that have been successfully developed with this framework.

the similar issue of balancing processing time with the time required for data access and movement among elements of a distributed or parallel application.

The solution to the out-of-core processing problem is typically divided into two parts: (1) algorithm analysis, to understand the data access patterns and, when possible, redesign algorithms to maximize data locality; and (2) storage of data in the secondary memory using a layout consistent with the access patterns of the algorithm, amortizing the cost of individual I/O operations over several memory access operations.

To achieve real-time rates for visualization and/or analysis of extreme scale data, one would commonly seek some form of adaptive level of detail and/or data streaming. By traversing simulation data hierarchically from the coarse to the fine resolutions and progressively updating output data structures derived from this data, one can provide a framework that allows for real-time access of the simulation data that will perform well even on an extreme scale data set. Many of the parameters for interaction, such as display viewpoint, are determined by users at runtime, and therefore, precomputing these levels of details optimized for specific queries is infeasible. Therefore, to maintain efficiency, a storage data layout must satisfy two general requirements: (1) if the input hierarchy is traversed in coarse-to-fine order, data in the same level of resolution should be accessed at the same time; and (2) within each level of resolution, the regions in close spatial proximity are stored in close proximity in memory.

Space-filling curves [9] were also used successfully to develop a static indexing scheme that generates a data layout satisfying both the above requirements for hierarchical traversal (see Fig. 19.2). The data access layer of ViSUS employs a hierarchical variant of a Lebesgue space-filling curve [5]. The data layout of this curve is commonly referred to as an *HZ-order* in the

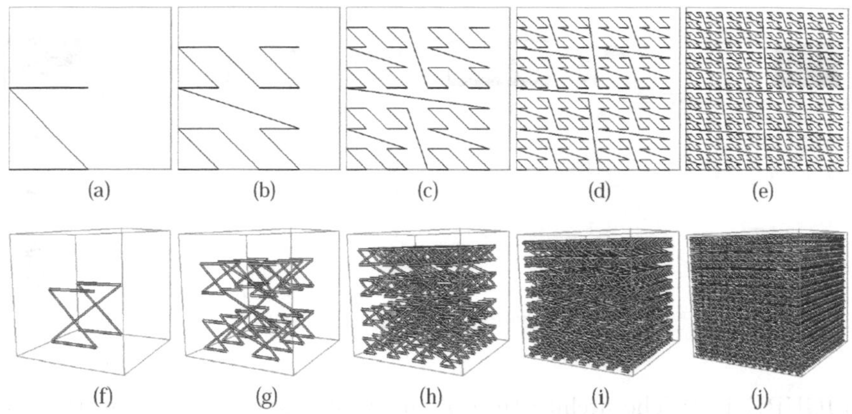

FIGURE 19.2: (a)–(e) The first five levels of resolution of the 2D Lebesgue's space-filling curve. (f)–(j) The first five levels of resolution of the 3D Lebesgue's space-filling curve.

literature. This data access layer has three key features that make it particularly attractive. First, the order of the data is independent of the out-of-core block structure, so that its use in different settings (e.g., local disk access or transmission over a network) does not require any large data reorganization. Second, conversion from the Z-order indexing [4] used in classical database approaches to ViSUS's HZ-order indexing scheme can be implemented with a simple sequence of bit string manipulations. Third, since there is no data replication, the performance penalties associated with guaranteeing consistency are avoided, especially for dynamic updates and increased storage requirements, typically associated with most hierarchical and out-of-core schemes.

Parallel I/O for Large Scale Simulations. The multiresolution data layout of ViSUS, discussed above, is a progressive, linear format and, therefore, has a write routine that is inherently serial. During the execution of large scale simulations, it is ideal for each node in the simulation to be able to write its piece of the domain data directly into this layout. Therefore, a parallel write strategy must be employed. Figure 19.3 illustrates different possible parallel strategies that have been considered. As shown in Figure 19.3a, each process can naively write its own data directly to the proper location in a unique underlying binary file. This is inefficient, though, due to the large number of small granularity, concurrent accesses to the same file. Moreover, as the data gets large, it becomes disadvantageous to store the entire data set as a single, large file and typically the entire data set is partitioned into a series of smaller more manageable pieces. This disjointness can be used by a parallel write routine. As each simulation process produces a portion of the data, it can store its piece of the overall data set locally and pass the data on to an aggregator process.

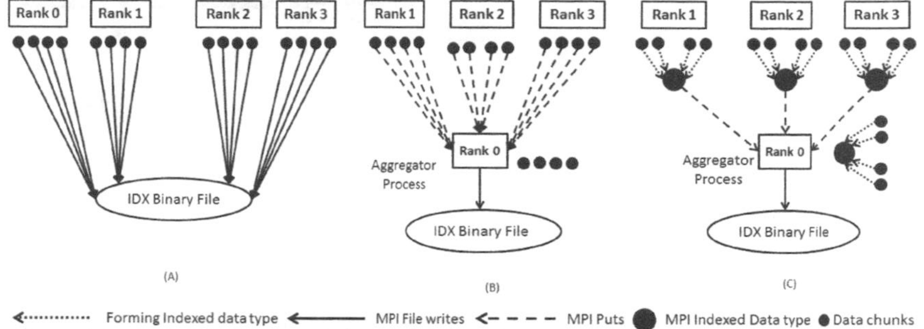

FIGURE 19.3: Parallel I/O strategies: (a) Naive approach where each process writes its data in the same file; (b) an alternative approach where contiguous data segment is transmitted to an intermediate aggregator that writes to disk; and (c) communication reducing approach with bundling of noncontiguous accesses into a single message.

The aggregator processes can be used to gather the individual pieces and composite the entire data set. Figure 19.3b shows this strategy, where each process transmits a contiguous data segment to an intermediate aggregator. Once the aggregator's buffer is complete, the data is written to disk using a single large I/O operation. Figure 19.3c, illustrates a strategy where several noncontiguous memory accesses from each process are bundled into a single message. This approach also reduces the overhead due to the number of small network messages needed to transfer the data to the aggregators. This strategy has been shown to exhibit a good throughput performance and weak scaling for S3D combustion simulation applications when compared to the standard Fortran I/O benchmark [2, 3]. In particular, recent results[2] have shown empirically how this strategy scales well for a large number of nodes (currently up to 32,000) while enabling real-time monitoring of high-resolution simulations (see 19.3).

LightStream Dataflow and Scene Graph. Even simple manipulations can be overly expensive when applied to each variable in a large-scale data set. Instead, it is ideal to process the data based on need, by pushing data through a processing pipeline as the user interacts with different portions of the data. The ViSUS multiresolution data layout enables efficient access to different regions of the data at varying resolutions. Therefore, different compute modules can be implemented using progressive algorithms to operate on this data stream. Operations like binning, clustering, or rescaling are trivial to implement on this hierarchy, given some known statistics on the data, such as the function value range, etc. These operators can be applied to the data stream as is, while the data is moving to the user, progressively refining

[2]Execution on the Hopper 2 system at NERSC.

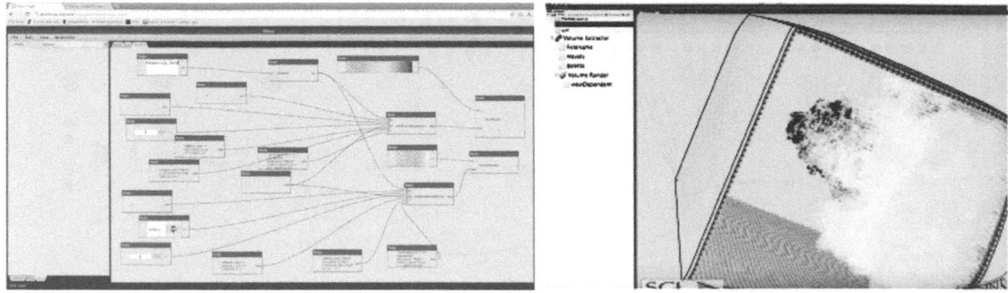

FIGURE 19.4: The LightStream data flow used for analysis and visualization of a 3D combustion simulation (Uintah code). Several data flow modules chained together to provide a light and flexible stream processing capability (left). One visualization that is the result from this data flow (right).

the operation as more data arrives. More complex operations can also be reformulated to work well using the hierarchy. For instance, using the layout for 2D image data produces a hierarchy, which is identical to a subsampled image pyramid on the data. Moreover, as data is requested progressively, the transfer will traverse this pyramid in a coarse-to-fine manner. Techniques like gradient-domain image editing can be reformulated to use this progressive stream and produce visually acceptable solutions [10, 7, 8]. These adaptive, progressive solutions allow the user to explore a full resolution solution as if it was fully available, without the expensive, full computation.

ViSUS LightStream facilitates this stream processing model by providing definable modules within a data flow framework with a well-understood API. Figure 19.4 gives an example of a data flow for the analysis and visualization of a scientific simulation. This particular example is the data flow for a Uintah combustion simulation used by the Center for the Simulation of Accidental Fires and Explosions (C-SAFE) at the University of Utah. Each LightStream module provides streaming capability through input and output data ports that can be used in a variety of data transfer/sharing modes. In this way, groups of modules can be chained to provide complex processing operations as the data is transferred from the initial source to the final data analysis and visualization stages. This data flow is typically driven by user demands and interactions. A variety of "standard" modules are part of the base system, such as data differencing (for change detection), content-based image clustering (for feature detection), or volume rendering with multiple, science-centric transfer function. These can be used by new developers as templates for their own progressive streaming data processing modules.

ViSUS also provides a scene graph hierarchy for both organizing objects in a particular environment and the sharing and inheriting of parameters. Each component in a model is represented by a node in this scene graph and inherits the transformations and environment parameters from its parents. 3D volume or 2D slice extractors are children of a data set node. As an example

FIGURE 19.5: The same application and visualization of a Mars panorama running on an iPhone 3G mobile device (left) and a Powerwall display (right). Data courtesy of NASA.

of inheritance, a scene graph parameter for a transfer function can be applied to the scene graph node of a data set. If the extractor on this data set does not provide its own transfer function, it will be inherited.

Portable Visualization Layer—ViSUS AppKit. The visualization component of ViSUS was built with the philosophy that a single code base can be designed to run on a variety of platforms and hardware ranging from mobile devices to Powerwall displays. To enable this portability, the the basic draw routines were designed to be OpenGL ES compatible. This is a limited subset of OpenGL used primarily for mobile devices. More advanced draw routines can be enabled if a system's hardware can support it. In this way, the data visualization can scale in quality depending on the available hardware. Beyond the display of the data, the underlying GUI library can hinder portability to multiple devices. At this time, ViSUS has made use of the Juce[3] library, which is lightweight and supports mobile platforms such as iOS and Android in addition to major operating systems. ViSUS provides a demo viewer, which contains standard visualizations such as slicing, volume rendering, and isosurfacing. Similarly to the example LightStream modules, these routines can be expanded through a well-defined API. Additionally, the base system can display 2D and 3D time-varying data. As mentioned above, each of these visualizations can operate on the end result of a LightStream data flow. The system considers a 2D data set as a special case of a slice renderer, and therefore, the same code base is used for 2D and 3D data sets. Combining all of the above design decisions allows the same code base to be used on multiple platforms, seamlessly, for data of arbitrary dimensions. Figure 19.5 shows the same application and visualization running on an iPhone 3G mobile device and a Powerwall display.

[3]See `http://www.rawmaterialsoftware.com`.

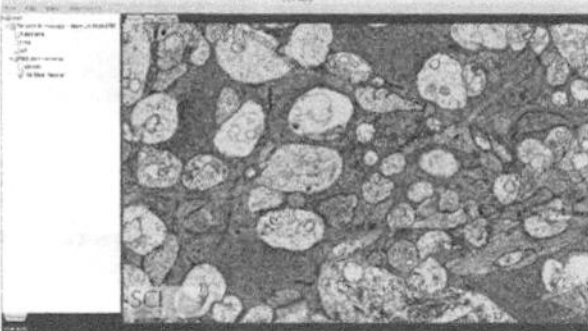
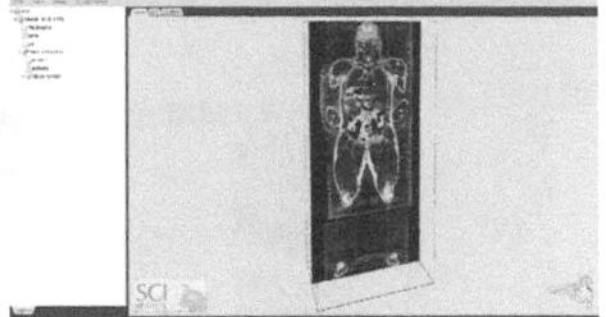

FIGURE 19.6: The ViSUS software framework visualizing and processing medical imagery. On the left, the Neurotracker application providing the segmentation of neurons from extremely high-resolution Confocal Fluorescence Microscopy brain imagery. This data is courtesy of the Center for Integrated Neuroscience and Human Behavior at the Brain Institute, University of Utah. In the center, an application for the interactive exploration of an electron microscopy image of a slice of a rabbit retina. This data set is courtesy of the MarcLab at the University of Utah. On the right, a 3D slicing example using the Visible Male data set.

19.3 Applications

As shown in the upper and right portions of the infrastructure diagram in Figure 19.1, ViSUS has the versatility to be used in a wide range of applications. This section highlights a representative subset of these applications. The general philosophy behind the ViSUS applications is the deployment of light tools that are task driven as a complement to general-purpose solutions provided by other existing systems.

Neurotracker and Other Medical Applications. The Neurotracker is an application built on the ViSUS framework that targets the segmentation of neurons from extremely high resolution Confocal Fluorescence Microscopy brain imagery.[4] The core data processing of the Neurotracker uses the ViSUS I/O library to get fast access to the brain imaging data combined with multiresolution topological analysis used to seed the segmentation of neurons. A marching image segmentation routine extracts filament structures from the topological seeds. The user interface of the the Neurotracker is built with the ViSUS GUI components specialized for segmentation.

Figure 19.6 also highlights two additional medical imaging applications. In the center image of Figure 19.6, an example of the ViSUS framework is used for interactive exploration of an electron microscopy image of a slice of a rabbit retina. In all, this 2D data set is over 3.4 gigapixels in size.[5] On the right of Figure 19.6 is a 3D data slicing example of the Visible Male[6] data set comprised of over 4.6 billion color voxels.

[4]Data is courtesy of the Center for Integrated Neuroscience and Human Behavior at the Brain Institute, University of Utah (`http://brain.utah.edu/`).

[5]Data is courtesy of the MarcLab at the University of Utah (`http://prometheus.med.utah.edu/\textasciitilde{}marclab/`).

[6]See `http://www.nlm.nih.gov/research/visible/visible\_human.html`.

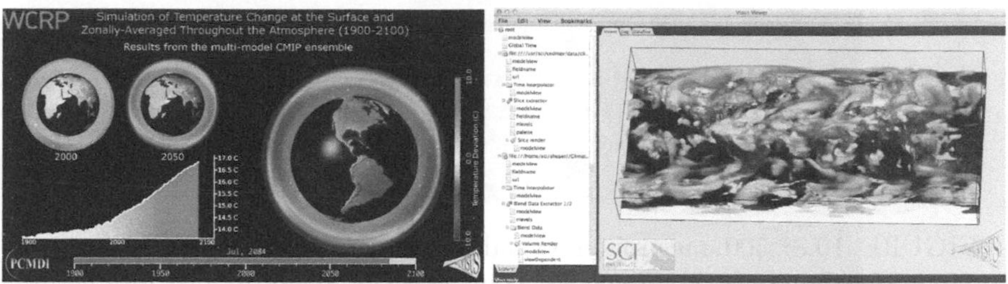

FIGURE 19.7: Remote climate visualization with ViSUS. The ViSUS framework providing a visualization for a temperature change ensemble simulation for the Earth's surface for the December 2009 climate summit meeting in Copenhagen (left). A visualization of global cloud density for a more recent climate simulation (right).

Web server and plug-in. ViSUS has been extended to support a client–server model in addition to the traditional viewer. The ViSUS server can be used as a standalone application or a web server plug-in module. The ViSUS server uses HTTP (a stateless protocol) in order to support many clients. A traditional client–server infrastructure, where the client established and maintained a stable connection to the server, can only handle a limited number of clients robustly. Using HTTP, the ViSUS server can scale to thousands of connections. The ViSUS client keeps a number of connections (normally 48) alive in a pool using the "keep-alive" option of HTTP. The use of lossy or lossless compression is configurable by the user. For example, ViSUS supports JPEG and EXR for lossy compression of byte and float data, respectively. Because the ViSUS server is an open client–server architecture, it is therefore possible to port the plug-in to any web server which supports a C++ module (i.e., apache, IIS). The ViSUS client can be enabled to cache data to the local memory or to a disk. In this way, a client can minimize transfer time by referencing data already sent, as well as having the ability to work offline if the server becomes unreachable. The ViSUS portable visualization framework (Appkit) also has the ability to be compiled as a Google Chrome, Microsoft Internet Explorer, or Mozilla Firefox web browser plug-in. This allows a ViSUS framework based viewer to be easily integrated into web visualization portals.

Remote Climate Analysis and Visualization. The ViSUS software framework has been used in the climate modeling community to visualize climate change simulations comprised of many variables over a large number of timesteps. This type of data provides the opportunity to both build high-quality data analysis routines and challenge the performance of the data management infrastructure. Figure 19.7 shows ViSUS rendering 10TB of data used by a finalist in the Data Transfer Challenge of Supercomputing 2009. Initially, this work showed the possibility of transferring, transforming, analyzing, and rendering a large data set on geographically distributed computing resources [1]. At the same time, this work was also used to introduce more

FIGURE 19.8: 500 megapixel visualization of a 2D panorama data set of Mount Rushmore (left). The color shift between images in a panorama mosaic (middle). An application using a LightStream data flow to provide approximate gradient domain solution as a user interacts with the data (right). Data set courtesy of City Escapes Photography.

forms of analysis and visualizations, providing novel insights into the dynamics of global carbon cycle, atmospheric chemistry, land and ocean ecological processes, and their coupling with climate. This work was used to present findings regarding the Earth's temperature change based on historical and projected simulation data at a December 2009 climate summit meeting in Copenhagen. The right of Figure 19.7 shows the ViSUS framework providing a visualization of global cloud density for a more recent simulation.

Panorama Multiscale Processing and Viewer. The ViSUS framework along with the LightStream data flows can be used for real-time, large panorama processing and visualization. In Figure 19.8, the left image is a visualization of a 500 megapixel 2D image of Mount Rushmore.[7] In this flattened image, there are detected color shifts due to the fact that the image is a mosaic of many individual images. The middle of Figure 19.8 provides a visualization of the original picture data for a panorama of Salt Lake City, which is comprised of over 600 individual images for a total image mosaic size of 3.2 gigapixels. As mentioned in 19.2, the LightStream data flow can be used to operate on the multiresolution data as the panorama is being viewed and provide an approximate gradient domain solution [10, 7, 8] for each viewpoint. In this way, a user can explore the panorama as if the full gradient domain solution was available without it ever being computed in full. This preview is shown on the right in Figure 19.8.

Real-Time Simulation Monitoring. Ideally, a user-scientist would like to view a simulation as it is computed, in order to steer or correct the simulation as unforeseen events arise. Simulation data is often very large. For instance, a single field of a timestep from the S3D combustion simulation in Figure 19.9 (left) is approximately 128 GB in size. In the time needed to transfer this single timestep, the user-scientist would have lost any chance for significant steering/correction of an ongoing simulation or at least take the opportunity

[7]Data courtesy of City Escapes Photography (`http://www.cityescapesphotography.com`).

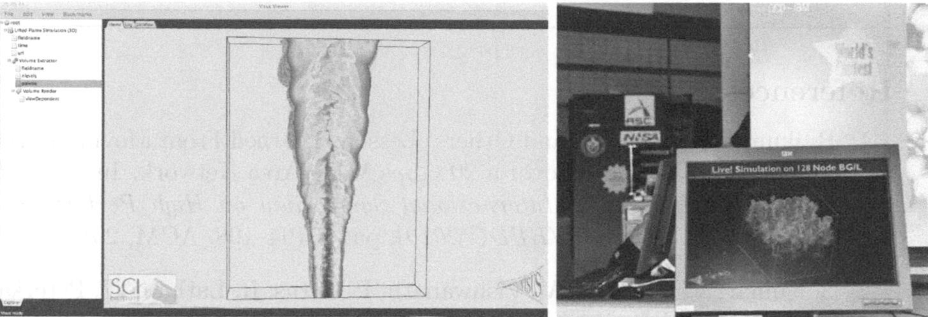

FIGURE 19.9: Remote visualization and monitoring of simulations. The left image shows remote visualization from an S3D combustion simulation run on a supercomputer at NERSC in Oakland, CA, with results displayed on a desktop machine located at the SCI Institute (Salt Lake City, Utah). The right image show two ViSUS demonstrations of LLNL simulation codes (Miranda and Raptor) visualized in real-time while executed on the BlueGene/L prototype at the IBM booth of the Supercomputing Exhibition.

to save computing resources by early termination of a job that is not useful anymore. By using the parallel ViSUS data format for simulation checkpointing [2, 3], this data can link directly with an Apache server using a ViSUS plug-in running on a node of the cluster system. By doing this, user-scientists can visualize simulation data as checkpoints are reached. ViSUS can handle missing or partial data, therefore the data can be visualized even as it is being written to the disk by the system.

ViSUS's support for a wide variety of clients (a standalone application, a web-browser plug-in, or an iOS application for the iPad or iPhone) allows the application scientist to monitor a simulation as it is produced, on practically any system that is available without any need to transfer the data off the computing cluster. As mentioned above, Figure 19.9 (left) is an S3D large-scale, combustion simulation visualized remotely from an HPC platform.[8]

This work is the natural evolution of the ViSUS approach for targeting practical applications for out-of-core data analysis and visualization. This approach has been used for direct streaming and real-time remote monitoring of the early large-scale simulations such as those executed on the IBM BlueGene/L supercomputers at LLNL [6] shown in Figure 19.9 (right). This work continues its evolution toward the deployment of high performance tools for *in situ* and post-processing data management and analysis for the software and hardware resources of the future including exascale DOE platforms of the next decade.[9]

---

[8]Data is courtesy of Jackie Chen at Sandia National Laboratories, Combustion Research Facility `http://ascr.sandia.gov/people/Chen.htm`.

[9]Center for Exascale Simulation of Combustion in Turbulence (ExaCT) (`http://science.energy.gov/ascr/research/scidac/co-design/`).

References

[1] Rajkumar Kettimuthu and Others. Lessons Learned From Moving Earth System Grid Data Sets over a 20 Gbps Wide-Area Network. In *Proceedings of the 19th ACM International Symposium on High Performance Distributed Computing (HPDC 2010)*, pages 194–198. ACM, 2010.

[2] S. Kumar, V. Pascucci, V. Vishwanath, P. Carns, R. Latham, T. Peterka, M. Papka, and R. Ross. Towards Parallel Access of Multi-dimensional, Multiresolution Scientific Data. In *Proceedings of 2010 Petascale Data Storage Workshop*, November 2010.

[3] S. Kumar, V. Vishwanath, P. Carns, B. Summa, G. Scorzelli, V. Pascucci, R. Ross, J. Chen, H. Kolla, and R. Grout. PIDX: Efficient Parallel I/O for Multi-resolution Multi-dimensional Scientific Datasets. In *Proceedings of IEEE Cluster 2011*, September 2011.

[4] J. K. Lawder and P. J. H. King. Using Space-Filling Curves for Multi-Dimensional Indexing. In *Advances in Databases*, volume 1832 of *Lecture Notes in Computer Science*, pages 20–35, 2000.

[5] Valerio Pascucci and Randall J. Frank. Global Static Indexing for Real-time Exploration of Very Large Regular Grids. In *Supercomputing '01: Proceedings of the 2001 ACM/IEEE Conference on Supercomputing (CDROM)*, New York, NY, USA, 2001. ACM.

[6] Valerio Pascucci, Daniel E. Laney, Ray J. Frank, F. Gygi, Giorgio Scorzelli, Lars Linsen, and Bernd Hamann. Real-Time Monitoring of Large Scientific Simulations. In *SAC*, pages 194–198. ACM, 2003.

[7] Sujin Philip, Brian Summa, Peer-Timo Bremer, and Valerio Pascucci. Parallel Gradient Domain Processing of Massive Images. In Torsten Kuhlen, Renato Pajarola, and Kun Zhou, editors, *Eurographics Symposium on Parallel Graphics and Visualization*, pages 11–19. Eurographics Association, 2011.

[8] Sujin Philip, Brian Summa, Valerio Pascucci, and Peer-Timo Bremer. Hybrid CPU-GPU Solver for Gradient Domain Processing of Massive Images. In *2011 IEEE 17th International Conference on Parallel and Distributed Systems (ICPADS)*, pages 244–251, December 2011.

[9] Hans Sagan. *Space-Filling Curves*. Springer-Verlag, New York, NY, 1994.

[10] B. Summa, G. Scorzelli, M. Jiang, P.-T. Bremer, and V. Pascucci. Interactive Editing of Massive Imagery Made Simple: Turning Atlanta into Atlantis. *ACM Transactions on Graphics*, 30:7:1–7:13, April 2011.

[11] J. S. Vitter. External Memory Algorithms and Data Structures: Dealing with Massive Data. *ACM Computing Surveys*, 33(2):209–271, March 2000.

Chapter 20

The VAPOR Visualization Application

Alan Norton
National Center for Atmospheric Research

John Clyne
National Center for Atmospheric Research

20.1 Introduction 415
20.1.1 Features 416
20.1.2 Limitations 417
20.2 Progressive Data Access 417
20.2.1 VAPOR Data Collection 418
20.2.2 Multiresolution 419
20.3 Visualization-Guided Analysis 420
20.4 Progressive Access Examination 422
20.4.1 Discussion 423
20.5 Conclusion 424
References 427

VAPOR is an open source visual data analysis package developed by the National Center for Atmospheric Research, with support from the National Science Foundation. VAPOR provides a highly interactive, platform independent, desktop exploration environment, capable of handling some of the largest numerical simulation outputs, yet requiring only commodity computing resources. The cornerstone of VAPOR's large data handling capability is a wavelet-based progressive access data model that enables the user to tradeoff I/O, memory, and computing resource requirements for data fidelity. This chapter provides an overview of VAPOR, placing an emphasis on the VAPOR strategy for handling large data. To illustrate the effectiveness of this technique, the chapter concludes with a brief case study of two data sets: one from Magneto-Hydrodynamics, and another from numerical weather prediction.

20.1 Introduction

VAPOR is a visualization package that was designed from the outset as a means to enable the interactive exploration of massive, time-varying, gridded data sets, primarily those resulting from high-resolution numerical simulations [2, 1]. For numerous computational scientists, the greatest limitation in the visualization of large data is often the I/O. Other computational costs incurred during analysis (e.g., computer graphics rendering, flow integration, calculation of derived quantities) are frequently dwarfed by the time it takes to retrieve the data from a disk.

VAPOR addresses this issue by attempting to minimize the amount of data read from the secondary storage. This is accomplished through the use of both a progressive access data model, designed to support efficient region-of-interest (ROI) access, and a strong reliance on the caching of previous data retrievals in random access memory (RAM). Hence, VAPOR's advanced visualization capabilities, and the ability for the user to make speed/quality trade-offs to maintain interactivity, enable rapid feature identification or identification of spatiotemporal regions of interest. Once identified, these reduced-size ROIs can be quantitatively or qualitatively explored with progressively finer detail.

While suitable for use in numerous computational science domains, VAPOR is primarily designed to address the needs of the earth and space sciences communities, and, in particular, weather, solar, oceanic and climate science, and related disciplines. As a result, VAPOR supports features typically not found in other packages (e.g., handling of geo-referenced data), but may lack capabilities found in more general visualization packages (e.g., support for unstructured computational grids).

Finally, VAPOR was designed to run on a commodity desktop (or laptop) computing platform with a shared memory architecture. As discussed below, interacting with sizeable data sets is enabled in VAPOR with only modest computing resources.

20.1.1 Features

This section provides a brief overview of some fundamental capabilities of the VAPOR GUI. These capabilities are aimed at enabling visualization-guided analysis. VAPOR is integrated with NumPy [3] environments to provide more quantitative analysis, and to provide the ability to manipulate variables and derive new quantities.

- The VAPOR GUI was designed to run and enable highly interactive performance using only a desktop or laptop computer equipped with hardware-accelerated graphics. Through the use of two user-specified data reduction parameters (see 20.2) and the reuse of previously computed and cached results, the system can provide interactive performance on commodity platforms.

- The GUI provides common advanced visualization capabilities associated with state-of-the-art interactive visualization, making use of the GPU for improved performance whenever possible. These capabilities include volume rendering, isosurfaces, contour planes, steady and unsteady flow lines, image-based flow visualization, inclusion of geometric models, transfer function editing, calculation of derived variables, coordinate axis annotation, color bars, etc.

- VAPOR supports an embedded NumPy calculation engine. Derived variables are easily expressed as Python expressions or Python scripts. These scripts are executed only on demand. When a derived variable is requested, Python calculates the derived variable at the desired accuracy and the calculation is constrained to the requested ROI. The new variable is immediately available to all data operators in the GUI. Moreover, the derived quantity is stored in cache, if space is available, to improve access speed for subsequent references.

- VAPOR provides several forms of steady and unsteady flow integration and visualization. The GUI offers numerous data-driven methods for seeding integration locations. For example, random seed locations may be biased toward high-magnitude spatial regions of an arbitrary variable. The flow integrator does not require uniform temporal or spatial sampling, and the GUI allows the selection of limited spatial and temporal extents.

20.1.2 Limitations

Several features that are commonly found in visualization packages, aimed at large data, were deliberately not included in the VAPOR design. Many of these limitations or omissions stem from VAPOR's focus on the earth and space science communities, and its emphasis on desktop computing.

- Computational grids are limited to structured, regular tessellations, though the sampling between grid points need not be uniform.

- There is no support for distributed memory architectures. A shared-memory programming model is assumed. Thus, the resolution of a grid is constrained by the available shared-memory address space.

- All data in a VAPOR session are presumed to arise from a single numerical experiment and are constrained to a single sampling grid; variables sampled at different rates, or with different coordinates cannot be mixed in a single session.

20.2 Progressive Data Access

VAPOR's method for handling gridded data sets, resulting from high-resolution numerical simulations, differs from many other scientific visualization applications. Its approach is motivated by the widening gap between compute and I/O performance. For example, in many analysis operators, especially when time varying data are involved, the single biggest bottleneck is the rate at which data can be retrieved from a disk.

Another technology trend factored into VAPOR's design is the advancement in output display resolution. While very high-resolution tiled display devices have been deployed at a number of research facilities, the number of pixels available to a typical researcher working in his or her office has not changed significantly over the years, relative to the rate of progress of other technologies.

With these thoughts in mind, the primary approach to enabling interactive processing of large data, employed by VAPOR, is progressive access. The aim of this data model is to allow the user to make trade-offs between speed and accuracy, providing a form of *focus-plus-context* [4]. Users are able to accelerate the scientific discovery process by formulating hypotheses in an interactive mode, using less resource-intensive approximations of their numerical model outputs. These hypotheses can later be validated on data, to increase resolution accuracy, at the cost of reduced interactivity.

The progressive access data model employed by VAPOR is based on the energy, or information, compaction properties of the discrete wavelet transform. An overview of wavelets, their suitability for compression of scientific data sets, and the mathematical framework for much of the discussion in this section is described in Chapter 8.

20.2.1 VAPOR Data Collection

To take advantage of VAPOR's progressive access capabilities, a data set must first be translated into VAPOR's progressive access data format—VAPOR Data Collection (VDC). A variety of command line tools supporting common file formats and user-callable libraries are provided to facilitate translation. Each variable is transformed, one timestep at a time, from the spatial to the wavelet domain. The resulting wavelet coefficients are sorted based on their information content—coefficients with larger magnitudes contain more information—and distributed to a small, finite number of bins. The number of bins, as well as the number of coefficients stored in each bin, is determined by the user, subject to the constraint that the aggregate number of wavelet coefficients in the bins equals the number of coefficients output by the wavelet transform, which in turn equals the number of grid points.

The reconstruction of a variable from wavelet space requires the application of the inverse discrete wavelet transform. If all of the bins are used for reconstruction, no information is lost (up to a floating point round-off). The user may choose, however, to only include a subset of the bins in reconstruction, producing an approximation of the original data. While any combination of bins might be used to reconstruct an approximation of the original variable, the best results are obtained when the subset of bins used in reconstruction are those containing the largest magnitude coefficients. This subsetting of wavelet coefficients is called *level-of-detail* (LOD) selection.

Each wavelet coefficient bin, for each variable, at each timestep, is stored in a separate file. This distribution of bins into different files provides some flexibility in large data management. Files corresponding to bins with less information content might be kept on tertiary storage (e.g., tape), while smaller but higher information content bins can be kept on secondary storage (e.g., rotating disk). By default, a VDC distributes wavelet coefficients into four LOD bins that are sized to offer compression rates of 500:1, 100:1, 10:1, and 1:1, when the bins are combined for reconstruction. Thus, using lower resolution LODs can result in significantly reduced I/O and storage requirements.

To improve both the computational performance of the forward wavelet transform (transformation to wavelet domain) and inverse wavelet transform (transformation to spatial domain) operations themselves, and the performance of data ROI subsetting, data volumes are decomposed into blocks prior to the forward transformation. Blocks are individually and independently transformed, sorted, and stored to files. The block dimensions are a user parameter. The default for the VDC of a 3D variable is 64^3, which empirically strikes a reasonable balance between computational efficiency on cache-based microprocessors, disk transfer rates, and provides enough degrees of freedom for high rates of compression. This latter point warrants further explanation. Because each block is treated independently, and the distribution of wavelet coefficient bins are fixed for all blocks, the maximum compression rate possible is a function of the block size. Larger blocks offer a better compression rate at the expense of an increased computational cost and storage access when a ROI is retrieved.

20.2.2 Multiresolution

The principal benefits of LOD selection are reduced I/O transfer times—which often dominate analysis—as well as the opportunity for a reduction in disk storage requirements if the lower information content wavelet coefficient bins are not stored. However, the reconstructed data contains the same number of grid points as the original data, whether generated from an approximating LOD, or losslessly reconstructed using all wavelet coefficients. Hence, the CPU or GPU computational cost of processing a variable, as well as the RAM requirements, are no less than those required for a conventional data representation, and may easily overwhelm the resources of a desktop computing platform regardless of the LOD.

Fortunately, an intrinsic property of the wavelet transform is multiresolution; an N^d signal is decomposed into dyadic hierarchy, where each level in the hierarchy approximates the next, finer level using only the fraction $\frac{1}{2^d}$ of the sample points (see 8.4 for a detailed discussion). This property affords VAPOR users two forms of quality control: the first form is LOD selection, based on wavelet coefficient prioritization; and second, resolution or *refinement* control, based on the grid sampling rate. LOD selection primarily impacts I/O, while refinement control has implications for both primary storage (RAM) and computation. Coarser grids have a smaller memory footprint, and can substantially reduce memory requirements, as well as the computational and graphical expense of many visual and nonvisual analysis operations, whose costs are proportional to the number of grid points.

The next concrete example illustrates the concepts of LOD and multiresolution, and their respective impacts on computing resources. Assume a computing mesh with 1024^3 grid points that are transformed into a VDC, resulting in three levels of detail corresponding to compression rates of 1:1 (no compression), 10:1, and 100:1. The wavelet coefficients for each LOD reside in separate files on a disk named *lod2*, *lod1*, and *lod0*, respectively. Moreover, the number of coefficients stored in each file would be approximately $1024^3 - \frac{1024^3}{10} - \frac{1024^3}{100}$, $\frac{1024^3}{10} - \frac{1024^3}{100}$, and $\frac{1024^3}{100}$, respectively. As described earlier, reconstruction of our data using the coarsest approximation (100:1) requires reading the coefficients from *lod0*, while the second coarsest approximation is reconstructed from the coefficients from both *lod1* and *lod0*, and so on. The choice of LOD will determine how much data are read from a disk.

Due to the multiresolution properties of wavelets, a second form of data reduction can be had by performing an incomplete wavelet reconstruction, halting the inverse transform after the grid has been reconstructed to 512^3, 256^3, or 128^3 grid points, for example. A multiresolution approximation contains fewer grid points than the original data, thus resulting in reduced memory and compute resources required to store and operate on the approximation. Note that the grid resolution refinement selection is independent of the LOD. However, regardless of the refinement level, higher level LODs will contribute more information, leading to more a more accurate approximation.

20.3 Visualization-Guided Analysis

As an illustration of the visualization-guided analysis capabilities of VAPOR, this section describes the workflow used in addressing a research problem in Magneto-Hydrodynamics (MHD) [6]. The data set explored is output from an MHD simulation with a high Reynolds number, computed on a 1536^3 grid, with 16 variables, requiring 216GB per timestep. It was expected that, at this resolution, geometric structures, known as *current sheets*, would form. A current sheet is characterized by the magnitude of the electrical current achieving local maxima along a 2D surface. While current sheets were expected to appear, it was not known exactly what shapes these surfaces would take. There are theoretical reasons to expect that the current sheets could wind into a rolled-up structure, i.e., a current roll; however, this phenomenon was not observed in previous simulations. Direct volume rendering of one scalar variable from the data, at full-resolution, without data reduction, would be possible, but the demands on computing resources would be substantial. The

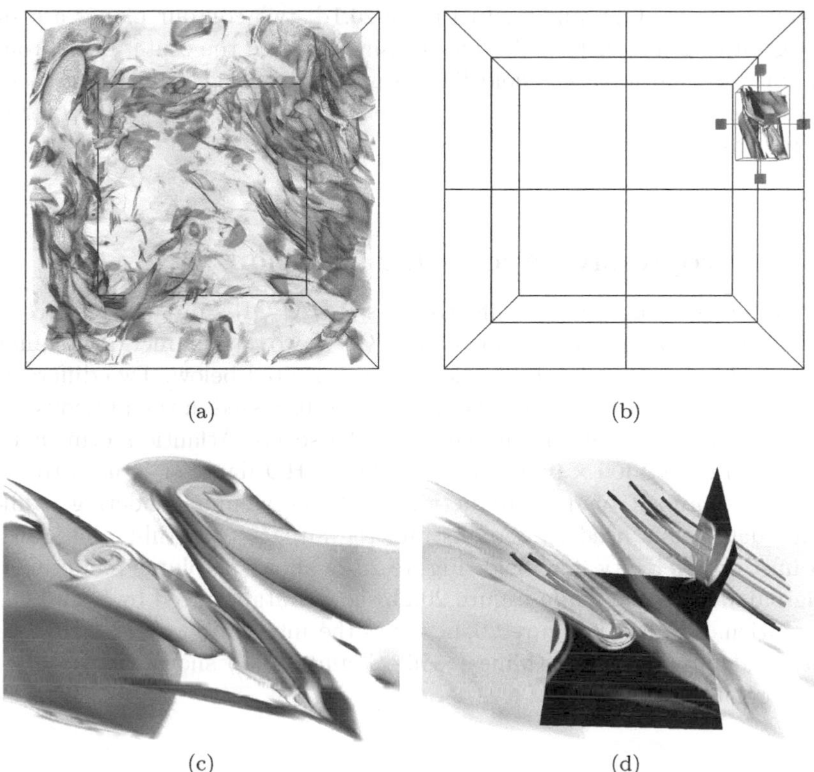

FIGURE 20.1: Exploration of the current field of a 1536^3 MHD simulation. A volume rendering of current magnitude generated using reduced data (a), an isolated ROI exhibiting a current "roll-up" (b), a close up of the phenomenon (c), and magnetic field lines passing through the center of the roll-up (d). The images in (b)–(d) were generated with the highest refinement level and LOD data.

image in Figure 20.1a is a full-domain volume rendering of the current field reconstructed from the VDC by reading only $\frac{1}{100}$ of the available wavelet coefficients and performing a partial inverse wavelet transform to produce a grid with $\frac{1}{64}$ the resolution of the original mesh (384^3 grid points). Both the reduction in grid resolution (saving computation and memory costs) and in the wavelet coefficients used to reconstruct the variable from wavelet space (saving I/O costs) were necessary for interactivity.

Through visual inspection of the highly compressed data involving the interactive manipulation of viewpoints, transfer functions, and cutting planes, a small ROI was identified, containing a current roll shown isolated in Figures 20.1b and 20.1c. Once isolated, additional visualization and analysis tools can be interactively applied in this smaller region, using both increased grid resolution and LOD quality. In Figure 20.1d, two contour planes are shown, along with the magnetic field lines passing through the center of the roll. The seeding location for these field lines was selected by picking locations on the contour planes in regions of high current magnitude, near the center of the current roll cross section.

20.4 Progressive Access Examination

To further illustrate the effectiveness of VAPOR's progressive access data model, a qualitative comparison of VAPOR's two data reduction techniques—LOD and refinement level selection—are presented below. Two different data sets are used for comparison: the MHD data discussed in the previous section, and a numerical weather simulation of the severe Atlantic storm, Erica [5], computed on a $1300 \times 950 \times 50$ grid. The MHD data set is used to demonstrate the impact of data reduction on direct volume rendering, while the Erica data set is used to demonstrate the impact to pathline integration in an unsteady velocity flow field. Figure 20.2a shows a volume rendering of the original MHD data, while Figure 20.2b shows data reduced by both LOD and resolution. Similarly, Figure 20.4a shows the integration over 20 timesteps of five randomly seeded pathlines, while Figure 20.4b shows the results when using reduced data.

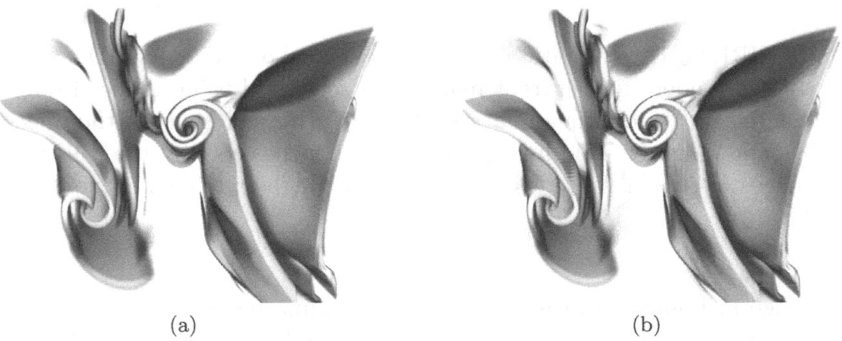

(a) (b)

FIGURE 20.2: Volume rendering of an isolated ROI showing the magnitude of the current field from a 1536^3 simulation. The original, unreduced data are shown (a) along with data reduced by a combination of both LOD and resolution (b). The corresponding reduction factors for LOD and resolution coarsening are 10:1 and 8:1, respectively.

Figures 20.3 and 20.5 compare images generated with reduced MHD and Erica data, respectively. The top rows show images generated from data at the highest refinement level, but with LODs corresponding to compression rates of 10:1 (a), 100:1 (b), and 500:1 (c). Similarly, the bottom rows show images generated from data at the highest LOD, but with refinement levels corresponding to grids at $\frac{1}{8}$ (d), $\frac{1}{64}$ (e), and $\frac{1}{512}$ (f) of the original resolution.

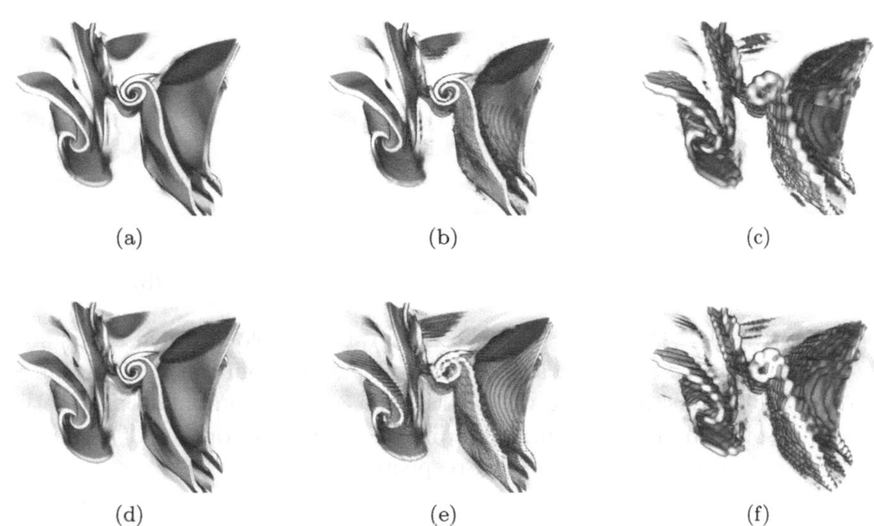

FIGURE 20.3: Direct volume rendering of reduced MHD enstrophy data (volume rendering of original data shown in Figure 20.2a). The images in the top row were produced with the native grid resolution, but varying the LODs with reduction factors of 10:1 (a), 100:1 (b), and 500:1 (c). The bottom row used the highest LOD for all images, but varies the grid resolutions with reduction factors of 8:1 (d), 64:1 (e), and 512:1 (f). Reduced LOD primarily benefits I/O performance, while reduced grid resolution primarily benefits memory, computation, and graphics.

The reader can subjectively evaluate the quality of these images, keeping in mind the substantial reductions in data involved. While the leftmost images would obviously be preferable for publication purposes, many analysis operations might be suitable for enabling qualitative understanding, using highly compressed data for visualization.

20.4.1 Discussion

In general, the accuracy of the reduced data depends strongly on the data's properties, in particular the degree of coherence between neighboring samples. However, the two examples presented above illustrate some principles generally applicable in the visualization of large data sets. When performing volume rendering (or similar visualizations, such as isosurface rendering, that map data values to graphics primitives), there is little benefit to having grid

(a) (b)

FIGURE 20.4: Pathlines from the time-varying velocity field of a simulation of the Atlantic storm Erica. Pathlines generated from the original, unreduced data are shown (a), along with data reduced by a combination of both LOD and resolution (b). The corresponding reduction factors for LOD and resolution coarsening are 10:1 and 8:1, respectively.

resolutions whose screen projection sampling rate exceeds, or even approaches, that of the display device itself. Visual quality rarely improves by increasing the refinement level once the projected voxels subtend a screen area smaller than the screen pixel size. In such a case, little information is lost if resolution is reduced, to afford interactivity.

Similarly, when performing an unsteady flow integration (or other operations requiring significant CPU processing on multiple volumes of data), it is again valuable to perform the initial analysis and visualization interactively. Interactivity can be obtained by using both lowered resolution and level of detail, and also by reducing the time sampling rate. By performing visualization and analysis on a small subregion, the quality impact of resolution, time sampling and compression level can be assessed. The desired visualizations can be previewed interactively at a lower accuracy by setting up the appropriate parameters for a subsequent noninteractive session of sufficient accuracy, to precisely illustrate the features of interest.

20.5 Conclusion

VAPOR provides a desktop environment for the interactive exploration of high-resolution numerical simulation outputs. The interactive exploration of data sets, whose size would otherwise overwhelm desktop computing re-

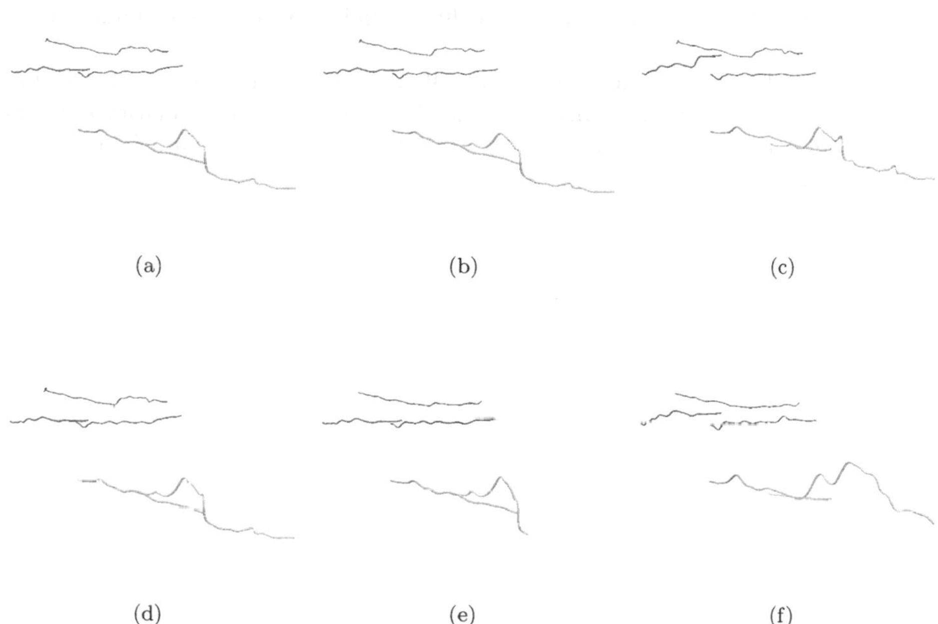

FIGURE 20.5: Pathline integration of five randomly seeded pathlines using reduced storm simulation data. Pathlines generated with original data shown in Figure 20.4a. The images in the top row were produced with the native grid resolution, but varying the LODs with reduction factors of 10:1 (a), 100:1 (b), and 500:1 (c). The bottom row used the highest LOD for all images, but varies the grid resolutions with reduction factors of 8:1 (d), 64:1 (e), and 512:1 (f).

sources, is enabled by the use of a variety of user-controllable data reduction techniques. These include:

- compression ratio (LOD selection), which is in direct proportion to the I/O performed;
- refinement level (resolution), which is proportional to processing time (including graphics rendering time), and also, it can have a significant impact on memory overhead, and;
- a time sampling rate and region size, both of which are in direct proportion to both processing time and I/O time.

To maintain interactivity, users can explicitly manipulate the parameters associated with all of these data reduction offerings provided by VAPOR.

When final, high-quality results are required, the parameters can be set for possibly noninteractive visualization and performed without user supervision, provided the computing platform has sufficient resources to handle the full fidelity data.

In addition to these user-controllable mechanisms, VAPOR also makes extensive use of data caching to avoid the unnecessary recalculation of previous results, and more significantly, to minimize the reading of data from secondary storage.

References

[1] John Clyne, Pablo Mininni, Alan Norton, and Mark Rast. Interactive Desktop Analysis of High Resolution Simulations: Application to Turbulent Plume Dynamics and Current Sheet Formation. *New Journal of Physics*, 9(301), 2007.

[2] John Clyne and Mark Rast. A Prototype Discovery Environment for Analyzing and Visualizing Terascale Turbulent Fluid Flow Simulations. In *Proceedings of Visualization and Data Analysis 2005*, January 2005.

[3] Eric Jones, Travis Oliphant, Pearu Peterson, et al. SciPy: Open Source Scientific Tools for Python, 2012. `http://www.scipy.org/`.

[4] John Lamping, Ramana Rao, and Peter Pirolli. A Focus+Context Technique Based on Hyperbolic Geometry for Visualizing Large Hierarchies. In *Proceedings of the SIGCHI conference on Human Factors in Computing Systems*, CHI '95, pages 401–408, 1995.

[5] Ching-Hwang Liu and Roger M. Wakimoto. Observations of Mesoscale Circulations within Extratropical Cyclones over the North Atlantic Ocean during ERICA. *Monthly Weather Review*, 125(3), 1997.

[6] P. D. Mininni, A. Pouquet, and D. C. Montgomery. Small Scale Structures in Three-Dimensional Magnetohydrodynamic Turbulence. *Physical Review Letters*, 97(244503), 2006.

Chapter 21

The EnSight Visualization Application

Randall Frank

Computational Engineering International, Inc.

Michael F. Krogh

Computational Engineering International, Inc.

21.1 Introduction ... 429
21.2 EnSight Architectural Overview ... 429
21.3 Cluster Abstraction: CEIShell ... 432
21.3.1 Virtual Clustering Via CEIShell Roles ... 433
21.3.2 Application Invocation ... 434
21.3.3 CEIShell Extensibility ... 434
21.4 Advanced Rendering ... 434
21.4.1 Customized Fragment Rendering ... 435
21.4.2 Image Composition System ... 437
21.5 Conclusion ... 440
References ... 442

21.1 Introduction

The vision behind EnSight has always been a full-featured, interactive, high performance visualization tool capable of scaling to the largest data sets. At the center of that vision is the ability to effectively leverage advanced computer systems both at the desktop/display and in computational clusters. This chapter reviews the EnSight framework for distributed application launch and rendering. The distributed launch system facilitates custom deployment of scalable EnSight visualization solutions. It has evolved to meet the increasingly varied requirements of its users, including heterogeneous environments with potentially complex and restrictive access controls. The rendering system is specifically geared toward improved user interaction and advanced rendering in distributed, high performance scenarios by exploiting advances in desktop graphics card technologies and direct interaction methodologies.

21.2 EnSight Architectural Overview

To place these commentaries in context, it is useful to provide a brief overview of the EnSight visualization framework. EnSight [6] was designed around a core, client–server architecture. The server provides the means for reading data from storage through a built-in or user-defined reader. It is responsible for the extraction of geometric primitives from the data, which are transmitted to the client for display. The client performs all of the graphical rendering and manipulation tasks using these primitives. Geometry and commands move between client and server applications over an abstract communications channel capable of leveraging TCP/IP, MPI, or other transport layers via user-written plug-ins.

The core server includes the ability to read multiple data files as a single data set utilizing a single process—thread-based, parallel processing. Larger data sets stress scalability limits of the thread-based parallelism utilized in the basic client–server application scenario. For such data sets, the framework supports distributed memory processing (e.g., on cluster systems). The server of servers (SOS) and collaboration hub (C-HUB) applications are introduced in these scenarios. The applications provide the logic and processing necessary to allow multiple client (C-HUB) and server (SOS) instances to work collectively. The SOS can be used without the C-HUB to allow multiple servers to interact with a single client, but the use of the C-HUB, to utilize multiple clients, necessitates the use of an SOS.

In most visualization scenarios, the extracted geometric primitive data is a small fraction of the size of the raw data. The magnitude of this data reduction, coupled with client performance benefits realized though the exploitation of modern programmable graphics cards, makes it practical to use far fewer client instances than server instances for distributed visualization. For a large number of cases, only a single client instance is required, resulting in the common configurations shown in Figure 21.1.

The SOS provides distributed processing capabilities layered on top of the core server, providing cross-server aggregation and dynamic object remapping. The SOS also allows for optional automatic decomposition of a single data set over multiple servers with dynamic ghost element generation. These two features allow for N to M conversions between the degree of natural decomposition present in the data set (N) and the number of servers (M). The SOS provides a parallel asynchronous data reduction network (similar to the MPI all-reduce functionality). This network allows servers and user-defined readers to perform global computations, exchange particle trace locations, etc. The SOS manages the merging and chunking of geometry packets from the various servers, including the ability to compute new geometric decompositions that target specific numbers of client applications. The revised geometric decomposition is routed to the various clients via the C-HUB to load balance rendering work over the clients.

The C-HUB serves to link together multiple instances of the client applica-

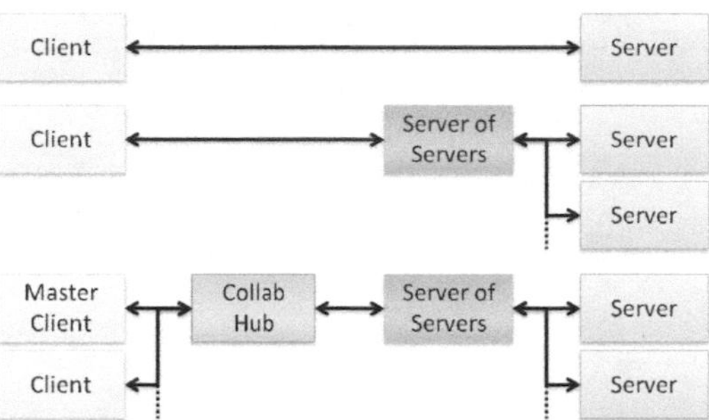

FIGURE 21.1: The basic EnSight application configurations. Top to bottom: traditional client–server (Ensight Standard), single client multiple servers (Ensight Gold), multiple clients and servers (Ensight Distributed Rendering—DR).

tion and synchronize their interactions. It provides channels for the transmission of user interaction and scripted commands. More importantly, it generates routing information that the SOS uses to decompose geometry packets to match the client configuration. The C-HUB routes the geometry packets generated by the SOS to individual clients with selective duplication of geometry to specific client instances. When combined with dynamic head tracking and support for multiple independent viewing frusta, the system enables flexible tiled and CAVE configurations as shown in Figure 21.2.

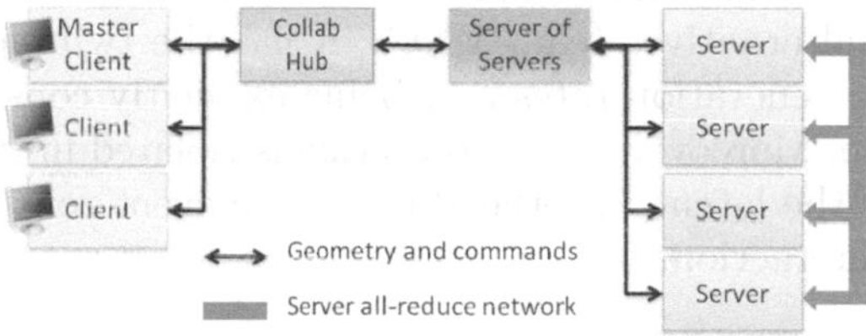

FIGURE 21.2: EnSight parallel rendering for CAVE and planar tiled displays. The geometry is duplicated on all of the clients.

The client provides an independent pixel fragment compositing engine network, which allows the output of multiple graphics cards to be combined into a single image. The source to this system was released by CEI as open source [2] and a variation is currently available as the paracomp [1] library. Combining dynamic geometry routing with one or more instances of the compositing system allows clients to render to tiled displays, where each panel of the display is driven by a separate compositing network as shown in Figure 21.3.

In a multiclient rendering configuration, the client process accepting user input is referred to as the *master* and the other clients are referred to as

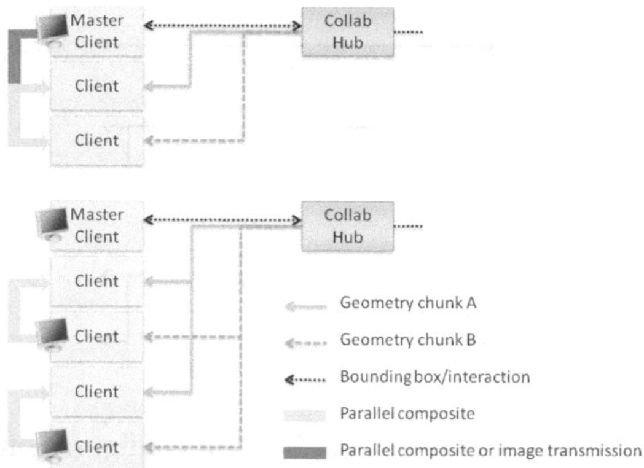

FIGURE 21.3: EnSight rendering with parallel compositing. Chunks of geometry can be selectively duplicated to parallel composite to a single display or support multiple parallel compositing instances for tiled display output.

slaves. The master client's participation in the parallel compositing system is optional. The network topology of the session's cluster and desktop hardware tends to drive the decision. When the master and slave clients are located on a common network (e.g., all on the same cluster), the final image delivery is handled by the compositing library. In other situations, a remote image transmission service is provided by the C-HUB. The C-HUB image delivery service supports bandwidth-limited, asynchronous remote image transmission with frame differencing and compression. When compared to generic remote image delivery alternatives (e.g., VNC), it has the twin advantages of being driven at image generation rates and being explicitly constrained to the rendered image size. Moreover, user interaction is isolated in the locally executed client, avoiding the latency overhead of remote event routing for many forms of direct user interaction.

21.3 Cluster Abstraction: CEIShell

The obvious challenge presented by these varied configurations is how to realize a specific instance of the framework (an EnSight session) within a given site's computational and networking constraints. At a minimum, an EnSight session is composed of a single client and server. Optionally, a session might also include additional servers and rendering clients, an SOS, and a C-HUB. Traditionally, EnSight has used a configuration file scheme based on a secure shell (ssh), or equivalent, for launching the various application instances in an EnSight session. This approach works well in static and relatively homogeneous computing environments. However, increasingly complex security requirements and system heterogeneity, coupled with an increase in scale and system complexity, have predicated a move to a new launch system for the EnSight framework. User requirements for the new system include support for queuing systems, improved launch scalability, dynamic application configuration, VPN communications, session persistence, and enhanced debugging capabilities.

To meet these requirements, CEIShell was developed. Simplicity and extensibility were key themes in the design of the system. The simplicity of CEIShell contributes greatly to its overall robustness and reduces the amount of work necessary to customize it to specific platforms, queuing systems, firewall policies, etc. The system can be extended to support site-specific network and job configuration protocols. CEIShell is a small C++ application that weaves four concepts together: tree-based communications with other CEIShells, communications with locally running applications, a tag-based resource specification, and local job management. Perhaps the largest benefit of CEIShell is the improvement in session robustness, stemming from the fact that EnSight can probe potential session resources (e.g., validating environmental parameters, OpenGL configurations, etc.) before attempting to launch the session.

A collection of CEIShells is invoked as a rooted, acyclic, tree network. Typically, command line parameters indicate parent–child relationships as well as specific communication protocols and their parameters. The actual commands used to launch the multiple CEIShells on the various computers are provided by site-specific customizations. Issues such as remote computer access, network security, cluster queuing system commands, and user credentials dictate how applications, including CEIShell, are actually invoked on behalf of the user. The commands to launch a specific instance of a CEIShell network are commonly wrapped in a custom script that is parameterized by the site. For example, the number of compute nodes, the number of rendering nodes, and the users credentials might be parameters mandated by the site maintainer. The EnSight client provides a customizable GUI that can be used to invoke a site's CEIShell network scripts along with any required parameterization.

21.3.1 Virtual Clustering Via CEIShell Roles

Along with the specification of CEIShell's parent–child communication parameters, a site customization script indicates what roles each CEIShell provides. Roles are simply ASCII tags applied to each CEIShell instance. Each shell may have any number of roles. The CEIShell itself does not dictate the use of roles other than to use them to perform a resource search. A simple but useful example of roles would be to give nodes in a cluster standardized names. For example, one might allocate 10 nodes via a batch system on a cluster but assign the roles "node00" through "node09" to the nodes. When EnSight encounters a reference to "node04" in a resource or other configuration file, the CEIShell network can remap the name via the roles to a specific, dynamically allocated node. In this use-case, the CEIShell becomes a virtual cluster that always appears the same to a calling application, regardless of which nodes are actually allocated. EnSight uses specific predefined tags to indicate where individual EnSight framework components are to be realized when it is not provided with a more specific tag from a configuration file. For example, the role "SOS_SERVERS" indicates that the CEIShell could run an EnSight server for a distributed EnSight session. EnSight can inquire how many CEIShells have been assigned a particular role and use that information to determine, at runtime, the number of EnSight servers or rendering nodes to use.

21.3.2 Application Invocation

Once EnSight decides what components and how many of each it needs to invoke, it requests the CEIShell network to actually launch the components on its behalf. This approach simplifies the core EnSight application logic by insulating the application from the details of remote resource access and launching details, and provides a site with the flexibility needed for tight integration with its computing environment.

The CEIShell is designed to aid in complex session debugging tasks. It redirects the output from the applications it launches and other CEIShells back to the root CEIShell. From there, it can redirect this output to the main application (e.g., the EnSight client) where a dialog might be displayed with highlighted warnings and errors. CEIShell can also route communications on behalf of the application using an intrinsic VPN mechanism. This system allows application components to talk to each other without allocating additional communication channels by multiplexing application communication over the intra-CEIShell communication channels. This can be essential if application components reside multiple network hops away from each other, precluding the establishment of direct communication channels.

21.3.3 CEIShell Extensibility

CEIShell is extensible in several ways. It knows nothing about a particular application, including EnSight, so it can be used for any distributed network application. Because roles are not specific to CEIShell, individual applications may dictate their semantics and exploit them for other tasks. For example, a site might decide to launch X11 servers with an EnSight session-specific life cycle on specific nodes, in the CEIShell network, using their own custom roles. CEIShells communicate through transport plug-ins as specified through generic URLs that indicate which transport and parameters to use. These transports are implemented as dynamic shared libraries that implement basic communication primitives. A site may provide its own custom communications transport abstraction. CEIShell ships with transports for named pipes and TCP/IP sockets as well as a source code to an MPI transport plug-in.

21.4 Advanced Rendering

One of the most recognizable aspects of a visualization framework is its interactive rendering system and associated user interface. The ease with which useful visual representations of large, complex data sets may be constructed is one measure of the effectiveness of such interfaces. The EnSight rendering infrastructure has undergone several major revisions designed to improve this measure. Recent revisions have been motivated by the need to support distributed-memory volume rendering and improve direct user interaction with extremely large data sets. These changes facilitated a move to a more modern, Qt-based GUI (see Fig. 21.4) that supports improved direct interaction mechanisms—drag and drop, context sensitive menus, etc.

In order to improve application responsiveness with large data sets, it became necessary to be able to perform as many actions as possible without forcing geometry to be re-rendered, as rendering-induced latency tends to increase with the rendered polygon count. Additionally, direct interaction and feedback methods introduced in the revised EnSight GUI added the requirement to support object-level picking and high performance object recognition, with both geometry and annotation elements in the rendering system. The resulting rendering system supports dynamic selection changes with object silhouette highlighting and object picking, without forcing additional rendering operations. The approach is to base the rendering system on an image fragment compositing system implemented with OpenGL shaders. This mechanism maps naturally to distributed parallel compositing, allowing a single, unified system to support both simple desktop and distributed parallel rendering.

The dynamic shader system uses OpenGL multiple render-target enabled framebuffer objects, making it possible to break up the rendering pipeline into independent, retained layers for the 3D geometry and 2D annotations. Hardware picking and layered rendering make it possible to eliminate the traditional EnSight modes (plots, annotations, viewports, etc.), replacing them with context-sensitive direct interaction, which accelerates and simplifies user interaction. The system was developed as an extension to the existing parallel compositing system. All of the geometry rendering operations are implemented in a framework that is based on the ability to generate complete OpenGL shader programs. The shader programs represent the current rendering state dynamically, from a collection of shader program fragments.

21.4.1 Customized Fragment Rendering

The dynamic fragment system follows the Chromium [7] OpenGL model without the network components. It uses a state tracking system [4] to determine what rendering functionality is active and computes the actual fragment program from the state. As the core application makes OpenGL calls, a model of the current rendering state is formed and tracked, avoiding the overhead of

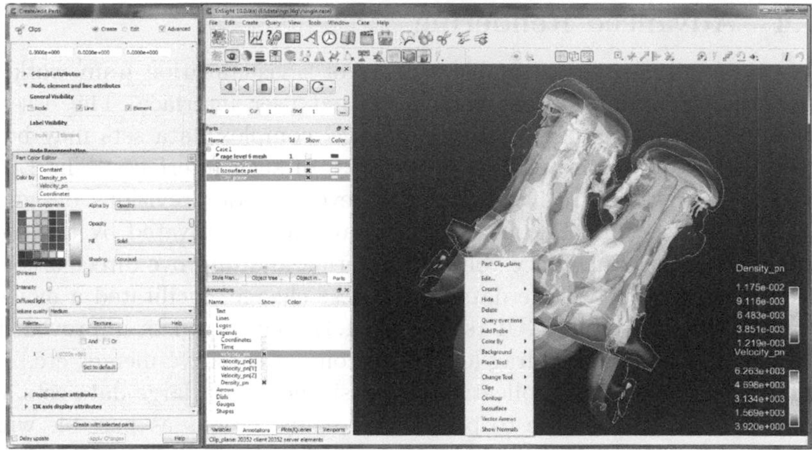

FIGURE 21.4: The EnSight 10 GUI includes dynamically generated silhouette edges for selections and targets, hardware-accelerated object picking, and context sensitive menus. These form the basis of a comprehensive direct interaction, drag and drop interface.

redundant state changes. When the modeled state is bound, the engine looks at the current OpenGL state and generates a complete fragment shader by combining fragments of GLSL source code that model the various OpenGL rendering features (e.g., lighting, texturing models, etc). The shader is compiled and cached for future reuse. In a typical session, Ensight will encounter 15–20 independent OpenGL rendering states and will generate and cache a shader program for each such state.

There is a formal OpenGL extension system, which allows for the redefinition and extension of rendering capabilities. The system allows for custom OpenGL rendering as CEI OpenGL extensions. For example, EnSight uses a custom texturing interpolation function to implement its "limit fringes" feature. The implementation is specified as a hand-coded GLSL fragment. The application then passes a unique OpenGL texturing enumerator to the state tracker via the state tracked `glTextureEnv` calls. Through this mechanism, the rendering engine can be extended to support rendering styles, not intrinsic to OpenGL, in a form that can be additively combined, on-the-fly, with other OpenGL rendering features. Structured and unstructured volume rendering and depth peeling have been implemented as such extensions, allowing the system to support a hybrid transparent polygon/ray tracing rendering system.

A key feature of this new OpenGL abstraction is to embrace fragments composed of *deep pixels*. A deep pixel includes, color, depth and secondary information related to the object types, symmetry, etc. The EnSight rendering system includes two independent depth buffers and at least two color buffers. For fast, pixel accurate, transparency surface rendering, EnSight uses a depth peeling [5] algorithm. The paired depth buffers are used for depth peeling and the secondary color buffers are used to store ancillary metadata information.

Every visible object in EnSight has the ability to store one or more rendering tags. Tags are rendered into secondary color buffers. They include information about the basic state of the object (e.g., selected, highlighted, etc.) and they hint at how a given object should behave in various interactive contexts. For example, tags encode the relative importance of an object to the current set of operations, making it easier for the user to select the most appropriate options for a given operation. They can also be used to inform antialiasing post-processing operations for edges and boundaries [8]. The dynamic highlighting and picking systems are based on these tags. If the user clicks on a pixel, the actual target object can be determined directly from the tag buffer by reading back a few pixels without referencing the actual geometry.

Dynamic highlights can be updated in real-time response to single pixel cursor motion, exploiting custom shaders to perform the necessary topological analysis without re-rendering any geometry. Tags are also used to encode the context of an object. For example, if a given object is currently being rendered using a symmetry operation, the specific mode of symmetry is included. This information is used by the picking system to generate coordinates in the original data space with simple coordinate inversion operations. The overall user experience is greatly improved with the unique types of instantaneous feedback afforded by the deep pixel tags.

The dynamic fragment framework outlined previously is used to implement a ray casting volume rendering system capable of supporting arbitrary polyhedral elements [3] while maintaining a high-efficiency rectilinear grid renderer that can be used when appropriate. The algorithm allows for the rendering of volumes with embedded, transparent surfaces by terminating the volume rays at the depth peels, and later continuing them at the depth surface of the peel. Coupled with natural ray restart at the volume element block boundary, the system supports both embedded transparent surfaces and volume data presented as unsorted, concave blocks. Very little pre-processing (only simple, unconstrained blocking) and no presorting of volume domain elements is necessary. Another way to conceptualize the system is as a space-leaping volume renderer that leverages depth peels as leaping boundaries, switching back and forth between ray casting and traditional polygon rendering, as dictated by the geometry. The results of the ray casting segments are blended into the peel, effectively collapsing the combined region into a single image fragment. This formulation allows volume rendering to fit naturally into the distributed compositing system. Examples that exercise both the structured volume shader and the unstructured volume shader are shown in Figures 21.5 and 21.6, respectively.

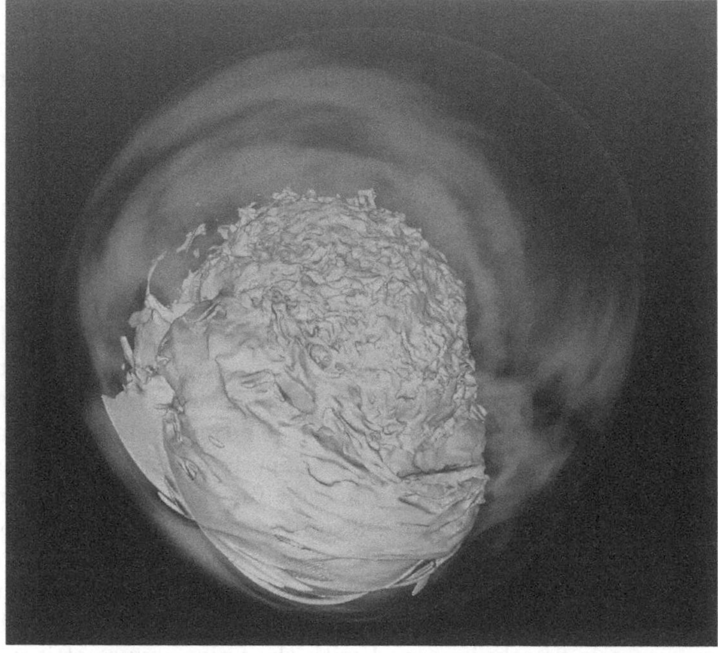

FIGURE 21.5: Supernova remnant density field rendering, demonstrating distributed structured volume rendering, with embedded polygonal surfaces. Data courtesy of Dr. John Blondin, North Carolina State University.

21.4.2 Image Composition System

EnSight has always utilized a multi-pass rendering system. It supports threaded rendering over multiple tiles with tile chunking, antialiasing, and other functions being handled in the various passes. The rendering engine maps the various passes into textures and framebuffer objects generated with each pass and in many cases retained from frame to frame. Key retained fragments include the annotation layer and depth peels with their associated metadata tags. These image fragments are rapidly composited to form the final image though a filtering engine that is used to generate dynamic selection highlighting and image-based, full-scene antialiasing.

Selection highlighting maps the current list of selected objects in a scene into a state table, and applies it to the pixel object tags via texturing. Topological edge detection operators generate constrained silhouette edges for the current selected and targeted objects, which are then blended into the final image. This last step is implemented as a single fragment program and can be regenerated from the retained textures and framebuffer objects without re-rendering the core geometry, enabling dynamic changes in the graphical entity selection, independent of the data set size and its associated rendering expense. The annotation plane is one such fragment layer that contains all 2D

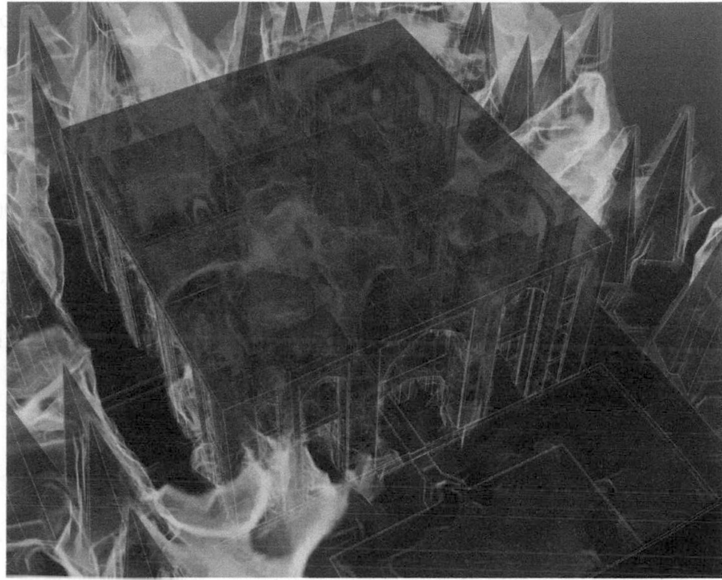

FIGURE 21.6: Simulated air flow through the Kashan-Iran Fin Garden building and surrounding vegetation. Unstructured volume rendered velocity through semi-transparent architectural geometry. Data courtesy of Studio Integrate.

annotations and plots. These plots and annotations can be moved, edited, and rendered independently from the rest of the 3D scene—a feature increasingly important when volume rendering large, distributed data sets.

For distributed rendering, EnSight DR has used an open source parallel compositing system utilizing the parallel pipeline algorithm [9] to implement sort-last parallel rendering. A unique feature of this compositor is its extension to include per-pixel fragment sorting. In the system, image fragments are generated by the various slave rendering nodes. Each node renders multiple fragments, and in most cases, one per depth peel. These fragments include color, depth, and pixel metadata. Fragment data is compressed with run-length encoding, as fragments from later peels often contain a very small fraction of valid pixels. The bandwidth savings stemming from this compression, especially in deeply pipelined situations (e.g., animations), can be significant. The compositing engine maintains a per-pixel sort based on the depth value of each pixel. Once all of the fragments have been received and sorted, they are blended together to form a final image subrectangle at each of the nodes participating in the composite operation before collecting the rectangles together to form a final image. This per-pixel sorting in the compositor allows for distributed transparency using distributed depth peeling, avoiding the need for any pre-processing and sorting.

An example of imagery, generated with the compositor and deployed with EnSight remotely in a fully parallel, hardware-accelerated configuration, can be seen in Figure 21.7. The image contains three temporal snapshots taken from a billion element 3D simulation of the implosion of an Inertial Confinement Fusion (ICF) capsule illustrating the development of turbulence in the fuel as the capsule implodes. The white isosurface represents the interface between the inertial drive shell and the central fuel and is over 600 million triangles in the final frame. The complex blue and yellow isosurfaces represent vortex tubes inside the fuel which begin as relatively coherent structures but rapidly develop a fully turbulent character. The snapshots show a close-up view of a small region near the pole of the capsule as it becomes fully turbulent. Additional details may be found in a report by Thomas and Kares [11].

21.5 Conclusion

The EnSight visualization framework rendering and distributed launching system have progressed considerably in more recent releases to better meet user requirements and exploit technological and methodological advances. Additional changes in algorithms and implementations will continue to be reformulated to track the nature of new data sets. The deployment system targets are changing as well. New system targets necessitate changes in task subdivision and location to reflect changes in the balance of computational resource and bandwidth—for example, the exploitation of "cloud" and other distributed resources.

Looking further ahead, a unique challenge facing EnSight is how to best meet increasingly refined requirements of scientific inquiry applied to complex data set domains [10]. The composition data sets is changing to reflect new simulation technologies. For example, there is an increased use of ensemble methods and aggregate data sets, which include the results of multiple simulation methods and parameterization. Additionally, many solvers are adopting meshing technologies, which make heavy use of arbitrary polyhedral elements. These trends in data set composition, combined with contemporary simulation-coupled decision-making processes, represent some of the most exciting challenges on the horizon, which could potentially redefine the traditional notion of "visualization."

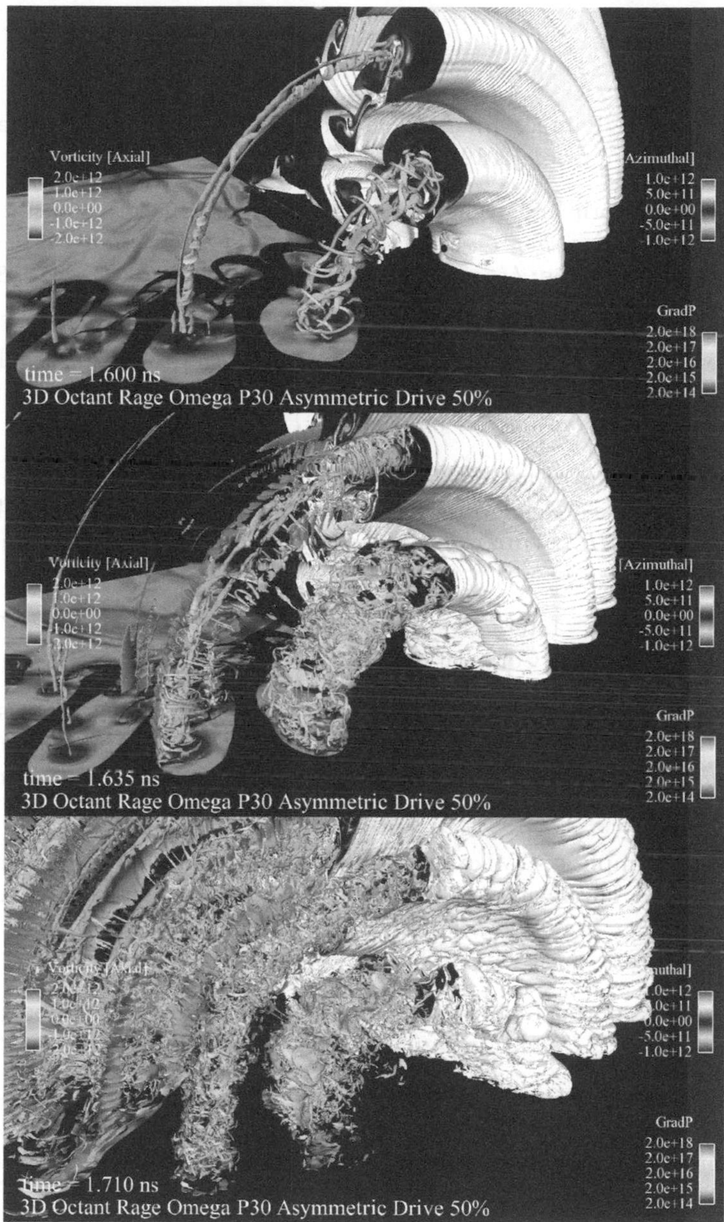

FIGURE 21.7: Inertial Confinement Fusion implosion visualization. The images were computed using EnSight servers located at Lawrence Livermore National Laboratory and EnSight clients running on a rendering cluster located at Los Alamos National Laboratory. Images courtesy of Robert Kares, Los Alamos National Laboratory.

References

[1] Paracomp Parallel Image Composition Library. `http://paracomp.sourceforge.net`, 2012.

[2] Byron Alcorn and Randall Frank. Parallel Image Compositing API. In *IEEE Workshop on Commodity-Based Visualization Clusters*, October 2002. `https://computing.llnl.gov/vis/images/pdf/IEEE_Viz_2002_rjf.pdf`.

[3] Fbio F. Bernardon, Christian Azambuja Pagot, Joo Luiz Dihl Comba, and Cludio T. Silva. GPU-Based Tiled Ray Casting Using Depth Peeling. *Journal of Graphics, GPU, and Game Tools*, pages 1–16, 2006.

[4] Ian Buck, Greg Humphreys, and Pat Hanrahan. Tracking Graphics State for Networked Rendering. In *Proceedings of the ACM SIGGRAPH/EUROGRAPHICS Workshop on Graphics Hardware*, HWWS '00, pages 87–95, 2000.

[5] Cass Everitt. Interactive Order-Independent Transparency. Technical report, NVIDIA Corporation, 2001.

[6] A. Grimsrud and G. Lorig. Implementing a Distributed Process between Workstation and Supercomputer. In *Applications of Supercomputers in Engineering: Algorithms, Computer Systems and User Experiences*, pages 133–144, 1989.

[7] Greg Humphreys, Mike Houston, Ren Ng, Randall Frank, Sean Ahern, Peter D. Kirchner, and James T. Klosowski. Chromium: A Stream-Processing Framework for Interactive Rendering on Clusters. In *SIGGRAPH'02*, pages 693–702, 2002.

[8] Rusty Koonce. Deferred Shading in Tabula Rasa. In Hubert Nguyen, editor, *GPU Gems 2*, pages 429–457. Addison-Wesley, 2008.

[9] Tong-Yee Lee, C. S. Raghavendra, and John B. Nicholas. Image Composition Schemes for Sort-Last Polygon Rendering on 2D Mesh Multicomputers. *IEEE Transactions on Visualization and Computer Graphics*, 2:202–217, September 1996.

[10] Mike Long and Randall Frank. The Eyes' Mind: Visualization in the Linux Supercomputing Era. *Desktop Engineering*, September 2007.

[11] Vincent Thomas and Robert Kares. LA-UR-11-03527. Technical report, Los Alamos National Laboratory, 2011.

Index

in situ, 178
  data staging, 297
  processing, 152–173, 287, 289, 290, 297, 299

active pixels, 318
active-pixel encoding, 318
adaptive mesh refinement, 295
ADIOS, 162–169, 296–298
aggregator, I/O, 344

brushing and linking, 110

callback, 320
CEIShell, 368
Chromium, 29
Chromium Renderserver (CRRS), 36
co-rendering, 30
co-scheduling, 293
compositing, *see also* image compositing
compression
  lossless, 39
  lossy, 39
Compute Unified Device and Systems Architecture (CUDASA), 213
Compute Unified Device Architecture (CUDA), 199
concurrency, 285, 298
context and focus, 110, 356
contracts, 18
convolution kernel, 265
cores, 10

data classification, 204
data flow networks, 12
data parallelism, 11, 176
deep pixels, 372
depth peeling, 372
Digital Light Processing (DLP), 216
direct volume rendering, 49–51, 200
distributed cache (DC), 52
Distributed Parallel Storage System, 35
distributed visualization, 23
distributed volume rendering, 373
distributed-memory parallelism, 10
dynamic load balancing, 21

embarrassingly parallel, 14
energy groups, 288, 295
ensembles, 288
EnSight, 365–376
exascale, 62, 162, 165, 283–301
external faces algorithm, 330
extreme-scale data, 249

fault-tolerant computing, 287
filters, 12

ghost
  cells, 20, 182, 330
  data, 15, 228
  faces, 19
GLEAN, 297, 298
GPGPU
  definition, 198
  programming languages, 199
    CUDA, 199
    OpenCL, 200
GPU clusters, 204
graphics pipeline, 196
graphics processing unit (GPU), 195–219, 226, 233, 268, 286

halo exchange, 228
hidden-surface removal, 321
High Angular Resolution Diffusion Imaging (HARDI), 211
hybrid parallel
  architecture, 224
hybrid parallelism, 11, 222
hybrid pipeline partitioning, 30

I/O
  collective, 249, 253, 254
  noncollective, 249, 251, 253
  striping parameters, 251
IceT, 315–323
image compositing, *see also* compositing
  radix-k, 63, 70
  2-3 swap, 63, 69
  binary-swap, 63, 67, 317, 318
  binary-swap algorithm, 67
  definition, 65
  direct-send, 63, 66
  hardware, 68
  IceT library, 68, 315
  optimizations, 64, 67, 71
  parallel, 63–77, 318, 367, 374
  performance experiments, 73
  SLIC library, 68
  tree-based methods, 67
inactive pixels, 318
index-based computation, 292
indexing, 105
  bitmap indexing, 105
information visualization, 110
integral curves, 80
  Finite-Time Lyapunov Exponents, 81
  GenASiS, 84
  hybrid parallel, 236–238
  integration time, 80
  Lagrangian Coherent Structures, 81
  Lagrangian method, 80
  NIMROD, 85
  parallel space–time data structure, 94
  parallelization, 83
  parallelizaton over seeds, 236
  parallelize over data, 83, 87, 235, 237, 242
  parallelize over seeds, 83, 85, 235, 240
  parallelize over seeds–data hybrid, 88
  seed point, 80
interlace, 318
interpolation
  artifacts, 19

Joule certification, 254

kd-tree decomposition, 205

large display visualization, 215
latency, 25
level of detail (LOD), 147, 177, 197, 319, 326, 356

massive data, 249
maximum intensity projection (MIP), 204
maximum intensity difference accumulation (MIDA), 204
message passing, 11
metadata, 19
  contracts, 187
MIP Map, 211
multi-pass rendering, 373
multiple render targets, 372
multiresolution methods, 127

nodes, 10, 222
non-volatile memory, 286, 297
nonlinear gradient interpolation, 203

occlusion culling, 210
Open Computing Language (OpenCL), 200
OpenGL, 47, 196, 319, 320
OpenGL fragment shader, 372

OpenMP, 11, 212, 223, 225, 274, 298, 313
optimization
  using metadata, 19
out-of-core methods, 292

parallel coordinates, 110
ParaView, 286, 325–339
particle accelerator
  linear, 119
  plasma-based, 121
particle-in-cell simulation, 118
pipeline execution
  execute, 13
  pull, 13, 36, 181
  push, 13, 36, 181
  update, 13
pipeline parallelism, 188
post-processing, 289, 290, 292
power consumption, 284, 297
preintegration, 203
Processing Element, 10, 222
progressive data access, 356
pthreads, 11, 223, 225, 274, 313
pure parallelism, 13, 248

QDV, *see also* query-driven visualization, 103–125
  B-tree, 105
  connected component analysis, 113
  connected component labeling, 113
  connected components, 112
  histogram-based parallel coordinates, 111
query interface, 110
query-driven visualization, *see also* QDV

radix-k, 70–71, 317, 318
ray casting, 202
ray tracing
  cluster-based, 52
region-of-interest, 353
relevance function, 204
remote and distributed visualization, 23
remote visualization, 23, 367
rendering, 45
  particle-based, 207
  round-robin distribution, 46
  sort-first, 46, 205, 316
  sort-last, 47, 205, 316, 318
  sort-middle, 46, 316
  sprites, 48
rendering geometry, 47
  send-geometry, 47
RFB protocol, 36
run-length encoding, 318

scalable rendering, 41
Scalable Rendering Threshold (SRTP), 41
scatter-gather, 14
schema, 293
scientific data models, 293
send-data partitioning, 28
send-geometry partitioning, 29
send-images partitioning, 26
shader, 196
  domain shader, 197
  fragment shader, 197
  geometry shader, 197
  High Level Shading Language (HLSL), 198
  hull shader, 197
  OpenGL Assembly Language, 198
  OpenGL Shading Language (GLSL), 198
  Phong shading, 208
  tessellation shader, 197
  vertex shader, 197
shared-memory parallelism, 11
sinks, 12
slice-based rendering, 202
sources, 12
space-filling curve, *see* Z-order curve
  HZ-order curve, 343
  Lebesgue curve, 343
spatial resolution, 288

spatial decomposition, 315
splats, 208
static load balancing, 21
steaming
  functional parallelism, 188
stencil kernel, 265
streaming, 21
  data parallelism, 188
  Demand Driven Visualizer, 190
  lazy evaluation, 190
  OOC construction procedure, 186
  piece, 187
  QSplat, 185
  sparse traversal, 183
streamlines, 80, 239
structured volume rendering, 373
supercomputer, 10

task, 10, 222
task parallelism, 11
TCP-based communication, 36
temporal parallel coordinates, 112
temporal resolution, 288
threads, 11, 222
tiled display, 215, 216, 218
topological methods, 294
traditional parallelism, 222
transfer function, 204
tree communication scheme, 259
tricubic B-spline texture filtering, 203
trillion cells, 249, 253, 254, 258

UDP-based communication, 36
uncertainty quantification, 287, 288, 292

VAPOR, 353
Virtual Network Computing (VNC), 36
Visapult, 30, 34
visibility ordering, 321
VisIt, 286, 304–314
  pipeline partitioning, 40
visualization
  definition, 1
visualization pipeline, 1, 328
ViSUS visualization framework, 341
volume rendering, *see also* direct volume rendering, 49–51
  hybrid parallel, 224

wavelet coefficients
  approximation, 133
  detail, 133
wavelet transform
  Haar, 133, 136

Z-order curve, 129